WORLD RAILWAYS *of the* NINETEENTH CENTURY

A Pictorial History

in Victorian Engravings

THE JOHNS HOPKINS UNIVERSITY PRESS
Baltimore & London

Printed in China on acid-free paper

9 8 7 6 5 4 3 2 1

The Johns Hopkins University Press
2715 North Charles Street
Baltimore, Maryland 21218-4363
www.press.jhu.edu

Library of Congress Cataloging-in-Publication Data
Harter, Jim.
World railways of the nineteenth century : a pictorial history in Victorian engravings / Jim Harter.
p. cm.
Includes bibliographical references and index.
ISBN 0-8018-8089-0 (hc. : alk. paper)
1. Railroads—Pictorial works. I. Title: World railways of the 19th century. II. Title.
TF145.H28 2005
385′.09′034—dc22 2004021142

A catalog record for this book is available from the British Library.

A list of sources for the illustrations appears on the last three printed pages of the book.

This book is dedicated to the memory of my father,
JOSEPH WALTER HARTER, 1908–1971, my first mentor in railway matters.
It is also for my stepfather RALPH KERR, C.E. (ret.), of Albuquerque,
who since joining our family has contributed greatly to us all.

CONTENTS

PREFACE

THIS book is about the development and expansion of railways in the nineteenth century. It is an amazing story. Simply stated, the steam locomotive was the greatest technological breakthrough in transportation since the invention of the wheel. Its revolutionary potential inspired the collective imagination, resulting in many economic, social, and technical changes. Concurrent with these was a great acceleration of events all over the planet.

Arguably, this was the central occurrence of its age, around which many other developments unfolded. Successive changes since then have led to our own modern era. In the twentieth century, trucks, busses, cars, and airplanes emerged to challenge railroad dominance. As these technologies improved, railways correspondingly declined, losing not only most of their traditional turf but much of their place in our collective memory. Over time we have slowly forgotten the story of their early development; thus it is important to periodically retell it. In that spirit I submit the present volume. Providing the most comprehensive overview of nineteenth-century railroading ever published, it offers what is perhaps the definitive story.

Most of us today may find it hard to imagine the overwhelming importance of railways in the year 1900. At the time they touched most people very directly and employed a much larger segment of the population. They symbolized many things: great fortunes, power, excitement, glamour, style, opportunity, long-distance travel, and, for many in remote locales, connection to the outside world. The only competition railroads had were animal-drawn vehicles, slow steam road machines, and the most primitive of gasoline-powered horseless carriages, which in most places had to operate on fragmentary networks of rudimentary roads.

Railways were by far the major carrier of people, mail, and all kinds of freight. The last included manufactured goods, produce, meat, milk, grain, gold, other valuables, lumber, bulk liquids, and minerals such as iron ore and coal. Also in use were small narrow-gauge lines that served the respective spheres of industry, mining, agriculture, timber, construction, and the military. Many towns and cities had tramway systems using various modes of traction. In addition to these, large cities like New York and London also employed mass transit systems.

The first steam locomotive was built in England in 1803. It resulted from the marriage of two technologies: the tramway, which traced its origins back to ancient Greece, and the steam engine, which had its beginnings in the previous century. The progress of this new technology was initially slow and awkward, but by 1825 a successful railway had been built. It functioned as a common carrier, open to anyone willing to pay for its use. It carried both freight and passengers at speeds far greater than had been previously possible. And unlike rivers and canals, which could freeze over in the winter, the railway was open for business all year.

This new technology spread quickly to America and most of Western Europe, and later to such countries as Canada, Russia, and India. Elsewhere in the world it took much longer to catch on. South America, Africa, and most of Asia lagged far behind in this development. South America and Africa never built large systems, and aside from India, Asia did so only after the West turned to highways and petroleum use. Data from 1885 demonstrate this lag. That year, while North America had 137,400 and Europe 114,200 route miles of track, respectively, Asia had built only 15,200 miles; South America, approximately 10,000; and Africa, 3,700 miles.

Central to the history of any technology is the sequence of forms or expressions that it takes during development. The evolution of the steam locomotive was the most essential element in early railway history. Its development is covered here, in considerable detail, with attention given to variations among different countries. Aside from the locomotive, developments were also taking place in operational practices, rolling stock design, bridge and tunnel building, railway station architecture, and alternative forms of traction. All of these subjects are addressed in this book.

This volume is unique, in that it is the first on the subject ever to be fully illustrated with Victorian wood engravings. Although the technique for wood engraving had been developed in the previous century, it was only around 1860 that it reached a level of accuracy sufficient to provide good renderings of railway scenes and equipment. In the beginning, engravings were done from artists' sketches, but later, many were done from photographs. Disseminated by means of the printed media, these images had a strong influence on the popular imagination. Because engraving was expensive and time-consuming, it became obsolete once the photographic halftone technique was perfected in the 1890s. Today, little is known about the engravers themselves, only that this once flourishing industry had all but disappeared by 1900.

While serving in the utilitarian sense by conveying essential visual information, engravings also function on an aesthetic level through the beauty of their composition and craftsmanship. Given new life in this book, they evoke the spirit of that wonderful bygone era. In varying degrees of quality, they depict scenes rarely available in surviving photographs. Occasionally, they also provide graphic representations of ideas that never materialized or show action scenes, which at the time were impossible to photograph. Also depicted are many beautiful examples of locomotives, passenger coaches, freight cars, streetcars, and so on. These often show detail better than surviving photos from the period.

These scarce engravings came from many sources, primarily books and journals published in America, England, France, and Germany. For those having an interest in the origin of these pictures, an image source list has been provided. The contents of this book have been organized first by topic, and then, insofar as possible, chronologically. When an image fits more than one category, I have chosen the one most appropriate to the editorial needs of this book. It is hoped, on my part, that the text and pictures of each chapter will help to amplify each other. Whenever possible, I have provided images corresponding to the specific text discussion.

Keeping an international audience in mind, I have tried to use railway terminology appropriate to each circumstance. Thus, when writing on America, words like *railroad, passenger car, freight car, switching, engineer,* and *fireman* have been used. For subjects elsewhere in the world, however, where English terminology or the foreign-language equivalent was more prevalent, I have used words like *railway, carriage, goods wagon, shunting, driver*, and *stoker.*

In addition, whenever the foreign names of railways have been available, like France's Nord and Ouest, they have been used rather than their English equivalent. This book has also been written with an eye to the political realities of that time. Thus certain parts of Central Asia are covered in the chapter on the Russian Empire, and material relating to early railways of Ottoman-controlled Europe can be found in the chapter on Asia.

This volume should provide those familiar with railway history much new material to contemplate. For newcomers it may offer an exciting door into a world of endless fascination; one that can be investigated further in the pages of hundreds of other railroad books. Finally, it might be noted that all of the visual material in this book lies in the public domain and thus can be used for clip art.

Note on the illustrations: With a few exceptions, engravings from original nineteenth-century publications have either been cut out and pasted up for this book or, when a size change has been required, have been professionally photographed accordingly. While this process has been destructive to the original sources, and costly to the author, it was nevertheless necessary in order to ensure the highest degree of image quality, thus fulfilling the author's intent that these images will be passed down, virtually intact, to future generations. In two cases prints were made from transparencies provided by the Mary Evans Picture Library of London. In a relatively few instances, engravings arrived in an unsuitable condition for this volume, but because they filled some necessary niche, were retouched, either by hand or (later) by computer. Retouching varied according to each image but usually involved improving line quality and sharpening detail. Different background skies were substituted in a few instances to improve graphic quality. A few images have been trimmed somewhat to fit existing space.

As this book was originally conceived only as an image source volume, no effort was made in its early stages to document the source of engravings. Furthermore, many of these images were purchased from print dealers and more often than not had incomplete data regarding the source. In some instances, publications are credited but without dates, because this is all that is known.

ACKNOWLEDGMENTS

MANY individuals have helped in various ways to bring this volume into being. I would first like to acknowledge my mother, Bennett Kerr, John and Marty Marmaduke, Ingo Swann, and Jean Letschert, all of whom gave indispensable aid to this project. I would also like to thank John H. White Jr., former curator of transportation at the Smithsonian Institution, who was very generous in answering a number of my inquiries. I am similarly indebted to Phil Atkins of the National Railway Museum, York, England, and Bernard Bonnet and Christian Bourdon of the Association Française des Amis des Chemins de Fer, Paris. For answering other queries, thanks go to Jean-Marc Combe of the Musée Français du Chemin de Fer in Mulhouse; G. P. Zakhrevskaya of the Central Museum of Railway Transport, St. Petersburg, Russia; Charles H. Stats, Railway & Locomotive Historical Society, Oak Park, Illinois; and Albert Eggerton of the Smithsonian Institution, Washington, D.C.

I am also indebted to my editor, Robert J. Brugger; Julie McCarthy; Maria denBoer; Susan Lantz; James Hendricks; John Digby; Joan Hall; Eric Edelman; Kathleen Dameron; Christine Merllie Young; Mike Harter; Barbara Harter Whitton; John Topham; Hugh F. Foster Jr., MG (ret.); Gene Wilde, Carl, and Tino at Miller Blueprint Co. of Austin; Hilary Evans of the Mary Evans Picture Library, London; Ruth Fiori; Terrelita Maverick; Phillip Krumm; Dennis Stacy; Jennifer Sunseri; and Larry Hollern.

I also received help from the following book and print dealers: Hans Fellner of Arts & Livres Librairie, Luxembourg; Librairie Alain Brieux, Paris; Robert Gavora, Talent, Oregon; A. Amitin Book Co., St. Louis, Missouri; Kathleen Manning of Prints Old & Rare, Pacifica, California; Robert Estrada, Albuquerque, New Mexico; Arnold B. Joseph, New York; Paul Sarna, New York; Shirley Solomon, New York; Shirley Sorenson, San Antonio, Texas; and Harry S. Newman of The Old Print Shop, New York.

WORLD RAILWAYS *of the* NINETEENTH CENTURY

CHAPTER

1

British Railways 1800–1850

RAILWAY technology traces its origins at least as far back as ancient Greece, where carts were pushed or pulled between parallel lines of stone slabs. This system later came to be used by the Romans. It also has been suggested that wooden rails were used across the Isthmus of Corinth for the overland transport of boats.

In northern Europe, primitive rail use dates at least to the fifteenth century. Over the next hundred years, the operation of carts over wooden rails in German mines became a common practice. This technology appears to have first reached Britain late in the sixteenth century in the area of Nottinghamshire. Beginning in 1602, rail lines were built in the colliery districts around Newcastle, and these later became known as "tramways." Typically, such lines were laid from the colliery to the nearest river, where the horse-drawn coal was loaded into boats. By providing a smoother and harder surface over which wheels would roll, this technology reduced resistance, allowing horses to pull much more weight than before.

Circa 1760, during a slump in the iron market, iron straps 5 feet long, 4 inches wide, and 1¼ inches thick began to be laid over the wooden rails. Designed for easy removal should the price of iron go up, these straps were seldom removed because they further reduced friction and improved track durability. Later, other types of cast iron rails were invented.

The prevailing practice in England and Wales was for operating wheels with flat surfaces on rails having outer flanges. This changed in 1789, when William Jessop built his Loughborough & Nanpanton Tramroad. This 3-mile line, connecting communities on the Leicester and Charnwood Forest canals, employed a new type of flat rail called the "edge rail." Being more vertical than horizontal, it was so called because the wheels ran on its upper edge.

Each rail consisted of short sections several feet in length, abutting one another, which were fixed by bolts to wood or stone. When viewed from the side, the lower edge of each section curved downward in a long ellipse suggesting the shape of a fish belly. This "fish-bellied" design provided greater strength. Reviving earlier experiments, flanged wheels were employed for use on this primitive track.

There was considerable tramway construction during this period. By the year 1800, in the Newcastle area alone, about 200 miles of tramway had been built above ground and perhaps an equal distance in the mines below. Two other tramways worth noting were the Surrey Iron and Oystermouth lines. The former, opening on July 26, 1803, became the world's first public railway. Inspired by the operation of barge canals, it was set up to allow cartage contractors, or anyone else with a horse and rail car, to move over the line upon payment of a toll. Extending for 9½ miles, the line connected Wandsworth, on the Thames, with Croydon. It was originally intended to extend all the way to Portsmouth. The Oystermouth Rwy., also known as the Swansea & Mumbles, opened in April 1806. On March 25, 1807, it became the world's first line to inaugurate passenger service.

Railways, as we know them today, resulted from combining this tramway technology with that of steam power. The steam engine also had roots in antiquity. Hero of Alexandria, in the first century CE, described the aeolipile, a primitive steam-reaction turbine. Although experiments with primitive steam devices continued in later centuries, it was not until James Watt made improvements on an engine built by Newcomen that steam power reached a level of practicality. Watt received patents for his improvements in 1769. Early steam engines were put to such uses as driving factory machinery, pumping water, and powering lifts. Because they were enormous in size, operated at low levels of steam pressure, and were very inefficient in fuel consumption, they were not readily adaptable for providing power for mobile vehicles. This would require smaller and more efficient engines. Watt later made other improvements on the steam engine, but he avoided experiments involving higher pressure.

Nonetheless, as early as 1769, efforts were made to adapt steam for traction purposes. That year the Frenchman Nicholas Cugnot built a carriage that employed two single-acting vertical steam engines operating alternately on a driving axle. In 1786 Symington built a carriage powered by a Watt condensing engine. However, neither of these experiments was very successful. In Philadelphia, Oliver Evans developed a high-pressure steam engine that he adapted for industrial use in 1802. Two years later he built an ungainly vehicle, the *Oructor Amphibolis,* which, in essence, was a steam-powered scow with wheels. It could move on either land or water. In later years steam road vehicles continued to achieve greater refinement. Had there been less vehement opposition to their use over British roadways, they conceivably might have competed against the railways.

Richard Trevithick built the first steam locomotive at Shropshire in 1803. It had one cylinder, a flywheel, and operated at relatively high steam pressure. Through experimentation Trevithick found that if supported by sufficient weight, drive wheels had adequate adhesion to draw heavy loads on smooth rails. One of his most important discoveries was that by conveying exhaust steam from the cylinder to the chimney by means of a pipe, which he turned upward, the blast of steam created an artificial draft, causing the fire to burn much better. This innovation he called the "blast pipe."

On February 24, 1804, Trevithick's engine was tried on the Penydarran Tramway in Wales. It drew cars containing 10 tons of iron and about 70 people to Merthyr Tydfil, a distance of 9 miles. This experiment was considered a success from a mechanical point of view, although, commercially speaking, horse-drawn power remained less expensive. The primary problem was that the weight of the engine was too heavy for the existing track; thus any cost savings from using steam was offset by the expense of track replacement. In 1805 Trevithick built his second engine for a colliery at Gateshead. Later, in 1808, he set up a circular demonstration track in London, upon which ran his fourth locomotive, *Catch Me Who Can.* A short time later an increase in the cost of horse fodder, owing to the Napoleonic wars, made steam traction a more viable option.

In 1811, John Blenkinsop, proprietor of the Middleton Colliery, had Matthew Murray, an engineer of Leeds, build a 5-ton locomotive, the *Blenkinsop.* Lighter than Trevithick's engines, it was uniquely designed with cogwheel driving gear. Acting on a center rack-rail it provided the necessary traction for drawing heavy loads as well as for ascending grades. The locomotive had four supporting wheels, but the driving force was conveyed by connecting rods to cranks, and then by means of spur-wheels and gear to the main cog or driving wheel. Unlike Trevithick's engine, the *Blenkinsop* had two cylinders, thus eliminating the need for a flywheel. The initial run in August 1812 was successful. The engine went into immediate service hauling coal 3½ miles daily from Middleton to a wharf in Leeds. It consumed 75 pounds of coal and evaporated 8 cubic feet of water per hour. On a level track it could draw 94 tons at 3½ m.p.h. and 20 tons at 10 m.p.h. Other similar locomotives were built and worked well for many years. They were considered a commercial success.

At New Bottle, in 1813, William Brunton devised a four-wheel locomotive that relied for traction on two piston-powered "legs" operating on the ground. The motivation for this arrangement was apparently the same lack of confidence in smooth rail adhesion that had inspired Blenkinsop. Brunton's *Steam Horse* remained in operation until it exploded in 1815.

In May 1813 Mr. Blackett of the Wylam Colliery and William Hedley completed construction of the locomotive, *Puffing Billy.* After studying Blenkinsop's system on the Middleton line, they decided to return to Trevithick's idea of operating smooth driving wheels on smooth rails. After determining in a test that this provided sufficient adhesion, they built an engine, which they mounted on their test frame. Tried out on the 5′ gauge Wylam line, it was unsuccessful because of a lack of steam. Blackett and Hedley provided a new boiler and cylinders, adding two of Trevithick's blast pipes, which gave the engine a puffing sound. Following the completion of *Puffing Billy,* they built a second engine, *Wylam Dilly.* Both locomotives were successful, but their weight caused the cast iron plate rails over which they rode to break frequently. To overcome this problem each engine was placed on two four-wheel bogies, with each of the eight wheels receiving power from gears. The locomotives were returned to four-wheel operation when the plate-way was replaced by iron track in 1830.

On July 25, 1814, the *Blucher,* a locomotive built by George Stephenson, was successfully tested on the Killingworth Colliery line near Newcastle. Two days later it conveyed a train of eight wagons of coal, weighing 30 tons, up a gradient of 1 in 450 at 4 m.p.h. Stephenson, born alongside the Wylam tramway in 1781, was a self-taught engineer who had grown up in poverty. However, he showed uncommon intelligence and self-determination. In 1816, he was consulted over a proposed rail line to connect the coal-rich town of Darlington with the port of Stockton. Upon his advice, the directors decided to construct the line with iron rails rather than using wood. Stephenson was appointed engineer and surveyor of the line, and after gaining the confidence of the directors, he successfully persuaded them to employ a steam locomotive.

There was considerable opposition to this line, however, and it did not receive approval by Parliament until 1821. On September 27, 1825, it was finally completed. Stephenson's 0-4-0 *Locomotion,* the first engine having coupled wheels, pulled an inaugural train consisting of 6 coal wagons, a covered coach carrying the directors and proprietors, 21 wagons loaded with passengers, and, lastly, 6 more coal wagons. Traveling down a gentle incline, the train reached a top speed of 15 m.p.h. A crowd of 12,000 watched the train's departure at Darlington, and 40,000 were present for its arrival at Stockton, where it received a 21-gun salute.

Although the railway was initially intended only for hauling merchandise, the company soon allowed a contractor to offer daily passenger service between Stockton and Darlington. The *Experiment,* an enclosed passenger coach designed by Stephenson, began operations on October 10, 1825. Drawn by a horse, it made the daily 12-mile trip between Stockton and Darlington in about 2 hours. The success of this venture created a demand for other rail carriages, and these were eventually attached to goods trains.

Such a development had been foreseen years earlier by Thomas Gray, a writer in Nottingham. Gray had visited a colliery line and became obsessed with the vision of a much greater rail network in the country, one capable of carrying passengers as well as goods. He became such a persistent advocate of this dream that many in England insisted he was mad. Nevertheless, his prophetic treatise on railways was inspirational to some, and went through five printings in his lifetime.

Another important step in locomotive design occurred with the rebuilding of the Stockton & Darlington's *Royal George* in 1827. Constructed the previous year by Wilson & Co., of Newcastle, it had an unusual four-wheel, four-cylinder, 0-2-2-0 configuration. Following a collision, it was rebuilt by the railway's engine superintendent Timothy Hackworth. He removed two cylinders, and in a new innovation, gave it six coupled wheels and provided a tender at each end. This resulted in both greater power and greater adhesion. The maximum performance for this engine was drawing 32 cars of coal weighing 130 tons at 5 m.p.h. It worked on the line until 1842.

George Stephenson's son, Robert, was born in 1803. From an early age his father instilled in him self-discipline and the desire for knowledge. Robert trained under his father's guidance in engineering skills, and attended the University of Edinburgh for one term. In 1822, after returning from school, he apprenticed himself to his father, who besides constructing the Stockton & Darlington line, had begun the manufacture of steam locomotives in Newcastle. Two years later, exhausted from overwork, he took a position leading a party to South America for the inspection of silver and gold mines. He later spent time with Richard Trevithick. It was only in 1828 that Robert returned to work at Newcastle.

During his son's absence, George Stephenson had taken a position as principal engineer for construction of a rail line between Liverpool and Manchester. These two cities, having a combined population of 350,000, were among England's most important. The railway to be built between them was the first conceived for improving commerce between two cities. However, gaining parliamentary approval for it was difficult. In opposition were landowners, canal proprietors, and stagecoach operators. Once construction of the line commenced, it required considerable efforts and engineering expertise. In the course of the project 63 bridges were erected, numerous cuttings and embankments made, a tunnel constructed, and a swamp crossed. Including legal and parliamentary expenses, land purchases, surveying, and the like, costs for the line's construction totaled 739,183 pounds.

Initially the company was undecided about the type of traction to be used for the line. Deciding that horses were inadequate, it sent out a number of engineers, including Stephenson, to investigate other railways and to consider the relative merits of steam locomotives versus stationary engines. Stationary engine systems required setting up engines at certain intervals called stations. Trains of wagons could thus be drawn by means of long ropes from station to station. While slow, these systems were dependable and were especially useful on grades. The company finally chose steam locomotives, but also employed stationary engines on two slopes. By this time the company had also begun to consider the idea of offering passenger service. This would require passenger trains to operate at speeds at least comparable to those of competing stagecoaches. At this point the company decided to offer a competition with a prize of 500 pounds, in hopes of attracting men of science to design an improved locomotive.

What resulted was the famous Rainhill trials of October 1829, where five locomotives competed for honors: Robert Stephenson's *Rocket,* Timothy Hackworth's *Sans Pareil,*

Braithwaite and Ericsson's *Novelty*, Timothy Burstall's *Perseverance,* and Brandeth's *Cyclopod.* Certain specifications were required for each engine. Four- or six-wheel locomotives could be entered, the former not exceeding 4½ tons and the latter not exceeding 6 tons. Each engine was required to consume its own smoke, and on a level plane pull a train of carriages for an extended distance averaging at least 10 m.p.h., while not exceeding a steam boiler pressure of 50 p.s.i. For the trials, each locomotive was required to run 10 round trips over the 1¾-mile distance of the course at full speed, the object being to reach or surpass the required rate of speed. Following this, the whole sequence was repeated once more in order to prove that the engine could complete a journey from Liverpool to Manchester and back.

Brandeth's *Cyclopod,* a four-wheel vehicle powered by two horses on a treadmill, was disqualified before the trials. Hackworth's *Sans Pareil,* a hastily built four-wheel engine, exceeded weight restrictions, wasted fuel, and had other problems. The Braithwaite and Ericsson engine was quick, unconventional in design, and borrowed ideas from Goldsworth Gurney's experimental steam carriages. Brightly painted and polished, it employed two hand-operated bellows, and was clearly the crowd's favorite. However, it suffered two mechanical breakdowns and was withdrawn. Burstall's *Perseverance,* weighing only 2.85 tons, was eliminated early in the trials.

In contrast, the Stephenson engine, which had received input from both father and son, was well engineered and performed admirably. It was simple in design, yet more powerful than previous locomotives. Its most important feature was the multitubular boiler. In this system many copper tubes of small diameter replaced the single large U-tube mounted in the boiler barrel. Hot smoke and gasses exiting from the firebox traveled toward the chimney through these tubes. This change, first suggested by Henry Booth, secretary of the Liverpool & Manchester Co., greatly hastened steam conversion by providing more surface area for heating. Another innovation on the *Rocket,* pioneered the previous year in the *Lancashire Witch,* was positioning the cylinders at an inclination where the wheels were driven directly by the piston rod. Previously the cylinders had always been placed vertically. A single set of drivers was used, supported in the rear by carrying wheels. Blast pipes, one from each cylinder, were extended into the smokestack. The *Rocket* was the only engine to complete the trials successfully. It averaged 16 m.p.h. over the course and at one point reached a top speed of 29 m.p.h. For a brief period it became a prototype for the manufacture of other locomotives. However, technology was evolving so quickly that the *Rocket* was soon obsolete.

The Liverpool & Manchester line was inaugurated on September 15, 1830, with great fanfare. However, on that day the first public railway fatality occurred. A visiting dignitary, William Huskisson, M.P., was struck by the *Rocket* and died shortly afterward. Despite this inauspicious beginning the line became immediately profitable and quickly gained a good volume of both freight and passenger traffic. The changes that it brought were immediate. Men could travel from either city to the other, conduct business, and return home the same day. Market prices for coal and other merchandise were reduced as transportation costs decreased. Contrary to the fears of many, the line naturally stimulated commerce, and property values along its route increased. Double-tracked over its entire route, it became a model of inspiration for future railway construction and was visited by engineers from around the world.

Following the completion of the Liverpool & Manchester line, the Stephensons became involved in the construction of the London & Birmingham Rwy., and later in building the Birmingham & Derby, North Midland, York & North Midland, Manchester & Leeds, and Northern & Eastern lines. During this same period they were also extensively involved in planning a 347-mile rail network in Belgium. In the meantime they continued to manufacture locomotives both for the domestic market and for export. They continually undertook new experiments, and when successful, the improved engines became prototype designs for manufacture.

The 2-2-0 *Planet,* built for the Liverpool & Manchester line in 1830, had two horizontal cylinders within the frames beneath the smoke box. This design provided both a shorter route for the spent steam into the blast pipe and a direct drive to the double-cranked rear axle. The 0-4-0 *Samson* was built in a fashion similar to the *Planet,* but its four coupled wheels gave it better adhesion. Also on this locomotive was another important innovation: the world's first whistle. (Previously some engines had carried bells or trumpets.) This design became so popular that Stephenson could no longer keep up with demand, and had to license manufacturing rights to the firms of Matthew Murray, Hawthorne of Newcastle, and Taylor & Co. The 2-2-2 *Patentee,* built in 1833, had a larger boiler and firebox. It was basically an improvement on the Planet class 2-2-0, correcting the instability inherent in that design. It had 12″ × 18″ cylinders and operated at a boiler pressure of 50 p.s.i. For the next few decades, the 2-2-2 became the most common type of English engine. The 0-6-0 *Atlas,* built for the Leicester & Swannington line in 1834, became an important prototype for later goods locomotives.

In addition to companies manufacturing Stephenson's designs, other locomotive builders entered the market and soon began building other types of engines. This group included Timothy Hackworth; Sharp, Roberts & Co. of Manchester (this company soon changed its name to Sharp, Stewart & Co.; later in the century it would shift its locomotive production to a facility in Glasgow); Edward Bury & Co. of Liverpool; Kitson-on-Leeds; Fenton, Murray & Jackson of Leeds; Jones, Turner & Evans of Newton; Nasmyth, Gaskell & Co. of Manchester; Mather, Dixon & Co. of Liverpool; the Haigh Foundry Co. of Wigan, Stothert & Co. of Bristol; Rennie & Co. of London; and Rothwell & Co. of Bolton-le-Moors. By 1840 Britain had approximately 30 locomotive manufacturers.

Edward Bury, who for a time was locomotive superintendent of the London & Birmingham line, invented a distinctive cylindrical firebox with a domed crown. This innovation was soon widely copied in America and on the Continent. Bury's firm gained the concession for the North Western Rwy. in 1836, but was out of business by 1850.

The first locomotive-tender combinations, soon known as "tank engines," were built by Forrester & Co., for Ireland's Dublin & Kingstown Rwy., in 1835. Having no tender, these engines carried both fuel and water onboard. Unlike regular locomotives, they could be operated in either direction, and thus did not require turning around. The original two were used for working mail-boat trains at the pier at Kingstown. Circa 1847, tank engines were running on the London & Blackwall Rwy., and on a branch line between Cheadle Hulme and Macclesfield. Originally, water was stored either at the rear or in side tanks, but saddle tanks were later developed. Tank locomotives became popular, from about 1850 onward, for both short-distance runs and switching.

In 1830 the Duke of Wellington, then leader of the British government, was asked to appoint some engineers to lay out four or more main railway lines to be built by the state. The duke's reply was that he did not like railways. Mr. Goulburn, chancellor of the exchequer, took no further action in the matter, fearing strong opposition from powerful vested interests. Thus, in the absence of governmental direction, private enterprise quickly filled the vacuum and supplied the impetus to Britain's subsequent railway development.

People did not like railways for a number of reasons. Some feared change. Others thought railways might threaten their means of livelihood, invade their property, or bring some other disruption to their lives. Railways were also bringing changes to the landscape, threatening its beauty and harmony. In *The Railways of Britain,* Jack Simmons states:

> The railways have done much to change the face of the landscape. Consider, as a simple example, the building of the great embankments that appear in the prints of Bourne, and the changes they wrought in the undulating country that they strode over. When the railways were being built in the 1830's and 1840's, many people attacked them for the destruction they were wreaking in the countryside—a great part of England was then a planned landscape in a sense in which it will never be again. Both in country and in town, these were the years in which England attained its greatest physical beauty. By 1830 the work of the 18th century landscape artists had had time to come to maturity: the trees they had planted had grown, and all the changes they had made had mellowed. Millions of pounds had been invested, altogether, in this improvement. We cannot be surprised, therefore, that the cutting of railways through a landscape so recently brought to perfection was vigorously opposed.

Between 1833 and 1837, England's first railway boom occurred, with construction beginning on many new lines. The first to serve London was the London & Greenwich Rwy. Its first section, from London to Deptford, opened on December 14, 1836. Greenwich was reached about two years later. This line became known for a viaduct having nearly 900 arches. Another early railway, the London & Croydon, inaugurated on June 5, 1839, gained access into the city by operating over the Greenwich tracks beginning at Corbett's-Lane. Both companies had termini near what would later become the South Eastern line's London Bridge Station.

The London & Birmingham Co. inaugurated its first portion, from Camden Town to Boxmoor, on July 20, 1837. The railway completed an extension into Euston Square three months later. The first portion of the London & South Western line ran from a station at Nine Elms Road to Woking. It opened on May 21, 1838. Other early lines serving London were the Eastern Counties, inaugurated on June 20, 1839, and the Blackwall Rwy., which opened on July 6, 1840.

The opening of the Great Western Rwy. occurred on June 4, 1838. A 2-2-2 Stephenson engine, the *North Star,* pulled the inaugural train from a London terminus at Bishop's Bridge Road to Maidenhead. By 1841 this line extended all the way to Bristol. Isambard Kingdom Brunel, a true engineering genius, was its "architect." He broke from the 4′8½″ gauge precedent established by George Stephenson and built his railway with a 7′¼″ gauge. This came to be called the "broad gauge." Stephenson's gauge, originally derived from the 4′8″ roadways common in northeastern England, then became known, for a time, as "narrow gauge."

Through careful reasoning Brunel determined that wider locomotives could operate at higher speeds: first, a wider gauge provided greater stability, thus there was less tendency for the train to wobble or leave the track; and, second, a larger heating surface was provided for making steam. A few other lines followed Brunel's example, and this ignited a great controversy that took years to resolve. The primary problem of this two-gauge system was the great inconvenience at the change of gauge points where goods or passengers had to be transferred from the cars of one line to another. This practice, called *transshipment,* created delays and extra expenses, and increased the potential for both damage to merchandise and theft. Until 1842 transshipment was virtually universal. Railways usually limited their operation to their own equipment; thus transshipment occurred anytime goods passed from one rail line to another. The establishment of the Railway Clearing House that year made possible through shipment of goods in the original car from origin to destination. However, through shipment was impossible when gauges changed. In 1846 Parliament passed the Gauge Act requiring all further railway construction in England and Scotland to be in the 4′8½″ gauge, which from that time on was called the "standard gauge." The standard for Ireland was set at 5′3″, a compromise average of various gauges already existing there.

A second period of rapid growth resulted in the railway mania of 1845–46. In November 1845, the *Times* observed that over 600 new railways were being promoted, while preliminary work had begun on perhaps that many more. From 1845 to 1848, in three sessions of Parliament, about 650 articles of railway legislation were passed. These acts included authorization for construction of about 9,000 miles of line. Because laissez-faire capitalism prevailed at the time, the government provided little guidance. Thus in many cases railway construction was primarily motivated by greed and directed to those areas where it was thought the greatest potential profits lay. This led to duplication of services in some areas, while other parts of the country were deprived of railways for years. Speculation became rampant, culminating in a crash, where, on average, the value of railway stocks fell by 50% and many fortunes were lost. Among the risky investment possibilities of this period were so-called atmospheric railways. These are discussed in greater detail in chapter 24.

One man intimately connected with the events of the time was George Hudson, who became known as the Railway King. Having inherited a fortune, he rose to become lord mayor of York in 1837 and then chairman of the York & North Midland line. Through shrewdness, charisma, and Herculean energy he eventually consolidated this line and others into the Midland Rwy., a system that extended for 1,016 miles. This was no small accomplishment, considering that the total distance of British railways built by 1850 was only about 6,500 miles. Hudson once made 100,000 pounds in a single day. He became a vastly admired celebrity and, despite a lack of personal refinement, was courted by the great and famous. However, at some point his dealings became unscrupulous, and in 1849 public revelations forced his downfall.

In *Our Iron Roads,* Frederick S. Williams quotes an unnamed writer of the time on the subject: "The Truth is, Mr. Hudson is neither better nor worse than the morality of 1845. He rose to wealth and importance at an immoral period; he was the creature of an immoral system; he was wafted into fortune upon the wave of a popular mania; he was elevated into the dictatorship of railway speculation in an unwholesome ferment of popular cupidity, pervading all ranks and conditions of men; and whatever may be the hue of the error he committed, it is rather too much to expect of him that he should be purer than his time or his associates."

The first royal mail was carried on the Liverpool & Manchester line on November 11, 1830. In 1837 the General Post Office negotiated contracts with most of the major lines of Britain for the transport of mail. The first Traveling Post Office (TPO) began operation on the Grand Junction Rwy., between Birmingham and Liverpool, on January 6, 1838. In its initial form, it consisted of a converted horse wagon, in which raw mail was sorted en route to its destination. On the night of February 4, 1840, England's first special mail train was inaugurated on a run between London and Twyford. Beginning circa 1840 methods began to be devised for picking up and dropping off mail while trains were still in motion. Circa 1844 special mail trains on the South Eastern line began carrying the Indian Mail between Folkestone and London.

Improvements in locomotive design continued at a rapid pace. Robert Stephenson, whose father had patented steam locomotive springs in 1816, got his own patent for steam brakes in 1833. Beginning in the 1830s, experiments were undertaken to increase the size of driving wheels on passenger engines. By using larger drivers, top speeds could be considerably increased, although this was done at the expense of having less power for starting from dead stops. This trend probably reached its ultimate expression in the *Ajax,* built in 1838, which had drivers 10 feet in diameter. During trials in the 1830s, locomotives reached speeds in excess of 60 m.p.h., and this trend would continue.

In June 1841 Robert Stephenson introduced his long boiler design. The increased length of the boiler tubes resulted in a more productive use of the heat from the firebox without raising the locomotive's center of gravity. Another Stephenson improvement was link motion, invented in 1842 by William Howe, an apprentice at the Newcastle works. A shifting link, fitted to the valve gear, provided a way of cutting off steam at different stages of the stroke. A long cutoff provided smooth, powerful starts, while a short cutoff worked better at higher speeds. The link conserved steam pressure, water, and fuel by allowing the natural expansion of the steam, initially provided to the cylinder, to finish out the stroke. Although others were experimenting with similar systems, the Stephenson device won out and eventually gained almost universal acceptance.

In 1837, Daniel Gooch, age 21 and a former Stephenson employee, became locomotive superintendent of the Great Western Rwy. He began by upgrading the haphazard collection of existing engines, built by a number of different manufacturers. Soon he took over designing the company's locomotives, using a Stephenson-built Patentee class engine as his prototype. Gooch, along with his brother John, who held a similar position with the South Western Rwy., devised their own cutoff device. In the early 1840s a speed competition emerged between the broad-gauge Great Western and some of its narrow-gauge rivals. This contest came to be called "the battle of the gauges," and in it, Gooch played a prominent role. In 1845, his 2-2-2 *Ixion* set a new speed record. Pulling a 71-ton train, it averaged 55 m.p.h. in the 53-mile stretch between London and Didcot, and reached a top speed of 61 m.p.h.

The establishment of the Swindon works in 1846 allowed the Great Western to build most of its own locomotives. In doing so, it followed the lead of the Grand Junction Rwy., which had constructed a similar facility at Crewe in 1843. The first new engine from Swindon, completed in 1847, was the 2-2-2 Great Western. Having driving wheels of 8-foot diameter, it attained an average speed between Didcot and London of 67 m.p.h. and reached a top speed of 78 m.p.h. A number of similar engines were built on this pattern. The following year an extra axle was added at the front, creating the world's first 4-2-2 type. The first of these, the *Iron Duke,* became a prototype for 29 other engines built from then until 1855. These locomotives, having cylinder dimensions of 18″×24″ and operating with a boiler pressure of about 100 p.s.i., gave the Great Western an enviable reputation for speed as well as a certain glamour. Beginning in 1849, the company's London-Didcot express train became known as the *Flying Dutchman.*

Another important engine designer of this period was Thomas Crampton. He had previously served as Gooch's assistant before taking out a patent, in 1843, for a locomotive of long boiler design. It had large drivers in the rear, behind the firebox, and an additional set of carrying wheels in front, and was thus designated as a 6-2-0. In this plan he also placed the cylinders and valve gear outside the wheels, thus allowing easier maintenance. This system was supposed to solve the problem of having large driving wheels while at the same time maintaining a low center of gravity.

Two standard-gauge engines of this design were built in 1846 for export to Belgium. In trials on the London & Northwestern line, the *Namur* was reputed to have exceeded the top broad-gauge speed to that time. In 1848 Crampton's *Liverpool* was tested for a period of time on the same line. Although it reached a top speed of 78 m.p.h., it did some damage to the track and gave Crampton engines a poor reputation in England. After a few modifications, most problems of this design were eliminated. While few were sold in Britain, they achieved great popularity on the Continent, and between 1849 and 1864, about 300 were built in France and Germany.

The Grand Junction Rwy., inaugurated in 1837, connected Liverpool to Birmingham. It was constructed by Thomas Brassey, who later gained fame as an international railway builder. After consolidation of the Grand Junction, London & Birmingham, and Manchester & Birmingham lines into the London & Northwestern Rwy. in early 1846, locomotive building was continued at Crewe. Under Alexander Allan and Francis Trevithick, innovations were made in the company's standard-gauge engine designs, which greatly improved performance. Allan invented his own form of link motion and, following the precedent of the previous engine superintendent, William Buddicom, placed cylinders and valve gear on the outside. Trevithick's 2-2-2 *Cornwall,* completed in 1847, had drivers 8′6″ in diameter, the largest ever used on a standard-gauge engine. In trials it reached a top speed of 79 m.p.h.

Railways came later to Scotland. The first tramway was the Kilmarnock & Troon line, which opened in July 1812. The first steam locomotive was operated there in 1817. The next line to open was the Monkland & Kirkintilloch Rwy. in 1824. Later in the decade construction began on a number of other lines. Among these, the most important was the Garnick & Glasgow line, which connected to the Monkland & Kirkintilloch. The primary purpose for railways of this period was for hauling coal, although some passengers were carried as well. Construction of major lines in the country did not begin until the 1840s. On February 21, 1842, the 46-mile Edinburgh & Glasgow Rwy. was completed. The Caledonian Rwy. opened a route between Glasgow and Carlisle in 1848. That same year a route between Glasgow and Aberdeen was completed by the Scottish Central and Scottish North Eastern lines. On February 16, 1850, service began on the world's first railway ferry between Granton and Burntisland. Operating over the Firth of Forth, the *Leviathan* connected portions of the Edinburgh, Perth & Dundee line.

During the same period a few small railways were constructed in Wales, mainly for transporting coal and other minerals. Among these was a tramway of a tiny 1′11½″ gauge used for transporting slate. This was the Festiniog Rwy., a line that later became famous. A similar line was the 4′ gauge Padarn Rwy., which opened in 1843. It transported slate from Dinorwic Quarry, Llanberis, to Port Dinorwic on Menai Strait. A much longer railway, the South Wales line, opened on June 18, 1850. Beginning near Hagloe, where it connected with the Forest of Dean Rwy., it proceeded to Chepstow, crossed the River Wye, and continued to Cardiff. From Cardiff it went northwest through a coal-rich region, returning to the coast at Briton Ferry, and continuing to Swansea. The Taff Vale Rwy., first authorized in 1836, was constructed between Merthyr Tydfil and the Cardiff Docks. Including a few branches the line was 32 miles long. Like many other railways, it later built numerous extensions.

The first line in Ireland was the Dublin & Kingstown Rwy., which opened on December 17, 1834. During the 1840s construction began on the Great Southern & Western line, and by April 1849, it extended for 145 miles, connecting Dublin and Mallow. In May 1850 the Cork, Black Rock & Passage Rwy. was inaugurated. The "Irish Mail" train service, from London, was inaugurated on August 1, 1848, using the Chester & Holyhead line to reach the port of Holyhead. However, until Stephenson completed his bridge over Menai Strait in 1850, this journey had to be completed by ferry. Traffic had previously been routed through Liverpool.

By the end of 1850, a through route for carrying traffic between London and Edinburgh had been pieced together. Known as either the Great Northern or East Coast route, it involved the cooperation of the Great Northern, Scotland's North British line, and, in between, various small companies that were amalgamated into the North Eastern Rwy. in 1854. The completion of two of Robert Stephenson's bridges in 1850, one crossing the Tyne at Newcastle and a second over the Tweed at Berwick, greatly facilitated this enterprise. Service out of London initially began at the Great Northern's York Road Station; however, when King's Cross opened in 1852, it became the new terminus. This new route acquired a little more glamour with the inauguration of the *Flying Scotsman* express on August 8, 1850.

By 1850 the general outline of a rail network had been established in England and was taking form elsewhere in Britain. This rapid spread of railways accelerated industrial development, resulting in ever increasing commercial activity and prosperity. A sizable portion of this activity involved the importation of raw materials and exportation of manufactured goods. The creation of railways meant that facilities for these industries no longer had to be located near navigable waterways. During this period England's position as the world's leading industrial, mercantile, and colonial power was greatly consolidated.

The creation of these lines also provided unprecedented opportunities for economical passenger travel. This resulted in a greater mobility on the part of the public and an accelerated flow of ideas, ultimately leading to a growing sophistication of England's populace. All of these developments were synchronous with gradual social changes of various kinds, including a noticeable enlargement of the middle class. Thus passenger travel greatly contributed to railway earnings. In fact, until 1852, passenger revenues exceeded those of goods.

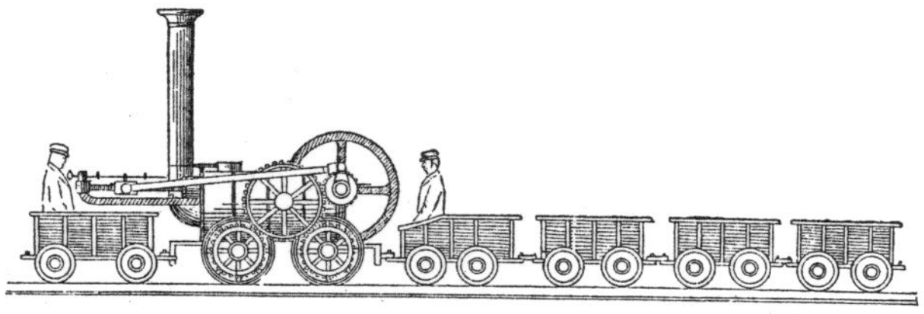

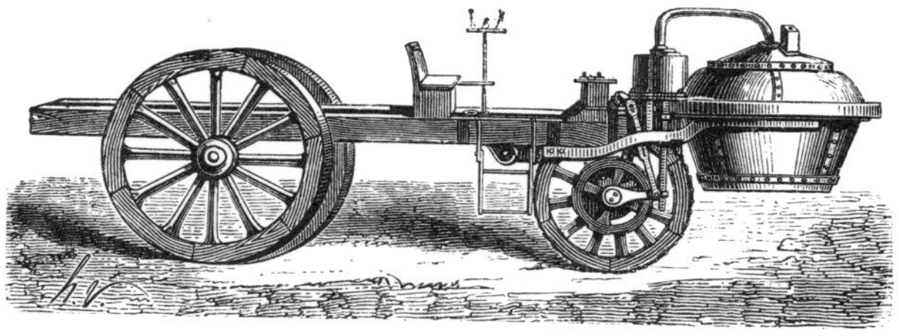

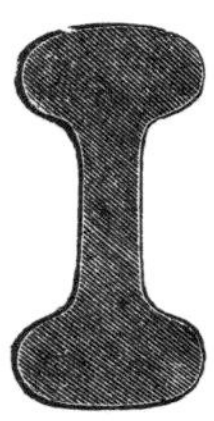

TOP LEFT: William Brunton's *Steam Horse*, New Bottle, England, 1813. TOP RIGHT: Trevithick's engine *Catch Me Who Can* being demonstrated at Euston, London, 1808. UPPER MIDDLE LEFT: Trevithick's first engine at Penydarran Tramway, Wales. MIDDLE LEFT: Blenkinsop's rack railway, Middleton Colliery, Leeds, 1812. MIDDLE RIGHT: Cugnot's steam carriage, 1769. LOWER MIDDLE LEFT, BOTTOM LEFT: Early rails. BOTTOM RIGHT: World's first locomotive built by Trevithick, 1803.

PLATE
1

PLATE 2

Top left: Blackett and Hedley's *Puffing Billy*, Wylam Colliery, 1813. **Top right:** The *Blenkinsop*, Middleton Colliery, 1812. **Middle left:** Braithwaite and Ericsson's *Novelty*, 1829. **Middle center:** 0-6-0 *Experiment*, Stockton & Darlington Rwy., circa 1828. **Middle right:** 0-4-0 *Samson*, Leicester & Swannington Rwy., 1833. **Bottom left:** Function of Blenkinsop's cog/rack system. **Bottom right:** Stephenson's *Rocket*, Liverpool & Manchester Rwy., 1829.

Top left: Opening of the Stockton & Darlington line, 1825: Top right: The *Royal George*, world's first engine having six coupled wheels, Stockton & Darlington Rwy., 1827. Middle left: Stephenson's *Rocket*, 1829. Middle right: Stephenson engine with four coupled wheels, Killingworth Colliery, 1815. Bottom: Stephenson's *Locomotion*, first engine of Stockton & Darlington Rwy., 1825.

PLATE 3

PLATE 4

Top left: Timothy Hackworth's *Sans Pareil*, 1829. Top right: 0-4-0 *Comet*, Leicester & Swannington Rwy., 1832. Middle left: 2-2-0 *Liver*, Liverpool & Manchester Rwy., 1832. Middle right: 0-2-4 *Earl of Airlie*, Carmichael & Co., Dundee; Dundee & Newtyle Rwy., 1833. Bottom left: 0-4-0 *Agenoria*, Foster, Rastrick & Co., Stourbridge; Shutt End Rwy., 1829. Bottom right: Early printed ticket, circa 1845.

Liverpool & Manchester Rwy., 1831: Top: First-class passenger train. Middle: Second-class passenger train. Bottom: Second-class passenger coach designed by George Stephenson.

PLATE

5

PLATE 6

Top left: 2-2-2 passenger engine with domed firebox, Great Western Rwy., circa 1838. Top center: Sections of various kinds of track including Vignoles, Brunel, and Barlow. Top right: George Stephenson. Bottom: Robert Stephenson's 2-2-2 *Patentee*, Liverpool & Manchester Rwy., 1833.

PLATE 7

2-2-2 *North Star* designed by Daniel Gooch, 1837. Gooch worked for the Stephenson Co. at the time, but later gained fame as locomotive superintendent of the broad-gauge Great Western Rwy. This engine may have originally been intended for Russian use, but instead ended up at the Great Western, pulling its inaugural train from Paddington on June 4, 1838.

Stephenson long boiler engines: TOP: 4-2-0 passenger locomotive "A," 1846. BOTTOM: 2-2-2 passenger engine, 1842.

PLATE
8

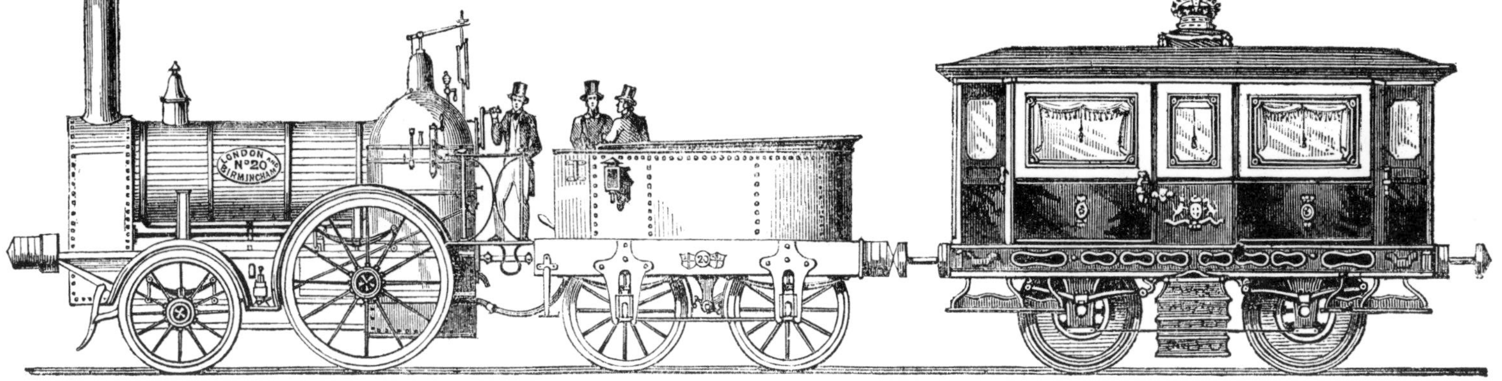

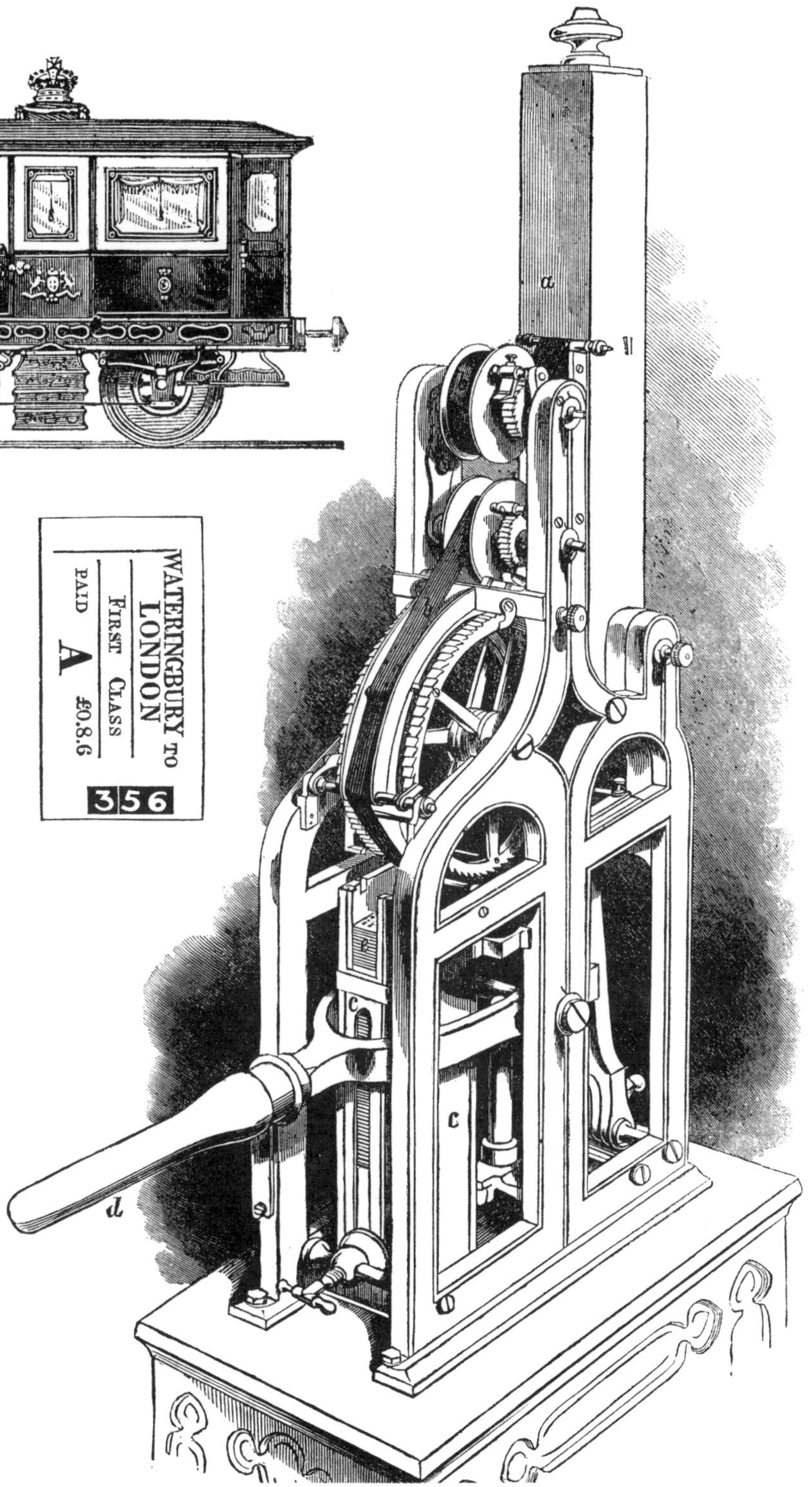

PLATE 9

Top: Royal train, London & Birmingham Rwy., 1843. The 2-2-0 Planet type engine, built by Edward Bury, has his distinctive circular style firebox with domed crown. **Bottom left:** Interior of royal carriage, South Western Rwy., 1844. **Bottom right:** Railway ticket machine designed by Thomas Edmonson, circa 1845. Prior to this time all tickets were written out by hand.

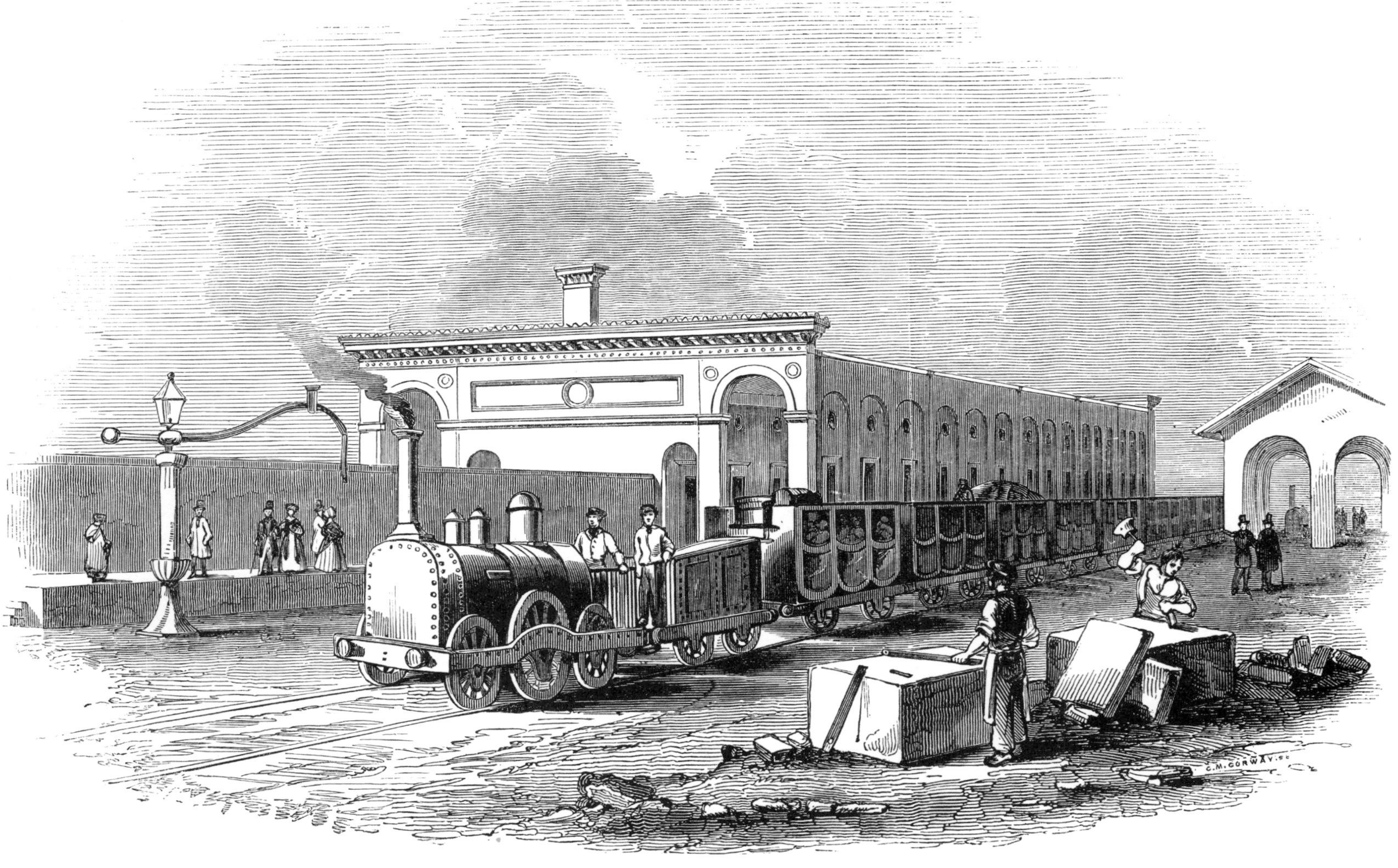

Top: Needham Market Station, designed in the Tudor style by Frederick Barnes, Norwich & Ipswich Rwy., 1849. **Bottom**: Cambridge Station, designed by Francis Thompson, Eastern Counties Rwy., 1845.

PLATE 10

Top: Bardney Bridge, Great Northern Rwy., circa 1849, depended on piling for support. Middle: Alderbottom Viaduct, East Lancashire Rwy., 1848, employed bracing on top of stone piers. Bottom: Ogden Viaduct, East Lancashire Rwy., 1848, was typical of the stone arch type.

PLATE

11

CHAPTER

(2)

British Railways 1851–1900

THE skeletal form of railway development was established in England by midcentury, but it would require 30 more years before the system was substantially fleshed out. Aside from a lull during the Crimean War, there was considerable railway activity during the 1850s as new lines emerged, others rapidly consolidated, and all strove to gain greater access to territory and business. Later, during the boom of the mid-1860s, this activity became even more frenzied.

Perhaps the easiest way to gain a general understanding of this expansion would be to imagine being in London, facing northward, and looking at a compass and a map. Directly northward, the Great Northern became the most dominant railway. Its main trunk line extended north from London to Doncaster, and its branches reached westward into the Midlands, eastward into Norfolk, and north along the coast. Connecting with the North Eastern and North British lines, it provided the east coast route to Scotland. King's Cross Station, built in 1852, became its London terminus.

In the area north of Doncaster and extending into southern Scotland, the North Eastern Rwy. established its supremacy. This line was amalgamated in 1854 from the York & North Midland; York, Newcastle & Berwick; Leeds Northern; and Malton & Driffield railways. It later incorporated the Newcastle & Carlisle in 1862, and a year later, the Stockton & Darlington.

Looking to the northeast from London, the next region is East Anglia, comprised of the counties of Norfolk, Suffolk, and Essex. In 1862 the Great Eastern Rwy. was consolidated from the Eastern Counties, Eastern Union, East Anglian, Norfolk, and New Market lines, immediately giving the company a near monopoly in the region. Having its London terminus at Bishopsgate, and later at Liverpool Street Station, the railway served such cities as Norwich, Harwich, and Yarmouth.

Southeast of London, in Kent, a fierce competition emerged between the South Eastern and London, Chatham & Dover lines. This rivalry was costly to both companies, as their main lines ran closely parallel to each other, both terminating at the port of Dover. The South Eastern's London terminus was initially London Bridge Station, but in the 1860s the company extended farther into the city, opening both the Charing Cross and Cannon St. termini. The London, Chatham & Dover line terminated at Victoria Station, but later it also used the Ludgate Hill and Blackfriars stations. In 1874 the company opened a station at Holborn Viaduct.

The London, Brighton & South Coast Rwy., amalgamated from the London & Croydon and London & Brighton lines in 1846, dominated the area immediately south of London. Victoria Station was also its terminus, which it opened on October 1, 1860. Somewhat to its west, the London & Southwestern line carved out territory, extending a trunk line to Devon, but also sending feeder lines southward to serve channel ports like Plymouth, Portsmouth, and Southhampton. The company's London terminus was at Waterloo Station. The Great Western Rwy. was strong west of London. Besides reaching west-southwest all the way to Penzance in Cornwall, it extended many branches in western England and Wales, and eventually built northwestward into Liverpool. Its London terminus was Paddington Station, built in 1854.

The London & North Western Rwy. expanded in the West Midlands and into the northwest, eventually creating, in association with the Caledonian line, a west coast route between London and Scotland. The company's London terminus was Euston Station, built in 1838. The Midland Rwy. continued to expand in the Midlands, but also reached southward to Bristol; from this route a branch extended to the Welsh port of Swansea. By arrangements, first with the London & North Western, and then with the Great Northern, it gained access to London. However, the Midland finally opened its own London terminus, St. Pancras Station, in 1868. In 1876, after building north through the Pennine Mountains to Carlisle, it connected with both the North British and Glasgow & South Western lines, creating a third London-Scotland route.

Although this picture gives a general overview of the expansion of the major English railways, it ignores the reality of the competition between them at points where their territories overlapped, and also omits many smaller lines. One worth mentioning is the Furness Rwy. It carved out a small but profitable territory for itself along England's remote northwest coast in the Cumberland region. Another important development of this period was the rapid construction of London's underground passenger system. This story is told in chapter 28.

As expansion proceeded from midcentury into the 1860s, all kinds of turf wars were fought. These were played out like an elaborate game of chess between powerful figures like Captain Mark Huish of the London & North Western, Edmund Denison of the Great Northern, Charles Russell of the Great Western, and James Allport of the Midland. These individuals formed various alliances of convenience, and making all kinds of side deals, used smaller railways like pawns to block the moves of their opponents.

Huish, who had created an alliance of railways known as the Euston Square Confederacy, was an especially ruthless player, and his most hated opponent was the Great Northern. When this railway first appeared, in the late 1840s, all significant traffic between the Midlands and London was passing through Euston Station. It was known that the Great Northern would soon open its own London terminus, and Huish, not wanting to give up his profitable monopoly, decided to act. He made arrangements with potential connecting lines to block any aid to his enemy. In one instance the Manchester, Sheffield & Lincolnshire, which had already granted the Great Northern running powers to Grimsby, reneged on this promise, blocking the tracks there and refusing to water trains at Retford. Huish then contacted the North British Rwy. to discourage any similar support. This company refused to commit itself, one way or the other.

Fortunately, at this critical moment, two allies supported the Great Northern. The Lancashire & Yorkshire granted it running powers from Askern to Knottingley, and the York & North Midland provided a connecting link from Knottingley to Burton Salmon. With these developments, the North British line added its support. However, the Great Northern was still blocked on other fronts, and was thus forced into a pooling agreement in 1851. But this truce was very brief. The year 1851 featured the London Exhibition and many visitors were streaming into the city. By this time the Great Northern had opened a temporary terminus in the city and quickly it and Huish's Confederacy became engaged in serious passenger competition.

However, this contest was not as one-sided as it appeared. Hamilton Ellis in *British Railway History: 1830–1876* makes the following observation:

> The Great Northern possessed several technical advantages over its Confederate "friends" at that time. All its rolling stock was new, and in Archibald Sturrock it had a first-rate locomotive engineer, so that its trains were punctually run and, by the standards of the time, exceptionally comfortable. In such respects the London & North Western was at a disadvantage; its Southern Division was still worked chiefly by a fearsome legacy of small Bury locomotives and decrepit rolling-stock bequeathed by the London & Birmingham; not least among the sights of Exhibition year was that of the ancient Bury engines, in twos and threes and fours, struggling along with enormous excursion trains, with all schedules, regular passenger, excursion or goods, dissolved into chaos incredible.

At this point the Great Northern's fortunes improved. Queen Victoria soon used the line for making her trips to Scotland, and within a few years through-train operation on the east coast route to Edinburgh was worked out. The Great Northern also gained access to the South Yorkshire coalfield, and soon took over a great part of the London coal business. Space limitations prevent the inclusion of further episodes of this drama; however, it should be mentioned that it was not long before the Manchester, Sheffield & Lincolnshire and Midland lines abandoned Huish and went over to the other side.

The rivalry in Kent between the South Eastern and London, Chatham & Dover lines was equally intense but of a different nature. At first the South Eastern line had this territory all to itself. It originally had a somewhat circular route dropping south from London through East Croydon to Red Hill. From Red Hill the route took a 90 degree turn, proceeding east-southeast through Tonbridge to Ashford. From Ashford a branch extended northeast to Canterbury, while another continued to Folkestone and Dover. Between 1845 and 1852 other branches were added. These included a 51-mile route west from Red Hill to Reading in the Thames Valley, a line from Ashford to Hastings, and another between Tonbridge and Hastings.

The London, Chatham & Dover first entered the scene in 1853, receiving authorization as the East Kent Rwy., to build a line from a South Eastern connection at Strood through Chatham and Faversham to Canterbury. This was probably not immediately perceived as threatening because in effect the new line would feed business into the South Eastern system. However, this new railway had greater ambitions in mind. In 1855 it received authorization to build from Canterbury to Dover, and in 1858 to extend westward to St. Mary Cray. This town had become the western terminus of the Mid-Kent Rwy., first authorized in 1856.

By this time the East Kent line was plotting to find a way into London, and was looking at various small railways, either planned or under construction, between it and its objective. On August 1, 1859, it changed its name to the London, Chatham & Dover, thus announcing its true intentions. On July 22, 1861, the company opened its extension to Dover and soon negotiated running power rights over the Mid-Kent line from St. Mary Cray to Bromley, on the West End & Crystal Palace Rwy. from Bromley to Beckenham, and on the Victoria Station & Pimlico Rwy. into London. The inaugural train from Dover arrived in London on November 1, 1861, taking a route 10 miles shorter than that of its competitor, the South Eastern. Its London-Canterbury distance was 20 miles shorter.

This bold move initiated a destructive war between the lines that lasted for nearly 40 years. In order to remain competitive the South Eastern line constructed a more direct London-Dover route. Each railway tried to attract a greater share of the business by offering lower rates and/or better service. However, as each had to share a territory more ideally suited for a single railway, both companies were reduced to a state of poverty, gaining reputations for inefficiency, poor equipment, and dirty carriages. This conflict was finally resolved in 1899, when the two companies agreed to a "working union." By the 1870s, most of Britain's other railways had learned to live with each other. While still having occasional differences, they became more civil and cooperative toward one another.

In Scotland other lines struggled to expand their operations. In 1862 the North British Rwy. completed its Waverly route from Edinburgh to Carlisle. In 1865 it absorbed the Edinburgh & Glasgow Rwy. That same year both the Scottish Central and Scottish North Eastern lines were absorbed into the Caledonian. Later, both it and the North British Rwy. constructed many branch lines in the areas surrounding Glasgow and Edinburgh and also immediately to the north. The Glasgow & South Western Rwy. built mainly in the area its name suggests. The Great North of Scotland line, which began in 1852, constructed north and west from Aberdeen. Farther to the west, in 1861, the Highland Rwy. undertook the most formidable task of railway building encountered to that time in Britain. Under the engineer Joseph Mitchell, a line was built through the Grampians, with a lengthy ascent from Blair Athol to Druimuachdar Pass. This route eventually connected Perth and Inverness.

In Wales the Cambrian line became the dominant railway. It built in the coastal region along Cardigan Bay and also reached inland, with branches extending to Whitechurch and Wrexham in the north and Brecon in the south. In south Wales during the 1850s construction began on two lines primarily intended for the extraction of coal. These were the Rhymney and Brecon & Merthyr Tydfil Junction lines.

Although railway progress lagged in Ireland for a time, the two main rail centers became Dublin and Belfast. Eventually the main railways serving Eire were the Dublin, Wicklow & Wexford; Cork, Bandon & South Coast; Great Southern & Western; Waterford, Limerick & Western; and the Midland Great Western. The Great Southern & Western ran from Dublin to Cork and the southwest. The Midland Great Western operated between Dublin and Galway. Connecting Eire to Northern Ireland was the Great Northern of Ireland Rwy. This line had come into being in 1877, when it was consolidated from the Dublin & Belfast Junction; Dublin & Drogheda; Ulster; Irish North Western; and other railways. Besides extending from Dublin to Belfast, it also reached Londonderry and Bundoran. Also important in Northern Ireland were the Belfast & Northern Counties and Belfast & County Down lines.

Between 1860 and 1870 total route miles in the United Kingdom increased from 10,443 to 15,537. The increase in England and Wales was from 7,583 to 11,043; in Scotland from 1,486 to 2,519; and in Ireland from 1,364 to 1,975. Total passenger traffic in Britain also grew substantially during this period, increasing from 193,435,678 to 330,004,398 (excluding commuter traffic). Among major British railways at the time, the London & Northwestern was the busiest. In 1873 it operated 1,935 locomotives, 2,722 passenger carriages, and 39,581 goods wagons, carrying a total of 42,511,777 passengers and 27,208,042 tons of goods.

At midcentury the locomotive business was generally

good, although it had changed somewhat from 1840, when there were about 30 builders. Some early firms had disappeared while others emerged in their stead. Later entrants in this competition include Beyer, Peacock & Co. of Gorton, East Manchester, in 1854; Ruston & Hornsby Ltd., Lincoln, 1854; Hudswell, Clarke & Co., Leeds, 1860; Manning, Wardle & Co., Leeds, 1860; Fletcher, Jennings & Co., Whitehaven, 1861; Avonside Engine Co., Bristol, 1864; Dubs & Co., Glasgow, 1864; Hunslett Engine Co., Leeds, 1864; Atlas Works, Bristol (first under Fox, Walker & Co., 1864, and from 1880 on under Peckett & Sons, Ltd.); Black, Hawthorn & Co., Gateshead-upon-Tyne, 1865; and W. G. Bagnall Ltd., Stafford, 1887.

The company shops established at Crewe and Swindon in the 1840s later prospered, becoming known as railway towns. In *Our Iron Roads*, Frederick S. Williams comments on these: "Some towns have been the distinct creation of railway establishments. So it is with Crewe. Here stood one farmhouse: now there is a railway establishment that employs 5,000 or 6,000 men; and between 2,000 and 3,000 locomotives have been here erected. So with Swindon. Some 45 years ago, a little party of gentlemen sat down on the greensward to take their luncheon . . . Where that luncheon was that day, eaten by Isambard Brunel and Daniel Gooch, the platform of the Swindon station is today, and hard by are Swindon Junction and the vast locomotive establishments." At the time Crewe and Swindon served the London & North Western and Great Western lines, respectively. However, the former firm also built engines at Wolverton until 1863, and the latter initiated production at Wolverhampton in 1859. Such success inspired competing railways to do likewise. Among these were the North Eastern, which began production at Gateshead-upon-Tyne in 1849; the Midland at Derby in 1851; the Great Eastern at Stratford in 1851; the Manchester, Sheffield & Lincolnshire at Gorton in 1858; and the Great Northern at Doncaster in 1867. Other information on locomotives from 1851 to 1900 can be found in chapter 5.

One interesting development was the introduction of cheap worker trains. In the mid-1850s the Great Eastern line was expanding its system within London. Under an act of 1854 authorizing the various extensions, the company was required to transport workmen, at a special rate, between the stations of Edmonton and Walthamstow and its Bishopsgate terminus. In March 1865, the London, Chatham & Dover line initiated special workmen's trains between the inner suburbs of Kent and Victoria Station. Inbound service only ran between 5:00 and 6:00 in the morning. Afternoon outbound service ran after 5:15 on weekdays and 2:30 on Saturdays. This precedent was later followed to some degree by other railways, especially the London, Brighton & South Coast and Great Eastern lines. Another group of companies, like the London & South Western, provided uneven service or sold reduced-fare tickets in some stations, but not in others. Many workers, for example, walked miles from more distant stops into Shepherd's Bush Station to board special trains. Companies more conservative in their philosophy, like the London & North Western and Great Western, had a less charitable attitude toward the whole idea, seeing it as smacking of radicalism.

Other technical and service innovations improved railway operations during this period. In 1852 the Great Northern Rwy. made the first attempt to heat passenger carriages by introducing "foot warmers." In 1857 the Midland line laid the world's first steel rails at Derby. Besides being stronger and safer than iron, they also could support heavier rolling stock. In 1860, on the Chester & Holyhead Rwy., watering troughs were installed for the first time in livestock wagons. The initial use of gas lighting occurred in 1863 on passenger coaches of the North London line. In 1871, the Caledonian Rwy. became the first company to experiment with compressed air brakes. Later, in 1878, the vacuum brake was introduced in England. After some years of competition this system became favored over air brakes, and in the twentieth century, after railway nationalization, was universally adopted.

At a meeting of the British Railway Association in 1870 the engineer Robert F. Fairlie read a paper entitled "The Gauge for the Railways of the Future." This ignited a new gauge controversy just as the old one (the Great Western had begun conversion of its broad-gauge routes to standard in 1868) was finally being resolved. Fairlie had recently designed a unique double boiler tank engine for the 1′11½″ gauge Festiniog Rwy. in Wales. This rustic line was originally begun in 1832 and operated as a gravity and animal traction route before converting to steam in 1863. Still existing today, it extends for a distance of 13¼ miles and ascends 700 feet in altitude from the port city of Portmadoc to some slate quarries. Including a one-mile extension to Duffws, it had, in the nineteenth century, four other stations besides the two termini. Passenger service began for the first time in 1864.

The Festiniog line's original engines were provided by G. England & Co. Fairlie's locomotive, an 0-4+4-0T articulated type, was built by the Hatcham Iron Works and delivered in 1869. Given the name *Little Wonder*, it soon gave very impressive performances—in one instance, drawing a train of 180 tons up the 1 in 85 gradient on the line. This feat drew attention not only to Fairlie's locomotive, but also to this successful and economical narrow-gauge operation. Many engineers came to visit the railway, and it quickly became the prototype for a new trend in railway construction.

Narrow-gauge railways were often more economical than standard-gauge ones for the following reasons: (1) less land had to be graded for their construction; (2) lighter track could be used; (3) curves could be sharper, thus giving engineers more flexibility in planning routes, which in turn reduced the necessity for many tunnels and bridges; (4) those tunnels and bridges still necessary could be smaller and require lighter materials; and (5) as lighter engines and rolling stock were used, their cost was also less. The primary disadvantages of this system included transshipment costs and reduced trainload capacity. Thus narrow gauge was ill suited for areas of high traffic density, but ideal for isolated railways with lower traffic volume and minimal connections to other lines. They were also excellent for industrial, agricultural, and military use.

Although narrow-gauge lines were subsequently built in many countries, they were especially important for underdeveloped nations, as they offered, perhaps, the only affordable means for having railways. Aside from industrial use, however, there was considerable opposition in Britain to any narrow-gauge construction. As a result, only a small number of these lines were built, and these were largely limited to Wales and Ireland.

By 1880 Britain's railway system had largely been built, and from this point on there was relatively little new construction. One new line incorporated that year was the Hull & Barnsley Rwy. Once inaugurated, it competed with the North Eastern Rwy. in hauling coal from South Yorkshire to the port of Hull. It connected with other railways in the Yorkshire coalfield area and gained running powers over an adjacent section of the Midland Rwy.

In 1884 a rail line was authorized between the coal-rich Rhondda Valley and Barry, southwest of Cardiff. Nearby railways and dock companies opposed this venture, but it went ahead. Port facilities were built at Barry, and 26 miles of track laid to the coalfield. This railway, the Barry, became a very successful operation.

On February 1, 1886, the Mersey Rwy. was inaugurated. Running beneath the Mersey River it connected Liverpool with Birkenhead. The tunnel was drilled through solid rock far beneath the bed of the river. Built mainly for serving local commuter needs, this line had a number of underground stations. Its 0-6-4T Beyer-Peacock engines were the largest used up to that time for passenger service in Britain.

On September 1, 1886, the Severn Tunnel, Britain's longest at 4¼ miles, was opened by the Great Western line. Although planned by Charles Richardson in 1865, its initial construction was delayed until 1872. During the tunneling, the passage was flooded twice. The second incident, caused by a tidal wave in 1883, set the work back considerably. Once completed, the tunnel provided a direct connection between London and Wales.

Construction also began in 1880 on a much more ambitious tunnel under the English Channel from Shakespeare Cliff, near Dover, to Sangatte in the Pas de Calais, France. Such a route had first been proposed in 1802 for use by stagecoaches. In the 1860s this idea was revived as a way to establish a rail connection between Britain and the Continent. Channel tunnel companies were incorporated in England in 1872 and in France in 1875, and from 1880 to 1883 construction proceeded on both ends. However, opposition to this project was strong, for both economic and military reasons, and it was finally dropped. Other important nineteenth-century British tunnels were the London & North Western's Stand Edge Tunnel and the Woodhead Tunnel on the Manchester, Sheffield & Lincolnshire. These each exceeded 3 miles in length.

By 1886 the rail network in Britain had grown to a total of 19,169 route miles. Operated over this system were 15,196 locomotives, 33,656 passenger carriages, and 464,139 goods wagons. The six largest lines were the Great Western, 2,346 miles; London & Northwestern, 1,828 miles; Great Eastern, 1,558 miles; Midland, 1,339 miles; Great Northern, 1,045 miles; and North British, 1,015 miles. Together, these six accounted for almost 48% of Britain's total route mileage.

Although London had a substantial underground rail system by this time, the city's mainline carriers, going back to the earliest times, had also shared in the local business. In its March 9, 1888 issue, *Railroad Gazette* reported that the Great Eastern had no less than 45 passenger stations and 28 goods stations within the limits of London. Many of the passenger stations were of considerable size, and nearly all were built of brick, with solid stone platforms and iron and glass roofs, and had comfortable waiting rooms. The typical Great Eastern suburban train consisted of 15 coaches, each from 27 to 31½ feet in length. Each carriage could seat either 32 first-class or 50 second- or third-class passengers.

In 1889 a tragic rail accident at Armagh, Ireland, took 80 lives, including many children. It occurred when a packed excursion train stalled while ascending a steep grade. The train was divided into two parts, each to be pulled over separately. However, efforts to secure the rear section were inadequate, and it ran backward down the incline, colliding with another train. The resulting furor quickly led to passage of the Regulation of Railways Act of 1889. This legislation essentially required that all passenger trains have a continuous braking system operating throughout the train, and that if any coach or coaches should become disconnected, their brakes would automatically be applied. As a further safety measure, it also required all lines to operate on the system of "block-working." This was an interlocking method of traffic control in which the line was divided into many small sections. No train could proceed into the next "block" until any ahead of it had moved out.

In 1890 British railways carried over 817 million passengers. The same year goods traffic figures showed that over 126 million tons of coal were carried and 383,000 tons of fish. In 1891 there were 516 rail companies in the United Kingdom, of which 266 were operated or leased by other companies. The Great Western line alone leased 36 railways and shared ownership in 9 others. This company finally completed its conversion from broad gauge to standard in May 1892. By the end of 1897 Britain had 21,433 route miles of track.

During the 1890s there was comparatively little new construction. The most ambitious expansion was undertaken by the Manchester, Sheffield & Lincolnshire line. Under the direction of Sir Edward Watkin, this railway managed to gain access to London by an arrangement with the Metropolitan line, and built a terminus at Marylebone Road. This extension was opened on March 15, 1899, the same day that the company changed its name to the Great Central Rwy. It immediately began contending for traffic between London and its home area of the north and east Midlands. Unfortunately, the company's timing was bad. Britain's economy had entered a downturn and three other London connected railways were competing for the same traffic. From this time until 1923, when the Great Central was grouped into the London & North Eastern system, the company's investors never received a single dividend.

In Scotland the West Highland line completed routes to Fort William in 1894 and reached Mallaig just after the turn of the century. This opened an area of the country that had never before known rail service. In 1898 completion of the Carr Bridge on the Highland line reduced by 26 miles the route from Inverness to the south.

The construction of small feeder lines in country districts was made possible by the Light Railways Act of 1896. A few of these were completed just after the turn of the century. However, they were expensive in relation to the amount of traffic generated, and by then, railways were beginning to consider a new mode of transport, the petrol bus, for serving rural needs.

The introduction and expansion of railways in Britain during the nineteenth century brought about many far-reaching changes. These became so much a part of the fabric of their time, interconnected with everything else, that they are not always easily differentiated. Perhaps most apparent was the great stimulus railways provided for business and industrial activity. Not only did they bring raw materials to factories and manufactured goods to markets, they introduced a new level of efficiency, resulting in a general price decline. This generated an increase in purchasing power for the entire nation. As industries and other businesses expanded there was a greater demand for workers. This offered a new range of choices for the rural poor, creating in time an exodus to the cities.

Railways also accelerated the flow of information and ideas. As mail service improved and telegraph networks were established, government, commercial, and personal communications flowed at an ever faster pace. Mail trains also carried newspapers from Fleet St. to increasingly distant destinations. While this provided an ever-expanding base of readership, it also had the effect of influencing public opinion on an increasingly national level. Thus in this and other subtle ways, railways contributed to a greater cohesiveness and unity in the country.

One early manifestation of this unifying power was seen in 1852, when time was first standardized. Until the 1840s all time in Britain had been a local matter. However, as rail companies grew, scheduling became increasingly difficult. The London & Birmingham tried to accommodate local variations in its schedule, but the Great Western, and shortly thereafter the Caledonian, standardized their schedules according to Greenwich time. Although there was initial resistance in some regions, the idea of having a single national time caught on. This had a synchronizing effect on activities in the country.

As Britain's rail network expanded the entire population became increasingly mobile. While few among the lower classes traveled at first, changes were gradually made to accommodate them. Improvements were made in service and rolling stock, fares were lowered, and cheap worker trains were introduced. As circumstances got better, working-class families could afford to visit distant relatives at Christmas or ride excursion trains to places like Brighton. Improvements in suburban transit and the introduction of street railways allowed both employers and employees to live at ever-increasing distances from work. Sanitary conditions improved as congested slums were abandoned in favor of better, but still basic, suburban dwellings.

In the changed set of realities brought in by the railways there was an increased intermingling of the various social classes. As the old landed aristocracy lost power to the new industrial bourgeoisie, there were many from humble beginnings who rose to prominence, among them people like George and Robert Stephenson. This new social freedom helped to emancipate the lower classes to some degree, and Britain slowly began moving away from aristocracy toward meritocracy.

However, all of this progress was not gained without paying a price. Railway and industrial expansion had a shadow side, one bringing ugliness, social dislocation, bad architecture, and dangerous levels of smoke pollution onto the landscape. After introduction of the motor car in the twentieth century, environmental deterioration accelerated. In giving birth to the steam railway, Britain absorbed the most direct forces that it unleashed, becoming in effect, its primary experimental laboratory.

Railway inaugurations: Top: Arrival of first train at Barnstaple, North Devon Rwy., 1854. Bottom: Inaugural train passing Neath Abbey Station, Swansea & Neath Rwy., 1863. This line was built to provide a direct rail link for transportation of coal and iron ore from Merthyr Tydfil and Aberdare Valley to the port of Swansea.

PLATE 12

Construction of engines at the Stephenson Locomotive Works, Newcastle, England, 1864: Top: Operation of a steam hammer. Bottom: The fitting room.

PLATE 13

PLATE 14

Arrival of worker's penny train at London's Victoria Station; London, Chatham & Dover Rwy. This service, inaugurated in March 1865, operated between the inner suburbs of Kent and London, running between 5:00 and 6:00 a.m., and after 5:15 p.m. on weekdays and 2:30 p.m. on Saturdays.

Two views at Holbeck Junction, Leeds, Midland Rwy., 1868. The 2-2-2 engine in the foreground of the lower picture is one of Sharp-Stewart manufacture. *Editorial Note:* These two pictures originally constituted a very wide single engraving. To avoid a great reduction in size it was separated in the middle with the left side placed at the top and the right at the bottom.

PLATE 15

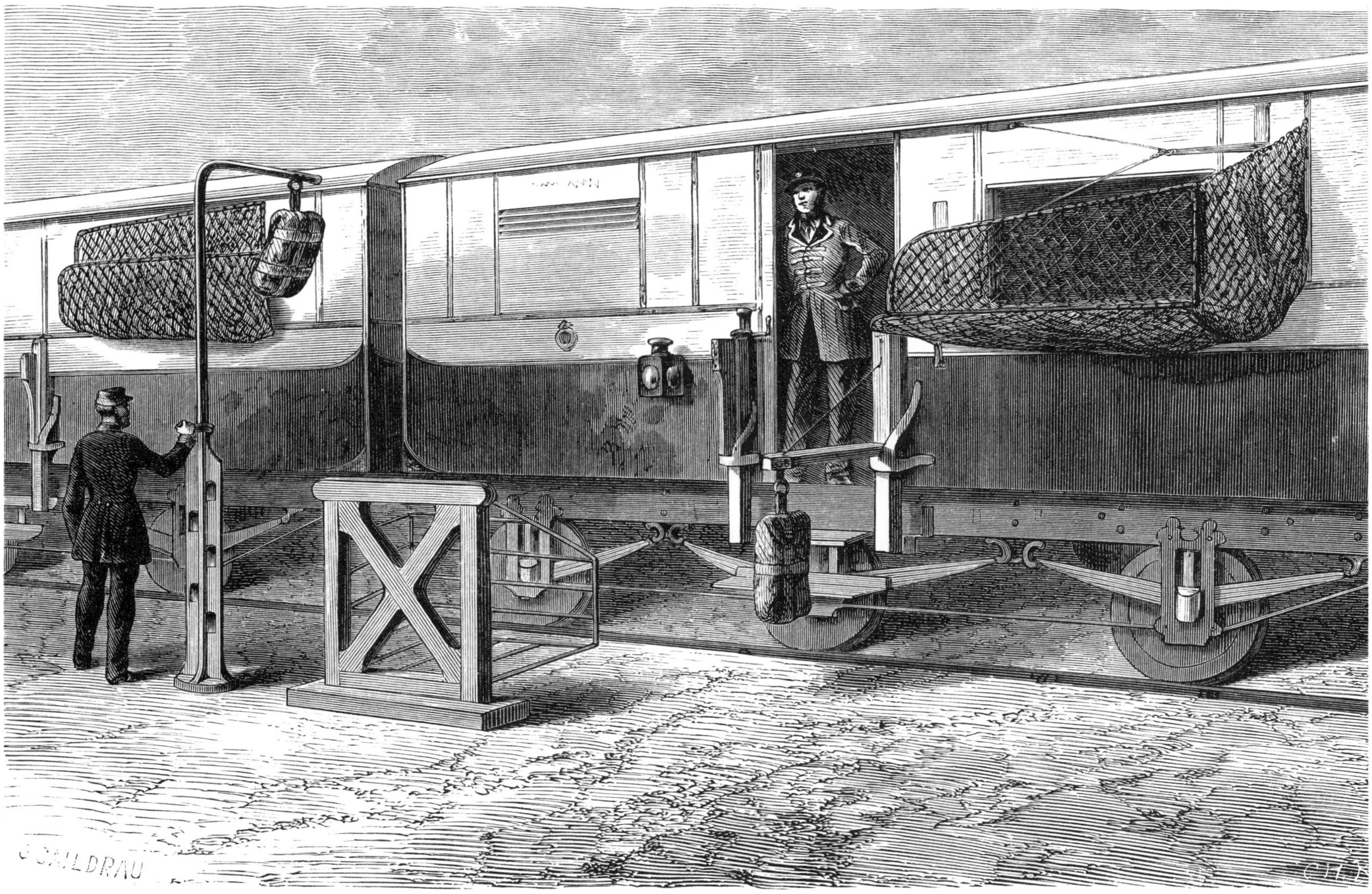

Top: Robert Fairlie's 0-4+4-0T *Little Wonder* pulls a mixed goods and passenger train on the famous 1′11″ gauge Festiniog Rwy. of Wales, 1871.
Bottom: British system designed for both picking up and dropping off mail bags while the train continues in motion, circa 1860s.

PLATE 17

Proposed facilities at Dover for a channel railway ferry, South Eastern Rwy., 1870. The Archcliffe Fort provided military security. The tunnel beneath the fortress apparently was a link to a junction with the London, Chatham & Dover line.

PLATE 18

A night scene in London: "Returning from the pantomime—the last train," 1871.

Passenger train on viaduct in London's Ludgate Hill district, London, Chatham & Dover Rwy., circa 1872. This engraving is from a number of sketches the famous artist Gustave Dore made in London.

PLATE
19

PLATE 20

"Arrival of the express train at York: Ten minutes for refreshment," 1872. This Great Northern train, the *Flying Scotsman*, departed King's Cross Station in London at 10:00 a.m. and arrived in York at 2:15 p.m. From there it continued on to Edinburgh via the east coast route.

Queen Victoria of Britain looks out from the royal train as it crosses the Tay Bridge near Dundee, Scotland, North British Rwy., 1879. Some months later this bridge collapsed in a storm, carrying down with it an entire train.

PLATE

21

Scenes at Glasgow during a railway strike, Christmas holidays, 1890: 1. Passengers at the Midland line's St. Enoch Station awaiting transit to the south. 2. Delayed luggage and parcels, Queen St. Station. 3. Barricade at Queen Street St. 4. Excursionists waiting at Central Station.

PLATE 22

CHAPTER

3

American Railroads before the Civil War

As progress was made in early English railways, news traveled across the Atlantic and was received with some interest in America. In comparison with England, America was still a primitive land. Its major cities were largely ports, and most of the commerce between them moved by water. Inland travel was much slower and more difficult. Roads tended to follow old Indian trails or cow paths; while they sufficed for travel by foot or on horseback, they generally were not well suited for wagon or carriage. As little political will existed at the time on the national or state level for involvement in "internal improvements," road or canal building was chiefly dependent on local efforts. While toll roads were built by enterprising individuals such developments were relatively few. Thus in the absence of an improved system, both local and regional economies, by necessity, remained somewhat rudimentary and largely self-reliant.

A breakthrough in this situation was the inauguration of the Erie Canal in 1826. It opened up, to some extent, trade with the Midwest frontier. This commerce previously had been routed via the Mississippi River system or across the Appalachian Mountains. New York City profited from this development, as it received traffic diverted from Boston, Philadelphia, and Baltimore. These cities sought to regain lost commerce by developing their own routes into the Midwest. However, canals were expensive, time-consuming to build, and, unless the geographical conditions were just right, totally impractical. Railways, for the first time, offered a possible alternative. Thus this new technology arrived in America at a critical moment and radically changed everything.

America's earliest lines were inclined plane tramways. First, perhaps, was one built at South Hadley, Massachusetts, in 1795. Another was built in 1805 at Beacon Hill in Boston. In Massachusetts in 1826, a 3-mile horse drawn tramway was built connecting the Quincy granite quarries to the Neponset River. Constructed a short time later was a line near Mauch Chunk (now Jim Thorpe), Pennsylvania. It ran from a coal mine to the Lehigh Valley Canal. Construction of the 16-mile Carbondale Tramway began in 1827. When completed, it extended from the Delaware and Hudson Canal at Honesdale, Pennsylvania, to coal mines owned by the canal company. It was on this line that America's first locomotive, the English-built *Stourbridge Lion,* was tried. It was tested in August 1829 by Horatio Allen, but found to be too heavy for the track. An early American proponent of steam railways was Colonel John Stevens. In 1825 he had a small demonstration engine built that operated on his estate in Hoboken.

In 1827 the merchants of Baltimore received a charter for building the Baltimore & Ohio R.R. Construction was inaugurated on July 4, 1828, and the first section of 15 miles, to Endicott's Mills (now Endicott City), was opened on May 22, 1830. This line was built to carry both freight and passengers. Initially horses were used, but on August 28, 1830, an experimental engine, built by Peter Cooper, was tried. In a famous contest, his locomotive, the *Tom Thumb,* pulled a loaded passenger carriage in competition with a similar carriage drawn by a horse. Cooper's engine gained the advantage, but suffering a mechanical breakdown, it lost the race. Nonetheless, this test seemed to resolve the issue decisively in the locomotive's favor.

Following news of the Rainhill trials, small delegations of Americans traveled to England to study the workings of Stephenson's Liverpool & Manchester and Stockton & Darlington lines. A number of English engines were ordered at the time, and these soon influenced the construction of locomotives in America. In 1830 the Charleston & Hamburg line in South Carolina was opened, employing the first American engine built for regular service, the 0-4-0 *Best Friend.* This engine was designed by Horatio Allen and E. L. Miller. Allen, formerly an employee of the Delaware & Hudson Co., had been among those traveling to England, and while there had placed the order for the *Stourbridge Lion* and three other engines. He became chief engineer of the Charleston & Hamburg line in August 1829.

That same year his unique 2-2+2-2 *South Carolina* was added to the line for freight service. Odd in appearance, it had a firebox in the middle and boilers extending both forward and backward. King-bolts were employed to allow flexibility in the middle between the two sets of drive wheels. In essence this created two trucks, or bogies, each consisting of a set of carrying wheels and a set of drive wheels. This allowed better performance on curves, and pioneered the idea of "articulation," a later practice of separating groups of drive wheels on larger locomotives. When completed on October 2, 1833, the Charleston & Hamburg line (later known as the South Carolina R.R.) extended for a total of 135 miles and was briefly the longest in the world.

Both the *Best Friend* and *South Carolina* were built by the West Point Foundry, located at the foot of Beach St. in New York City. The third engine built by this firm was the *De Witt Clinton.* It became the first locomotive of the Mohawk & Hudson R.R. This 17-mile line, inaugurated in 1831, later evolved into the New York Central System. In *The History of the First Locomotives in America,* by William H. Brown, Judge J. L. Gillis recalled his ride on this inaugural journey, which departed from Albany:

> The train was composed of coach-bodies, mostly from Thorpe & Sprague's stage-coaches, placed upon trucks. The trucks were coupled together with chains or chain-links, leaving from two to three feet slack, and when the locomotive started it took up the slack by jerks, with sufficient force to jerk the passengers, who sat on seats across the top of the coaches, out from under their hats, and in stopping they came together with such force as to send them flying from their seats.
>
> They used dry pitch-pine for fuel, and, there being no smoke or spark-catcher to the chimney or smoke-stack, a volume of black smoke, strongly impregnated with sparks, coals, and cinders, came pouring back the whole length of the train. Each of the outside passengers who had an umbrella raised it as a protection against the smoke and fire. They were found to be but a momentary protection, for I think in the first mile the last one went overboard, all having their covers burnt off from the frames, then a general melee took place among the deck-passengers, each whipping his neighbor to put out the fire. They presented a very motley appearance on arriving at the first station. There rails were procured and lashed between the trucks, taking the slack out of the coupling-chains, thereby affording us a more steady run to the top of the inclined plane at Schenectady.

America's earliest rails were flat, or "strap," rails. They were constructed by laying long wooden beams, typically 6″×12″ or 8″×12″, on heavy wooden ties, each a few feet apart. These beams were fitted together at each end, and long iron strips were laid along the inside top edge. These strips, about 2 feet long by ½ inch thick, were attached by spikes. However, in 1830, Robert Stevens conceived the stronger and safer "H" or "T" rail design still in use today. Stevens, along with his father Colonel John Stevens and brother Edwin were constructing the Camden & Amboy R.R. He designed this new rail while onboard a ship to England to buy locomotives. In time this iron rail replaced the strap rails and made possible the use of much heavier locomotives. The first portion of the Camden & Amboy, a 14-mile stretch between Bordentown and Hightstown, New Jersey, was opened in December 1830.

In midsummer 1831, the *York,* a locomotive built by Davis & Gartner of York, Pennsylvania, began service after winning a competition held by the Baltimore & Ohio line. It initially operated between Baltimore and Endicott's Mills, pulling a train of four cars. In 1832 the 21½-mile Saratoga & Schenectady R.R. was opened. Also that year the Baltimore & Ohio created the first sea-rail interchange in Baltimore. In 1833 the company's construction had been completed to Harper's Ferry, 81 miles distant. A 30-mile branch to Washington was also completed by the B & O that year.

In 1834 the 82-mile Philadelphia & Columbia line and 36-mile Portage R.R. were opened in Pennsylvania. These lines connected with existing canals to provide a system extending from Philadelphia to the Ohio River at Pittsburgh. That same year the Paterson & Hudson River R.R. and 34-mile Cayuga & Susquehanna R.R. were inaugurated. In 1835 three lines in Massachusetts opened for service: the 44-mile Boston & Worcester R.R., the 41-mile Boston & Providence line, and the 26-mile Boston & Lowell R.R. At the end of that year America had 1,098 route miles of track.

Besides the West Point Foundry, a number of other individuals and firms began, during the 1830s, to build locomotives. These included Phineas Davis of Baltimore, the companies of Mathias Baldwin, Long & Norris, Henry R. Campbell, and Garrett & Eastwick of Philadelphia, and Rogers, Ketchum & Grosvenor of Paterson, New Jersey. Baldwin's first locomotive, *Old Ironsides,* made its debut on the Philadelphia, Germantown & Norristown R.R. in November 1832. While American builders tended to follow English designs at first, a number of innovations shortly began, eventually leading to engines of a distinctly American style. One early improvement was made when the first "cow catcher" was added to the front end of the Stephenson-built *John Bull.* Later changes included adding cabs to protect the train crew and funnel stacks for catching burning cinders in the smoke. Inside cylinders were moved to the outside. This allowed easier maintenance and eliminated the troublesome crank-axle.

Among the most important of the early innovations was the four-wheel "bogie," or truck, introduced in 1833 by John B. Jervis. Prior to this time, lead or trailing carrying wheels, when used at all, had been limited to two in number. In Jervis's invention four carrying wheels were mounted on a separate platform located beneath the front part of the engine and connected by a central pivot. Swiveling on this pivot, the truck improved the ability of the locomotive to negotiate curves and also provided greater track stability. This invention made a lot of difference on the primitively constructed American lines of the time.

Alone among early American engine builders, Phineas Davis and his assistant Ross Winans pioneered their own designs from the very beginning, built primarily for the Baltimore & Ohio line. Davis developed the successful "grasshopper" type in the early 1830s. After Davis was killed in a rail accident in 1835, Winans continued to build locomotives of this style, but soon introduced the "crab" type, which had horizontal cylinders. By the early 1840s he introduced more powerful engines known as "mud-diggers."

In 1836, William Norris introduced the 4-2-0 *George Washington,* an engine that would earn a world reputation for Norris's company. In *The Locomotive and Philadelphia's Share in its Early Improvements,* Joseph Harrison Jr. cites the importance of this engine:

> This locomotive, when put upon the Columbia road, did apparently, the impossible feat of running up the old inclined plane at Peter's Island, 2,800 ft. long, with a rise of one foot in fourteen, drawing a load of more than 19,000 lbs. above the weight of the engine, and this too, at a speed of 15 m.p.h. This was no doubt impossible, if the simple elements of the calculation are only considered. But there was a point in this experiment, well known to experts at the time, which did make it possible, even by calculation; and this point consisted in the amount of extra weight that was thrown upon the drivers by the action of the draft link connecting the tender with the engine; the result being that about all the weight of the locomotive rested upon the drivers, less the weight of the truck frame and wheels in front. This most extraordinary feat, a writer on the subject says, "took the engineering world by storm, and was hardly credited."

Following this event the Norris firm received many orders from American railroads as well as others from abroad, including one from England. After Norris engines were shipped to Austria and Germany, they served as prototypes for later designs there.

In 1836 Henry R. Campbell designed the first 4-4-0 engine, a style later destined to be known as the "American" type. Campbell's intent was to "distribute the weight of the engine upon the rails more completely." However, his locomotive, once built, had problems in equalizing the weight on the driving wheels while accommodating the various undulations of the track. The Garrett & Eastwick firm followed up on this idea with its *Hercules,* built in 1837 for the Beaver Meadow R.R. This engine performed well and led to an order in 1839 from the Philadelphia & Reading R.R. for a 4-4-0 type, weighing 11 tons, 9 of which was to rest on the drive wheels. On February 20, 1840, this engine, the *Gowan & Marx,* pulled a train of 101 loaded four-wheel cars, having a weight of 423 tons, from Reading to Philadelphia, averaging a speed of almost 10 m.p.h.

The success of this locomotive was noted by a Russian mission visiting America at the time. Soon Eastwick, Harrison, a later partner of the firm, and Thomas and William Winans, sons of Ross Winans, were recruited to that country to build engines. Other companies also began production of the 4-4-0 type. The first Norris engine was completed by November 1839; however, Rogers did not build the type until 1844, and Baldwin, the year after. While these engines were initially designed for heavy freight use, they were soon employed for passenger service, and by the 1850s had evolved into that beautiful style for which they remain famous.

In 1840 the Hinkley & Drury firm opened in Boston. Later builders included the Taunton Locomotive Works, which opened in 1847; John Souther's firm in Boston, the Schenectady Locomotive Works; and the Portland Co., all opening in 1848, followed in 1849 by the Amoskeag Co. of Manchester, New Hampshire. Manufacturers opening during the 1850s included Danforth, Cooke & Co. of Paterson, Cincinnati Locomotive Works, Niles & Co. of Cincinnati, and William Mason of Taunton, Massachusetts. During these years and later, other machine shops entered the business; however, they typically lasted only a year or two, and had little interest in the kind of sustained innovation necessary to maintain the growth of a company.

Later in the 1840s, 0-6-0 and 0-8-0 engines came to be

more widely used for heavy freight. In 1847 Septimus Norris introduced the first 4-6-0 "ten-wheeler" type engine, a style that would later become the standard American freight locomotive. In 1848, Ross Winans, by then a veteran innovator in heavy-duty engines, built the first of his 0-8-0 "camel" locomotives for the Baltimore & Ohio R.R. This odd-appearing engine was also an attempt to address the problem of finding a suitable boiler design for burning the cheap, small-particle anthracite coal that was plentiful in the region. It had a long, narrow firebox that was serviced from two different fire doors, one in the usual place at the back and another on the top, which also had a coal hopper. For this reason the engineer's cab was placed on top of the boiler. This design was successful for its time and over 200 locomotives of this style were built. Other individuals soon made improvements on this design. That story, and other accounts of later locomotive development, can be found in chapter 5.

Traveling conditions gradually improved as time passed, but they were never ideal. Throughout the nineteenth century the quality of road engineering on American lines would remain inferior to that of their British counterparts. Typically, on American railroads grades were steeper, curves tighter, tracks of poorer quality, and bridges and roadbeds of more rudimentary construction. In contrast to British lines, which were often double-tracked, American lines were virtually all single-tracked. Until the 1840s, tunnels, due to their cost, were almost entirely avoided, and other ways were found to go over or around mountains. Travel over such routes was usually more difficult. Trains had rougher and potentially more dangerous rides, delays were frequent, and speeds had to be kept low to ensure safety.

The main reason for this was, obviously, lack of capital. America was largely an underdeveloped frontier nation, and so in railroad construction, as in many other endeavors, things were done as well as they could be under the circumstances. As time passed, however, both foreign and domestic rail investments increased and the American rail system improved greatly in quality. Later, as telegraphic dispatching came into use, largely eliminating the dangers of unexpected collisions, trains could be operated safely at higher speeds. Samuel Morse's telegraph was first tried between Washington and Baltimore on the B & O line in 1843. However, it was not until 1851 that a railroad first adopted the system for general dispatch use, and it took over two decades before the practice gained wide acceptance.

As in England, American postal authorities quickly saw the advantages of transporting mail over rail. Mail service was inaugurated on the Charleston & Hamburg line in November 1831. The Baltimore & Ohio began its mail service on January 1, 1832. Soon postal service was initiated on both the Camden & Amboy and Saratoga & Schenectady lines. The official beginning of the Railway Mail Service coincided with the inauguration of the B & O line between Washington and Baltimore on August 25, 1835. These developments proved so successful that on July 7, 1838, Congress passed a law establishing every railroad in America as a postal route.

Originally, sealed bags were given to railroads and these were placed in any car having space. Soon it was seen that although mail was technically freight, it should receive more expeditious service. Special mail cars were added to passenger trains, and, by the 1850s, trains consisting entirely of mail cars were occasionally operated. The practice of sorting mail en route, pioneered in England in the 1840s, was not initiated in America until the early 1860s.

While "railroad fever" quickly spread through America during the 1830s, it was brought down to earth by a financial crisis lasting from 1837 to 1840. One railroad, successfully built at the time, was described by William Sloane Kennedy in *Wonders and Curiosities of the Railway:*

> The pioneer railroad of the great western prairies was the Northern Cross road (now the Great Wabash). It originally extended from Meredosia, on the Illinois River, to Springfield, in the state of Illinois. Its origin was as follows. In 1837, the Legislature of Illinois appropriated ten million dollars for a magnificent system of internal improvements, and a large share of this was to go to railroads. Work on many of these was begun, but a great financial crash came, and the construction of all of them was suspended except that of the Northern Cross, which was nearly completed, and was speedily finished through to Springfield. The rails of the road were "strap" rails, ⅝″ thick, and were fastened to the wood by ten-penny nails sunk into the rail until the heads were flush with the upper surface. The first locomotive, built by Rogers, Grosvenor & Ketchum, was landed at Meredosia in the fall of 1838, and on November 8, the first puff of a locomotive was heard in the prairies of the West.

In 1840 America had a total of 3,328 route miles of track, with Pennsylvania being the dominant railroad state. By 1844 the Long Island R.R. had been extended from Brooklyn across the length of the island to Greenport. From there steamboat service was available both to Norwich, for rail connections to Worcester, and to Stonington, for rail connections to Boston. By 1850, America's system had reached 8,879 route miles, and New York had become the leader. Railroad building was still primarily focused in the Middle Atlantic and New England states, but this would soon change. In 1850 the first federal land grant bill was passed. It authorized grants of land to aid the Illinois Central and Mobile & Ohio companies in building a line from the Great Lakes to the Gulf of Mexico. These grants were to become great incentives for future railroad construction, first in the north central plains and later in the West, especially on the early transcontinental routes.

Although rail travel had improved greatly from its early days, it still typically required a lot of time and patience. Railroads did not go to many places, and traveling often required transfers between railroads as well as to stage and steamboat lines. A journey between Albany and Buffalo, for example, required the services of seven different railroads. *Disturnell's Railroad, Steamboat and Telegraph Book,* of January 1849, described a route from New York to New Orleans that included a rail trip to Washington, steamboat to Acquia Creek, train to Wilmington, North Carolina, steamboat to Charleston, rail to Augusta, Georgia, a combination of rail and stage to Montgomery, Alabama, stage to Mobile, and steamboat to New Orleans. One of several alternative routes included riding the B & O from Baltimore to Cumberland, and then taking a stage to Brownsville, Pennsylvania, where one could travel on by steamboat to Pittsburgh or on by stage to Wheeling. From either city one could then board a "commodious" steamboat to reach Cincinnati, St. Louis, and New Orleans. Such journeys typically took 11 to 15 days and cost between $33 and $50.

During the decade that followed, construction exploded, with railroads being built extensively in the Midwest, into Chicago, and to the Mississippi River. Also at the time many isolated lines were connected into larger systems, a number of trunk lines evolved, and many small railroads were absorbed by larger companies. One important development in 1860 was the completion of the 6′ gauge Atlantic & Great Western R.R. Extending from Salamanca, New York, to Dayton, Ohio, it provided an important link between the New York & Erie R.R., on one hand, and the Cincinnati, Hamilton & Dayton, and Ohio & Mississippi companies, on the other. All of these lines were of the same broad gauge, and together they formed a through route extending 1,200 miles, from New York to St. Louis.

Rail development was slower in the South. Compared to the North there was much less passenger traffic. Slaves rarely traveled, and population density was lower. As a result, southern lines were mainly built to facilitate the transportation of cotton. Typically constructed from the nearest coastal or river port, they delivered food and manufactured goods into the interior and hauled out agricultural produce and timber. These railroads were normally built in places where they did not compete with river traffic. Usually they were short in length and not intended to connect with other lines.

By 1860 America had 30,626 miles of track. This system provided a solid network in the northern part of the country as industrial growth rapidly took place there. It allowed raw materials from many different locations to reach factories, where they were converted into finished products, and these were then shipped again by rail to various destinations in the country or to ocean ports for overseas delivery. During this period rail service greatly improved, with trains operating faster, more frequently, and more often on time. Because of the railroad, the isolated regions of the country were slowly becoming part of an integrated whole, but the South was lagging behind in this development, and was perhaps somewhat resistant to it. This would give the North an important advantage in the Civil War (1861–65). This story is related in chapter 20.

PLATE 23

TOP LEFT: First engine tested in America, the *Stourbridge Lion*, built by Foster, Rastrick & Co. of Stourbridge, England, 1829. TOP RIGHT: Early American tramway, Quincy, Massachusetts, 1826. MIDDLE: Inaugural run of the *Best Friend*, Charleston & Hamburg R.R., January 15, 1831. BOTTOM: Contest of Peter Cooper's *Tom Thumb* with horse-drawn rail carriage, Baltimore & Ohio R.R., August 28, 1830.

Inaugural run of the engine *West Point*, Charleston & Hamburg R.R., March 5, 1831.

PLATE 25

Horatio Allen's 2-2+2-2 *South Carolina*, the world's first articulated locomotive, with portion of a train of cotton, Charleston & Hamburg R.R., autumn, 1831.

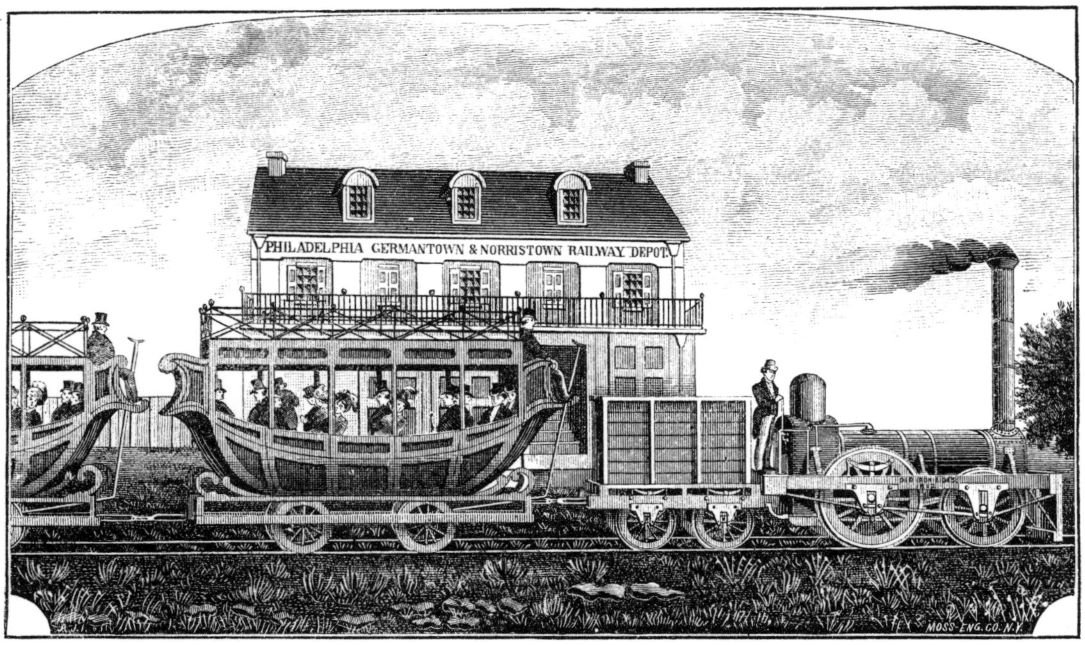

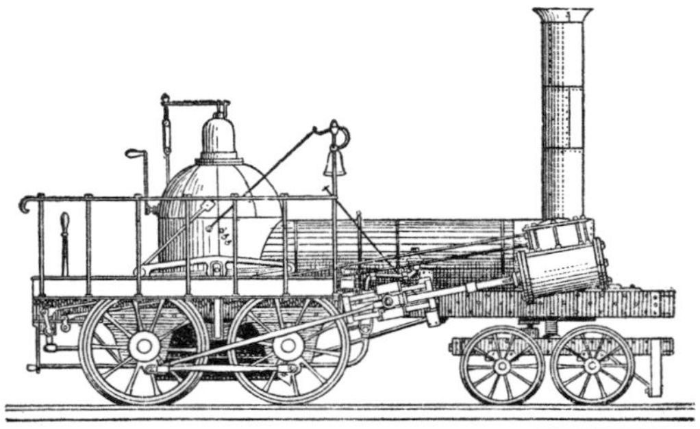

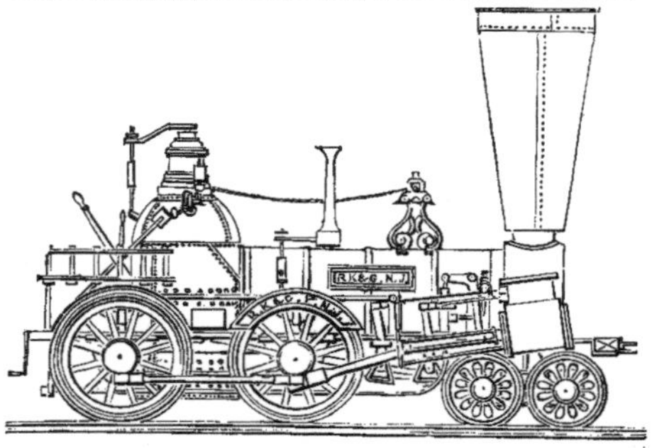

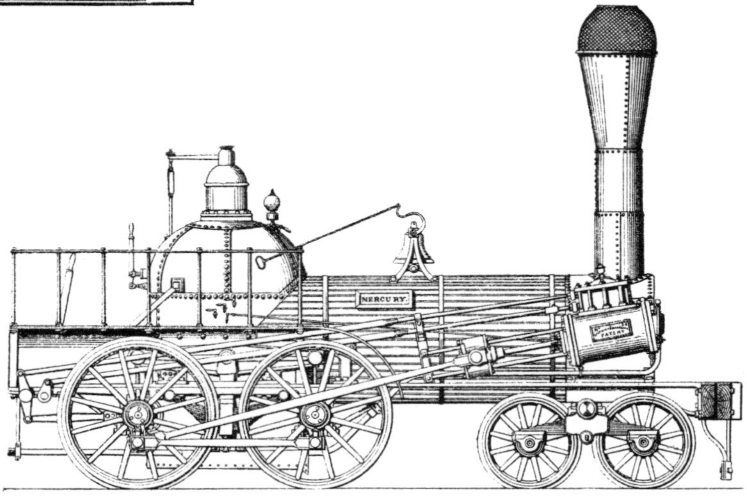

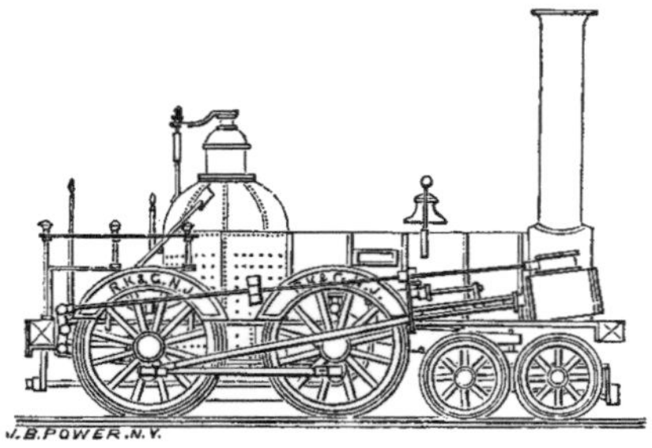

Top left, descending: Pennsylvania's first train pulled by Baldwin's *Old Ironsides*, 1832; 0-4-0 engine built by Braithwaite, London, 1838, (as later modified for Delaware, Lackawanna & Western Co.); Late Grasshopper style engine B & O R.R., 1838. Top right, descending: Eastwick & Harrison's 4-4-0 *Hercules*, 1837; Rogers 4-4-0, 1846; Eastwick & Harrison's 4-4-0 *Mercury*, 1842; Baldwin 4-2-0 engine of late 1830s; Rogers 4-4-0 engine, 1844; Rogers 4-2-0 engine, Tonawanda R.R., 1838.

PLATE
26

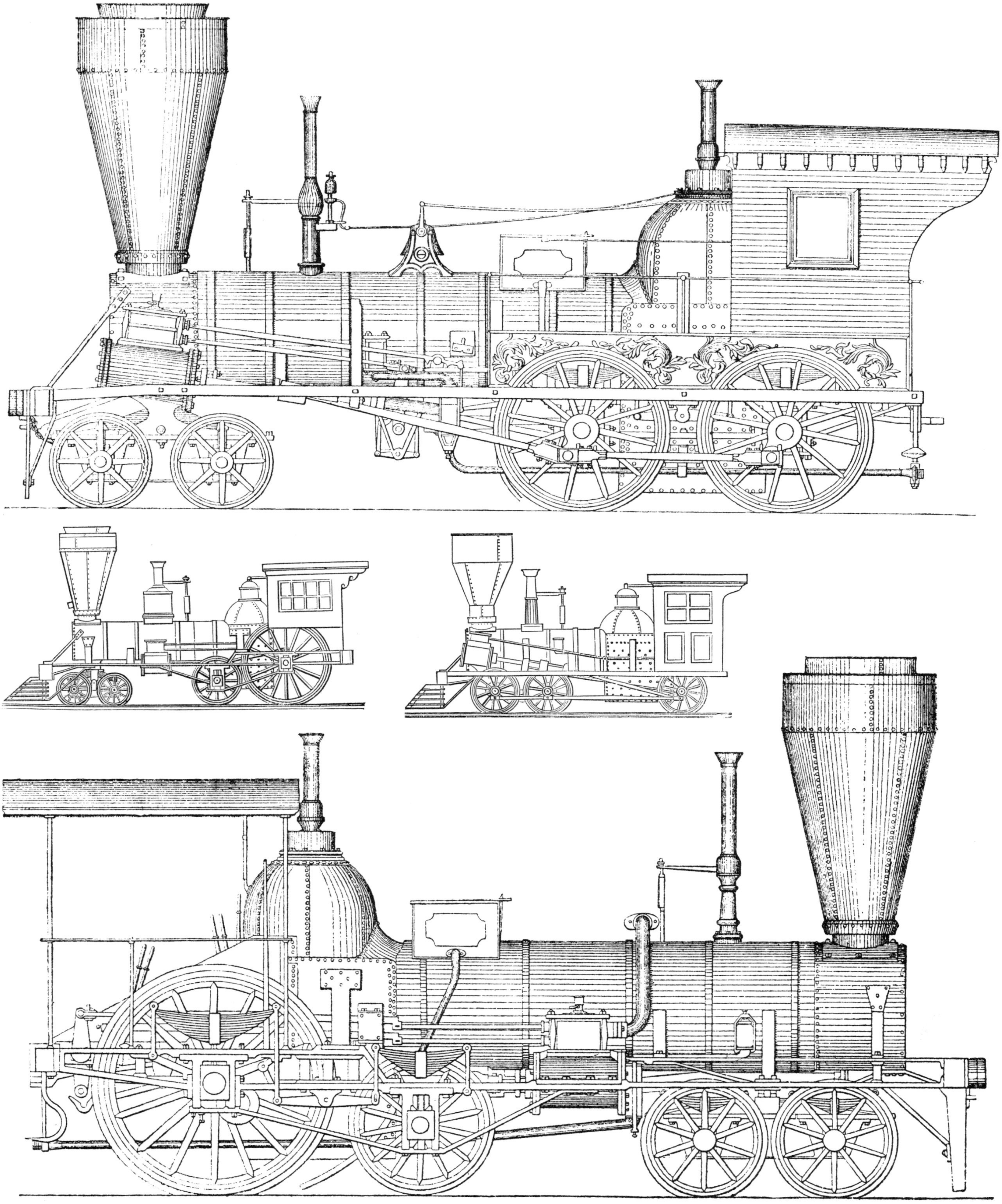

Baldwin locomotives of the 1840s: Top: 4-4-0 used for either freight or passenger service. Middle left: 6-2-0 express engine, 1848. Middle right: 0-6-0 freight locomotive, 1842. Bottom: 6-2-0 express engine.

PLATE 27

PLATE 28

Heavy freight engines of the 1840s: TOP LEFT: 0-8-0 *Dragon*, first to burn Cumberland coal and first to have a rocking grate. TOP CENTER: Rogers 4-6-0 "ten-wheeler" type, New York & Erie R.R., 1848. TOP RIGHT: Ross Winans 0-8-0 *Camel*, B & O R.R., 1848. BOTTOM: Winans 0-8-0 Mud Digger class engine *Cumberland*, B & O R.R., 1845.

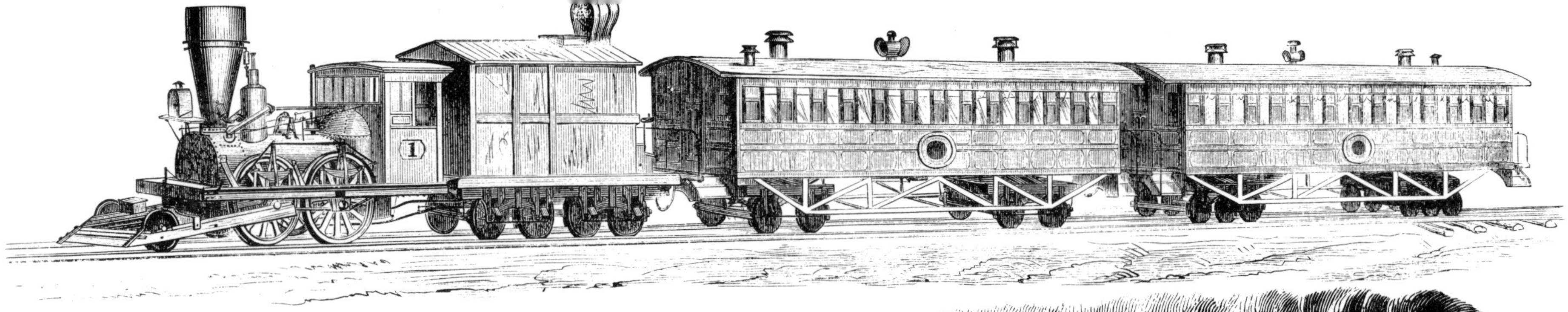

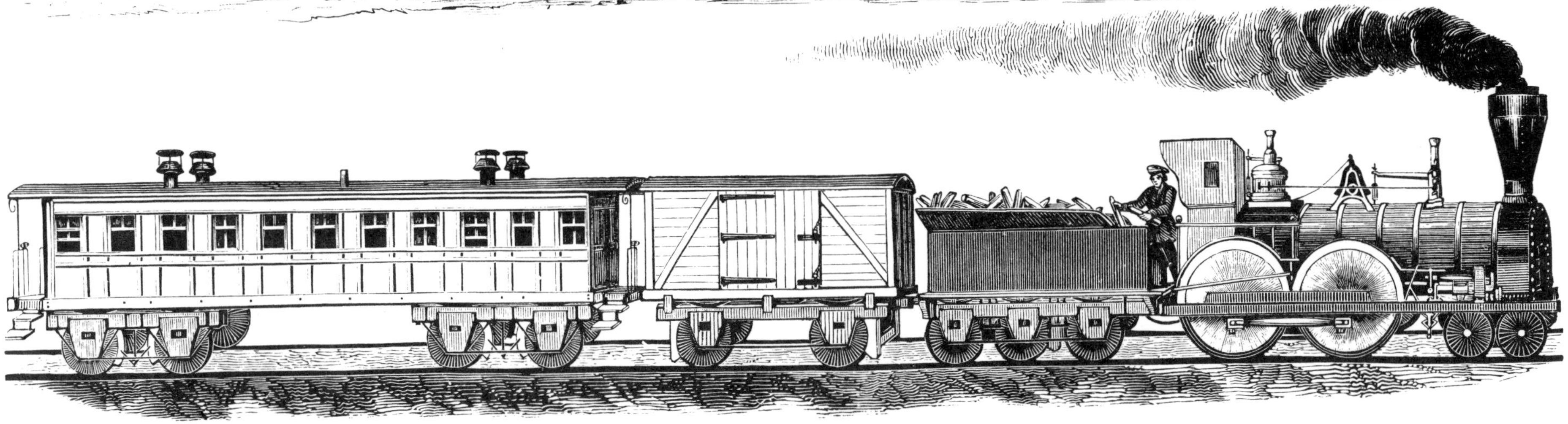

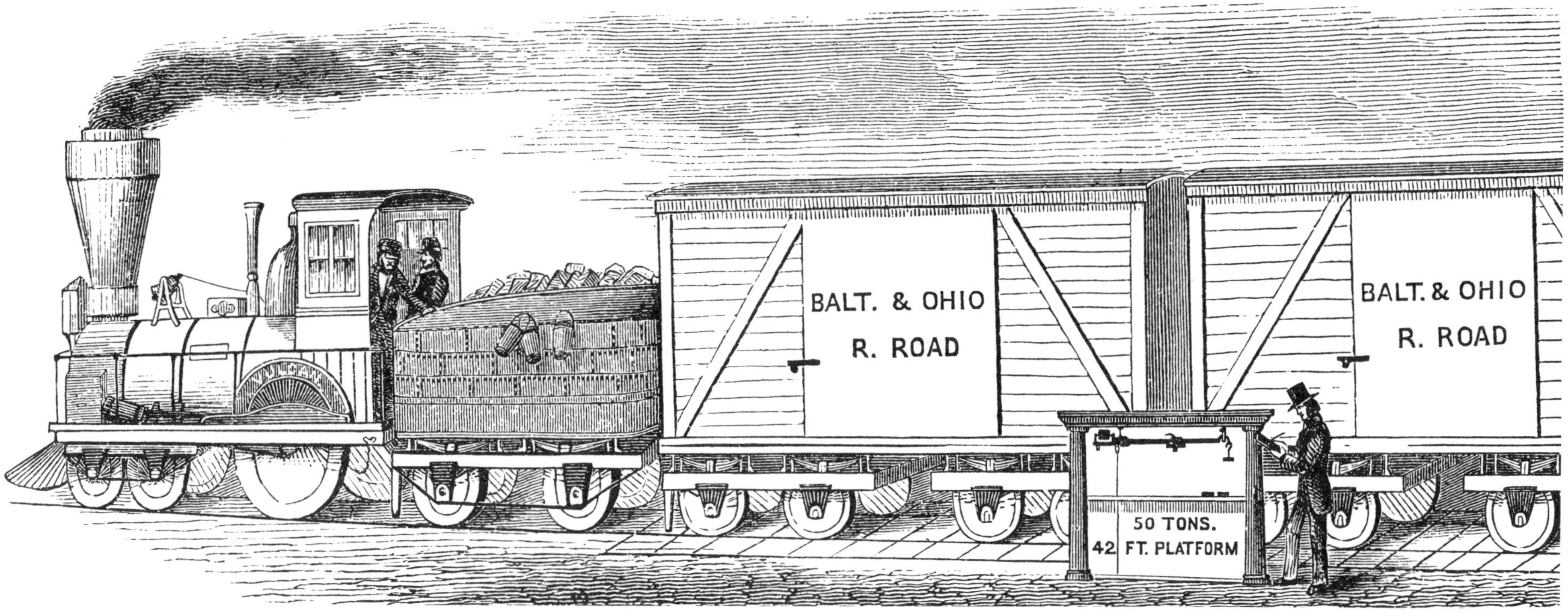

PLATE 29

TOP: Train pulled by famous English-built engine, *John Bull*, Camden & Amboy R.R., circa 1840s. MIDDLE: Passenger train, circa late 1840s. BOTTOM: Weighing boxcars, circa late 1840s.

Early American railroad stations: Top: Eastern R.R. Depot, Lynn, Massachusetts, 1857. Middle: Michigan Great Central R.R. Depot (note domed roundhouse), Detroit, 1854. Bottom: Boston & Providence R.R. Depot, Providence, Rhode Island, designed by architect Thomas A. Tefft in the Romanesque style, 1848.

PLATE 30

View of train at Bridgeport, Connecticut, New York & New Haven R.R., circa 1855.

PLATE 31

CHAPTER

(4)

American Railroads after the Civil War

In the aftermath of the Civil War the rail network of the North was more developed than ever before, while that of the South was in ruins. It would be several decades before the southern economy fully recovered. During this period the southern rail system was repaired and expanded, although during Reconstruction (1865–77), funds allocated for such purposes were sometimes plundered. Southern railroads also were hurt by the panic of 1873, and many lines ended up in receivership. During the 1880s railway mileage in the region almost doubled, growing from 14,778 to 29,263 miles. From then until the end of the century, many smaller lines were consolidated into larger systems like the Louisville & Nashville; Richmond & Danville (later reorganized as the Southern R.R.); East Tennessee, Virginia & Georgia; and Central of Georgia railroads.

In the North there was a general economic expansion after the war, and both rail and industrial growth continued. However, the main activity in new rail construction was moving increasingly westward. Thus comparatively few new lines were built in New England and the Middle Atlantic states, while rapid expansion continued in the Midwest. As in the South, there was also a considerable movement toward rail consolidation as the century waned.

By 1867 national attention turned to the West, where increasing progress was being made toward the completion of a transcontinental railroad. There had been interest in the idea of such a line for a long time. Besides opening the natural resources of the West for future utilization, it would provide a new route for trade with the Far East. Asa Whitney, a businessman who had traveled to China, petitioned Congress for such a line during the 1840s. This idea became more attractive after gold was discovered in 1848 and California gained statehood in 1850. Soon there was a national debate over which route such a railroad might follow. Southern congressmen and their supporters argued for a line through Texas to southern California. A central route, beginning in either Chicago or St. Louis, was advocated by people from the Midwest and most of the eastern seaboard. Those in the far north suggested a railroad from Lake Superior to the Pacific Northwest.

In 1853, a bill was passed by Congress authorizing U.S. Army survey crews to explore each of these possible routes. Shortly afterward, Mexico was persuaded to sell portions of land along its northern border with the United States, in what became known as the Gadsden Purchase. This land, later to become a part of southwestern New Mexico and Arizona, provided an easy way over the continental divide. Once the surveys were completed, the southern route appeared to offer the best possibilities. However, in the interim, the North-South division in Congress had grown so intense that no agreement was possible. Aside from the economic competition involved, many feared that the South intended to use the line to expand slavery across the Southwest to California. Thus while the issue continued to receive attention, it was left unresolved.

By the late 1850s pressure for a line was also received from California. Theodore Judah, builder of the state's first railroad, the Sacramento Valley line, made several lobbying trips to Washington. During this period he formed the Central Pacific Railroad with C. P. Huntington, Charles Crocker, Leland Stanford, and Mark Hopkins. When the South seceded in 1861, the transcontinental line became even more imperative, as it was seen as an important way of holding the remaining portions of the Union together. No longer having to contend with its southern members, Congress promptly acted to resolve the issue, and in 1862 President Lincoln signed the historic Pacific Railroad Act.

This legislation authorized construction of the line over a central route, giving the concession to two companies: the Union Pacific to start at Omaha on the Missouri River, and the Central Pacific beginning at Sacramento. Each was to build toward the other and would receive generous grants of land: a right of way, with land for other facilities, and five alternate sections on each side of the track. Generous loans were also provided, based on the terrain the lines would be built through. Congress followed up in 1864 with another important piece of legislation allowing the Northern Pacific Co. to build a line from Lake Superior to Puget Sound.

Construction on the Central Pacific route began on February 22, 1863. For the next several years, progress was slow for both companies. By 1863 Judah had become mistrustful of his partners and this ultimately led to a confrontation. Selling them his shares, he resolved to return to New York in hopes of gaining financial backing to buy them out. Unfortunately, this was not to be. While crossing the Isthmus of Panama, Judah contacted typhoid and died shortly after his New York arrival.

The chief officers of the Union Pacific proved to be both corrupt and incompetent, and it was only after they hired General Grenville M. Dodge to oversee construction that things began to change. Large crews of men and pack animals were employed as the line slowly made its way up the Platte River Valley. Although the terrain was relatively favorable, construction was difficult and time-consuming. The land was barren, and cross-ties, fuel, and water had to be brought in from elsewhere. Thus by the end of 1866 the railroad only extended a distance of 305 miles.

Many Union Pacific workers were immigrants from Ireland. Although sturdy and hardworking, they often had a fondness for alcohol and sometimes could be violent. Frequently taking advantage of them was a large group of prostitutes, gamblers, liquor salesmen, and criminals who followed in the rear. Construction crews also had to endure bitterly cold winters and occasional attacks by hostile Indians. While small groups or isolated individuals were most frequently targeted, an entire train was destroyed near Plum Creek Station on August 6, 1867, by a party of Cheyennes. Seven Union Pacific employees were killed, and another was scalped.

The Central Pacific line had greater difficulty in its initial recruitment of labor. Thus the company soon turned to the Chinese immigrant population, and discovered that these individuals made excellent workers. They were practical and stable, had good habits, and preferred tea to whiskey. Later, when the local source of manpower was exhausted, the railroad sent recruiters to China. The company also made efforts to pacify the Indians, and eliminated this threat to its workers. While gaining these advantages over the Union Pacific, it did have the disadvantage of building through the difficult terrain of the Sierra Nevadas.

While these companies were partners, they were also rivals, each wanting to build and gain control of as much of the total line as possible. As they drew closer they began to race each other. The Central Pacific finally won this competition, building 10 miles in a single day. However, the Union Pacific won the greater competition, as its portion of the total route was much longer. On May 10, 1869, the two lines met in an elaborate ceremony at Promontory Point, near the Great Salt Lake of Utah. Together they had laid a railroad stretching 1,780 miles. The presidents of both lines and many dignitaries were present when a final golden spike was driven and engines of each line steamed up to face each other. News of this event was quickly telegraphed to the rest of the country, and a new era was born.

Construction on the Northern Pacific route was delayed until 1870. Work commenced from both ends starting at Carlton, Minnesota (near Duluth), and from the Columbia River near Portland, Oregon. Unfortunately, the national financial panic of 1873 halted progress. By this time the lines had reached Bismarck, North Dakota, and Tacoma, Washington, respectively. Building did not resume until 1879. This railroad, which finally extended a distance of 2,260 miles, was completed in 1883 in a ceremony at Gold Creek, Montana. In the meantime, other transcontinental routes had been completed.

The second transcontinental line was completed at Deming, New Mexico, in 1881, when the Atchison, Topeka & Santa Fe and Southern Pacific railroads met. The Santa Fe had started construction in 1869 at Topeka, Kansas, and had closely followed the Santa Fe Trail. The Southern Pacific began laying track from Los Angeles in 1876 and crossed a very difficult stretch of desert. Extending its line farther east by 1882, it met the Texas & Pacific line, built from Shreveport, Louisiana, at Sierra Blanca, in far western Texas.

Continuing construction eastward from that point, the Southern Pacific took a route south of the T & P, and on January 12, 1883, near the confluence of the Pecos and Rio Grande rivers, met the Galveston, Harrisburg & San Antonio R.R. This opened a connection to Houston. In the meantime the Texas & Pacific finished an eastern extension of its route from Shreveport to New Orleans. Completed on October 15, 1882, this truly provided the South with a route to the Pacific.

Other important western lines included the Kansas Pacific, Denver & Rio Grande, and Great Northern lines. The Kansas Pacific R.R., a subsidiary of the Union Pacific, was built from Kansas City to Denver, and was completed on September 1, 1870. During its construction the legendary Buffalo Bill Cody was employed to provide meat for the workers. In one season his crew killed 4,862 buffalo plus many deer and antelope. Shortly, the shipment of Texas cattle provided the railroad with a lucrative source of income. The cattle were received at Abilene, after being driven up the Chisholm Trail. From Kansas City they were shipped to Chicago on other lines. The Kansas Pacific had a monopoly on this traffic until the Santa Fe built its railroad farther to the south, where it could intercept most of the business.

The Denver & Rio Grande line also had problems with the Santa Fe. Both companies were looking for a route westward that would avoid the high elevations of Colorado's Rocky Mountains. Raton Pass, on the New Mexico–Colorado border, provided a way to the south around this formidable range. Both railroads sent armed parties to take control of the pass. The Santa Fe group arrived first, and after intimidating its opponent, gained secure possession of the pass. Rail traffic through the pass was inaugurated on December 7, 1878.

A second confrontation between the lines occurred when the Santa Fe made a grab for the silver mining traffic in the Leadville area. To reach this district, it decided to extend a branch from its main line up the Royal Gorge of the Arkansas River. However, the Rio Grande company, having built south from Denver to Pueblo, was also preparing to go through the canyon. An interesting conflict ensued that involved both armed gangs and lawsuits. At one point the Rio Grande ran out of money, and its small narrow-gauge system was leased to the Santa Fe. However, when the lawsuit for control of the Royal Gorge was decided in favor of the Rio Grande, its former officers led a coup and regained control of the company. Santa Fe sympathizers were forcefully expelled throughout the Rio Grande system, and an armed gang was dispatched from Denver to Pueblo to retake company offices there. These were held for the Santa Fe by a tough gang from Dodge City led by Bat Masterson. Things were finally resolved peacefully; Masterson's group was bought off, and the Rio Grande took possession of the Royal Gorge.

Construction of this line was not easy. At some points, during surveying, engineers had to be suspended by ropes and lowered over the sides of cliffs. When blasting a roadbed through the mountainside proved too difficult at one location, a special bridge was devised. Two overhead rafters were constructed, the bases firmly braced at each side of the gorge. These supported the weight of a bridge that was built underneath. This line soon became the company's westward route over the Rockies. After completion of the Royal Gorge section, the road was extended over Tennessee Pass into the Colorado River valley. From there, construction continued until the line reached Salt Lake City. At this point it connected with the Western Pacific line, which extended to Oakland.

The Great Northern R.R., which began as the St. Paul & Pacific Co., was built by James J. Hill and completed in 1893. While situated north of the Northern Pacific line, it ran roughly parallel to that route, and terminated at Everett, near Seattle. By the turn of the century progress was being made on what would become America's final transcontinental route. This was the line of the Chicago, Milwaukee, St. Paul & Pacific R.R. Once completed, it connected Chicago and Seattle, and ran, for most of its route, south of the Northern Pacific.

While these longer routes served primarily to connect the eastern half of the United States with the West, other lines made important north-south connections, fed local traffic into the main rail systems, or operated independently in more isolated areas. By 1872 a number of such lines were being constructed in California. Among these were a Southern Pacific route between Los Angeles and San Francisco; a San Francisco & North Pacific line building from Petaluma up the Russian River to Cloverdale; the California & Oregon Extension route being constructed toward Oregon; and the San Joaquin Valley R.R. extending outward to Visalia. By 1897 this line connected Stockton and Fresno.

Many interesting railroad stories came out of the West. Two brief ones, related by *Railroad Gazette* in 1886, are good examples. In its issue of January 29 it ran the following story: "A weird and terrible tale comes from Montana, where the engineer and fireman of a train on the Utah & Northern road tell how a cab of their locomotive was invaded by a mountain lion, which sprang upon the footboard as the train was passing through a cut and at once took full possession. The invader continued to run the train until the engineer conceived the happy idea of blowing the whistle. This was too much for him, and so alarmed him, that he leaped from the cab, much to the relief of its rightful occupants."

The second story, from its March 12 edition, gives insight into the history of a steam engine. "An old locomotive, and the first ever run in California, originally built by the Norris Locomotive Works in Philadelphia in 1835 by William Norris, for the old State road from Philadelphia to Columbia, is still doing service among the Sacramento shops of the Central Pacific R.R. It was rebuilt at the locomotive Works of Septimus E. Norris, Lancaster, Pa., in 1865, and a tank placed on top of the boiler. It was subsequently sold to the Western Pacific R.R. and used for some time on the Market St. line in San Francisco. It next went into the possession of the Central Pacific R.R., and as stated above is still doing good service."

As rapid migration to the West took place, railroads proved to be absolutely essential both in bringing settlers to the region and in opening up its economic potential. They provided a means for transporting out timber, minerals, livestock, and farm products, while at the same time bringing in important manufactured goods. Thus, over time, they helped to integrate the West into the greater economic and political unity of the country.

Following the Civil War the nation went through a period of widespread corruption. In the general laissez-faire environment of the time, railroads could, to a large extent, do whatever they wanted. Dishonest company officers routinely enriched themselves by such means as manipulating stock and inflating construction costs. The Erie line, while under the control of Daniel Drew, Jim Fisk, and Jay Gould, gained an especially notorious reputation. When the Union Pacific was building the first transcontinental line, corrupt officers set up a false-front construction company, Credit Mobilier, which overcharged the Union Pacific by perhaps as much as $23 million. When this subterfuge was uncovered in 1872, a serious political scandal resulted.

In 1865 the nation's rail system totaled about 35,000 miles. By 1870 this figure had grown to approximately 53,000 miles, by 1880 it reached 93,296 miles, and by 1886 it was over 127,000 miles. In 1871 *Poor's Railroad Manual* reported that there were 1,300 to 1,400 American railroads either planned, under construction, or already completed. It also estimated that there were about 10,000 locomotives and 214,000 railroad cars of all kinds in the country. By 1886 the same publication had substantially revised these figures, reporting totals of 25,937 locomotives, 17,290 passenger cars, and 805,519 freight cars.

An important development after the Civil War was a move toward standardizing various railroad gauges. A majority of railroads at this time were using the English standard 4′8½″ gauge; other lines, especially in the South, were using other gauges, predominantly 4′10″ and 5′0″ and the Erie and allied companies, a 6′0″ gauge. By June 1, 1886, most of these lines had converted over to the standard gauge.

However, in 1871, a new gauge controversy arose. Narrow-gauge mania, which had begun in Britain the previous year, quickly spread to the United States and became a fad. Eventually, lines of various gauge sizes were built in many states, usually of 3′ or less. The 5½-mile Peakskill Valley R.R. had a gauge of only 2′ and the Mount Union a gauge of 2′3½″; the Pittsburgh & Castle Shannon line had a gauge of 3′4″ and the Brownsville & Rio Grande a gauge of 3′6″. Narrow-gauge lines probably achieved their greatest practicality in the mountainous regions of the West. Among such railroads were the Denver & Rio Grande; Denver, South Park & Pacific; and Eureka & Palisade.

Following the war, the Pennsylvania and New York Central & Hudson River systems established themselves as the leading railroads in the country. In 1873 the Pennsylvania worked 1,513 miles of track with 1,071 engines, 847 passenger cars, and 20,649 freight cars. That year it carried 15,057,153 passengers and 15,647,509 tons of freight. The line ran an average of 40 freight trains and 30 through-passenger trains a day between New York and Philadelphia.

By 1870 the total rail mileage in Illinois surpassed that of any other state. Chicago, its greatest city, had become America's primary rail center. It had rail connections in all directions, and served as the eastern terminus for four important "granger" lines. These railroads were the Chicago, Rock Island & Pacific; Chicago & Northwestern; Chicago, Burlington & Quincy; and Chicago, Milwaukee & St. Paul. The word *granger* referred to the National Grange, a farmers' organization that was very strong in this grain-growing region. Besides Illinois, this area included Minnesota, Iowa, Wisconsin, Missouri, Kansas, Nebraska, and the Dakota Territory.

Increasingly, after the Civil War, farmers in the region began to despise the railroads. Because these lines tended to serve separate territories, they became, in effect, monopolies. In areas where they had no competition, they usually kept both passenger and freight rates high. Another hated practice of the railroads was to downgrade grain when it was purchased from farmers, and then upgrade it when selling to the mills. However, in defense of the railroads, it should be noted that activity on these lines tended to be light most of the year. At harvest times traffic was heavy, but afterward, much of the rolling stock, which had required a substantial investment, lay idle.

The National Grange of the Patrons of Husbandry was founded in 1867. Initially intended to serve social and educational purposes, the organization soon became a vehicle for the expression of political discontent. In the early 1870s granger organizations gained influence in a number of state legislatures. Thus between 1871 and 1874, legislatures in Iowa, Illinois, Minnesota, and Wisconsin passed "granger" laws regulating railway freight rates. Soon other states passed similar bills. Granger laws were upheld in early Supreme

Court challenges, but in the Wabash case of 1886, it was held that states could not regulate rates for any shipments going outside the state.

Where rail competition existed it was usually intense and occasionally cutthroat. Large shippers sometimes manipulated competing railroads to gain their own advantage. In an effort to eliminate costly competition, railroads sometimes made secret agreements to fix rates and divide up the business. These and other practices are discussed further in chapter 15. Growing political pressure against abusive and discriminatory railroad practices finally resulted in the Interstate Commerce Act of 1887. This act set up the Interstate Commerce Commission, which was given oversight and regulatory control over all transportation between states. In addition to prohibiting rebates and pooling, it required that the rates of all interstate traffic meet a test of being "reasonable and just." At the time, however, the railroads still retained a lot of political power, and so did not encounter serious regulation until early in the next century.

While railroads had occasional labor disputes with their workers, going at least as far back as the 1850s, rail employees were generally happy with their companies and took pride in their work. By 1870, however, worker dissatisfaction began to surface. Many companies had grown larger, and in most cases, employee relations suffered as a result. Also, increasing rail competition forced companies to apply greater pressure on workers in an effort to both cut costs and boost performance. Following the financial panic of 1873, many lines cut employee wages, and in the next few years, made other reductions. Living expenses, however, did not go down. Because there was a plentiful supply of cheap labor, due to massive emigration from Europe, there was little the rail workers could do.

During this period no strong employee organization existed for voicing grievances. Those groups that had been formed were, like the Brotherhood of Locomotive Engineers, mainly created for fraternal purposes. Few incidents of labor unrest occurred until 1877, when the situation finally reached an explosion point. That year the average salary for engineers and conductors was $3 per day; $2 for firemen; and $1.75 for brakemen. The spark that ignited this tense situation was a secret agreement by the Pennsylvania, Baltimore & Ohio, New York Central, and Erie railroads to collude on rates and also to further reduce employee wages. The Pennsylvania line initiated the action, announcing a cut of 10% on June 1. The New York Central followed shortly, and then other companies, not a part of the original conspiracy, also announced salary reductions. These included the Lehigh Valley, Lackawanna, and Michigan Central railroads. On July 1, the Erie slashed its wages, and the Baltimore & Ohio, having previously cut pay only eight months earlier, announced its new reductions on July 16.

While employees of the other lines continued to work, the Baltimore & Ohio labor force reacted angrily. Strikes began immediately, with the firemen and brakemen leading the action. At Martinsburg, West Virginia, all rail traffic was halted and the state militia was called in. Because of public support, the soldiers were dissuaded from taking any serious action against the strikers. A short time later, however, federal troops sent by President Hayes arrived. In an attempt to disperse the crowd, they fired into it, killing 10 people.

Word of this incident spread quickly to other cities, igniting protests in such faraway places as St. Louis, Omaha, St. Paul, and Buffalo. At Camden Station in Baltimore an angry crowd gathered, and was fired upon by soldiers with Gatling guns. Legitimate protest in Chicago degenerated into a civil insurrection as armed bands of thugs, arsonists, and looters roamed the streets and caused considerable damage.

This outpouring of rage, however, found its most intense expression at Pittsburgh, where it was directed primarily at the Pennsylvania R.R. There, thousands of people were hurt and about 25 killed, as first police, and then sheriff's deputies, armed Pinkerton strikebreakers, and state militia, and finally federal troops were successively called in. Over 100 locomotives and about 2,000 cars were destroyed in the course of this rampage, which also claimed the Union Station and hotel, a roundhouse, and various company offices. Damage in the city itself was also considerable, and 10,000 troops were required to return the rail system to operation.

While considerable national sympathy initially existed for the workers, much of it disappeared during the riots. Afterward the railroads fired and blacklisted many of the participants, but at the same time grew a little more cautious about provoking their employees. Labor solidarity increased after this incident and union activity became more common. This soon resulted in the formation of the Trainmen's Union in 1883, the Order of Railroad Telegraphers in 1887, and the Brotherhood of Railway Carmen in 1891.

Eugene Debs, who previously had served as secretary of the Brotherhood of Railroad Firemen, organized the American Railway Union in 1893 and led a successful strike against the Great Northern R.R. In 1894, after the Pullman Co. in Chicago ordered a wage cut, Debs called a strike, instructing all his union members to walk off their jobs on any railroad carrying Pullman cars. As a result rioting erupted in Chicago, and hundreds of trains in the Midwest were stopped. Federal troops were quickly ordered in by President Cleveland, and after 12 people were killed, order was restored.

Following the incident Debs was jailed, but later ran a number of times as a Socialist Party candidate for the U.S. presidency. Although many strikers were once again fired and blacklisted, organized labor efforts in the industry continued to grow through the end of the century. This led to the establishment in 1894 of the Switchman's Union, and in 1898 of the Brotherhood of Railway and Steamship Clerks, Freight Handlers, Express, and Station Employees.

In 1863, the Pennsylvania R.R. began to replace iron rails on its line with steel. In the heated struggle of the Civil War, this development hardly attracted attention. However, it was significant as it made possible the operation of heavier trains with greater safety. In the period following the war, railroad operations and safety were improved by a number of technological developments. In 1873, an automatic coupling device was patented by Major Eli H. Janney, a Confederate veteran. Until this time couplers were mainly of the link and pin type. Coupling was done by hand and was dangerous. Accidents, which were common, typically sheared off or crushed fingers or whole hands. Both before and after Janney's invention, other safe coupling mechanisms were devised, but it was not until Congress passed the Safety Appliance Act in 1893 that they were made mandatory.

In 1869 George Westinghouse invented the first practical air brake system. By 1873 he further refined it and introduced a failsafe improvement that would function if cars on the train became disconnected. The timing of his invention could not have been better, as public indignation was mounting over the frequency of passenger train accidents. Thus many lines quickly adopted his system for passenger service, but, because of high costs, were slow in accepting it for freight.

Shortly after Westinghouse had introduced his invention, Lovett Eames devised a simpler vacuum brake system. This device was found effective, and was used on a number of railroads, including the New York elevated lines and Union Pacific. However, several persisting problems in the Westinghouse system were eventually corrected, and it soon gained preference over Eames' vacuum system. Despite this, though, most railroads continued to delay air brake installation in their freight car fleets. Finally, Congress tired of this procrastination, and in its 1893 safety legislation, compelled the railroads to act.

As we are so conditioned to life synchronized by standard time today, it is hard to imagine a time when it did not exist. But such was the case until November 18, 1883, when time zones set at the 75th, 90th, 105th, and 120th meridians of longitude were officially established. Before this, time had been a local matter. It varied from town to town, and this made the scheduling of trains, not to mention dispatching, difficult. In 1870 *Railroad Gazette* proposed a single standard time for the entire nation. Although there was little support for this suggestion, it stimulated thought on the matter. Ten years later the editor of the *Official Guide of Railways,* William F. Allen, gained acceptance for his idea of a system having four time zones.

Perhaps one of the most profitable lines in the world at the time, in proportion to its length, was New York's Marine R.R. It extended for just over a quarter of a mile, connecting Brighton and Manhattan beaches on Coney Island. Laid along the shore, it was vulnerable to storms and often had to be rebuilt. However, the line's income was such that rebuilding costs were insignificant.

In 1890 the total route distance of American railroads was 166,817 miles. Although the financial recession of 1893–96 slowed new growth somewhat, this figure still increased to 193,000 miles by 1900. Since the Civil War railroad efficiency had greatly improved in many ways. Railroad consolidation had greatly synchronized operations. Engines had become substantially larger and more powerful. From the 1860s on, steel began to be used in their construction, and was later increasingly utilized in building both freight and passenger cars. Car size had also grown in this period, and load capacity had increased accordingly. Trains were longer, faster, more frequent, and usually on time. Substantial progress had also been made in electrical traction, and electrification of many lines was imminent.

Street railways, first built in 1832, also had gone through their own evolution after the war, and later, as cable and trolley systems were laid, had greatly transformed the quality of the nation's urban life. The America of 1900, which railroads had largely created, was remarkably changed from the one of the Civil War period and bore little resemblance to the country of the late 1820s that had existed at their inception. That isolated, rural, and awkward land had given way, within 75 years, to an optimistic, dynamic, and relatively sophisticated nation, one having a prosperous agricultural and industrial economy and standing on the threshold of international greatness.

Top: Contest between rival groups for control of Albany & Susquehanna R.R., Harpersville, New York, 1869. **Bottom**: Scene in rail yard at Corry, Pennsylvania, 1867. This town, situated at a junction of the Pennsylvania and Atlantic & Great Western lines, was also a terminus for the Oil Creek R.R.

American station scenes, circa 1867: Top: Union Pacific depot, Omaha, gateway to the American West. Bottom: Depot at Stratford, Connecticut; New York & New Haven R.R.

PLATE
33

PLATE 34

Shooting buffalo from a train on the Kansas Pacific R.R., circa 1871.

Construction on the transcontinental rail line: TOP: Union Pacific workers laying track along the Platte River Valley, Nebraska, circa 1865.
BOTTOM: Workers of both companies join to construct final mile of track joining Union Pacific and Central Pacific rail lines, 1869.

PLATE
35

California scenes on the Central Pacific Railroad: Top: Terminus at Oakland Wharf, 1878. Bottom: Freight train crossing wooden trestle bridge in Sierra Nevada Mountains.

PLATE
36

PLATE 37

Double-headed train rounding "Cape Horn," above south fork of American River, Central Pacific R.R., 1878. This area, in the heart of California's gold country, was very close to where the initial discovery was made in 1848.

Colorado railroad scenes: Top: Train on west side of Marshall Pass, Denver & Rio Grande R.R., circa 1885. Bottom: North end of Raton switch-back near Trinidad, Colorado; Atchison, Topeka & Santa Fe R.R., 1879. This picture also shows the old Santa Fe Trail and a railroad tunnel under construction.

Double-headed passenger train in Las Animas Canyon, Colorado, Durango-Silverton branch, Denver & Rio Grande R.R., 1886.

PLATE 39

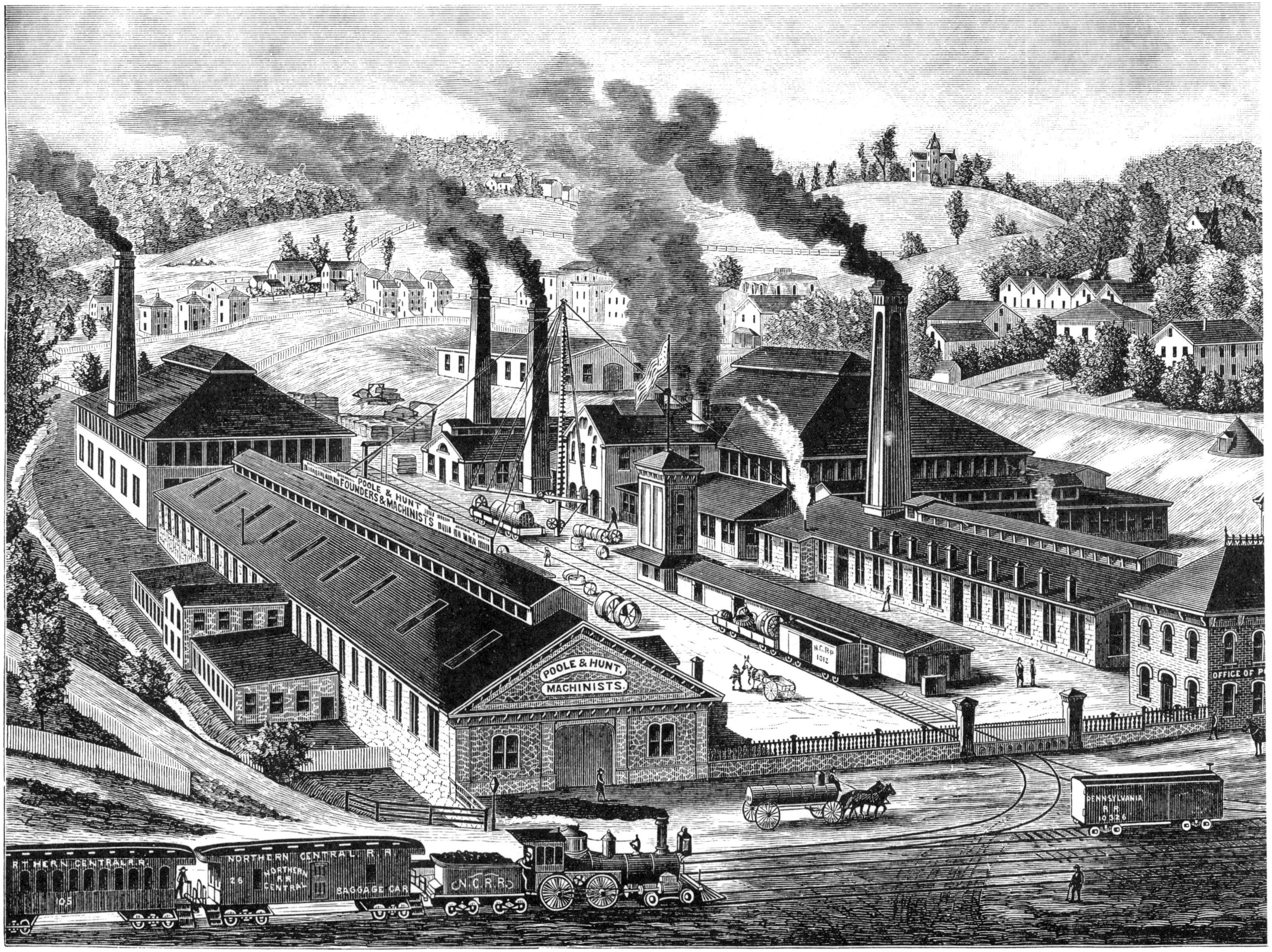

PLATE 40

A scene at Woodbury, Maryland, on the outskirts of Baltimore, showing line of the Northern Central R.R., circa 1870. The Poole & Hunt Co. had opened for business in 1853, initially planning to manufacture locomotives. When no orders were received they turned to other products, and later built some railroad cars and cable railway machinery.

PLATE 41

A scene during the great railroad strike of 1877: a fireman and engineer are forced by strikers to abandon their engine at Martinsburg, West Virginia, on July 17, Baltimore & Ohio R.R.

TOP: Manhattan Beach's prosperous Marine R.R. suffered periodic devastation by storm. This scene is from September 1889. BOTTOM: Eastbound and westbound trains exchange mail at Riverside, Nebraska, Union Pacific R.R., November 23, 1878.

PLATE 43

Logging train in a Michigan pine forest, 1886. There is a likelihood that the Shay geared engine depicted here is shop number 110, built for J. Cummer & Son of Cadillac, Michigan, on August 1, 1884.

Interior of observation car of the U.S. presidential train, showing President Benjamin Harrison seated on right, 1891. Andrew Jackson became the first American president to ride a train in 1833.

PLATE 44

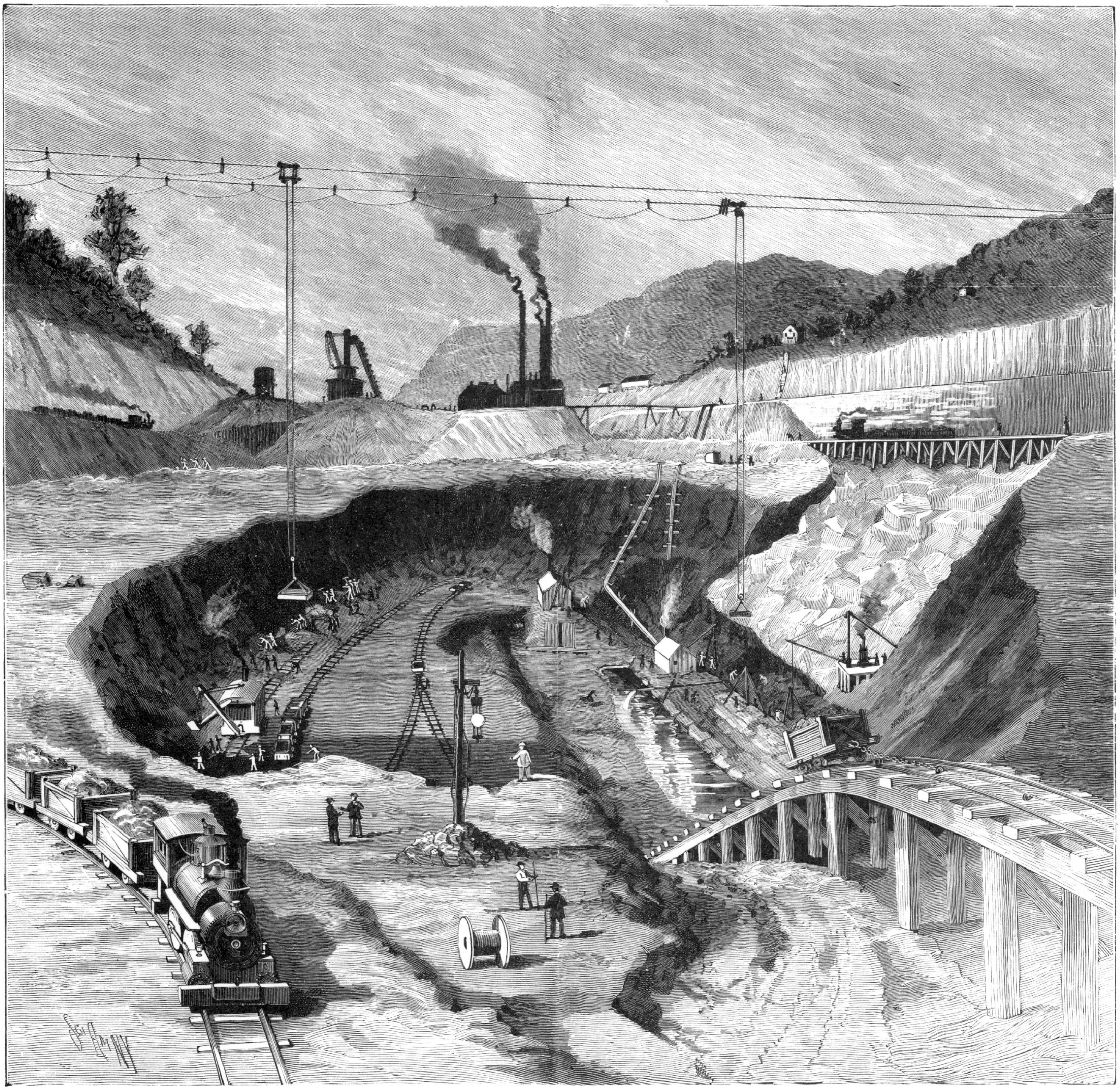

Temporary rail lines were often built by contractors to facilitate construction. This scene shows construction in progress on New York's Croton Reservoir dam, 1896. It would soon provide the water supply for the city of New York.

PLATE
45

CHAPTER

5

Steam Locomotive Development, 1851–1900

Once locomotive design was past its infancy, three concerns became paramount: power, speed, and economy of operation. By 1851, considerable technological improvements had been made, resulting in engines far superior to the initial ones. The perfect representative of this progress, perhaps, was the 4-2-2 Great Western express engine, *Lord of the Isles,* displayed that year at the London Exhibition. It had a certain beauty of style and was capable of speeds over 75 m.p.h. This locomotive was the most recently completed of Daniel Gooch's Iron Duke class, a group of engines that were setting phenomenal speed records at the time and making an impressive case for the advantages of the 7′ broad gauge over the 4′8½″ standard gauge.

The spirit of the period found its ultimate expression, perhaps, in 1854, when a 4-2-4 tank locomotive of the broad-gauge Bristol & Exeter line reached a speed of 81 m.p.h. This engine, designed by James Pearson, was somewhat remarkable. Its 9-foot diameter drivers were smooth and had no flanges. The carrying wheels at each end kept this locomotive on the track.

However, many important developments would occur in the latter part of the century, leaving engines like these far behind. By 1900, steam locomotives had reached entirely new levels of performance. The freight and passenger cars they pulled were larger and heavier, often constructed of steel or iron, rather than wood, like the cars of midcentury. Trains were generally longer and traveled at faster speeds, although they seldom ventured to exceed the speed records of the Iron Duke class. Improvements had also been made in adapting engines to specific kinds of tasks, whether for heavy freight hauling, express passenger service, commuter traffic, mass transit, mining, agriculture, or industrial use. This evolution was not easy, but was costly and time-consuming. It slowly proceeded as people took risks and made mistakes, but learned from these and persisted in their efforts.

In 1851 one of the most immediate needs of the railway industry was for a powerful freight locomotive capable of drawing heavy loads up steep mountain grades with tight curves. While 0-6-0 freight engines had first appeared in Britain in the 1820s and the 0-8-0 in America by the 1840s, none had yet reached a level of power, durability, and flexibility on turns well suited for heavy mountain work. Improvements in locomotive technology, however, would soon provide a breakthrough.

In Austria a decision had been made, amid the political upheaval of 1848, to construct the first trans-Alpine railway through Semmering Pass on the Vienna-Trieste Sudbahn. This move was viewed, by many, as reckless, for at the time it had not yet been demonstrated that an adhesion locomotive could successfully draw a train up the line's steep gradient of 1 in 40 for the lengthy distance of 14.3 miles. A rail publication in Stuttgart, remembering the famous Rainhill trials in England, suggested a similar competition to help the Austrian government out of its dilemma.

Thus in July 1851, the Semmering trials were held on an 8-mile completed portion of the line, with a 20,000 florin prize to go to the winner. Each of the four competing locomotives was restricted to a maximum load limit of 6¼ tons per axle and was required to make a dozen test runs in which it would do the following: (1) comfortably negotiate curves of 200-yard radius; (2) draw a 125-ton load up 4.7 miles of a continuous 1 in 40 gradient at a speed of at least 7 m.p.h.; (3) draw the same load downhill at about 19 m.p.h.; and (4) make a stop within 160 yards on the downhill grade.

The four locomotives entered were all unique in the technology they employed. The 0-8-0 *Vindobona* was entered by the Scot John Haswell, director of the Imperial Austrian Railway Works. Among the four, it offered the simplest design and was the first eight-coupled wheel type in Europe. An Englishman, Joseph Hall of the Maffei Locomotive Works of Munich, built the *Bavaria,* which employed chain coupling between the second and drive axles and between the trailing axle and a lead tender axle.

John Cockerill, an English industrialist who owned the Societe J. Cockerill in Seraing, Belgium, entered the 0-4+4-0 *Seraing,* which was essentially two 0-4-0 locomotives connected back to back. It is probable that the *Seraing* ultimately owed its inspiration to the world's first articulated engine, Horatio Allen's 2-2+2-2 *South Carolina,* built in 1832. However, this same back-to-back principle had later been patented by Johann F. Laussmann of the Bergisch & March Rwy. of Prussia. He had submitted his idea earlier to Cockerill in hopes of gaining backing, but instead found it used without his knowledge. His later protests achieved no result.

The 0-4+4-0 *Wiener Neustadt,* built by Wenzel Guenther, was another articulated engine, but unlike the Seraing, had a single long boiler. Its four cylinders were placed in the midpart of the locomotive, two on each side, facing each other.

The winner of the competition was the *Bavaria,* which drew its load uphill at a speed of 11.34 m.p.h. While the *Vindobona* was rejected for being decidedly inferior, the remaining contestants were purchased by the Sudbahn. However, none were found to offer continually reliable service.

Late in 1851, Wilhelm Engerth submitted a new design to the Austrian Ministry of Transport. His planned engine was essentially a semi-articulated 0-6-4T locomotive, with the permanently attached tender supporting the weight of the firebox. However, the wheels of the tender, or more concisely, rear carrying wheels, also functioned as a set of auxiliary drivers. These were connected to each other by side rods, with the front tender axle receiving power via gears from the main drivers. In November 1853 a locomotive built from these plans was tested at Semmering and found to be superior to the previous entrants. The line accepted Engerth's design as its prototype and eventually had 76 of these engines built. Subsequently, Swiss and French railways ordered many of these, but they were modified from the original design. The Schneider Works at Le Creusot built 30 Engerth 0-8-4T engines for the Nord Rwy. in 1856–57, and Grafenstaden, 10 more in 1857–58.

The problems railroads faced in the mountainous regions of the eastern United States were not unlike those on the Semmering line. Ross Winan's anthracite-burning 0-8-0 "camel" engines, introduced in 1848, had demonstrated some success in operating in such terrain. Beginning circa 1854, both Samuel Hayes of the B & O line and Zerah Colburn of the New Jersey Locomotive Works at Paterson began making their own improvements to this design. James Millholland of the Philadelphia & Reading began similar experiments, and in 1863 built the *Pennsylvania,* the world's first 0-12-0 engine.

In 1853 America's largest locomotive builder, Baldwin, discontinued its half stroke cutoff system, used since 1845, in favor of a variable stroke cutoff. About the same time, Stephenson's link-motion valve gear was introduced to America. This system, patented in 1848, provided a variable cutoff, but also allowed a "lead" or small amount of steam into the cylinder to cushion the impact of the returning piston. Baldwin, along with a number of other manufacturers, initially resisted adopting Stephenson's system. However, beginning in 1854, the company built a few engines having link motion when specified in the orders, and by 1857 had fully adopted the device. In France, in 1859, Henri Giffard introduced his cone feedwater injector, a device allowing water to be fed into the boiler despite the resistance of steam pressure. This innovation was quickly incorporated into American locomotive design.

During the 1850s America's 4-4-0 engines achieved a longer and leaner classic style, which remained intact, except for a few changes, into the 1880s. Used for both passenger and light freight service, they performed well and were built in far greater numbers than any other type. Great attention was paid to giving these locomotives a beautiful appearance. Typically they were colorfully decorated with miniature paintings and other ornamentation, and had fixtures of highly polished brass. Beautifully finished hardwood was sometimes used in their construction, and they usually received names, occasionally exotic or mythical ones. For many in the isolated rural communities through which these engines passed, they symbolized something romantic and magical, awakening fantasies of big cities and faraway adventure.

In 1851 manufacturers, such as Baldwin, built both 0-6-0 and 0-8-0 engines for heavier freight use. It was the 4-6-0 "ten-wheeler" type, however, that eventually became the standard American freight engine. Aside from the extra set of drivers, its general appearance was much the same as the 4-4-0 American type. Being used for freight, these locomotives usually received less artistic attention than those used for passenger service.

By the 1850s it could generally be said that two primary schools of locomotive design were emerging, the British (or European) and the American. The former type became characterized by plate frames, inside cylinders, rigid axles, unequalized springs, four- or six-wheel tenders, and a small open space for the train crew. Wind protectors were in use on some engines by this time, but cabs were rare until the 1870s. The crew area remained small, forcing both driver and stoker to stand. In comparison, Americans favored bar frames, outside cylinders, bogies, equalized springs, large cabs with seats, and eight-wheel tenders.

Although Baldwin began building the "ten-wheeler" type in 1852, it did not fully discontinue production of its 0-6-0 and 0-8-0 designs until the 1860s. By then other types of freight locomotives had begun to appear. The 2-6-0 Mogul type, first built in Russia by Eastwick, Harrison & Winans, was introduced to America in 1863. More massive engines appeared a short time later: the 2-8-0 Consolidation type in 1866 and the 2-10-0 Decapod type in 1867. The 2-6-0 had a significant advantage over the 4-6-0 type, in that the front driving wheels could be placed farther forward. This put a larger proportion of the engine's weight on the drivers, thus improving adhesion.

These new single lead axle types owed their introduction to Levi Bissel's invention of the two-wheel "pony" truck in 1858. This was a modification of a four-wheel "safety" truck he had patented in 1857. Previously front carrying trucks, which guided the locomotive gently around curves, had swiveled on a fixed central pivot. When speeds were low this design worked well enough. However, as speed increased, the trucks tended to oscillate, which could cause derailments by sticking when negotiating curves. Bissel overcame these problems by building the truck frame like a radius bar, which swiveled on a center pin behind the truck axles.

This invention also led to the introduction, in America, of the small 2-4-0 "pony" type engine. This style was typically used for light passenger service or switching. However, the most common switcher of the period was the 0-4-0 type. By the late 1870s heavier switching began to be done by 0-6-0 models.

In the early 1870s, small tank locomotives became popular in America for switching and other duties. While 0-4-0 and 0-6-0 varieties became common, other styles were developed. These included 0-4-2 and 0-4-4 engines by M. N. Forney (who also designed a 0-6-6); 2-4-2 and 2-4-4 "double ender" styles developed by Mr. Hudson, superintendent of the Rogers Co.; and the Mason "double truck" designs based on the Fairlie patent. Forney received his first patent in 1866, and Hudson the following year.

In the Europe of 1851, passenger locomotives tended to be either the 2-2-2 type in general use or the 2-4-0 type favored on some German and Austrian lines. The wheels on the single driver engines tended to be large, so as to maximize speed. However, these locomotives were slow in starting from a dead stop. The four-wheel locomotive bogie first came to Europe with the Norris engines. Having short wheelbases, they tended to oscillate at higher speeds on a straight course. John Haswell, who had built the *Vindobona* for the Semmering trials, made improvements on the Norris design, and eventually discovered a radius bar truck system similar to Bissel's. In 1871 the Austrian Kamper improved this bogie further by adding two swing links and inclining them slightly toward each other. These were suspended from a spherical point located in the middle. This balanced the load and greatly reduced oscillation.

When the Great Western's Daniel Gooch began experimenting with four lead carrying wheels in 1846, he did not employ bogies, but instead allowed lateral movement in the axles or omitted flanges on some wheels. Thus, his 4-2-2 engines, and the somewhat similar Crampton locomotives, were not ideal for lines with tight curves. Beginning in 1862, the Beyer-Peacock Co. of Manchester built engines using both the four- and two-wheel Bissel trucks. Further improvements in bogie design were made by William Adams in the 1860s.

Steel was first incorporated in the manufacture of locomotive parts in Germany and France as early as 1850. However, it was only in 1861 that Baldwin became the first American builder to use steel, initially introducing it for firebox construction. In 1863 the Pennsylvania R.R. introduced steel rails for the first time in America. This was an important development, as it made possible the construction of much heavier locomotives. Following the Civil War, coal was mined in much greater quantity and began to replace wood as the primary source of engine fuel. This change eventually made funnel stacks, used for trapping sparks and cinders, obsolete. These were gradually replaced with other styles.

Another potential source of fuel was oil. After the Civil War it began to be pumped in Pennsylvania, and a short time later, plentiful sources were discovered in Russia. Beginning circa 1868 Henri Sainte-Claire Deville began extensive experiments with the substance, testing it on two locomotives of the Est line. On September 6 of that year, Emperor Napoleon III rode on an oil-fueled train along with Deville and the company president, boarding at Châlons. While oil appeared to offer certain advantages, it was a more expensive fuel than coal. Thus there was little incentive to use it, except for areas in Russia where other fuels were scarce. There the engineer Thomas Urquhardt, of the Grazi-Tsaritsin line, continued Deville's work and improved the technology.

After Engerth's engines were introduced in the early 1850s, the next developments in heavy freight locomotives occurred in France. Improvements were made on Engerth's design by a number of French manufacturers, including Schneider of Le Creusot. This led to a greatly changed version by Edouard Beugniot in the late 1850s, which improved both traction and flexibility. Koechlin of Mulhouse built the first engines of this style for the PLM Rwy. in 1860.

In the late 1850s Engerth engines were extensively used for pulling coal trains on the Nord line. Jules Petiet, chief engineer of the railway, noticed that most of the short branches feeding the main line were 20 miles or less in length. As the fuel required for these short journeys was minimal, he thought it possible to improve service by operating tank locomotives. This would eliminate tender drag and improve adhesion by placing the additional weight of the reserve fuel on the drive wheels. Further, he also wanted to design a firebox that could efficiently burn the cheap local soft coal. His solution was to build an engine having a wide firebox with the grate above the frame. The boiler was larger in diameter but shorter in length than was typical. To counter a higher center of gravity, water tanks were situated underneath and on each side of the boiler barrel. The first locomotives of Petiet's design were 0-8-0 engines, some having tenders, built in 1859. More powerful Petiet locomotives were built from 1862 to 1866. Among these were Europe's first 0-10-0T engine and the world's first 0-6-6-0T 12-wheel divided drive type. This latter, however, was not articulated. Its four cylinders were placed at opposite ends of the engine, each powering three drivers. Even more unique was his 0-2-6-2-0T design of 1862, which had a set of drive wheels at each end with six carrying wheels in between.

Among the unusual features of Petiet's engines was a horizontal smokestack that exhausted at the rear, perhaps used so his tall locomotives could successfully pass under bridges or through tunnels. Petiet also included in some of his locomotives a tube of large diameter, situated along the back of the boiler, through which smaller pipes carrying steam to the cylinders were passed. Hot combustion gases were exhausted through this tube on their way to the smokestack, thus further heating the steam, drying it more and raising its pressure. This innovation anticipated the later development of the superheater. By burning cheaper coal and drawing four to five more cars than the Engerth engines, these locomotives demonstrated a clear improvement in operational efficiency.

In 1866 an 0-6-6-0 Petiet engine was used in the inauguration of the Enghien-Montmorency Rwy. Unlike previous Petiet locomotives, it featured a new boiler and firebox arrangement designed by the Belgian Alfred Belpaire. Belpaire's design was distinctive in appearance, having a flat top and squared sides. Like Petiet's earlier arrangement, it had a large boiler with a wide-grate firebox, and was intended for burning cheap coal. However, Belpaire's boiler was longer, and overall it was a more efficient system. It was displayed at the Paris Exhibition of 1867, and received considerable attention.

Following Petiet's creations and the introduction of America's new freight types, the next heavy locomotive of interest was the double boiler articulated tank design of Robert Fairlie. These so-called Double Fairlie engines were essentially an upgraded version of the Cockerill locomotive used in the Semmering trials. They had the appearance of two locomotives, the backs of which were joined to each other. Coming in 0-4+4-0 and later in 0-6+6-0 versions, they had two motor bogies, providing good flexibility on curves, and were especially useful in mountain terrain where their power and adhesion could be used to maximum advantage.

Fairlie, the locomotive superintendent of Ireland's Londonderry & Coleraine line, secured a patent for this design in 1864 and built the first prototype in 1865. His *Little Wonder,* sold to the Festiniog line in 1869, was so successful that many other locomotives of this style were soon ordered. Most were exported to such places as Mexico, Central and South America, and Russia. A few were tried out in the United States, but never gained acceptance. Fairlie also introduced the so-called Single Fairlie style. This was a single boiler tank engine having a motor bogie and trailing truck. The American manufacturer William Mason built about 150 "double truck" engines based on this design.

The engineer Jean Jacques Meyer also looked back to the Semmering trials for inspiration in creating his own articulated engine. Meyer, who had built locomotives at Mulhouse, sold his operation to Koechlin in the 1860s and moved to Paris. Returning to Guenther's plan for the *Wiener Neustadt,* he adopted a similar system and in 1869 built the 0-4+4-0 *L'Avenir.* A second engine, of 0-6+6-0 design, was displayed in 1873 at the Vienna World's Fair. However, neither of these engines performed much better than the 0-8-0 goods type already in wide use. The Guenther-Meyer type, as it became known, eventually found its home on mountain lines having tight curves. It had the advantage over the Fairlie in requiring less maintenance, but lacked the power to do its job satisfactorily. Greatly improved versions were built shortly after the turn of the twentieth century.

When Patrick Stirling, of the Great Northern line, introduced his gracefully styled 4-2-2 express engine in 1870, Britain's resistance to guide trucks began to crumble. Known variously as the "bogie single," "Stirling single," or "eight-foot single," his engine was more powerful than similar ones preceding it. Aside from the use of bogies, it also broke from common practice by using outside cylinders and omitting the steam dome. With only minor modifications this design was built for many years.

While 4-4-0 tank engines were built in Britain in the 1860s, it was not until 1873 that a standard express locomotive of this type was built. This was the creation of James Stirling, engine superintendent of the Glasgow & South Western line and brother of Patrick Stirling. Besides having a lead bogie and omitting the steam dome, this locomotive had coupled driving wheels of 7-foot diameter and inside cylinders. His engine initiated a new chapter in the history of British locomotives. The 4-4-0 type became widely adopted in the next decade, and many of these engines were of beautiful design. Beginning about 1886, however, there was resurgence in use of the single driver 4-2-2 type. This was attributed to Fred Holt's invention of steam sanding, which improved traction. However, about the end of the century heavier passenger carriages were widely introduced, and the 4-4-0 returned to prominence.

On the Continent a number of 4-4-0 locomotives were built in the 1850s; this type was adopted, in 1861, by the Baden State Rwy. That same year the locomotive builder Kessler supplied Switzerland with 12 4-4-0 tank engines. Afterward, this became a popular type in the country. Italy's first 4-4-0 engines were introduced in 1878 on the Alta Italia line. In France, where the 2-4-0 and 2-4-2 type passenger engines were popular, there was general opposition to the American type. When the Nord line began replacing Crampton locomotives in the 1870s, it did so with 2-4-0 engines. However, it had a few lines with severe curves, where it was thought that the 4-4-0 type might perform better. Thus in 1877 a few of these locomotives were introduced.

In 1889 a severe passenger train wreck occurred near Dijon, when a 2-4-2 engine derailed while going full speed downhill. Afterward it was decided to conduct test runs with various types of express locomotives to determine which gave the best performance. France's seven largest railways submitted their representative engines, with six being of the 2-4-0 and 2-4-2 types, and one, that of the Ouest line, being a 4-4-0. It was determined in the test runs (with speeds up to 75 m.p.h.) that the 2-4-2 engines caused the most stress to the track, while the 4-4-0 type provided the smoothest ride. Shortly, all of these lines adopted the 4-4-0. About the same time, the four-cylinder compound system was being introduced. Its combination with the 4-4-0 type proved very effective. By the end of the century, these French passenger locomotives were among the finest in the world.

While there was a general preference for inside cylinders in England, American engine design had evolved to favor outside cylinders. In its June 8, 1872 edition, *Railroad Gazette* presented the relative merits of these arguments:

> It certainly is somewhat curious that engineers in two countries speaking the same language and working to accomplish the same ends should, after forty years of experience and experiment, yet differ so widely, or at least have reached so little unanimity, in their practice. And yet it seems quite probable that the different conditions here and there may have caused this difference of practice. Here our rough roads made the flexible truck almost a necessity, and have been very destructive to crank-axles. With the better distribution of weight due to the use of a truck, we have found that the disturbing influence of outside cylinders was no very serious evil. The contention between those in favor of and those opposed to inside cylinders was very earnest, and almost rivaled in fierceness the "battle of the gauges" . . . With experience, however, it became apparent to those who took no part in the contest that outside cylinders had the best of it, or at least that locomotives thus equipped were less subject to failure and expense of repair, and thus the decision was quietly made in their favor. In Europe, however, the case is quite different. There the truck was not a necessity, and consequently the oscillation caused by outside cylinders was a greater evil than when a truck is used. The smooth roads there also lessened the liability to breakage of the crank axle. There can be no doubt that there are some advantages inherent in the use of inside cylinders. That a locomotive is less liable to lateral disturbance with inside rather than with outside cylinders, is certain.

Already during this period outside cylinders were increasingly used on the Continent. They also had been pioneered in the 1840s at Crewe, England, and as time passed they became increasingly acceptable to the country. Clement

E. Stretton, in *The Locomotive Engine and its Development,* relates that in the late 1880s the Great Northern Rwy. had two classes of express engines that were virtually identical, except that one had outside cylinders, while the other's were placed inside. In comparing the two he noted: "The practical result of working is that both types are very efficient. The inside cylinder class is less costly to build, the coal consumption is the same in both; and with reference to the question of speed it is found that the engines can run equally fast. For extremely fast running an inside cylinder engine is more suitable than an outside; it is therefore probable that in the future the inside cylinder class will become the standard." Not mentioned in this discussion was the fact that outside cylinders provided the advantage of easier maintenance.

Another American innovation requiring time for European acceptance was the enclosed cab. Italy was the first country to adopt this feature in the 1850s. Elsewhere, however, it was argued that cabs obscured too much of the engineer's view. So for the next few decades weather protection for the engine crew was minimal, although wind screens and, later, shortened cabs gradually gained favor.

In earlier locomotive technology, the successive improvements of the multitubular boiler and longer boiler design greatly improved both power and fuel economy. The later development of compounding led to a similar result. The basic principle of compounding involved allowing steam expended in the piston thrust in one cylinder to be exhausted to another, where it provided a second thrust at lower pressure. Originally devised for marine engines, this innovation was first attempted in locomotive design by the Eastern Counties Rwy. in 1850. An engine driver, John Nicholson, originated the idea for a two-cylinder compound system and submitted it to the locomotive supervisor, James Samuel. Two locomotives of this type were constructed, one for passengers and the other for goods. While results of this experiment were reported as highly satisfactory, no further engines of this type were built.

In 1866 Jules Morandiere of the Nord Rwy. designed an unusual compound locomotive having three cylinders and eight driving wheels. The wheels were coupled in two groups. A single inside high-pressure cylinder drove the forward axle of the rear group. Its spent steam was exhausted into outside low-pressure cylinders powering the front group. Designed for heavy passenger traffic on metropolitan lines, this engine was never built.

The next reported use of this innovation was the construction in 1876, by the Schneider Locomotive Works at Le Creusot, of three engines for the Bayonne-Biaritz Rwy. These were built from the designs of Anatole Mallet. Each engine had two outside connected cylinders of unequal size. The smaller high-pressure cylinder was 9½ inches in diameter, while the low-pressure cylinder had a diameter of 15¾ inches. Both cylinders had a stroke of 17¾ inches. The ratio of the capacity of the two cylinders was 1 to 2.78. These engines could be worked either compound or noncompound. When working compound, the steam from the boiler was admitted to the high-pressure cylinder only. It was then exhausted into a receiving chamber that carried it to the low-pressure cylinder. Thus the steam, having partially expanded in the high-pressure cylinder, continued its expansion in the low-pressure cylinder.

These engines gave such satisfactory results that similar ones were shortly ordered by the Paris-Orléans line, Spain's Norte Rwy., Russia's Southwestern Co., several German lines, and Austria's Kaiser Ferdinand Nordbahn. In Mallet's later engines steam was not admitted from the boiler directly to the low-pressure cylinder, but instead passed through a valve that reduced pressure on the larger piston and more closely balanced the work done by each cylinder.

Locomotive builders in several countries soon came up with a number of other compound designs, with the systems of the Frenchmen du Bousquet and de Glehn, the German August von Borries, Englishman Francis W. Webb, and American Samuel Vauclain eventually gaining the widest acceptance. Among these, Webb's system was perhaps the most unique. In his first engine of this type, the 2-2-2-0 *Experiment,* he essentially reversed Morandiere's idea. Two high-pressure cylinders were placed outside and connected to the rear drivers. A single low-pressure cylinder was placed between the frames and connected to the crank axle of the front drivers. Webb felt that the use of coupling rods created an unnecessary drag. Thus on his compound engines the two sets of drivers were allowed to run independently of each other.

Circa 1889, when he brought out a later version of his three-cylinder system in what became known as the Teutonic class, he omitted the Joy gear previously used for the inside cylinder. This resulted in a peculiar problem, noted in Gustav Reder's *The World of Steam Locomotives:*

> The low-pressure valve was operated with a loose eccentric which was kept in position for forward or reverse movement by stops. This, of course, prevented cut-off adjustment for the low-pressure cylinder. On starting, this loose eccentric sometimes created a scene which was the despair of the driver and which amused the travelers in the vicinity enormously; for, if

> the locomotive had backed up to the train waiting to depart, the inside eccentric obviously remained in the reverse position. On starting, it then often happened that both high-pressure cylinders caused the wheels to slip. The low-pressure cylinder was given a lot of steam and sent the axle it drove backwards, for that was the position of its gear, while the other driving axle tended to go forwards.

Shortly afterward Webb began experiments with a four-cylinder compound system.

Another important name in compound technology is that of Dr. Karl Gölsdorf. Born in 1861, he was the son of Louis Gölsdorf, chief mechanical engineer of the Vienna-Trieste Sudbahn. At age 30, Karl assumed a similar position with the Austrian State Rwys. and began building locomotives. Early in his career he developed a "starting valve," which simplified the complicated process of starting a compound engine from a dead stop. This device was later used on locomotives in Germany, Russia, America, Italy, Greece, Turkey, and Serbia. Gölsdorf's nineteenth-century engines were normally of the two-cylinder compound type, based on the Von Borries system. They were distinctive in design, usually having two steam domes with a large connecting pipe in between. Steam was collected in the rear dome and then passed to the forward one, in which was housed the regulator. Gölsdorf is also credited with building the first European locomotives of the 4-4-2 Atlantic type. Early in the twentieth century he introduced a four-cylinder compound system.

Compounding was an important improvement in locomotive technology. Besides offering increased power and economizing on fuel and water, time was saved as fewer stops were required to take on these items. In a Nord Rwy. test on December 25, 1891, a compound locomotive was found to save 14% in fuel and 23% in water over an ordinary engine having similar size and power.

In America, a further improvement was made on the anthracite-fueled engine designs of Winans and others: the Wootten boiler and firebox first introduced circa 1877. In a manner similar to the other systems, the engineer's cab was moved forward toward the middle, where it sat like a saddle over the boiler. This provided more space for the combustion area at the rear. The fireman remained there to fuel the engine, but was more vulnerable to bad weather. On some locomotives a fireman's cab was added to provide protection. These engines came to be called "camel backs" or "Mother Hubbards." They mainly operated in the Middle Atlantic states in close proximity to the anthracite coal mines.

The *Duplex,* the world's first 4-6-2 Pacific type, was an experimental locomotive built in 1886 by the Wilkes-Barre shops of the Lehigh Valley line. Designed by George S. Strong for heavy passenger service, it was of "camel back" style but had a double firebox, the casing being bifurcated like the legs of a pair of trousers. The adjoining fireboxes were housed in steel cylinders, which combined at the forward end into a single combustion chamber. Strong also designed a special valve gear for this engine. Other locomotives employing his innovations were soon built. However, these were all experimental prototypes and America's railroad industry ultimately rejected them.

By the 1880s Stephenson's link motion had largely been replaced by the Walschaert valve gear system. Walschaert, a Belgian, had invented his system in 1844, but it took many years before the advantages that it had over the Stephenson method were clearly understood. Link motion provided a "lead," or cushion of steam, in the cylinder to reduce the impact of the returning piston. However, as the cutoff was reduced for higher-speed operation, the lead was automatically increased to a corresponding degree. This resulted in sluggish performance once a certain speed had been attained. Walschaert's system allowed for greater control over the lead, thus eliminating this resistance factor. About the same time, lighter piston valves began to replace the old flat slide valves. This provided wider port openings and substantially improved cylinder performance.

In 1881 the Grant Locomotive Works of Paterson, New Jersey, built the famous *Fontaine* locomotive. Designed by Eugene Fontaine of Detroit, this engine was built for the Canada Southern Rwy. Its most unique feature was two sets of driving wheels, one above the other. It was reported to have reached a speed of 90 m.p.h. Fontaine's engine was an attempt to get around a problem inherent in North American railways: track instability. European railways, generally being of better construction, were safe for the operation of engines having large drivers and a high center of gravity. Such locomotives, however, would have frequently derailed on North American roads. Safe operation there required a lower center of gravity, and thus, out of necessity, drive wheels were kept smaller. While some compensation could be gained by operating at a greater number of revolutions, this was only safe to a point, as it put added stress on the cylinders, pistons, wheels, and valve gear.

Fontaine's unique solution to this problem was to put larger drivers above, resting on a smaller set of treads below, driving these by friction. These treads, in turn, were an outward extension of a wheel, larger in diameter, which ran on the track. Although this combination did seem to provide more speed with fewer piston strokes, while maintaining the necessary gravity, it did so at the cost of power. Fontaine's experiment received a lot of media attention at the time, but did not provide the breakthrough that was widely hoped for.

That same year Dr. Christian Raub of New York patented a plan for a perfectly balanced 2-4-2 tank engine called the *Central Power Locomotive.* This engine was designed in such a way that whether divided longitudinally or laterally, each half matched the other. Raub's objective was to build his locomotive on "correct principles and according to natural laws." To achieve this lofty goal he tried to align the center of gravity with the vertical median line of the engine, and locate the motive power at the middle of the engine in a plane extending through the center of gravity. In doing so he also hoped to eliminate any possible dead weight.

His locomotive had an unusual and advanced appearance for its time; it also offered some interesting features: (1) small boilers extending both forward and backward in the "Double Fairlie" fashion, each having a separate firebox in the cab; (2) a smoke chamber extending from both ends to the center, employed for collecting and returning smoke to a single centrally located stack; (3) a single large, centrally located steam dome for storing steam from both boilers, built in such a way that the lower portion of the smokestack passed through its center; and (4) vertically positioned cylinders, which transmitted power to cranks at each end of an intermediate driving shaft. Unfortunately, when later built by the Grant Co., Raub's engine proved to be an even greater failure than Fontaine's locomotive.

Also appearing in America about this time were locomotives designed specifically for logging operations. Usually primitive in construction, logging lines frequently had tight curves, little or no grading, and often, abrupt elevation changes. These circumstances, as well as the cargo of heavy logs, necessitated powerful engines that ran at very slow speeds.

Simplest and least powerful were engines designed for use on "pole roads," portable lines without sleepers built from logs laid out end to end. A typical road of the period was constructed of 30-foot pine poles, 8 to 12 inches in diameter. These were joined by use of a "lap," an extension cut from the large end of the log, with the small end of the next log fitted into it. A wooden pin held the two together. Pole road locomotives, similar in size to contemporary road traction engines, had double-flanged concave wheels and typically employed chain drive. Firms manufacturing engines of this type included the Tanner & Delaney Co. of Richmond and Adams & Price of Nashville. In 1885 engines of the latter company were geared at 4½ to 1 and had a top speed of 5 m.p.h.

A more widely used locomotive was the Shay geared engine, built by the Lima Locomotive Works of Lima, Ohio, from Ephraim Shay's patent of 1881. Ideal for use on sharp curves, these engines normally employed two or three vertical cylinders that were mounted on the right side. They powered a crankshaft, which transferred power to the wheels by use of beveled gears. The Lima Company manufactured variations of this design until 1945. In 1875 Charles D. Scott began experiments ultimately resulting in the Climax Geared Locomotive, first built in 1888 by Climax Manufacturing Co. of Corry, Pennsylvania. Unfortunately Scott realized little money from these efforts because his ideas were patented under the names of others. Luckier in this respect was Charles L. Heisler, whose first geared locomotive was built in 1891 by Dunkirk Manufacturing Co. of Dunkirk, New York. From 1894 on Stearns Manufacturing Co. of Erie, Pennsylvania, built Heisler engines.

As locomotive technology continued to evolve, so did fashions in design. In its July 6, 1883 issue, *Railroad Gazette* noted changing tastes at a recent exhibition in Chicago: "If compared with locomotives of twenty years ago, the plainness of finish of today would be a great contrast to the florid decoration which was then in vogue. The polished brass and bright colors have vanished, and now some of the progressive locomotive designers have attacked the decorative moldings of the dome, sand box, etc., and it seems certain that they, too, must go. We may, therefore, expect a new era in the design of American locomotives, and in the future their grace and beauty will depend only upon the skillful design of the machine itself, and not on any added decoration."

While express engine drive wheels were typically larger in England than on the Continent, experiments had been made there for years. In France, the 2-4-0 *L'Aigle,* designed by Blavier and Larpent, had drivers of 9′4″ diameter. Built by the Gouin Company of Paris, it was exhibited at the Paris Exposition of 1855. Perhaps the last extravagant experiment of this kind was Estrada's *La Parisienne,* a 0-6-0 express locomotive having drivers of 8-foot diameter. Displayed at the Exposition Universalle of Paris in 1889, it could reach a speed of 75 m.p.h. However, the engine's boiler could not produce a sufficient amount of steam for sustained performance.

Also exhibited at the Exposition was the world's first

compound articulated engine. It was a small 0-4+4-0 tank engine, designed by Anatole Mallet, for the portable 1′11⅝″ gauge railway system of Paul Decauville. As early as 1877 Mallet had taken an interest in the idea of compounding articulated engines. However, this problem was difficult, due to the movement of the separate driver bogies. In 1884 he received a patent for a design in which the front truck swiveled from a point at the engine's center, while the rear truck remained fixed. The high-pressure cylinders were rigidly mounted on the main frame, thus avoiding the use of ball joints for the high-pressure steam pipes. Ball joints were used on the pipes leading to the low-pressure cylinders. Having less pressure, they were not so vulnerable to rupture.

Mallet's Decauville locomotive was an immediate success. Soon, these and similar narrow-gauge engines were operating in Germany, Spain, Switzerland, and elsewhere. The first standard-gauge locomotives of this style were introduced on France's Herault Rwy. in 1891. Early 0-6+6-0 Mallet designs were not entirely successful; however, engines of this type, built in 1899 at Bryansk, for the Moscow-Kazan line, worked well. Afterward, Mallet's designs probably reached their full expression in the colossal American locomotives built in the early decades of the twentieth century. These were typically used in mountain areas to haul long trains of cars. The apogee, perhaps, was a 2-8+8+8-4 triplex version, which unfortunately, had severe problems due to constantly escaping steam.

In 1883 François Ricour, chief engineer of the Est line, began experiments with aerodynamic designs. In a 13-month test begun the following year, his "wind-cutter" engine consumed 12% less fuel than regular locomotives. Later, M. Baudry of the PLM Rwy. undertook similar experiments and came out with his C 61 class engine in 1898. It was built using the four-cylinder de Glehn compound system.

In *Railways of Western Europe,* O. S. Nock comments on this locomotive:

> The view of some English commentators of the day was that the aerodynamic fairings on the "C 61" class were not really justified. Neither they were, so far as ordinary running conditions went. But the PLM has somewhat exceptional physical circumstances to contend with in the Rhone delta. Between Avignon and the approach to Marseilles the line runs across a very flat and exposed countryside, subject to the furious wind of the Mistral, and the new locomotives were designed and described as "wind-cutters." Apart from this, however, they were most skillfully designed engines and, mainly due to their large low-pressure steam passages, which permitted a very free exhaust, they proved the most powerful and swiftest of all French 4-4-0's. One of them reached a maximum speed of 153 km./h., while another of them making up time with the "Rome Express," sustained 130 km./h. on level track, with a train weighing 170 tons. Both these spectacular feats of running were not publicized at the time because the legal speed limit in France was then 120 km./h.

Thuile's aerodynamic 4-4-6 engine, built by the Schneider Works at Le Creusot in 1900, had the driver's cab placed at the front. This provided a good preview of engine style changes coming up in the new century. That same year, "wind-cutter" locomotives, purchased by the Baden State Rwy., were introduced to Germany.

On May 10, 1893, an American 4-4-0 type, New York Central's No. 999, became the world's first engine to certifiably achieve a speed over 100 m.p.h. Pulling a four-car train near Batavia, New York, it reached a top speed of 112.5 m.p.h. Following this the locomotive was proudly exhibited in Chicago at the Columbian Exposition. Once it began regular service, its special drivers of 7′2″ diameter were replaced with smaller wheels. America's popular 4-4-0 type had reached its peak year of production in 1872. As the workload for these engines was increased, they were built heavier for better adhesion and were operated with higher steam pressure. Increasingly, 2-6-0 and 4-6-0 engines were introduced into passenger service. In 1888 the first 4-4-2 Atlantic type engine was built. It became increasingly popular as a passenger engine, gradually replacing aging 4-4-0 locomotives as the century waned.

In Britain workloads were also increasing. This was especially due to the growing size of passenger carriages and the greater amounts of metal used in their construction. As a result larger engines were finally introduced. The first 2-6-0 Mogul type made its appearance in 1878 on the Great Eastern line. The first 4-6-0 "ten-wheeler" began service in 1894 on the Highland Rwy. The first 4-4-2 Atlantic type was operated on the Great Northern line in 1898. The initial use of the 0-8-0 type occurred in 1889 on the Barry line in Wales, with the same company introducing the first 0-8-2T in 1897. These two engines were used for goods service.

At the end of the century, another innovation, promising new power and greater economy, reached a practical stage of development. This was the superheater, which had originally been anticipated by people like Richard Trevithick, the Frenchman Moncheuil, who patented the idea in 1850, and Jules Petiet. Dr. Wilhelm Schmidt, ably assisted by Geheimiath Garbe, successfully tested this idea for the Prussian State Rwys., producing his first superheater in 1897. In this system, steam drawn from the boiler, which still contained many droplets of water, was reheated by sending it through pipes in the lower part of the boiler barrel before it went to the cylinders. This process, which essentially dried out the steam and raised its pressure, was widely used after the turn of the century.

A major change in America's locomotive industry occurred in 1901 when the American Locomotive Co. was consolidated from the following firms: Brooks, Cooke, Dickson, Manchester, Pittsburgh, Rhode Island, Richmond, and Schenectady. The company's main plant was in Schenectady, and it acquired Rogers Locomotive Works in 1905.

French engines of early 1850s: Top: 0-4-2 engine for mixed use designed by Polonceau, Paris-Orléans Rwy. Bottom: 0-6-0 goods engine, J. E. Cail Co., 1851.

PLATE 46

PLATE 47

Top: Special 2-2-2 express engine built by Sharp, Stewart & Co., Manchester for Khedive Abbas I., Alexandria-Cairo Rwy., 1856. **Middle left:** 2-4-0 passenger engine, Sigl Co., Vienna, Warsaw-Terespol Rwy., 1867. **Middle center:** 0-8-4T Engerth goods engine, Vienna-Trieste Rwy., circa 1857. **Middle right:** 2-2-2 passenger engine *Cornwall*, as rebuilt in 1863, London & North Western Rwy. **Bottom:** 4-2-2 broad-gauge passenger engine *Lord of the Isles*, Great Western Rwy., 1851.

PLATE 48

4-2-4T broad-gauge express engine, built by Rothwell & Co., Bolton-le-Moors, for the Bristol & Exeter Rwy., 1853. In 1854 a locomotive of this type set a famous speed record of 81 m.p.h.

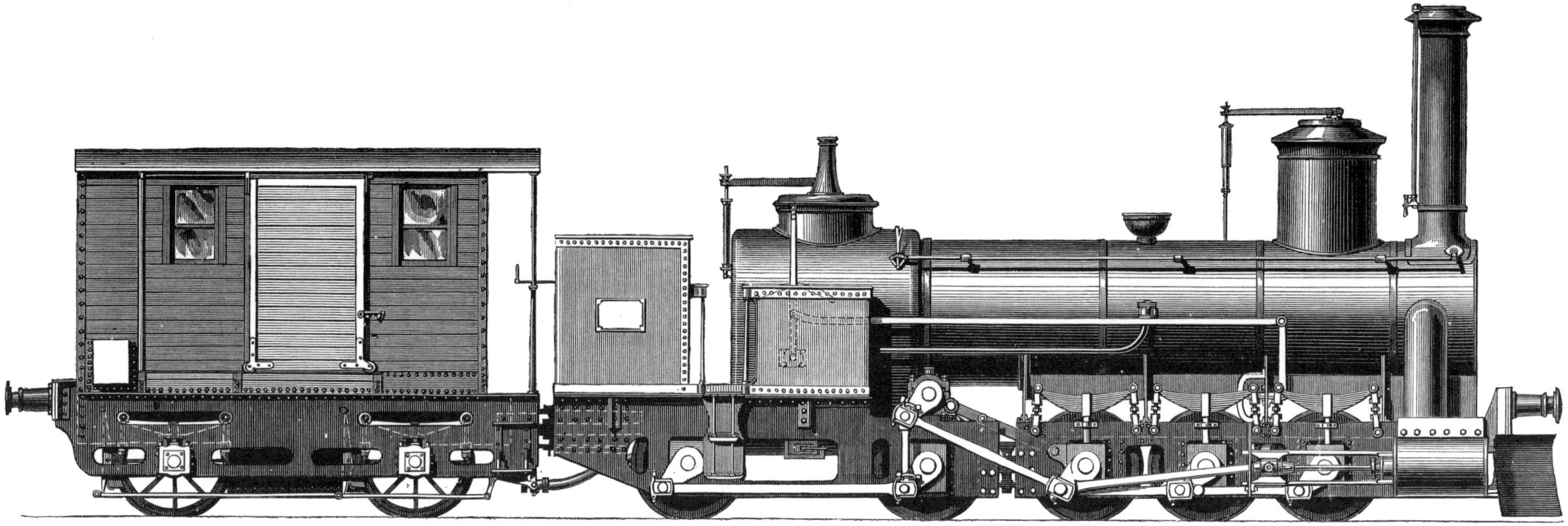

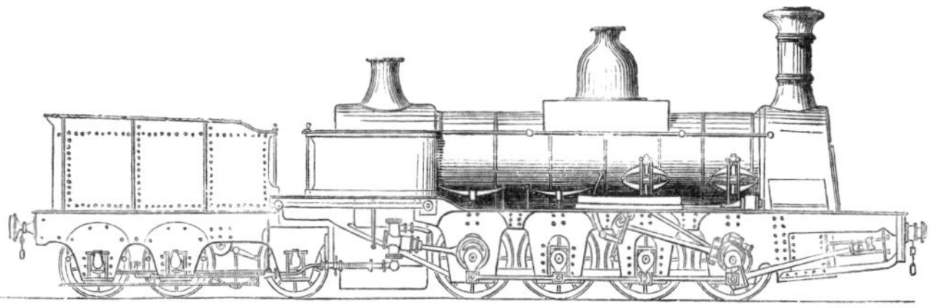

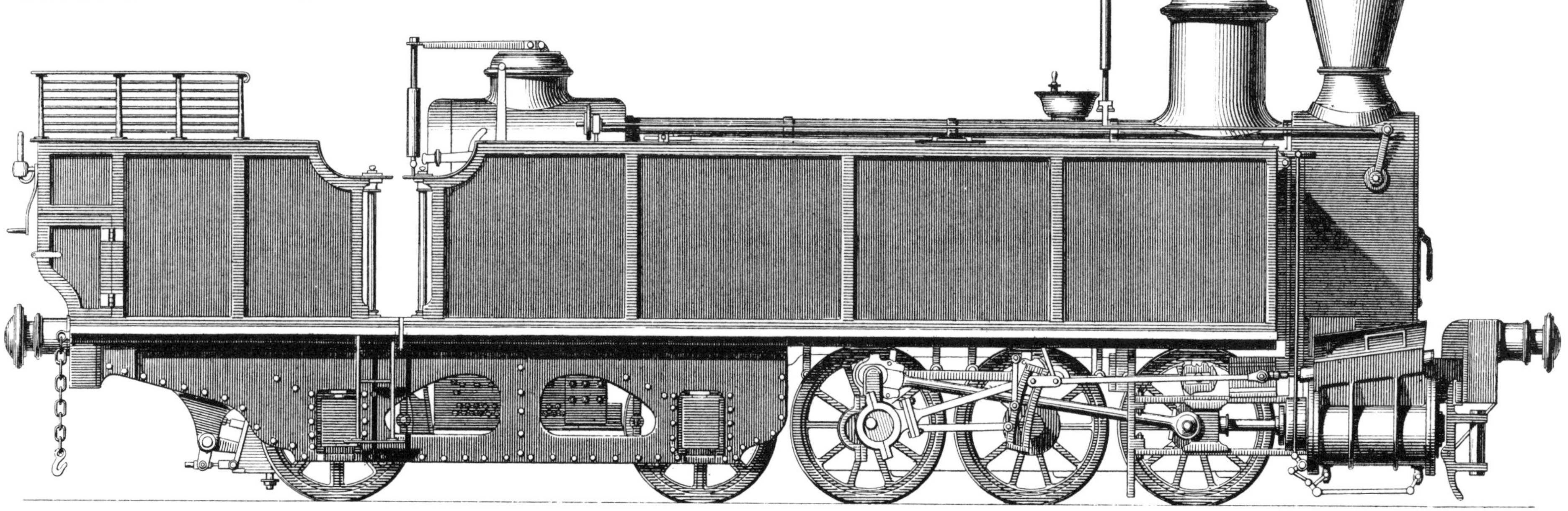

PLATE 49

TOP: The *Steyerdorf*, designed by Wilhelm Engerth and Pius Fink; Jassenova, Oravicza & Steyerdorf Rwy., Hungary, 1862. MIDDLE LEFT: 2-4-0 mixed service mountain engine, Richard Hartmann Co., Chemnitz, Saxony, 1862. MIDDLE CENTER: Beugniot goods locomotive used on Turin-Genoa line, circa early 1860s. BOTTOM: Early Engerth goods engine used at Semmering Pass; Vienna-Trieste Rwy., circa mid-1850s.

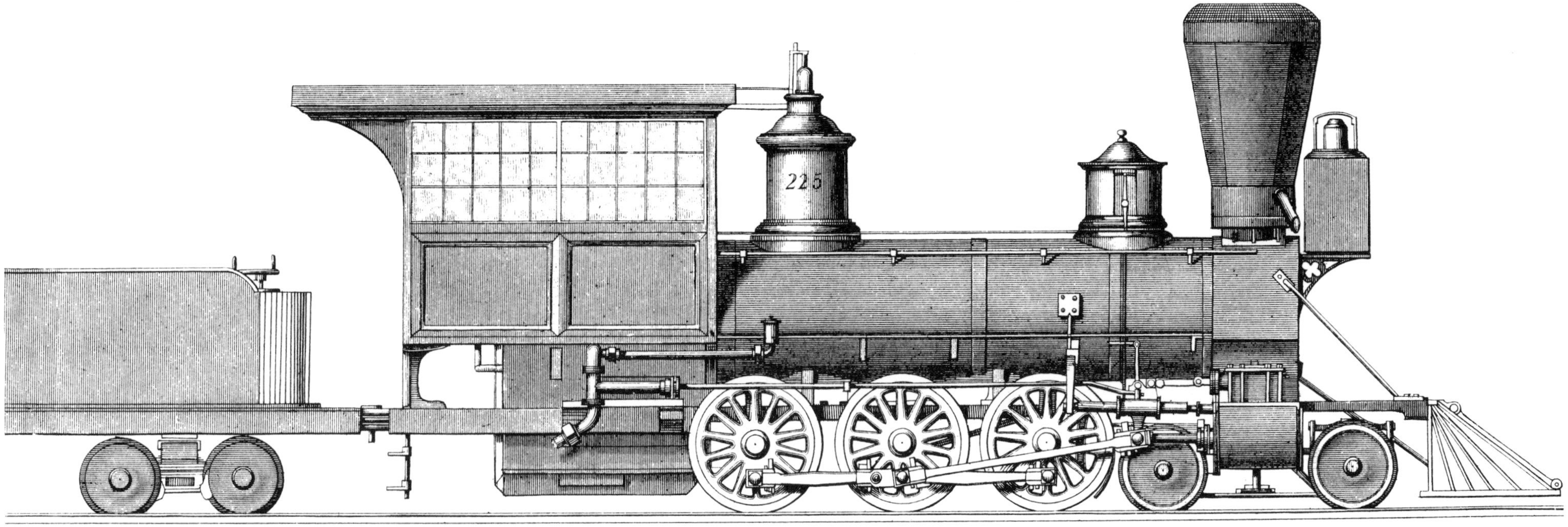

PLATE 50

Top: The first engine of the Manchester Locomotive Works, Manchester, New Hampshire, was this 4-4-0 "American" type of 1854. It was sold to the Central Military Tract R.R. in 1855 and renamed *Cossack*. This line was shortly absorbed into the Chicago, Burlington & Quincy R.R. Bottom: 4-6-0 "ten-wheeler" type designed by Tyson and built by Denmead & Son, Baltimore, B & O R.R., 1857.

PLATE 51

4-4-0 "American" type engine built by Danforth, Cooke & Co., Atlantic & Great Western R.R., 1862. This company's main line ran from Salamanca, New York, to Dayton, Ohio, with a branch to Cleveland. In connection with the Erie; Cincinnati, Hamilton & Dayton; and Ohio & Mississippi lines, it provided a through route of 6′ broad gauge from New York to St. Louis.

PLATE 52

American locomotives: Top: Baldwin 4-4-0 engine, circa 1855. Middle left: 4-4-0 engine of Amoskeag Co., 1851. Bottom: 0-6-0T engine built by H. J. Booth & Co., Union Iron Works, San Francisco, for California's Pittsburgh R.R., 1867. This $5\frac{1}{3}$-mile line ran from Pittsburgh on the San Joaquin River to the Pittsburgh & Union coal mines near Somersville. Coupled to the locomotive is a coal hopper car.

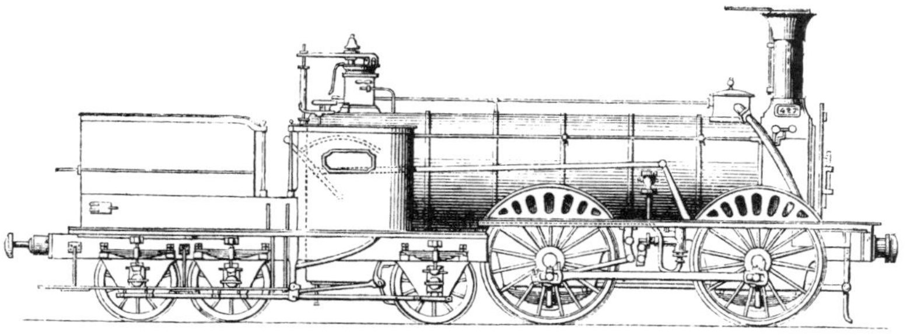

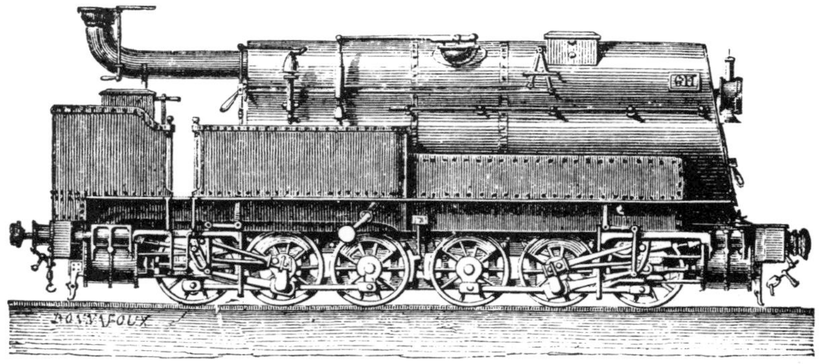

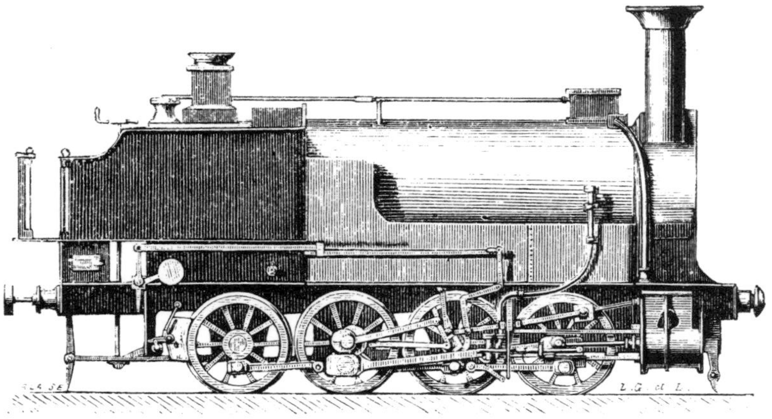

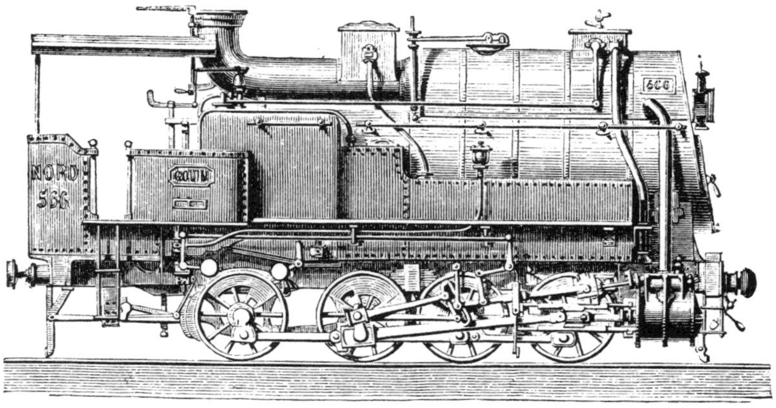

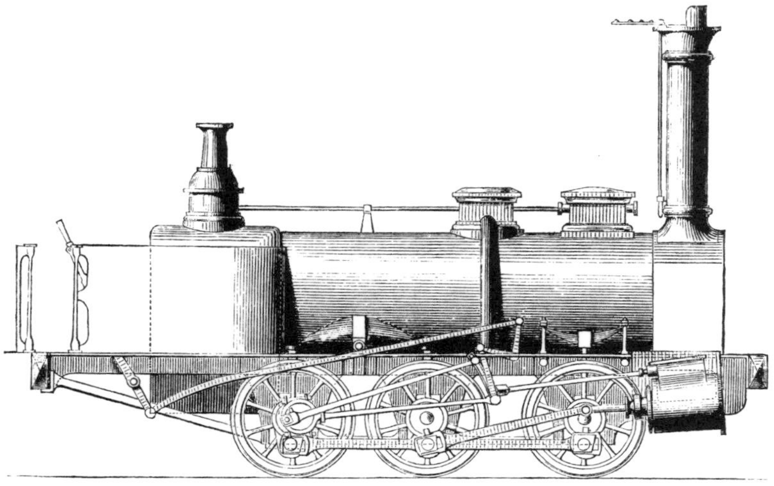

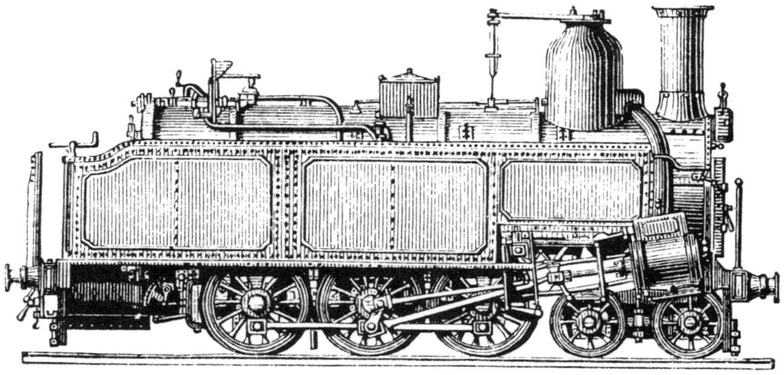

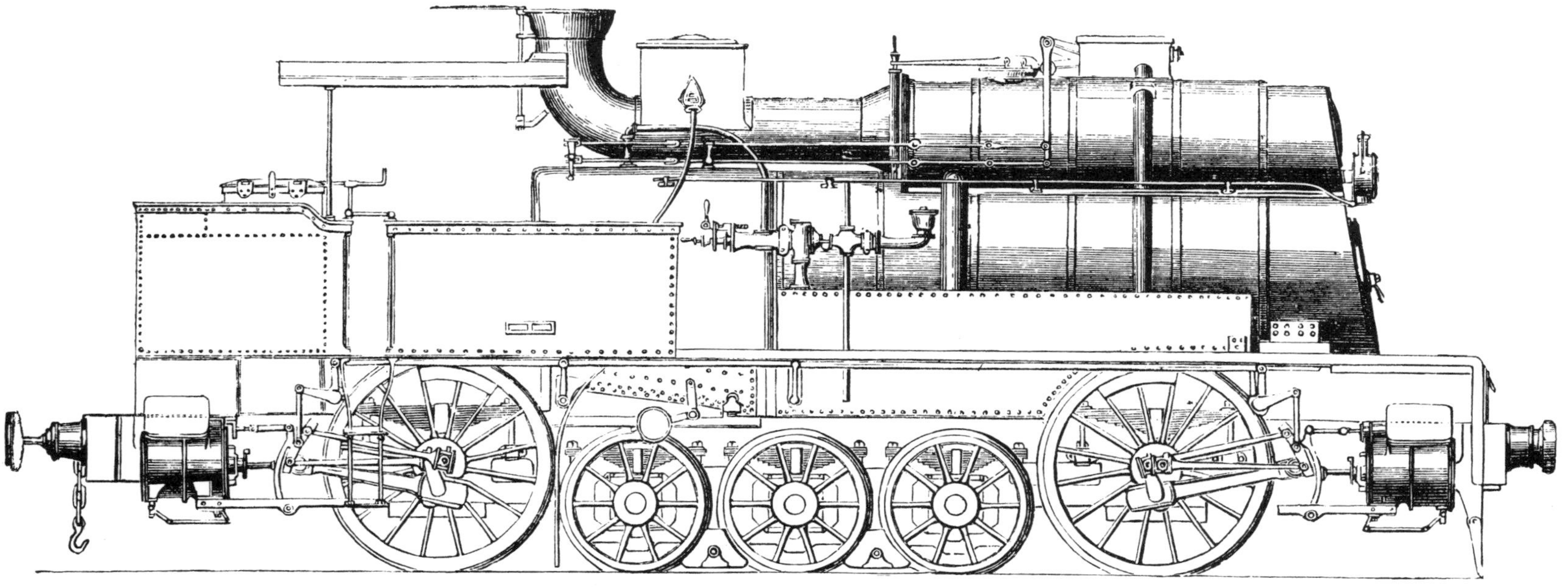

TOP LEFT: 0-4-6T Engerth mixed service engine, Nord Rwy., 1856. TOP RIGHT: Petiet 0-6-6-0T goods engine, Nord Rwy., 1862. UPPER MIDDLE LEFT: 0-8-0T heavy gradient engine, Nord Rwy., 1859. UPPER MIDDLE RIGHT: Petiet 0-8-0T, Nord Rwy., circa 1860. LOWER MIDDLE LEFT: 0-6-0T shunting engine, Paris-Orléans Rwy., circa 1850s. LOWER MIDDLE RIGHT: 4-6-0T goods engine, Societe St. Leonard, Liège, 1867. BOTTOM: Petiet 0-2-6-2-0T for mixed service, Nord Rwy., 1862.

PLATE 53

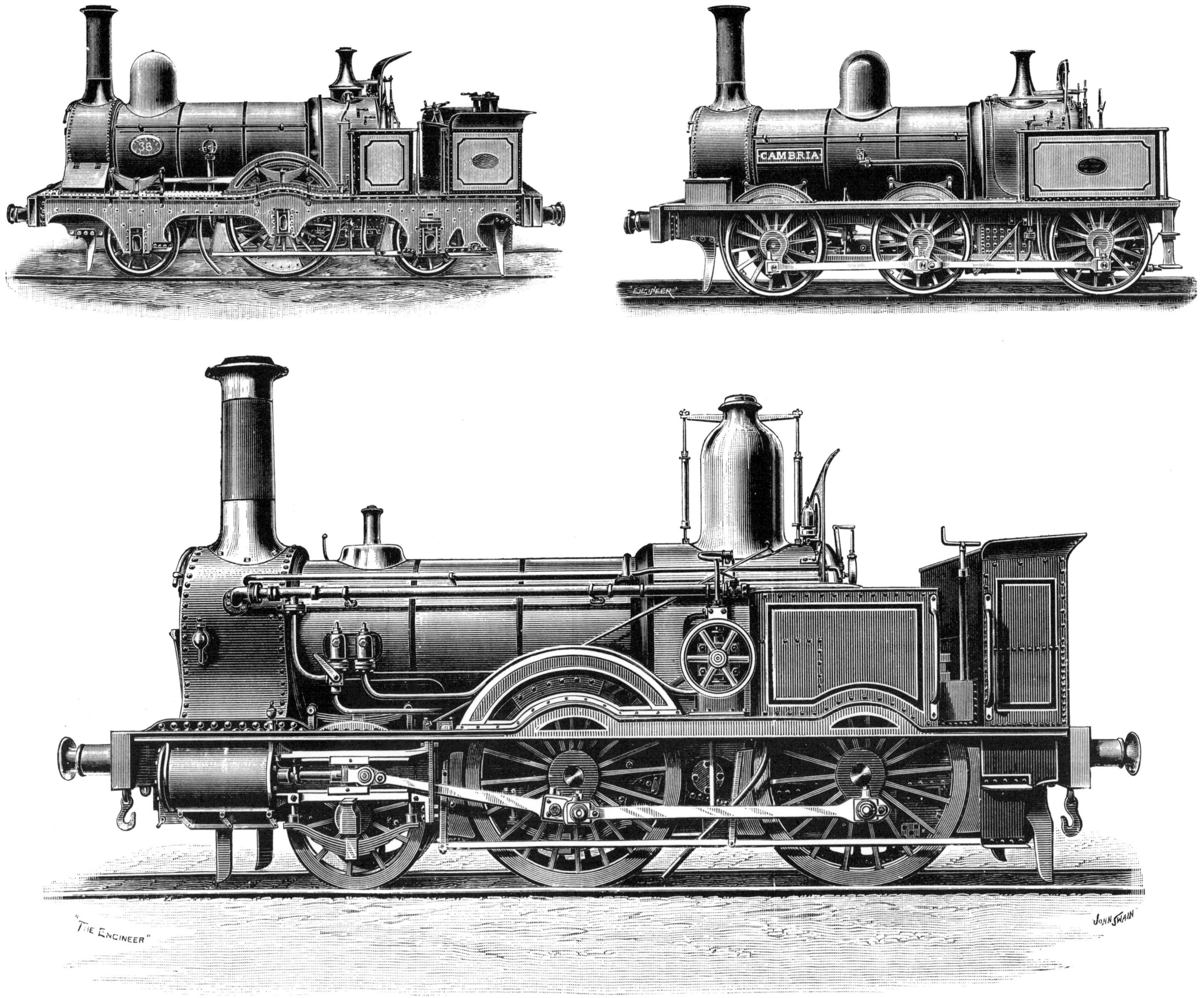

PLATE 54

British locomotives, 1863: Top left: 2-2-2T passenger engine, Sharp, Stewart & Co., Furness Rwy. Top right: 0-6-0 goods engine, Sharp, Stewart & Co., Cambrian Rwy. Bottom: 2-4-0T passenger engine for suburban traffic, Beyer, Peacock & Co., London & South Western Rwy.

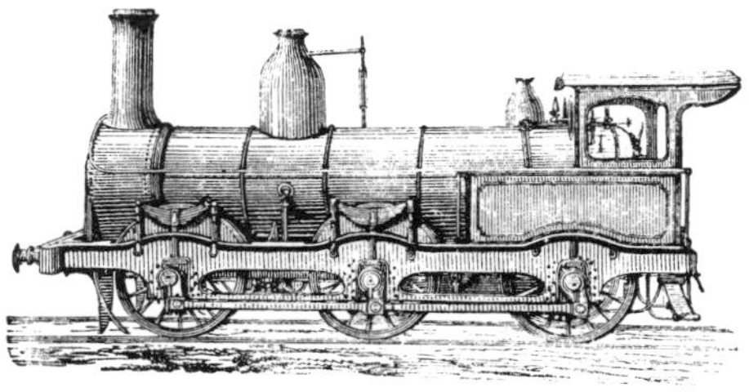

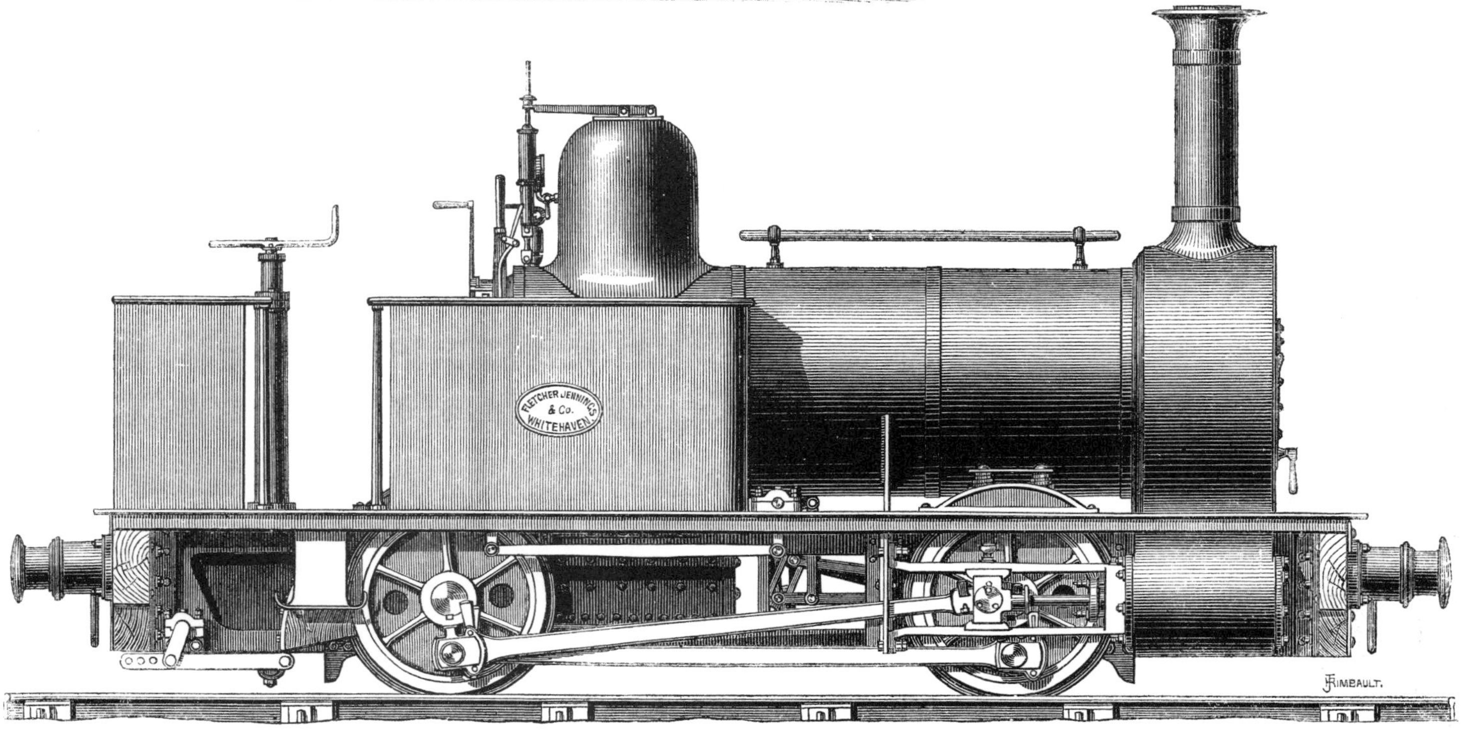

Top: 0-4-2T *Talyllyn*, built by Fletcher, Jennings & Co., Whitehaven, for 2′3″ gauge Talyllyn Rwy. of Wales, circa 1866. Middle left: 2-4-0 passenger engine; Sharp, Stewart & Co., Cambrian Rwy., 1863. Middle right: 0-6-0 Sharp goods engine, 1862. Bottom: Fletcher 0-4-0T *Dolgoch*, Talyllyn Rwy., circa 1866.

Top left: John Haswell's experimental four cylinder 4-2-0 *Duplex*, Austrian State Rwys., 1861. Top right: 2-2-2 *Problem*, London & North Western Rwy., Crewe, 1859. Middle left: 0-4-0T, Fletcher, Jennings & Co., Whitehaven, circa 1860s. Middle right: 0-4-0T, Manning, Wardle & Co., Leeds, circa 1860s. Bottom: 2-2-2 express engine, Neilson & Co., Glasgow, Caledonian Rwy., 1862.

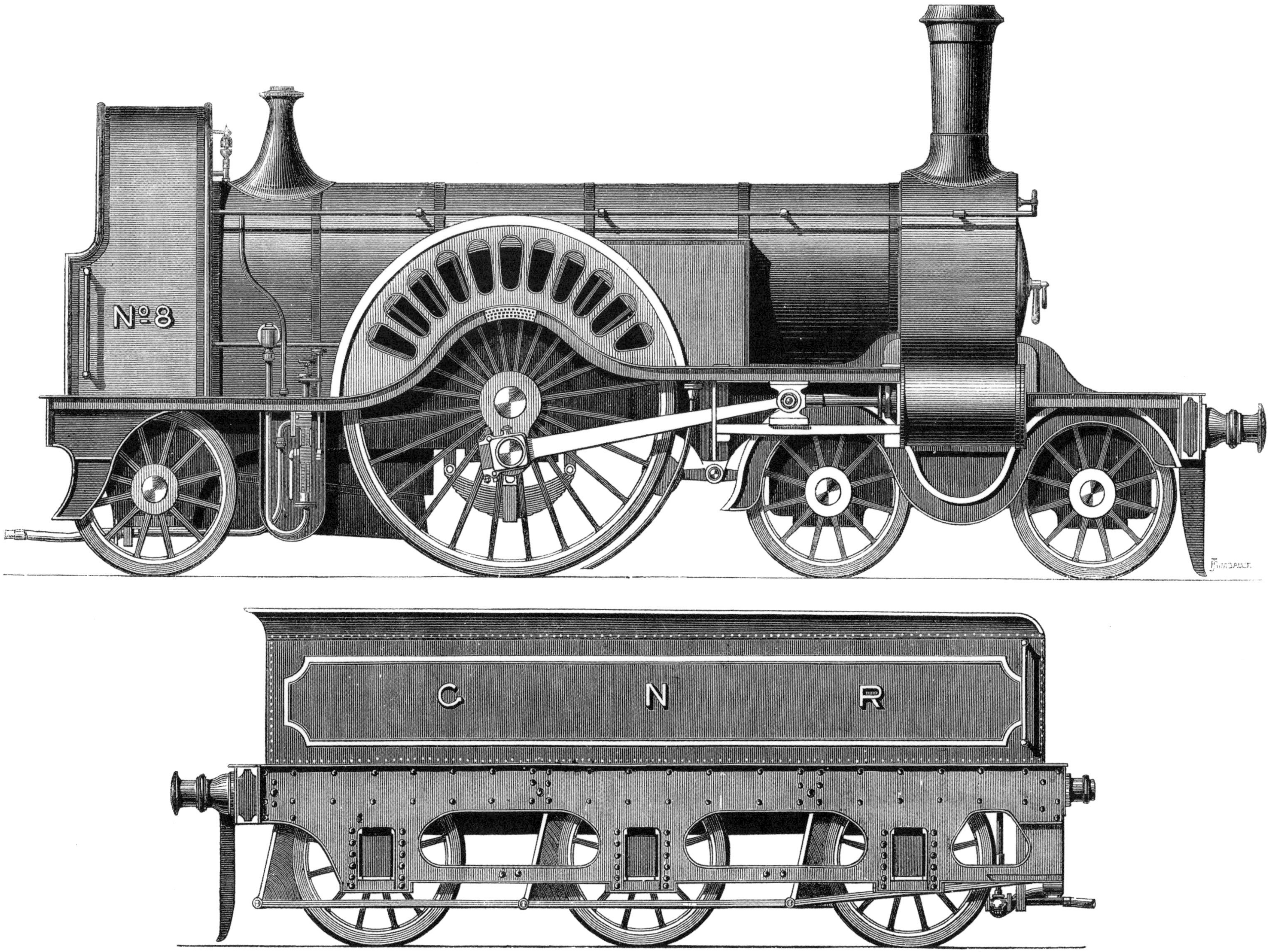

PLATE 57

4-2-2 express engine and tender designed by Patrick Stirling, Great Northern Rwy., 1870. Stirling's locomotive, known as the "bogie single," "Stirling single," or "eight-foot single," had drive wheels 8 feet in diameter, and was considered one of the most beautiful engines of the nineteenth century.

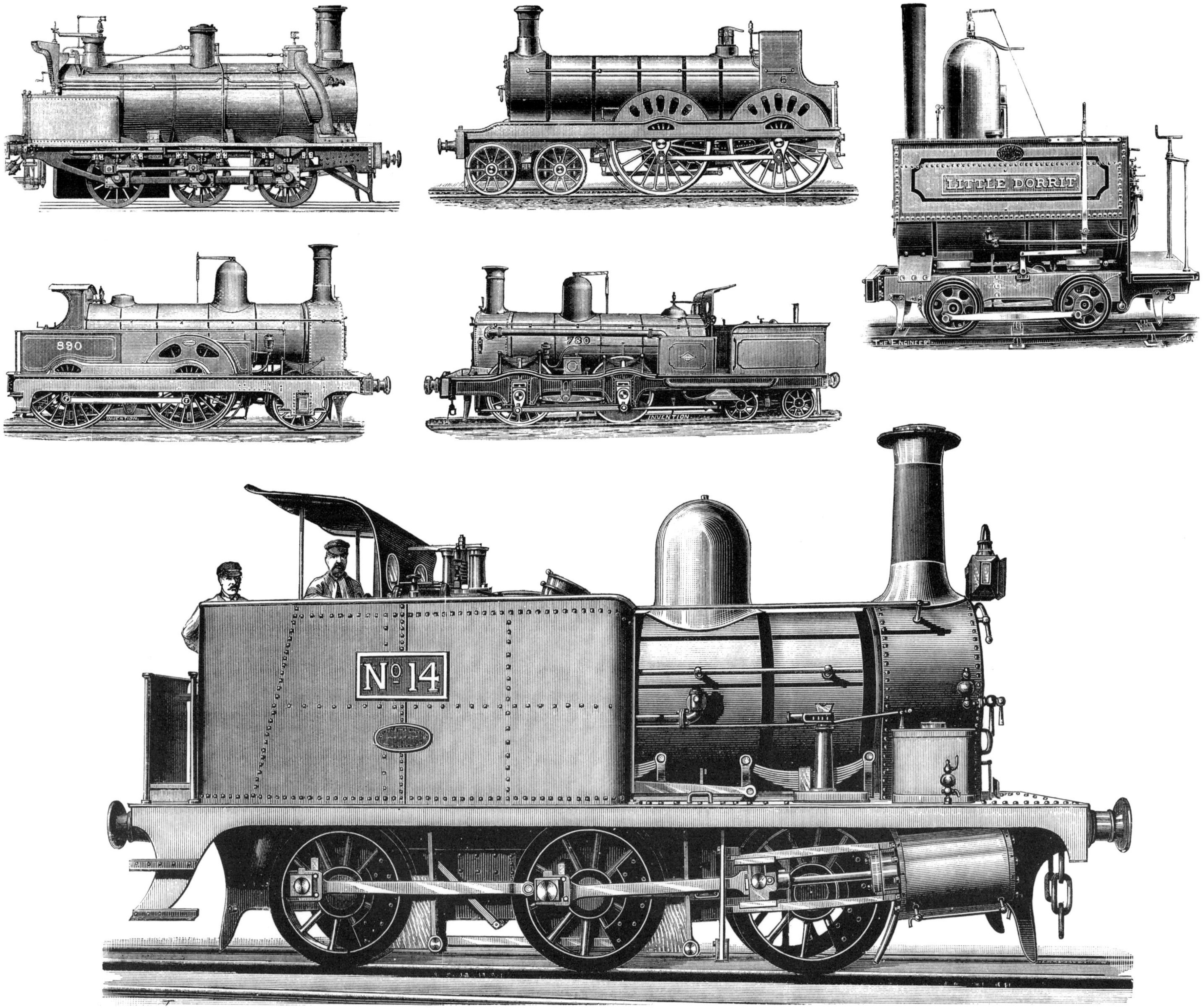

PLATE 58

Top left: 0-6-0 goods engine, Societe de Graffenstaden, Est Rwy., 1867. Top center: James Stirling's 4-4-0 express engine, Glasgow & South Western Rwy., 1873. Top right: Sharp 0-4-0T, Ebbw Vale Rwy., 1870. Middle left: Kirtley's 2-4-0 express engine, Midland Rwy., 1871. Middle center: 0-4-4T, Dubs & Co., Glasgow, Midland Rwy., 1870. Bottom: Sharp 0-6-0T goods engine, Cornwall Minerals Rwy., 1873.

PLATE 59

Small American engines: Top: Baldwin 2-4-0 passenger engine for 3′ gauge Denver & Rio Grande R.R., 1871. Bottom: 0-4-4T designed by Matthias N. Forney, 1870. As indicated by both "cowcatcher" position and flag direction, this engine was intended for operation with the boiler to the rear.

Small American engines, circa 1874: Top: 0-4-0 switch engine, Hinkley Co., Boston. Middle: 2-4-0 passenger engine, Grant Locomotive Works, Paterson, New Jersey, for Central R.R. of New Jersey. Bottom: Rogers 2-4-4T of "double ender" style developed by Mr. Hudson.

PLATE 60

Top: 2-8-0 Consolidation type freight engine, Pittsburgh Locomotive Works, circa early 1880s. Middle: Rogers 4-4-0 passenger engine, circa 1874. Bottom: 4-4-0 Baldwin passenger engine, circa 1874.

PLATE
61

TOP: 0-4+4-0T Meyer articulated engine *L'Avenir,* M. M. Cail Co., Fives-Lille, Great Luxembourg Rwy. (Belgium), 1872. BOTTOM: 4-4-0 express engine, G. Sigl Co., Wiener Neustadt, Vienna-Trieste Sudbahn, 1873.

Top: 4-4-0 passenger engine, G. Sigl Co., Wiener Neustadt, North Western Rwy. (Austria), 1873. **Bottom**: 0-4-0T for industrial use, $3'1\frac{1}{3}''$ gauge, Darmstadt Engine Works and Iron Foundry, Darmstadt, 1873.

PLATE 63

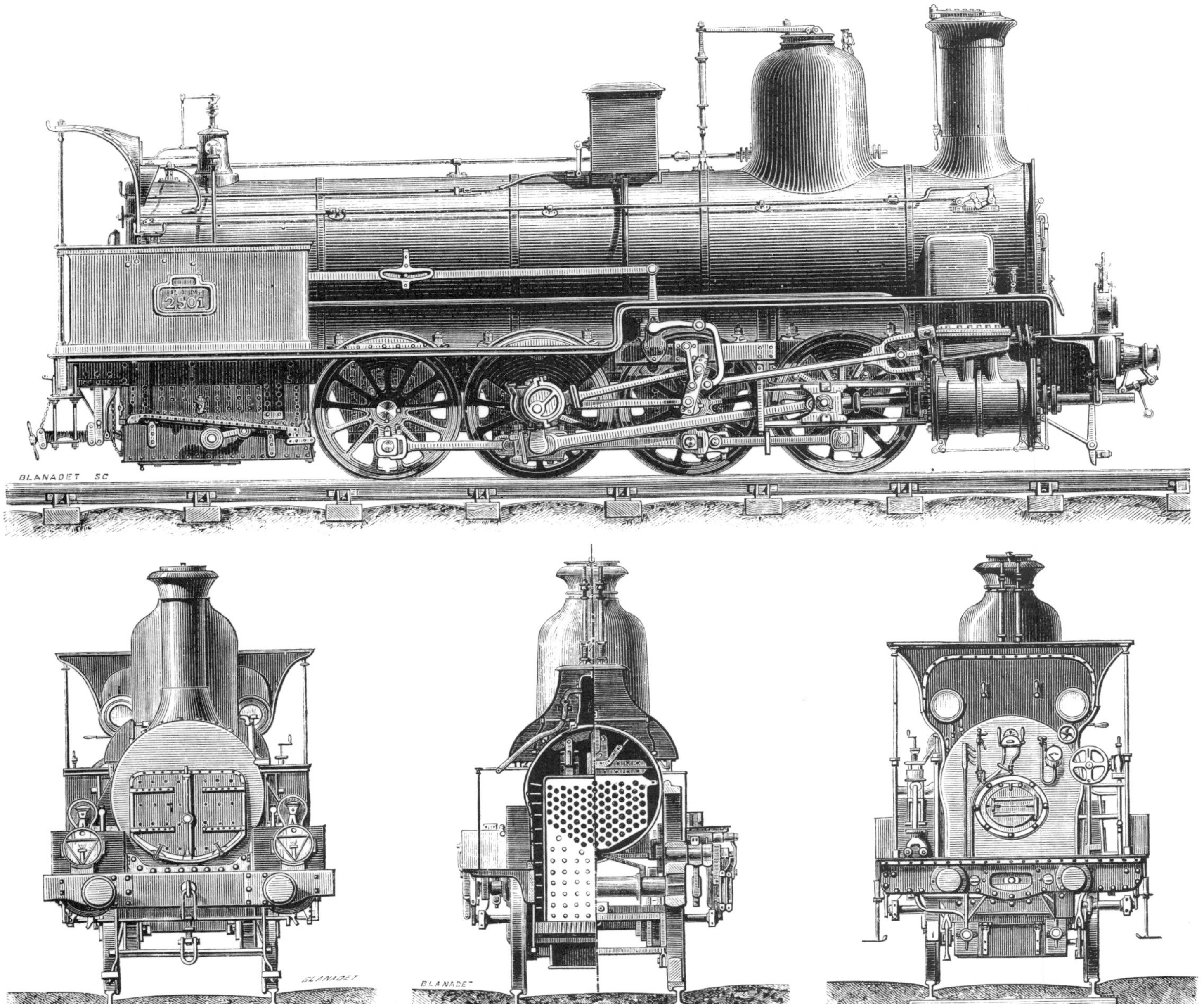

Locomotives of France's PLM Rwy., circa 1870s: Top: 0-8-0 goods engine. Middle left: Front view. Middle center: Cutaway section from rear. Middle right: Rear view showing inside of cab. Bottom: 2-4-0 Passenger engine.

PLATE
64

The articulated "double Fairlie" tank engines had two boilers and independent sets of drive wheels. Thus they could negotiate the tight curves common in mountainous terrain. TOP: 0-6+6-0T, Avonside Engine Co., Bristol, Chimbote-Huaraz Rwy., Peru, 1873. MIDDLE RIGHT: 0-6+6-0T, circa 1870. BOTTOM: 0-4+4-0T, Avonside Co., New Zealand Government Rwys., 1875.

PLATE 66

While differing considerably in appearance these tank engines were both applications of the "single Fairlie" concept where both drivers and carrying wheels functioned as bogies: Top: 0-6-6T "double truck" engine, Mason Machine Works, Taunton, Massachusetts, circa 1874. Bottom: 0-4-4T designed by G. P. Spooner for the 1′11″ gauge Festiniog Rwy. of Wales, circa 1877.

PLATE 67

Engines of the Rogers Locomotive & Machine Works, Paterson, New Jersey: Top: 0-6-0 switch engine, Georgia R.R., 1877. Bottom: 2-4-4T "double ender" style locomotive for suburban passenger service, Illinois Central R.R., 1880.

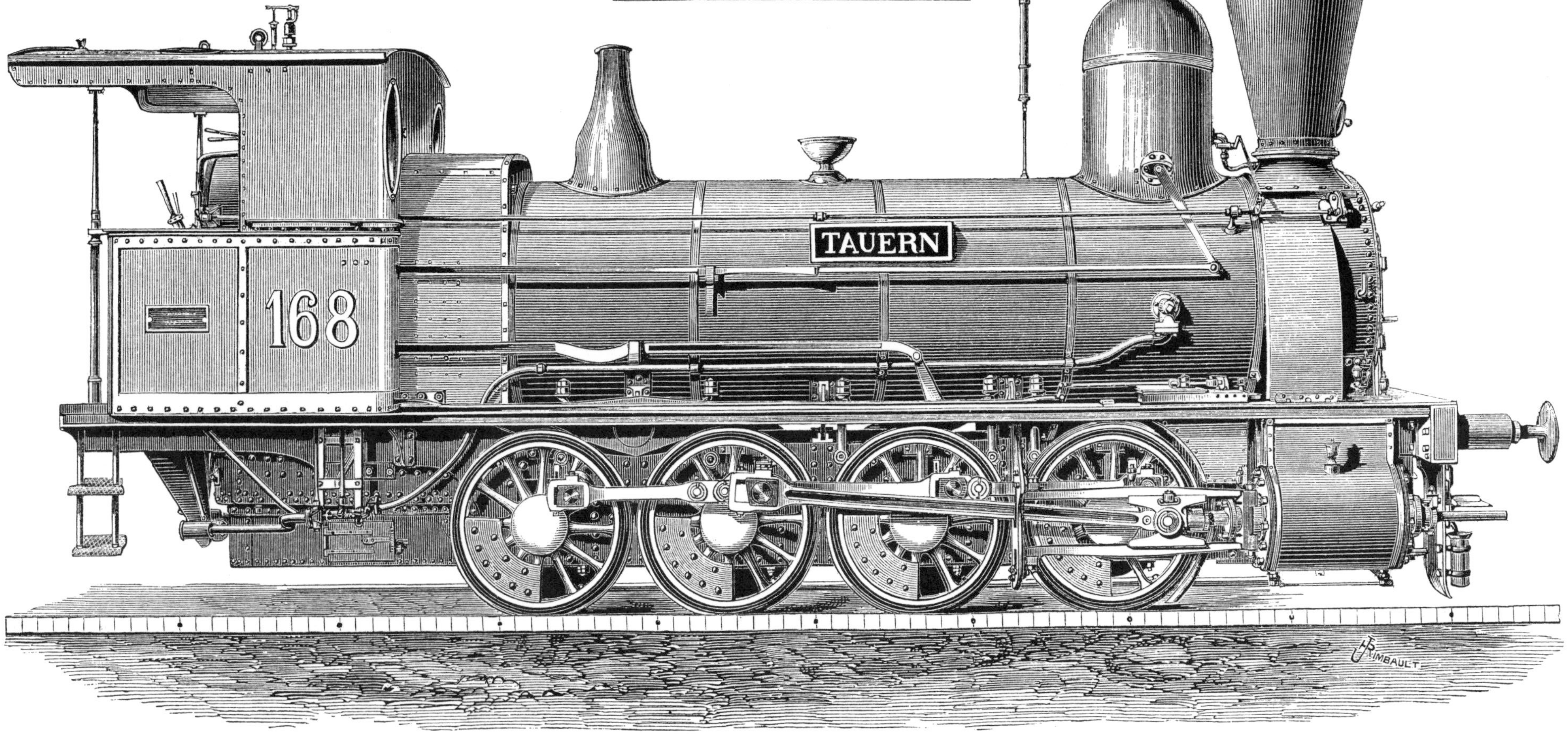

PLATE 68

TOP LEFT: 0-4-0T engine for 3′ gauge line, Black, Hawthorn & Co., 1876. TOP RIGHT: Eugene Fontaine's experimental engine, Grant Locomotive Works, Canada Southern Rwy., 1881. MIDDLE CENTER: 0-4-2T engine, H. K. Porter Co., Pittsburgh, Grand Haven R.R., Michigan, 1879. BOTTOM: 0-8-0 goods engine, Sachsiche Maschinenfabrik, Chemnitz, Saxony, Kaiserin Elizabeth Bahn, 1873.

Top: 2-2-2-0 four-cylinder compound engine designed by Alfred de Glehn, Nord Rwy., 1886. The drivers on this engine were uncoupled, with the front pair powered by inside HP cylinders and the rear by outside LP cylinders. **Bottom:** 2-4-0 passenger engine built by Krauss & Co. for Bavarian State Rwys., circa 1883. Note electric headlamp powered by dynamo-generator behind smokestack.

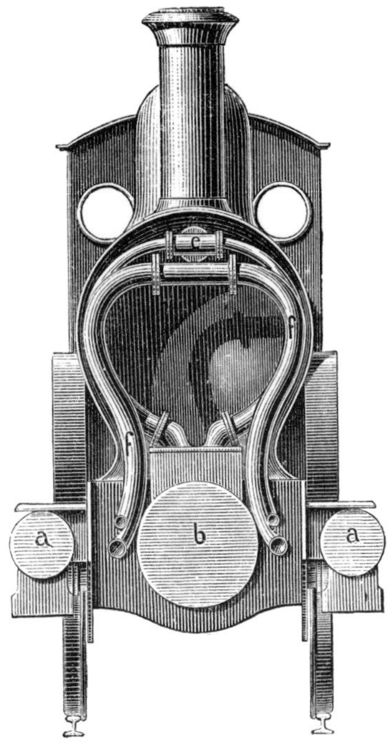

PLATE 70

2-2-2-0 compound engines of the London & North Western Rwy. employing F. W. Webb's three-cylinder system: TOP LEFT: Side view of Webb's original engine, 1882. TOP RIGHT: Cutaway view of same engine showing position of outside HP cylinders and inside LP cylinder. BOTTOM: The *Dreadnought*, 1884.

PLATE 71

Top: 4-4-0T passenger engine, Taff Vale Rwy., 1885. Bottom: 4-2-2-0T compound passenger engine, London & North Western Rwy., 1884. Beyer, Peacock & Co. originally built this engine in the 1870s as a 4-4-0T on the same pattern as the Metropolitan and District locomotives. Compounded by Mr. Webb in 1884, it pulled L & NW trains from Broad St., via Willesden, to the District line's Mansion House Station.

American saddle-tank locomotives: TOP LEFT: Light 0-4-2T logging engine, H. K. Porter Co., circa 1884. TOP RIGHT: 0-4-0T switch engine, Taunton Locomotive Co., circa 1874. MIDDLE LEFT: Porter 0-4-0T for use inside steel mills, circa 1888. BOTTOM: Porter open canopy 0-4-0T built for use on 20″ gauge line, Longfellow Mining Co., Clifton, Arizona, 1881.

PLATE 72

American saddle-tank locomotives, circa 1884: Top left: Porter light duty 0-6-0T. Top right: Porter 0-4-0T. Middle left: Porter 0-6-0T. Middle right: Porter light duty 0-4-0T. Bottom: Rogers "double ender" style 2-4-2T, Canarsie & Rockaway Beach R.R., Long Island.

PLATE 73

PLATE 74

Heavy American freight engines: TOP: 2-8-0 Consolidation type, Pennsylvania R.R., Class R, 1886. BOTTOM: World's first 2-8-2 engine (later known as Mikado type), Lehigh Valley R.R., 1884.

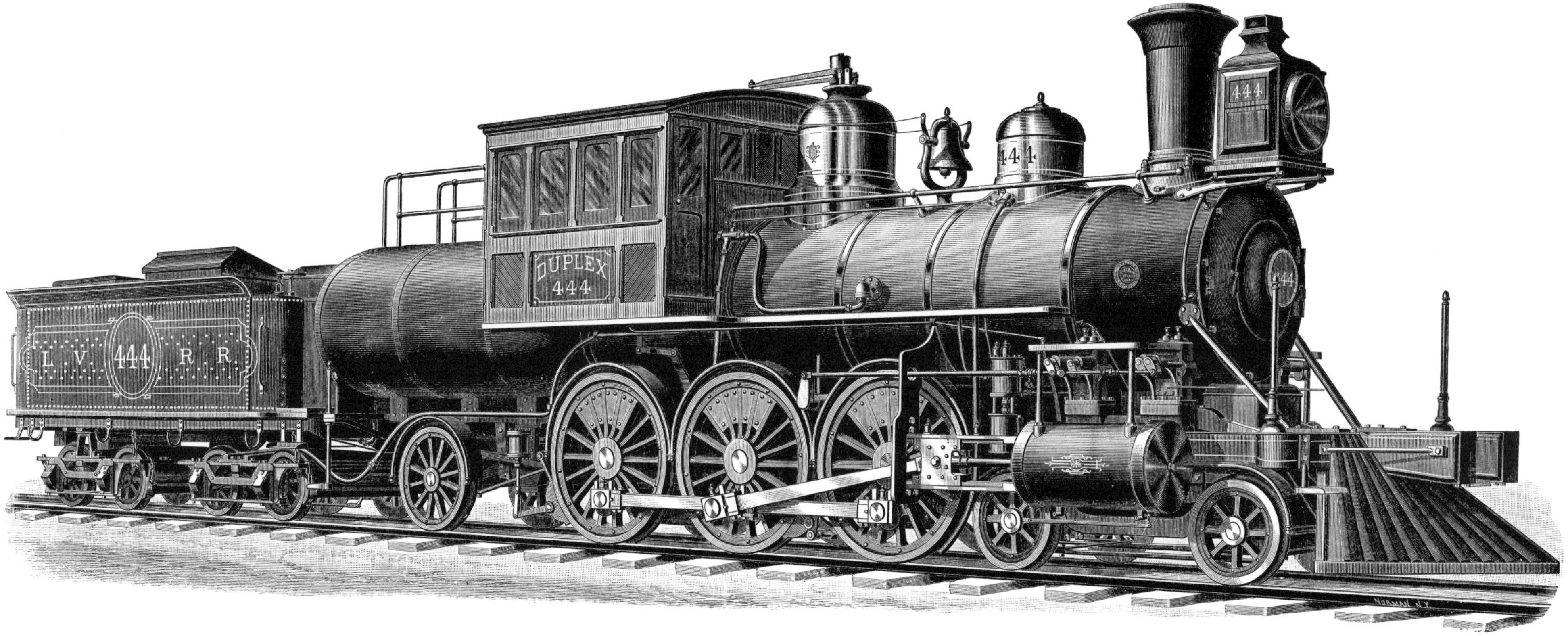

American engines of the "camel back" or "Mother Hubbard" style: Top: Rogers 4-4-0 engine with Wootten firebox, Union Pacific R.R., 1887. Bottom: World's first 4-6-2 (later known as Pacific type), fitted with Strong's patent boiler and built at Lehigh Valley's Wilkes-Barre shops, 1886.

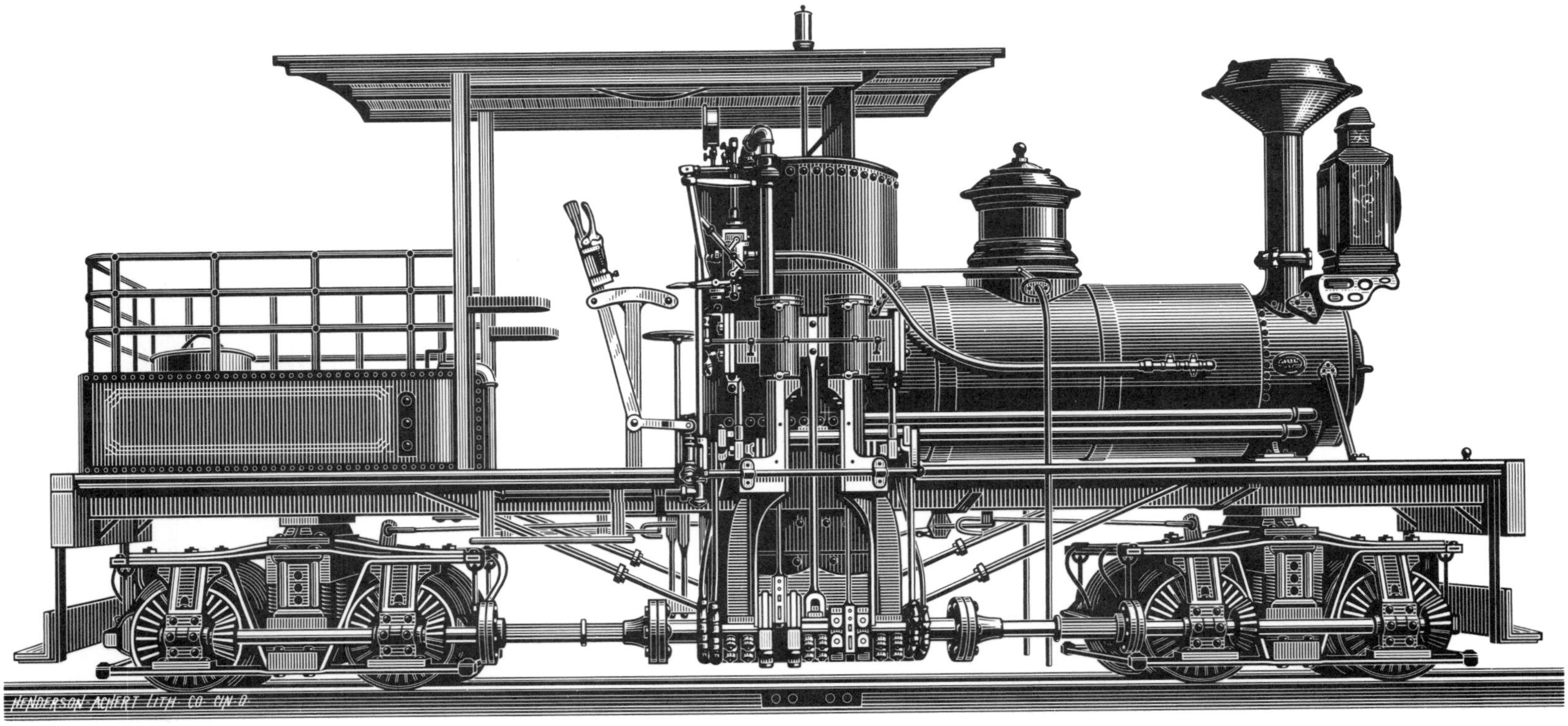

Specialized American logging locomotives: TOP: Wood-burning "T" boiler Shay engine, Lima Locomotive Works, Lima, Ohio. This engine was probably built between 1885 and 1888. BOTTOM: Chain drive "pole road" locomotive built by Adams & Price Co., Nashville, for Wallace, Sandford & Co., Wilson's Station, Alabama, 1885.

PLATE 76

Narrow-gauge locomotives built by the Decauville Co., Petit-Bourg, France: TOP: 0-4-0 engine, circa 1878. BOTTOM: 0-4+4-0T compound articulated engine, 1889. Exhibited that year at Paris's Exposition Universalle, it was Anatole Mallet's first articulated engine.

PLATE 78

TOP: 4-2-2 compound express engine (Worsdell & von Borries system), North Eastern Rwy., 1889. MIDDLE: 4-4-0 express engine, Glasgow & South Western Rwy., 1883. BOTTOM: 0-6-0 goods engine, Midland Rwy., circa 1889.

Top: 2-6-0 Mogul type wood-burning engine built by Swiss Locomotive Works, Winterthur for Finland's Wasa & Uleaborg Rwy., 1886. Bottom: 4-4-0 express engine, Ouest Rwy., 1889.

French express engines: Top left: 4-4-0 engine with Flaman boiler, Est Rwy., 1891. Top right: 4-4-0 compound locomotive, Ouest Rwy., circa mid-1890s. Bottom: Estrada's experimental 0-6-0 engine, 1886.

PLATE 81

Top left: 2-4-0 passenger engine, Borsig Co., Berlin, Prussian State Rwys., Class S 1, circa 1885. Top right: 0-6+6-0T Mallet articulated locomotive, Maffei Co., Munich, St. Gotthard Rwy., 1890. Middle center: 2-4-2 express engine, Austrian State Rwys., circa 1885. Bottom: 0-6-0 compound goods engine (Worsdell & von Borries system), Hannoversche Maschinenfabrik, Prussian State Rwys., 1889.

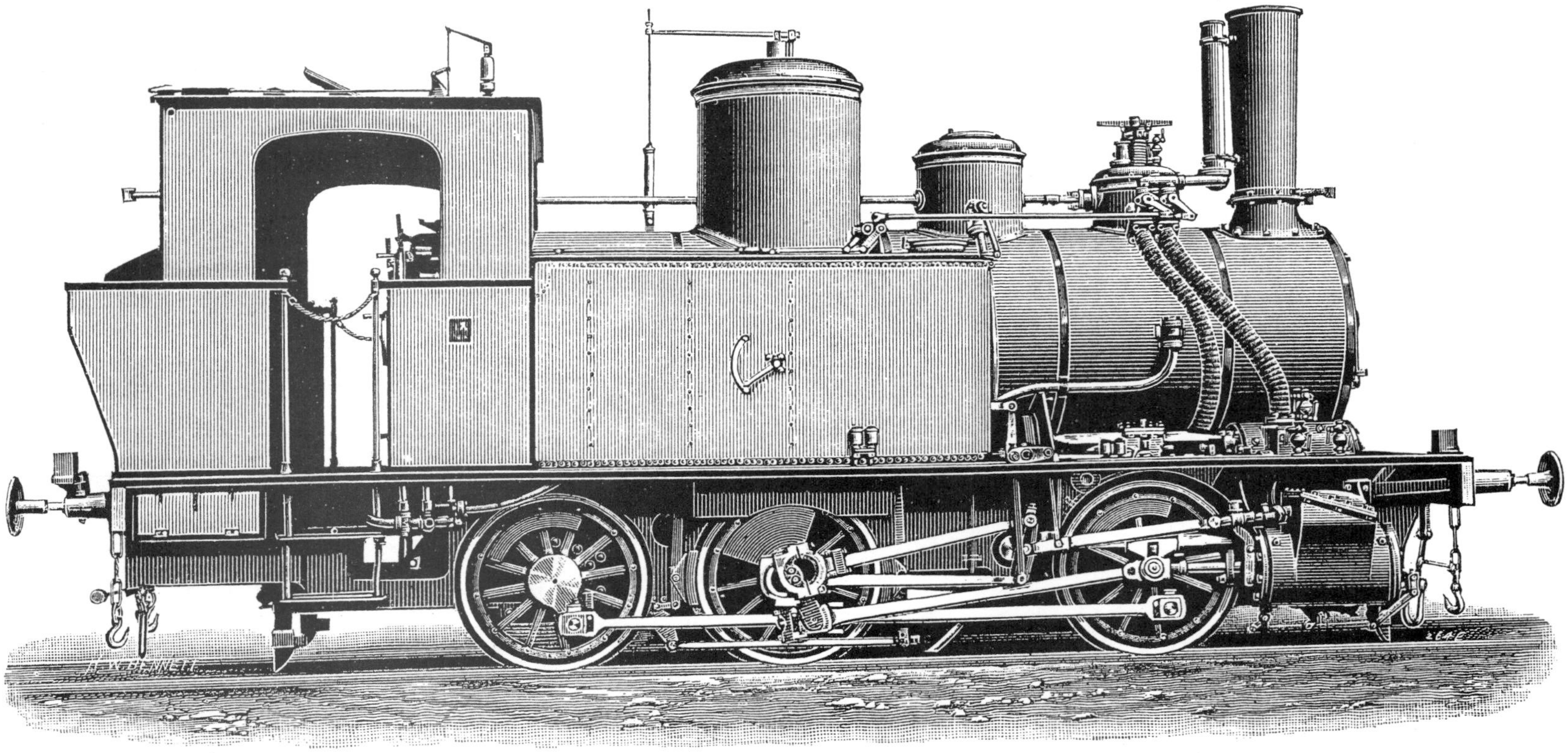

TOP: 0-6-0T combination rack-adhesion engine, Maschinenbau-Gesellschaft, Karlsruhe, for the Hoellenthal Rwy., circa 1890. BOTTOM: 0-8-0 goods engine built by Maffei Co. of Munich for the St. Gotthard Rwy., 1882.

PLATE
82

0-6-0 tank engines, 1890: Top: Locomotive for pulling light trains, Societe St. Leonard, Liège, Belgian State Rwys. Bottom: Goods engine designed by J. C. Park, locomotive superintendent, North London Rwy.

PLATE
83

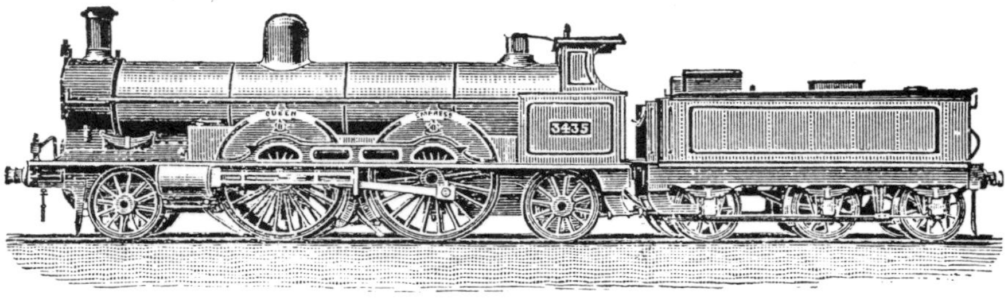

Top: Pechot-designed 0-4+4-0T articulated engine, Decauville Co., 1892. **Middle left:** 0-4-2T locomotive, Sachsische Maschinenfabrik, Cologne-Mindener Rwy., 1886. **Middle right:** 2-2-2-2 compound passenger engine, London & North Western Rwy., circa. mid-1890s. **Bottom:** Flaman triple boiler 2-4-2 Columbia type engine, Societe St. Leonard, Liège, Belgian State Rwys., 1888.

PLATE
84

Italian locomotives: Top: 0-6-0 goods engine, G. Ansaldo & Co., Sampierdarena, Adriatic Rwy., 1891. Middle: 0-8-0 goods engine, Mediterranean Rwy., 1889. Bottom: 4-6-0 engine, G. Ansaldo & Co., Mediterranean Rwy., 1891.

PLATE
85

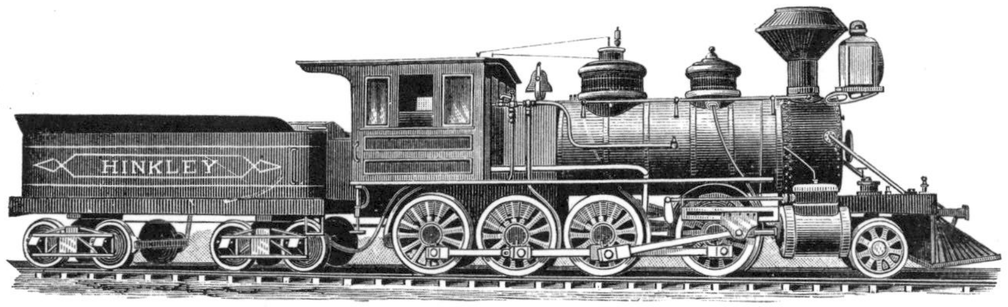

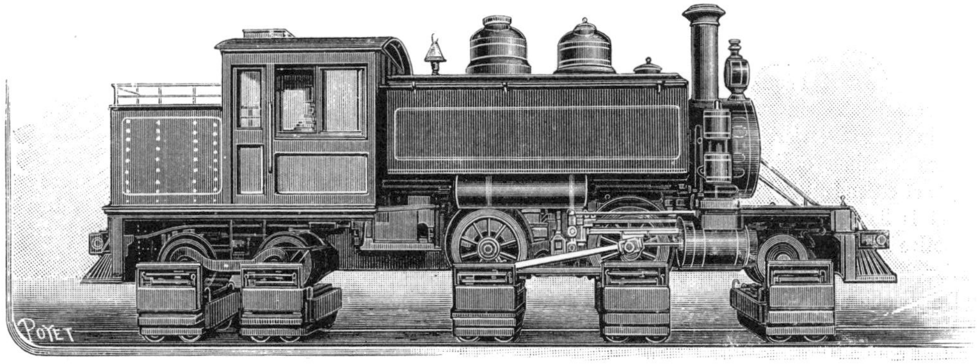

PLATE 86

Top: 4-4-0 passenger engine, Brooks Locomotive Works, Dunkirk, New York, for Lake Shore & Michigan Southern R.R., 1888. Middle left: 2-8-0 freight engine, Hinkley Co., Boston, circa 1883. Middle right: Weighing separate axles of a new 2-4-4T, Baldwin, Philadelphia, 1900. Bottom: 4-4-0 passenger engine, Old Colony R.R., 1893.

PLATE 87

The New York Central & Hudson River R.R.'s famous No. 999. This 4-4-0 engine, designed by William Buchanan and built at the company's shops in West Albany, set a world speed record on May 10, 1893, by pulling the Empire State Express at 112.5 m.p.h. between Batavia and Buffalo.

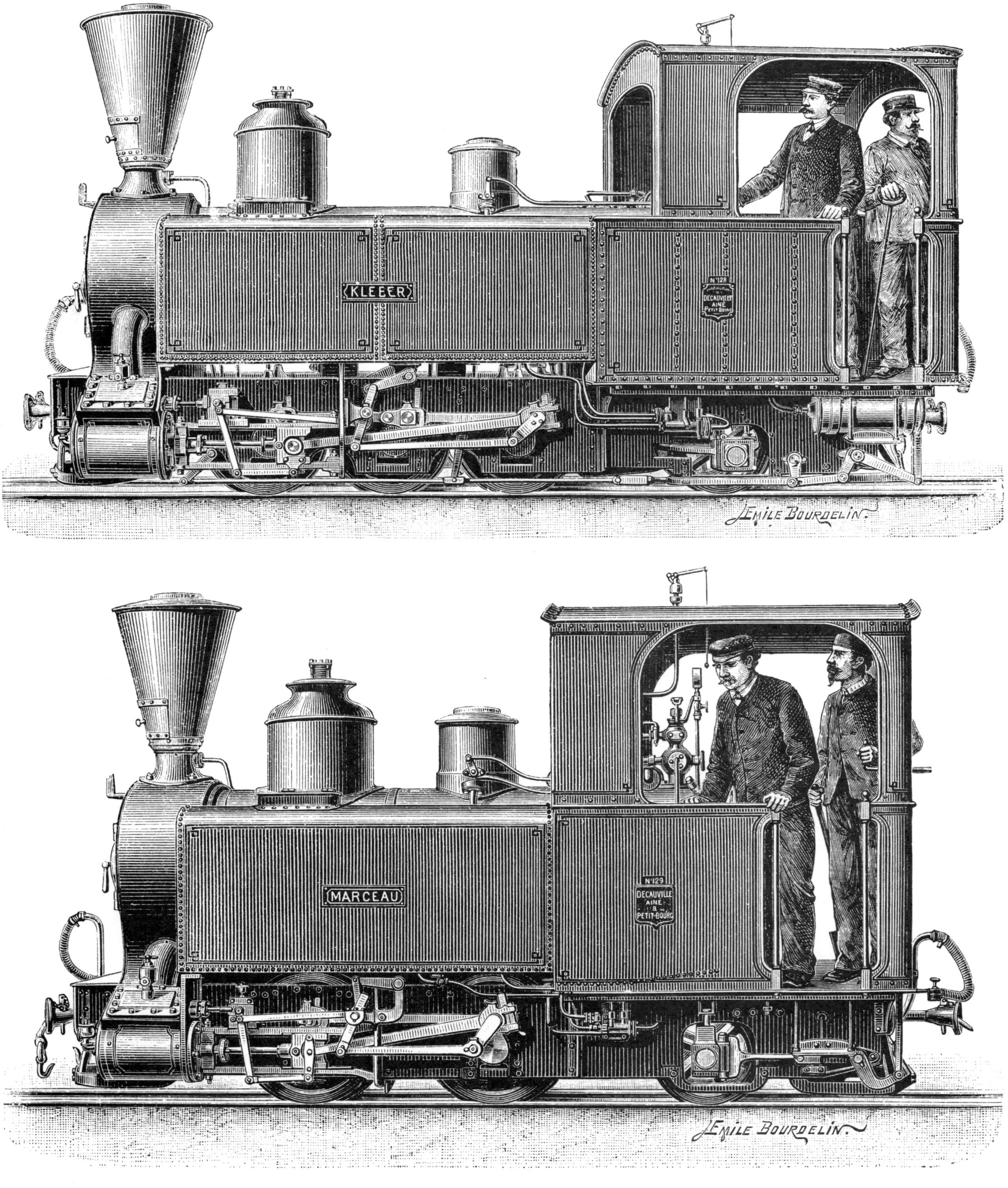

Decauville narrow-gauge tank locomotives, circa 1893: Top: 0-6-2T engine. Bottom: 0-4-2T engine.

Switzerland's Brunig Pass Rwy. consisted of two valley sections separated by a mountain section. About 1894 the Swiss Locomotive Works of Winterthur provided these two types of engines for working the line. Top: 0-4-0T rack engine for use in mountain section. Bottom: 0-6-0T adhesion locomotive for service in valley sections.

PLATE
89

PLATE 90

Heavy locomotives built by Swiss Locomotive Works, Winterthur, 1894: TOP: 0-4+4-0T Mallet compound articulated engine for Switzerland's meter-gauge Landquart-Davos Rwy. BOTTOM: 0-6-2T combination rack and adhesion engine for Turkish Empire's Beirut & Damascus Rwy.

PLATE 91

British narrow-gauge tank locomotives: Top left: 0-4-0T industrial engine, W. G. Bagnall Co., Stafford, 1896. Top right: 0-4-0T Bagnall engine built for sewage works at Beckton, 1894. Bottom: 2-6-2T built by Hunslet Engine Co., Leeds, for 3′ gauge, 31-mile Tralee & Dingle Rwy., Ireland.

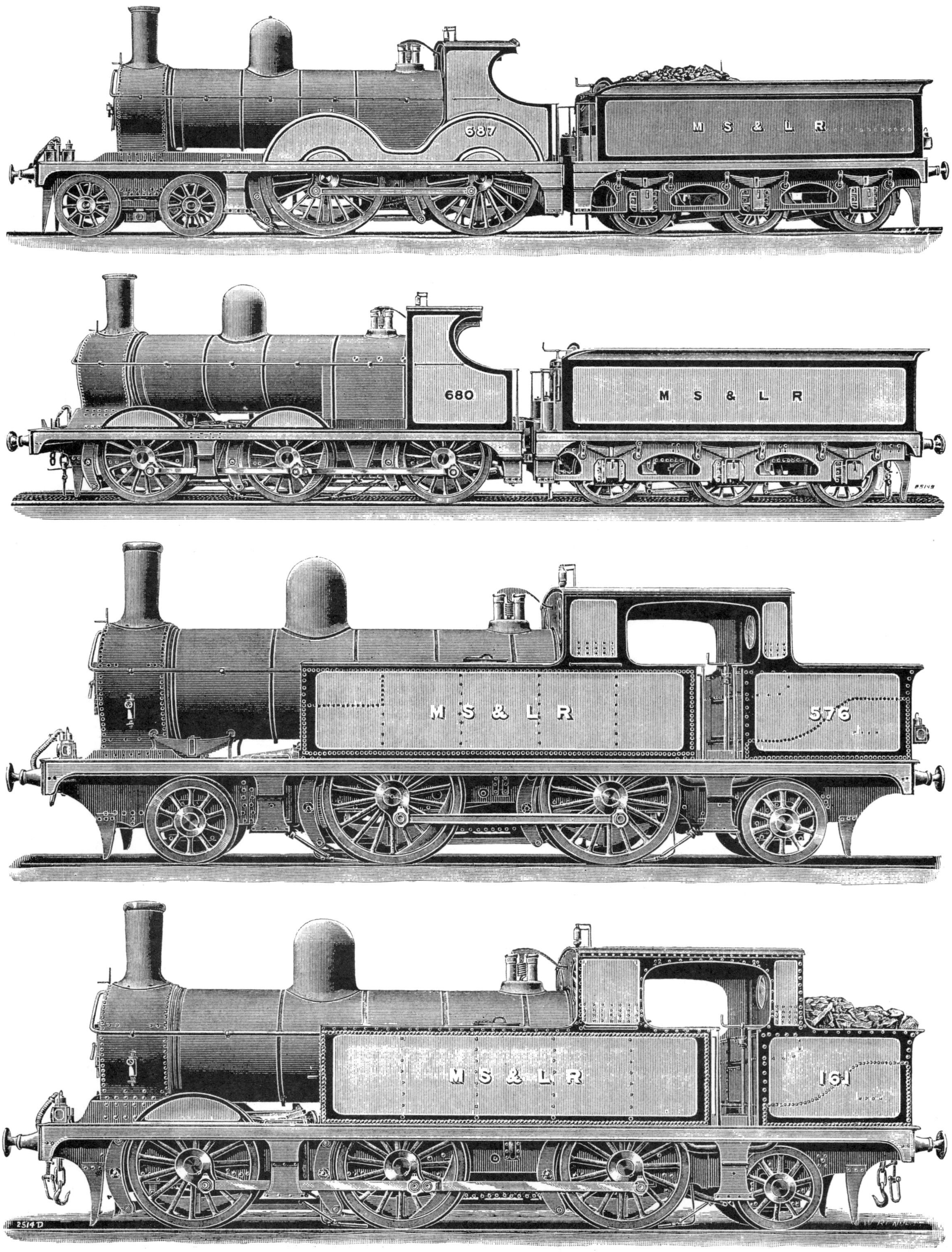

Standard locomotives of the Manchester, Sheffield & Lincolnshire Rwy., built at the company shops in Gorton, 1894: Top: 4-4-0 express engine. Upper middle: 0-6-0 goods engine. Lower middle: 2-4-2T for local traffic. Bottom: 0-6-2T for goods. Rapidly expanding at the time, the company became the Great Central Rwy. in 1899.

PLATE 93

British 0-4-4T passenger locomotives, 1895: Top: Engine for suburban service, Neilson & Co., Glasgow & South Western Rwy. Bottom: Branch line locomotive constructed at the North Eastern Rwy. shops at Darlington.

Top: 2-4-2 passenger engine with Corliss valve gear, Paris-Orléans Rwy., 1897. **Bottom**: 2-4-0 oil-burning locomotive built at the Great Eastern Rwy. shops in Stratford, 1894. This engine was intended for excursion service in the summer and fast goods work in the winter.

PLATE 94

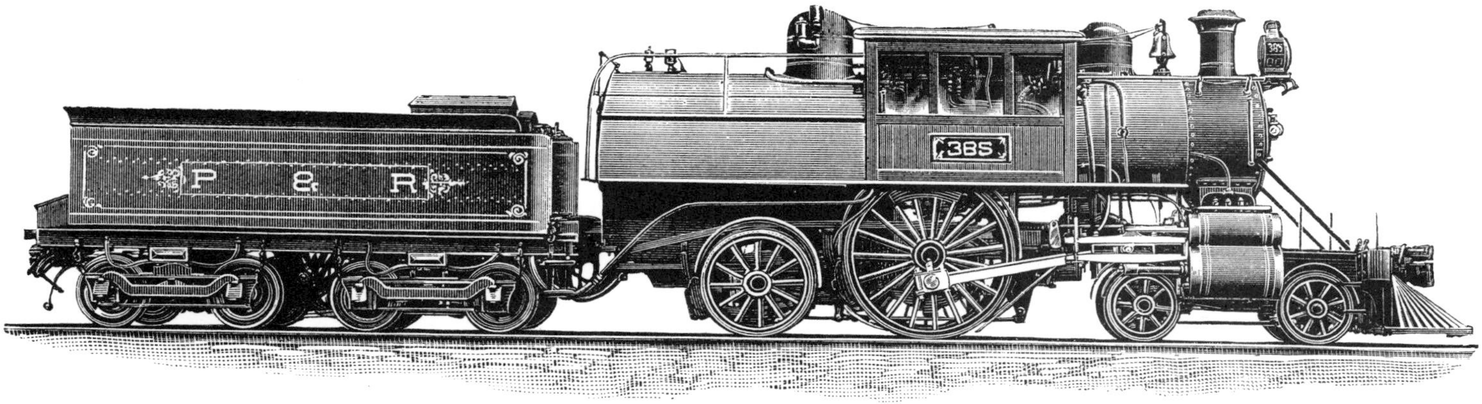

Top: 4-4-0 express engine, Brooks Locomotive Works, Lake Shore & Michigan Southern R.R., 1893. Middle: 4-2-2 express locomotive with Wootten firebox, Philadelphia & Reading R.R., 1895. Bottom: Brooks 4-6-0 express engine, Lake Shore & Michigan Southern R.R., circa 1895.

PLATE

95

Brooks locomotives from 1896: Top: 4-6-0 engine for mixed service, St. Lawrence & Adirondack Rwy. This 30.2-mile Canadian-American line operated between Valleyfield, Quebec and Malone, New York. Bottom: 2-6-0 freight engine, New York Central & Hudson River R.R.

PLATE 96

Top: Rogers 4-6-0 passenger engine, Illinois Central R.R., circa 1896. In Casey Jones' famous collision of 1900, he was operating a locomotive of this type. **Middle:** 4-4-0 express engine with super-heater, Borsig Co., Berlin, circa 1900. **Bottom:** New 4-4-2 Atlantic type engine receiving final inspection at Lokomotivfabrik Wiener-Neustadt, 1892. This was probably an early design of Dr. Karl Gölsdorf.

PLATE

97

PLATE 98

Heavy American freight locomotives, 1898: TOP: 115 ton 2-8-0 Consolidation type, Pittsburgh Locomotive Works. This engine was built for operation on the Union R.R. between Munhill and North Bessemer, Pennsylvania. BOTTOM: Brooks 4-8-0 "Twelve Wheeler" type, Great Northern R.R.

French compound locomotives: TOP: 4-4-6 aerodynamic front-cab express designed by Henri Thuile, Schneider Works, Le Creusot, 1900. MIDDLE: C 61 Class 4-4-0 express with Baudry aerodynamic design for improving train speeds against south France's strong Mistral winds, PLM Rwy., 1898. BOTTOM: 4-6-0 for goods or heavy passenger service, Ouest Rwy., 1900.

Top: Difficult to classify is this express engine by Krauss & Co., Munich, circa 1900. Normally it functioned as a 4-4-2 Atlantic type. However, it had retracting auxiliary drivers on the front bogie. These could be engaged for additional power for starting, heavy loads, or grades.
Bottom: 4-4-0 superheated express engine, Borsig Co., circa 1900.

PLATE 100

CHAPTER

6

French and Belgian Railways

PERHAPS France's history with railways began with Nicholas Cugnot's experiments with a steam carriage in 1769. Unfortunately, this invention was not successful, and soon it was largely forgotten. The first tramway in the country was constructed in 1778–79 at the mouth of the Loire River. It was built by the Englishman William Wilkinson for serving an ordnance factory. During the early nineteenth century, French engineers began to hear reports of the progress in steam traction being made in England. In 1818 de Gallois published *Annales des Mines,* a report on English rail progress. In the early 1820s plans began for construction of several tramways to connect the St. Etienne coalfield near Lyon with the Rhone and Loire rivers. The first of these, opening in May 1827, was a 10.56-mile line connecting St. Etienne with Andrezieux on the Loire. On March 1, 1832, passenger service was inaugurated on this line using horse traction.

Sometime after the Stockton & Darlington line opened in 1825, a French engineer, Marc Seguin, made a visit to study operations there. Seguin was one of four brothers, all engineers, who were nephews of the pioneer balloonist brothers, the Montgolfiers. Along with Edouard Biot they had been granted permission to construct a railway between St. Etienne and Lyon. During this and subsequent visits Seguin became acquainted with George Stephenson and placed orders for two of his engines.

When these engines were received it was found that they were deficient in producing steam. This caused Seguin to initiate experiments with a multitubular boiler design, about the same time Robert Stephenson began his own. Seguin's idea was patented December 12, 1827. For his first engine, successfully tested on November 7, 1829, he also employed an axle-driven fanning system to create a good draught. However, in the long run this innovation proved to be less successful than when blast pipes were used. Over the next several years Seguin built 12 additional locomotives at the Lyons-Perrache and St. Etienne railway shops. Work began on the St. Etienne–Lyon line on June 28, 1826, and beginning in 1830 both passenger and goods service was initiated as various portions of the line were opened. Finished in 1832, it was a well-engineered railway, extending for 36 miles and double-tracked throughout. The line had three tunnels, one of which, the "Terrenoire," was almost a mile long. In 1833 a third line in the area was completed, extending from Andrezieux to Roanne.

The first railway in the Paris area was the Paris–St. Germain line. It used both steam and animal traction. Promoted by the bankers Emile and Isaac Pereire, its first section opened on August 26, 1837, to Le Pecq. After the full line opened trains ran hourly from 7:00 a.m. to 10:00 p.m. In 1838 a Paris terminus was constructed at Rue de Stockholm, near Place de l'Europe. This station was built on a bridge above the tracks.

Concessions for two (passenger only) lines between Paris and Versailles were given on May 24, 1837. These railways were to run on opposite sides of the Seine. The right-bank concession was given to M. de Rothschild. The more successful of the two routes, it opened on August 2, 1839. Versailles Station was constructed as its Paris terminus. The left-bank concession was given to M. Fould. Its Paris terminus became the Gare de l'Ouest (now Gare Montparnasse). Also constructed for the line was the beautiful Val-Fleury Viaduct near Meudon. A notable engineering feat for the time, its foundations had to be sunk deep, past layers of clay, to reach hard bedrock, and this required extensive masonry work. Marc Seguin completed the route, and it opened on September 10, 1840.

Another early French line ran from Montpelier to Sète, a nearby Mediterranean port. Built by Mellet and Henry, the line opened in 1839, and used five engines constructed by the Fenton, Murray & Jackson firm of Leeds: the *Notre Dame des Tables, Cettoise, Rosine, Herault,* and *Montpellierenne.* That same year railways were opened connecting Epinal to the Burgundy Canal, and Alais to Beaucaire. All of these early lines were built in the 4′8½″ gauge, which became standard in the country. The Alsace Rwy., originally connecting Strasbourg to Basle and Mulhouse to Thun, was built to the highest English standards. The Mulhouse-Thun line was financed by Nicholas Koechlin, a local industrialist, and opened on September 1, 1839. The Strasbourg-Basle route, which opened September 19, 1841, became the first international railway, and for a brief time was the longest on the Continent.

Four inaugural trains were used when the 75-mile Paris-Orléans line opened on May 2, 1843. Among the dignitaries in attendance were the Duke of Nemours and the Duke of Montpensier. The following day, the two dukes attended the inauguration of the Paris-Rouen Rwy. In celebrating this occasion, a whole beef was roasted and served to 600 workmen who had just completed the Maisons-Lafitte Bridge.

Although most early French lines began with imported English engines, a small locomotive industry soon emerged in the country. In 1833, M. Edwards, manager of the Chaillot Iron Works in Paris, bought a 0-4-0 Samson class engine from the Stephenson Company. Using this as a prototype, the Chaillot firm began constructing its own locomotives. The company later changed its name to Derosne and Cail, and in 1849 began specializing in the production of a 4-2-0 version of the Crampton engine. A short time later, the firm's name was changed to J. E. Cail. It might be noted here that Thomas Crampton's fast locomotive, rejected in his home country of Britain, found a home in France and elsewhere on the Continent. Many of these engines were operated on the later French systems of the Nord, P.O., PLM, and Est.

The Schneider Company of Le Creussot built its first locomotive, the *Gironde,* for the Versailles Rwy. in 1838. Beginning about the same time in Mulhouse, the firm of Andre Koechlin, Stehelin & Huber built the *Napoleon.* This engine pulled the inaugural train on the Mulhouse-Thun line, and on that occasion, reached speeds of 50 m.p.h. This enterprise later became the Societe Alsacienne de Constructions Mechaniques (SACM).

In 1842, English engineers William Allcard and W. Buddicom were appointed to build engines for the Rouen Rwy. at its shop in Sotteville. Buddicom had previously been locomotive superintendent for England's Grand Junction line at Crewe and had pioneered engines of outside cylinder design. This pair did high-quality work, and later built many locomotives for the Ouest line. About the same time the Verpilleux firm of Lyons began making engines for the St. Etienne–Lyons Rwy.

While France had begun making progress in building railways, it lagged far behind England and the United States, and almost behind tiny Belgium. Still recovering from the aftermath of its revolution and the reign of Napoleon, the country was torn by deep ideological differences. Thus it was perhaps only natural that railways quickly became politicized. Followers of the Saint Simonite social philosophy favored them. This group advocated a utopian society based on free industrial enterprise. Others saw railways as "inherently aristocratic," and later as "essentially royalist" or "anti-republican." To some degree the country became paralyzed by a continually heated debate in the National Assembly over whether lines should be funded privately or publicly. Having no clear mandate, the government acted slowly in granting concessions, but was even slower to involve itself in railway construction. However, the underlying reality was that private-sector funds were insufficient for developing an adequate railway system. After state intervention in 1840 to rescue construction of the Paris-Orléans line, it became increasingly clear that more government involvement would be necessary.

Reaching a compromise, the National Assembly passed the Railway Law of June 11, 1842. This legislation provided the framework for a Paris-centered rail system having seven trunk lines radiating outward from the city. It also called for two cross-country routes, one from Dijon to Mulhouse and the other from Bordeaux to Marseilles. This system was to be, in essence, half-public and half-private. In *Railways, Past, Present, and Future,* G. Freeman Allen states: "In France the railway companies' spheres of influence were not left to their managers to resolve, belligerently, or any other way: they were delineated by the state. For the price of these sensible constraints, the companies had their track bed, its earthworks and engineering structures built for them with State and local authority finance; all they had to do was lay the track, erect the buildings, provide the rolling stock, then man and run the railway." The government provided about one-third of track maintenance costs after this, but also reserved the right to purchase railways at any time in the future. This law benefited the rail lines as well as the riding public, for the government proceeded, with great trouble and cost, to lay out the best possible national system for sustained train speed of any in Europe.

One interesting railway of the 1840s was the Paris-Sceaux

line. Going through hilly terrain, it employed a unique system devised by the engineer Arnoux, allowing trains to negotiate extremely tight curves. This railway receives more attention in chapter 29. In March 1847 the 59-mile Rouen–Le Havre Rwy. was inaugurated. Joseph Locke, a former pupil of Robert Stephenson, was chief engineer, and Mackenzie and Brassey served as contractors. France had only 350 route miles of track in 1842; this figure reached 1,250 by 1848. The revolution of that year, however, set railways back somewhat. Long-simmering grievances erupted, with employees demanding higher wages and calling for the expulsion of foreign rail workers (there were many British mechanics in the country at the time). Acts of sabotage were also directed against the lines, resulting in the destruction of stations, bridges, and other railway property. There was a general financial malaise as well, and in the radical political climate of the time, strong calls were made for nationalizing the railway system.

After the establishment of the Second Empire under Napoleon III things slowly improved. Railway financing became easier after bonds began to be issued. The government also actively promoted railway mergers. By 1858 six railways had been consolidated from the 77 existing in 1850, with each having a territorial monopoly.

Perhaps the easiest way to understand this new system would be to imagine oneself in Paris, with a map and compass, facing northward. Immediately ahead would be (1) the Nord Rwy., which, extending north from Paris, reached Calais, Lille, and the Belgian frontier. Owned by the Rothschild family, its Paris terminus was the Gare du Nord. Moving clockwise we next come to (2) the Est Rwy., which absorbed the Paris-Strasbourg Co. and other lines in the east. This line was of vast strategic importance because it connected Paris with areas along the German border. The original main line extended through Meaux, Château-Thierry, Epernay, Châlons-sur-Marne, and Nancy to Strasbourg, and ran north of the Vosges Mountains. The Gare de Strasbourg (later Gare de l'Est) was the company's Paris terminus. In 1859 the company added the Paris-Vincennes line, establishing its Paris terminus, the Gare de Vincennes, at Place de la Bastille. In 1860 the Est added the Paris-Reims-Charleville route. Later another main line was constructed from Paris through Troyes, Chaumont, and Belfort, to Mulhouse, running south of the Vosges Mountains.

Continuing our clockwise movement we come to (3) the PLM, or Paris-Lyon-Mediterranée Rwy., which stretched from Paris in a south-southeasterly direction through Dijon, Lyon, and Avignon to Marseilles. This railway was consolidated from the Paris-Lyon, Lyon-Mediterranée, Grand Central, Dauphine, and Lyon-Geneva companies. In time the PLM completed its Bourbonnais line, a second Paris-Mediterranean route running through Clermont Ferrand to Montpelier and the port of Sète. Another feature of the PLM was that it served France's three largest cities; thus it became the country's dominant carrier in terms of total route mileage, traffic volume, and revenue. The Gare de Lyon served as its Paris terminus.

Facing south-southwest was (4) the P.O., or Paris-Orléans Rwy., which despite its modest name, extended through Orléans to Blois, Tours, Poitiers, Angoulême, and Bordeaux. The company later built a second route directly south from Orléans through Vierzon, Châteauroux, Limoges, and Brive to Toulouse. Its Paris terminus was the Gare d'Orléans (now Gare d'Austerlitz) at Place Valhubert. The P.O. Company also operated the Sceaux line, its Paris terminus being at Place d'Enfert.

Even more southerly than the P.O. was (5) the Midi Rwy., which was constructed by the Pereires from Bordeaux to Toulouse and southeastward to Cebere at the Spanish border. Despite this company's relatively small territory, its strategic position allowed it to handle all of the French-Spanish traffic. At Bordeaux and Toulouse it connected with the P.O., and extending a branch from Narbonne to Sète it linked with the PLM. From Bordeaux the Midi later built a line south-southwest through Biarritz to reach Spain's Atlantic coast.

Returning to Paris, and once more moving clockwise, we come to (6) the Ouest Rwy., originally the left-bank Paris-Versailles line. It absorbed the St. Germain, Bretagne, and Normandie companies, all operating from Paris's Gare St. Lazare. After adding various extensions, the railway's main routes soon extended from Paris to Brest in Brittany, and to Granville, Cherbourg, and Le Havre in Normandy. During the 1860s construction began on Paris's Ceinture Rwy. This line eventually circled the outer perimeters of the city, providing a connecting link for the various railways. It also developed a substantial business in passenger and goods traffic. More can be found on this line and efforts to establish a Paris metropolitan system in chapter 28.

As in Britain, the large railways established company facilities to manufacture their own locomotives. The Est line constructed engines at Epernay, and the PLM in Paris. As rail lines expanded in France, increased economic activity followed. Cities having no railways often exerted political pressure to acquire them. In 1859, the government developed a policy that led to the creation of what was called the "new network." Formalized in the Franqueville Convention of that year, this plan requested that in exchange for state support of the existing system, railways should cooperate in extending branches to all cities in their territories having need of them. While the government provided no subsidy for building these lighter traffic routes, it did guarantee interest on capital the railways invested in construction. However, it also demanded that railways share profits, once they exceeded a certain margin.

While this policy led to the creation of many new routes, it hardly satisfied the clamoring requests for rail service. A law passed on July 12, 1865, authorized the construction, by local districts, of meter-gauge lines, so long as they were limited to short local routes connecting villages and small towns to larger communities. This led to the creation of a two-tiered rail system consisting of the six great companies and the new network of short meter-gauge feeder lines. However, some of these routes were built in standard gauge and connected to each other in a way that threatened to disrupt the existing order. In the 1870s, at the request of the government, some of these were absorbed into the big companies.

At the end of 1873 France had 11,511 miles of track. Between October 1, 1873, and October 1, 1874, 23 new railways or extensions of existing lines were built, totaling 326 route miles. Of these the 38-mile Orléans-Gien Rwy. was the longest.

In 1860 the fortunes of the PLM line changed for the better when France annexed Savoy and Nice from the Italian state of Sardinia. Existing railways there were absorbed into the PLM system. Among these was a work in progress, the railway tunnel beneath Mt. Cenis intended to connect Sardinia and Savoy. With the establishment of the Kingdom of Italy in 1861, this became a joint French-Italian project. Opened on September 17, 1871, it initiated many new commercial potentials for the two countries. More information on this tunnel and Fell's center-rail line over Mt. Cenis can be found in chapter 19.

The Est Rwy. suffered a much more unfortunate fate. Much of its system was destroyed in the Franco-Prussian War of 1870–71. Furthermore, when the Germans annexed Alsace-Lorraine, the company lost all of its lines there. Igney-Avricourt, 254.76 miles from Paris, became the railway's new eastern terminus, connecting there with Germany's new Imperial Elsass-Lothringen Rwy. Following France's defeat, the Est line became a handy scapegoat and took considerable blame for the loss. While confused railway operations had certainly contributed to the debacle, this had less to do with company management than with military incompetence. No clear military railway policy had been developed beforehand. Thus the railways and military had no previous experience in how to work together. Additional information on this conflict is provided in chapter 20.

Following the war, and suppression of the Paris Commune revolt, the Third Republic was installed. Finances were difficult for some time afterward. Railway service suffered and renewed calls were made to nationalize the system. A first step in this direction was taken in 1878 with the creation of the Etat (State) Rwy. At its inception, the line was given 1,625 miles of track consolidated from unprofitable local routes in the southwestern area of France. While it soon found itself in financial difficulty, this railway acquired running powers, from both the P.O. and Ouest lines, allowing it to create a Paris-Bordeaux route running through Chartres, Saumur, Niort, and Saintes.

France's government overextended itself in 1879 when it approved the Freycinet Plan. Under this legislation the state took upon itself the task of building a "third network" of about 6,500 miles of new track. The intent, essentially, was to put every Sous-Préfecture (small towns typically having populations greater than 2,000) on the standard gauge. As this plan was being implemented, however, the financial costs became too great. Under the conventions of 1883, the government divested itself of this obligation, placing responsibility for these routes in the hands of the large companies.

While route mileage in France increased by 30% between 1880 and 1888, gross income for the entire system remained almost unchanged. The *Railroad Gazette,* ever an observer of the international railway scene, made the following comments in its issue of November 16, 1888:

> The causes of this result are somewhat complicated. Something is due to the depression of trade which has been quite seriously felt since 1882; more, probably, is due to the failure of the companies to reduce rates (in order to encourage increased traffic). The French companies have so long enjoyed a monopoly in their several districts that there has been much less progress in railway administration in France than anywhere else. The government has exercised a great deal of control over the roads; but this control has been in the hands of engineers as distinct from business men, and has had almost no effect in compelling railways to keep pace with the times. It is doubtful whether the reduction in French railway charges has amounted to ten percent in the last 25 years. Further than this, the new roads have been built piecemeal, half under the control of the state, half under that of the companies. The distribution of lines in different parts of the country has been affected by political reasons rather than by commercial ones, so that the new lines have been a source of expense without as

yet having developed the traffic which was expected from them.

Important names in French railway history include Paulin Talabot, Auguste Perdonnet, and Eugene Flachat. Talabot (1799–1855), an engineer, began his career by constructing the Alais-Beaucaire and Avignon-Marseilles lines. Besides becoming president of the PLM Co., he also built railways in various parts of Italy, Portugal, and Algeria. Along with Coste, Perdonnet (1801–67) published the first French book on railways in 1828. Three years later he developed the first university curriculum for railway study. He later became chief engineer of the left-bank Paris-Versailles line, and was subsequently a director of the Est Rwy. His *Traité Élémentaire des Chemins de Fer* became a classic text on railways and was republished in a number of editions. Flachat (1802–73) gained his initial reputation during construction of the Paris–St. Germain line. He later became involved in locomotive, station, and bridge design, greatly influencing future trends in France. Flachat also assisted Talabot in construction of Bulgaria's Alais Rwy. In 1842 he became president of the left-bank Paris-Versailles line, and in 1845 assumed the same position with the Nord Rwy.

Following in the footsteps of Marc Seguin, a number of other French engineers began making important contributions to steam locomotive technology. In 1840 Benoit Paul Emile Clapeyron, of the right-bank Paris-Versailles line, began valve gear experiments, adjusting for a short cut off. In an effort to improve adhesion, Verpilleux in 1852 designed a four-cylinder engine with a permanently coupled tender. The second pair of cylinders powered the tender wheels. Both Polonceau and Forquenot of the P.O. line made improvements in passenger engine design, with the latter making an effort to standardize both designs and components. De Pambour invented the first practical steam pressure gauge. In 1859 Giffard devised the cone feedwater injector. About the same time Petiet began constructing his large innovative goods engines for the Nord Rwy. Train speeds also improved during this period. A speed record was set in 1855 when a Crampton-style engine averaged 60 m.p.h. between Paris and Marseilles, pulling the Imperial passenger train. Later, in 1877 Anatole Mallet introduced his two-cylinder compound engine. This was shortly followed by a three-cylinder system from Sauvage of the Nord line, and a four-cylinder system by du Bousquet and de Glehn.

From the mid-1840s on many engines were built in the large company shops; others were constructed by a number of locomotive manufacturers. While this group included a few of the early builders, some new companies were the Hallette Co. of Arras; the Cave, Claparede & Gouin firm of St. Denis; the Societe de Grafenstaden; Clement Desormes of Lyon; Parent & Schaken, originally in Lyon, but later at Fives-Lille; M. M. Cail of Fives-Lille; Weidenecht of Paris; and later on, the Decauville Co. at Petit-Bourg, which specialized in small narrow-gauge engines.

The Decauville story is especially interesting. Paul Decauville spent time studying the famous Welsh Festiniog line before devising his own 1′11⅝″ gauge portable railway system. Appearing in 1875, it was ingeniously conceived, having easily transportable engines and rolling stock, as well as track that could be quickly laid or dismantled. Ideal for domestic industrial and agricultural needs, this system also gained popularity in other parts of the world. Decauville railways were used in early excavations of the Panama Canal, in transporting sugarcane in Java and Australia, and for carrying passengers in Argentina and China. This system was also quickly adapted for military use. This subject receives more attention in chapter 20.

The August 15, 1885 issue of *Scientific American* gives the following description of the Decauville factory: "The Decauville works are at Petit-Bourg, only an hour from Paris. They cover about 20 acres on the bank of the Seine, and adjoin the PLM. The great hall is 525 ft. long by 525 ft. deep. The material is brought in at both ends (at one end the rails and steel for the road, and at the other, the sheet metal and iron for the cars), and the manufactured products are taken out at the middle, and loaded into the cars of the PLM Co. In July 1884, the works of Petit-Bourg attained their greatest development, with a thousand workmen, and 350 machines, which do the work of 3,000 men. Among others, there are four painting machines, which do the work of 60 painters. Three thousand cars and 93 miles of road are produced each month."

Circa 1871 the Belgian George Nagelmackers inaugurated France's first sleeping coach service. Later he introduced dining carriages to the country. On October 4, 1883, his Compagnie Internationale des Wagons-Lits inaugurated the Orient Express, a passenger train of unsurpassed luxury. It initially ran from Paris to Varna on the Black Sea. From there passengers could board steamers for Constantinople (Istanbul). In 1888 an overland route through Bulgaria to Constantinople was established. The company later operated other international luxury trains from Paris. For more on this story the reader is referred to chapter 13.

In 1883 French passenger trains operated at an average speed of 37 m.p.h. In 1886 express trains between Paris and

Boulogne averaged 42 m.p.h. Although locomotives were capable of pulling trains at a more rapid pace, speeds were held down for reasons of economy, reducing fuel costs as well as maintenance. During the 1890s train speeds improved as railways turned to the new innovation of compounding and the 4-4-0 type engine. This development is covered in chapter 5.

Circa 1889, French railways began to use both three-axle carriages of the English type and American-style bogie coaches. Total route mileage in France was reported as 20,743 in 1890 and 21,788 in 1893. In 1886 the PLM system alone extended for a total of 4,783 miles of track.

As the century waned the P.O. Rwy. decided to open a new terminus at a more convenient location in Paris. This became the Gare d'Orsay, which in recent times has been converted into an art museum. A tunneled extension was made from the Austerlitz station beneath the Seine to the new location. As air quality was an issue, it was decided to electrify the tunnel. Thus when the Gare d'Orsay opened in 1900, trains running between it and the old terminus were conveyed by electric locomotives. Outside of this area all traction was by steam.

Although mining tramways had existed in Belgium for over two centuries, the first horse traction line was the Hornu–St. Ghislain Rwy., inaugurated in May 1830. Besides connecting these two towns, this 11½-mile route provided a link between a coal mine and the Mons Canal. The line opened as Belgium was engaged in a war of independence from Holland.

Once the country had gained freedom, there was grave concern that the Dutch might act to impede commerce between Belgium's port city of Antwerp and the Meuse and Rhine rivers. While the Meuse flowed through such inland Belgian cities as Namur, Liège, and Maastricht, it had to pass through southern Holland before emptying into the North Sea. The Rhine also had to reach the sea through Holland. It connected Belgium with Dusseldorf, Cologne, and Mainz in Germany.

Beginning in 1831, plans were conceived for constructing an inland railway from Antwerp to connect with these two rivers. Besides protecting Belgian commerce, such a railway would open a more southerly route into Germany. Thus it had the potential for attracting traffic from France and England.

In contrast to the other nations of the European Continent, the Belgian government quickly saw a need for railways and acted decisively to begin their construction. On May 1, 1834, King Leopold I announced: "A railway system will be established throughout our kingdom. Its central point will be Malines and it will extend eastwards to the Prussian frontier, passing through Louvain, Liège, and Verviers. It will run north to Antwerp and west to Ostend via Termonde, Ghent, and Bruges. Its southern path will take it to Brussels and France."

Thus was the Belgian State Rwy., or Chemin de Fer de l'Etat, established. George Stephenson built its first line of 15 miles, connecting Brussels and Malines, which opened on May 5, 1835. A railway from Malines to Antwerp was inaugurated on May 3, 1836. In 1837 a route was opened between Ghent and Termonde. By 1839 it had been extended to Courtrai. Brussels' Nord and Sud stations were connected by rail in 1841, and by 1842 a line was extended south to Mons. The route from Verveirs to Aachen opened in 1843. There it met a German line that continued to Cologne. Thus by 1844 King Leopold's plan was realized. Railways were open from the port of Antwerp in the north to the French border in the south, and from the port of Ostend in the west to the German border in the east. In addition, maintenance workshops were built at Malines, where equipment of both lines could be serviced. A through Brussels-Paris route opened when France's Nord Rwy. reached the Belgian border in 1846.

The French literary giant Victor Hugo had his first railway experience on a visit to Belgium in 1841. At Mons he saw horses pulling a train of coal wagons. In Malines he joined a crowd watching the departure of a locomotive-drawn train for Brussels. Days later at Antwerp he took his first train ride, boarding a train for Brussels. Fortunately he left a record of this exhilarating 30+ m.p.h. experience, which was published in volume II of *Correspondance Familiale* (as translated in Graham Robb's *Victor Hugo: A Biography*): "The Motion is magnificent. You have to have felt it to know what it's like. The speed is unbelievable. The flowers at the track-side aren't flowers any more; they turn into blotches or red and white stripes. There are no points, only stripes. The corn is a huge mass of yellow hair; lucerne, long green tresses. Towns, steeples and trees dance about in a crazy jumble on the horizon. Now and then, a shadow, a shape, the upright figure of a ghost appears and disappears in a flash beside the door . . . Inside the carriage, people say, 'It's three leagues from here; we'll be there in ten minutes'." One can only imagine how such experiences later influenced art.

By 1840, the Belgian railway system extended for 204 miles, and by 1843, 350 miles. The government's investment in these early lines resulted in a heavy economic burden. Thus, beginning in 1844 the private sector was encouraged to become involved. By 1860 there were 465 route miles of track in the country, but a decade later, 1,850. By that time the government controlled less than one-third of the system.

The first locomotives used in Belgium were built by the Stephenson firm and were of the 2-2-2 Patentee class. These engines, the *la Fleche, Stephenson,* and *l'Elephant,* were operated on the Brussels-Malines line. In 1835, the English industrialist, John Cockerill, built his first locomotive, *Le Belge,* at his Belgian factory in Seraing. Beginning in 1844, he constructed engines based on the Norris designs, building 36 of the 4-2-0 type; and 37 of the 4-4-0 type. Many of these were exported. Cockerill gained more fame for his 0-4+4-0 *Seraing,* which was entered in the 1851 Austrian Semmering trials. Other Belgian locomotive manufacturers included la Haine-Saint-Pierre and Tubize, both located in La Haine; Couillet, near Charleroi; Evrard (later Compagnie Belge), at La Croyère near Tournai; Carel of Ghent; and Societe St. Leonard, of Liège.

Making important contributions to railway development were the Belgians Egide Walschaert and Alfred Jules Belpaire. Walschaert (1820–1901) held relatively modest positions in his life, rising only to become mechanical foreman of the Schaerbeck depot near Brussels. Nonetheless, he made an important innovation in 1844 with the introduction of his valve gear. It was years, however, before the value of this mechanism was fully understood, at which point it came into almost universal use. Belpaire (1820–93) became, in 1850, director of rolling stock for the Belgian State Rwys. His efforts to improve consumption of low-grade fuel resulted in his famous firebox design of 1864. Flat on the top and sides, but rounded where the surfaces met, it proved to be very efficient and was quickly adopted elsewhere.

Beginning circa 1870, the Belgian engineer George Nagelmackers initiated a sleeping car service modeled somewhat on the Pullman system. Although based in Brussels, his company operated internationally. After buying out his partner in 1876, he formed the Compagnie Internationale des Wagons-Lits. Beginning with its Orient Express in 1883, the firm began the operation of special luxury trains.

During the latter part of the century, light meter-gauge railways began to be built in Belgium. A network of these routes was established under the corporation, SNCV. The first SNCV line built was the 137-mile Ostend-Nieuport line, completed in 1885. Along the border with Holland other railways were built in the 3′6″ gauge, which allowed for interconnection with similar gauge Dutch railways. A number of these lines were mainly built for steam tram operation.

Total Belgian route mileage in 1880 was 2,550. The same figure at the close of 1892 was 3,590. That year the country had 2,866 locomotives, 4,314 passenger carriages, and 59,348 goods wagons. Of these figures the state controlled 2,030 miles of line, and had 2,091 engines, 2,841 passenger coaches, and 44,287 goods wagons. Narrow-gauge route mileage for 1892 was 584.

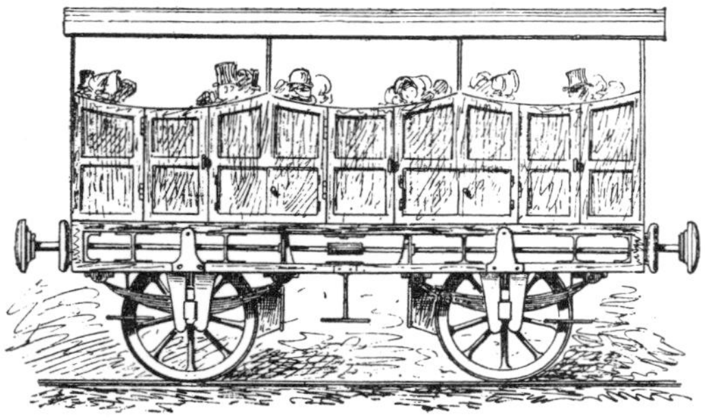

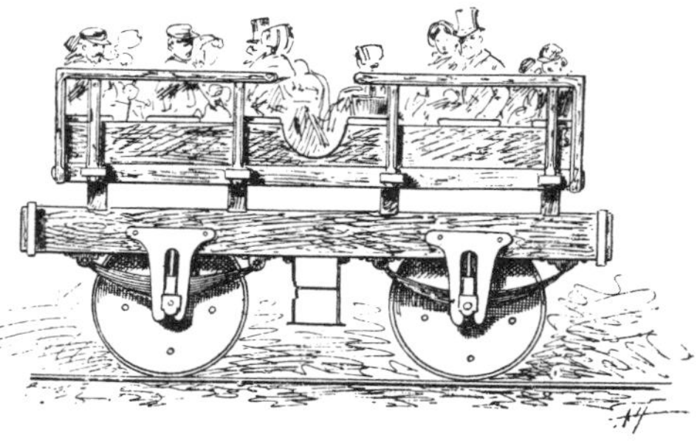

Top left: Train on Paris–St. Germain Rwy., circa late 1830s. Top right: Marc Seguin. Upper middle left: Cockerill's 2-2-2 engine *Le Belge*, 1835. Upper middle right: First-class carriage, Brussels-Malines Rwy., 1835. Lower middle left: Second-class carriage. Lower middle center: Third-class carriage. Lower middle right: Early French first-class carriage. Bottom left: Station of Paris–St. Germain Rwy. at Rue de Stockholm, 1838. Bottom right: Train on St. Etienne–Lyon Rwy., circa 1830s.

PLATE 102

Inauguration of first all-steam railway on the European continent, Belgium's Brussels-Malines line, May 5, 1835.

Station scenes in France and Belgium, 1846: Top: Gare du Nord, Paris. Upper middle: Gare d'Orléans, Paris. This later evolved into the Gare d'Austerlitz. Lower middle: Gare du Midi, Brussels. Bottom: Station of the Paris-Orléans line at Orléans.

PLATE 103

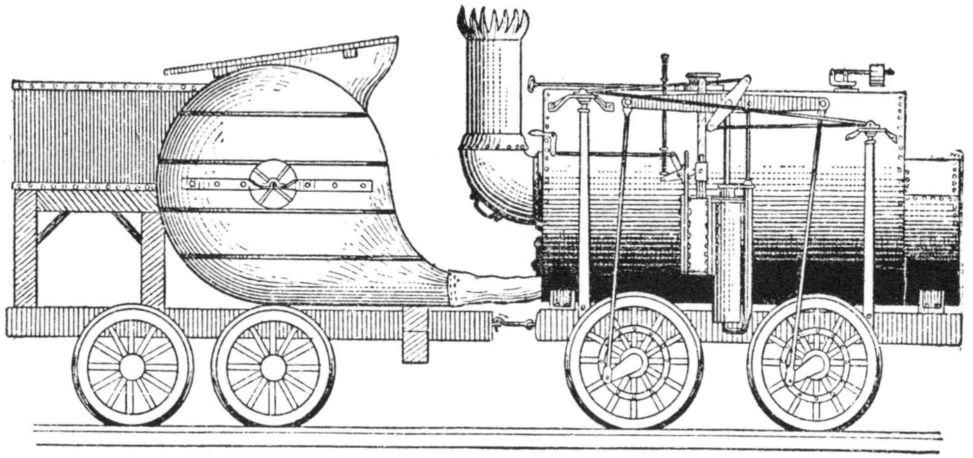

PLATE

104

Top left: Marc Seguin's engine of 1829. It had an axle-driven fanning system as well as multitubular boiler contemporary with Stephenson's *Rocket*. **Top right:** 2-4-0 Buddicom-designed goods engine, Ouest Rwy., circa 1840s. **Middle:** Early rails. **Bottom left:** 2-4-0 Stephenson-built engine, Paris-Versailles Rwy. (left bank), 1837. **Bottom right:** 2-2-2T Buddicom-designed passenger engine, Ouest Rwy., circa 1840s.

Top: 4-2-0 Crampton express locomotive, J. E. Cail Co., Paris, circa 1850s. **Bottom:** 2-2-2 passenger engine built on Stephenson's long boiler design, Derosne & Cail Co., Paris, 1846.

TOP: 2-2-2 passenger engine designed by Alexis Barrault, Derosne & Cail Co., Paris-Lyon Rwy., 1847. BOTTOM: 0-4-2 engine of mixed use type built for Nord Rwy. from 1867 to 1883. This is probably of the series 2487–2551 built from 1870 to 1875 at Fives-Lille.

PLATE
106

Festive French railway inaugurations: Top: Scene at inauguration of Toulouse-Sète section of Midi Rwy. at Toulouse, 1857. Bottom: Ceremony at Brest for a section of line extending to Guingamp, Ouest Rwy., 1865.

PLATE 108

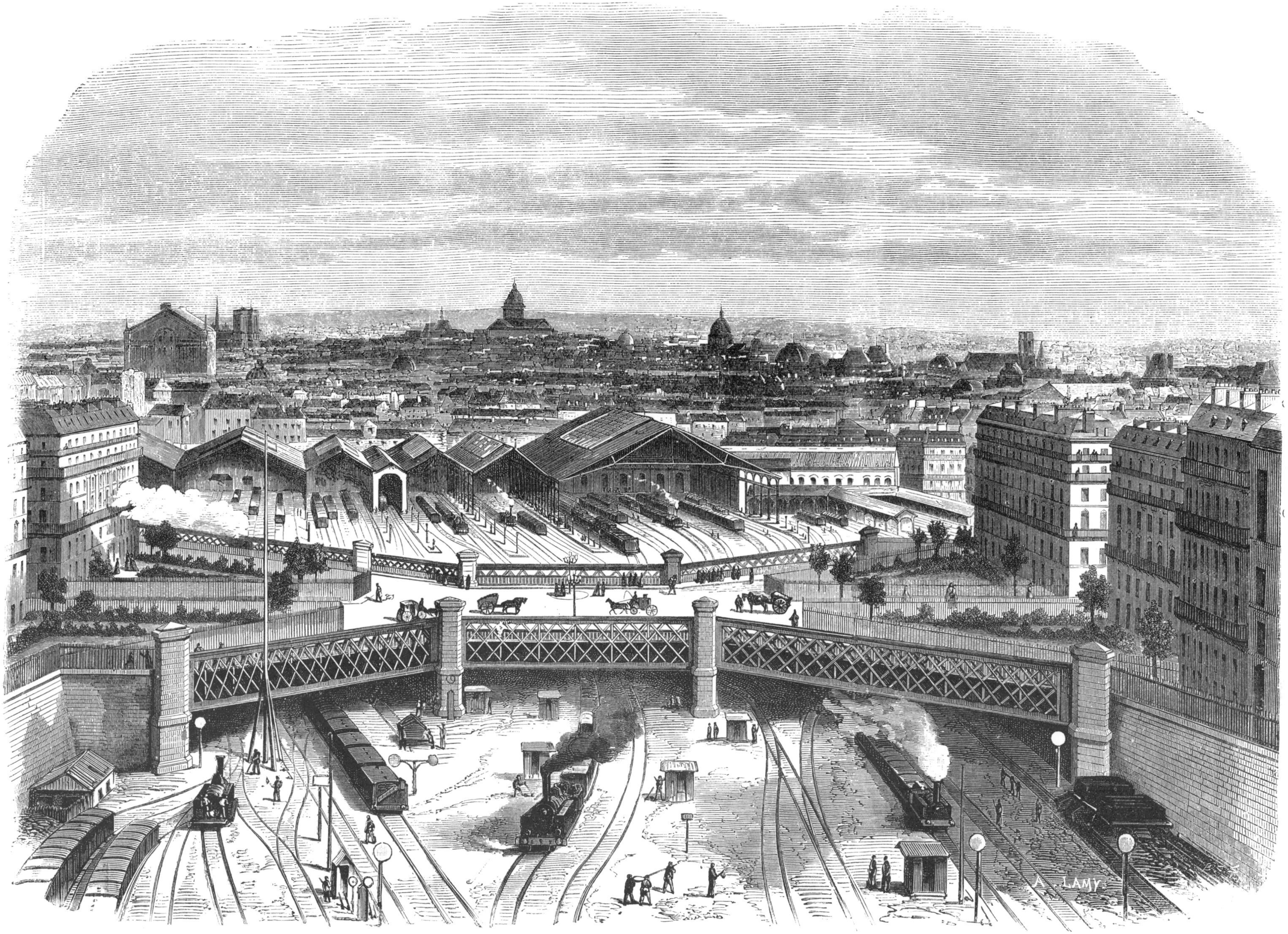

Back view of train sheds at Paris's Gare St. Lazare, Ouest Rwy., 1870. The iron bridge in foreground was the former site of Pl. de l'Europe. It connected Rues Madrid, Berlin, Constantinople, Londres, Vienne, and St. Petersburg. The terminal serviced 200 trains daily operating over seven different routes.

PLATE 109

A view of the British railway display at Paris's Exposition Universalle, 1878. The 2-4-0 express engine in the foreground was built by Sharp, Stewart & Co., Manchester, for the Paris-Orléans Rwy.

PLATE 110

Passengers on a German train disembark at the French border station of Igney-Avricourt for customs check before reboarding an Est Rwy. train. This town, 254.8 miles from Paris, became the Est company's terminus point after France's defeat in the Franco-Prussian war.

In an effort to prevent a rail strike in Paris from spreading elsewhere, the government sent in army troops, stationing them in outlying areas to ensure the continued operation of trains, 1898.

PLATE 111

PLATE 112

Arrival of a suburban commuter train in Paris, 1899. Note double-decker passenger carriages.

CHAPTER

7

Railways of the German and Austro-Hungarian Empires

WHILE use of tramways in German mines goes back a number of centuries, there was no substantial improvement in this technology until the early nineteenth century, when news of English progress reached the country. At the time Germany was not a united country, but consisted of a number of independent states. Among these, Prussia was dominant, in both military power and the political influence it exerted over the other states. However, some of these, like Bavaria, were resistant to Prussian hegemony and had close relationships with its rival, Austria.

In 1815 the first German locomotive was built in Berlin. Constructed for use in a Silesian mine, it proved inadequate for the job. In 1825, Friedrich List, a radical professor from Württemberg, fled to America. While in Pennsylvania, he became involved in railway developments there. In 1832 he accepted an offer to return to Germany, taking the position of U.S. consul in Leipzig. Much like Thomas Gray in England, he became a strong advocate for railways, and in 1833 published a plan for a Berlin-centered German rail system. Foreseeing political as well as economic benefits, List suggested that a strong rail network would help to facilitate German unification. However, these new ideas were met with considerable ridicule.

In 1834 List was recruited by another visionary, King Ludwig I of Bavaria, to assist in construction of his Nuremberg-Fürth Rwy. Later, after being involved in other railway projects, List was ignored, ostracized, and never given any acknowledgment for his contribution. Poverty-stricken and despondent, he committed suicide in November 1846. The Nuremberg-Fürth line, which had originally been promoted by Nuremberg citizens Plattner, Scharrer, and Ley, was built under the supervision of Paul von Denis and opened on December 7, 1835. Stephenson's *Der Adler*, a 2-2-2 Patentee class engine, pulled the inaugural train. In a practice that continued on this line until 1862, locomotives pulled passenger trains, while horses were used for goods traffic. This line was immediately successful, carrying 450,000 passengers its first year of operation.

On April 24, 1837, the first section of Saxony's Leipzig-Dresden Rwy. opened. This 75½-mile line was completed by 1839, and in its first year of operation served 412,000 passengers. Its first locomotive was the English-built *Comet*. Prussia's first railway was the 16½-mile Berlin-Potsdam route, which opened on October 29, 1838. In December of that year a line opened between Brunswick and Wolfenbuttel.

In 1839 first sections of a number of new railways were opened. These included the Magdeburg-Kalbe portion of the Magdeburg-Leipzig line, the Frankfurt Main–Hattersheim segment of the Taunusbahn route to Wiesbaden, and the Munich-Maisach sector of the Maximiliensbahn. The first two of these routes were completed in 1840. By this time construction was also taking place on Baden's Mannheim-Heidelberg route and the Berlin–Anhalter Bahn to Wittenberg. The first portions of these lines opened that same year. In 1842 the Rhine River and Belgian border were linked by completion of the Cologne-Aachen Rwy. When met by a Belgian line the following year, a route was opened from Cologne to the North Sea port of Antwerp.

In 1838 Prussia passed a railway law of sorts. This legislation offered no assistance to the private sector for constructing railways, nor did it indicate any immediate interest of the state in building them. However, it did not prevent construction, either. Thus what resulted briefly was a policy similar to that in England, where official approval had to be given before a line could be built. Soon concessions were granted for several new routes. These included the Berlin-Köthen, Berlin-Stettin, and Dusseldorf-Elberfeld lines. As Prussia's military began to perceive the strategic value of railways, the official position became more favorable to their development. Soon private railways began receiving limited state assistance. Circa 1843, official planning began for creating a larger rail system.

While Germany's first locomotives were imported from England, some were later ordered from the United States. From 1839 to 1841 the Norris firm of Philadelphia shipped 21 engines to Germany. These were probably all of the 4-2-0 type. In a trial on the Berlin-Potsdam route, one reached a speed of 27 m.p.h. while pulling a train of 16 cars carrying 353 soldiers with baggage and ammunition. Later, between 1844 and 1848 Norris exported 15 4-4-0 engines to Germany.

During this period domestic manufacture of locomotives and rolling stock began. The first successful German engine was the 0-4-2 *Saxonia*. It was built by Johann Schubert in 1839 at the Ubigau workshops of the Leipzig-Dresden line. In 1837 Johann Friedrich August Borsig began construction of his works in Berlin. However it was not until 1841 that he completed his first engines. Based on the Norris design, these were constructed for the Berlin-Anhalt Rwy. That same year in Munich, Josef Anton von Maffei completed his first locomotive, *Der Münchner*.

Borrowing Stephenson's *Patentee* design, Emil Kessler and Theodor Martiensen completed their first engine, the *Badenia*, at Karlsruhe in 1841. Unfortunately, Martiensen died in 1842 and the business failed later in the decade. Relocating to Esslingen, Kessler established Maschinenfabrik Esslingen to build locomotives for the Württemberg State Rwys. One early engine, the *Alb*, was intended for working in mountainous terrain. It was designed by Joseph Trick, who also became known for his "Trick port," an auxiliary conduit for the passage of steam into the cylinder.

In 1846, Georg Egestorff built the first locomotive at his Hannover factory. In 1870 this became the Hannoversche Maschinenfabrik. In 1847 Richard Hartmann began building locomotives in Chemnitz, Saxony, completing his first, the *Glück Auf*, on January 5, 1848. In 1870 this company became the Sachsische Maschinenfabrik. In 1848 Georg Henschel started locomotive construction in Kassel, completing the 4-4-0 *Drache*, in July 1849. In 1851, the famous Krupp firm, of Essen, pioneered the use of steel for manufacturing locomotive wheels. Shortly, steel would be utilized for making other engine parts.

Other companies to enter the locomotive business include Wohlert, Berlin, 1850; the Vulcan Works, Stettin (now Poland), 1859; Ferdinand Schichau, Elbing, Prussia, 1860; Darmstadt Engine Works and Iron Factory 1861; Krauss & Co., Munich, 1866; L. Schwartzkopf, Berlin, 1867 (it became Berliner Maschinenbau AG in 1870); Hagans, Erfurt, 1872; AG für Lokomotivbau Hohenzollern, Dusseldorf, 1875; Orenstein & Koppel, Berlin, 1876; and Markische Lokomotivfabrik, Schlachtensee, 1892. It should be noted that Orenstein & Koppel manufactured narrow-gauge rolling stock and equipment much like France's Decauville Co.

A thriving industry also developed for the manufacture of passenger carriages and goods wagons. For many years these did not differ in any significant way from English designs. Germany also adopted many English operational practices. Overnight passenger service was first inaugurated on the 178-mile Berlin-Hamburg Rwy. on April 1, 1847.

German railway engineers also borrowed from American methods. Thus grades were often steeper and curves tighter than those of contemporary English lines. Tunnel construction was avoided whenever possible. In fact, the first railway tunnel was only completed in 1840. It was built near Oberau, on the Leipzig-Dresden line, and was 1,683 feet long. As in America, German railways were built through large cities without any provision being made for the street traffic, other than level crossings.

While private companies were allowed to build railways in various areas of Germany, some states opted to undertake these projects themselves. Hannover began construction in 1842, Baden and Bavaria in 1846, and Württemberg in 1850. In 1846 the Baden State Rwy. completed an international route between Heidelberg and Basel. That same year, 10 of Prussia's 17 rail companies formed the Union of Prussian Rwys. This was done both to reach an accord on tariffs and to improve the synchronization of operations. In 1847 this union was expanded to become the Union of German Railway Administrations. This association helped greatly to integrate the workings of various railways and also to lessen the competition that existed among the state lines.

After the revolution of 1848 for strategic reasons Prussia constructed its first state railway, the Ostbahn, from Berlin to Königsberg, near the Russian border. This initiated a period of increasing government involvement in rail construction. Built soon afterward was the Lübeck-Buchen Rwy. It provided a more direct route for German-Scandinavian trade, which previously had gone through the port of Hamburg.

Although Germany had only 363 route miles of track in

1842, this figure climbed to 3,100 in 1848 and to 3,725 by 1850. While this system was still a somewhat fragmentary network, it already had stimulated industrial development and had begun to affect the country's social cohesiveness as well. As time went on there were important population shifts in the country, as more rural people sought employment in the industrial cities.

The main period of German rail expansion took place between 1851 and 1870, a time when the Prussian military became increasingly involved in railway planning. This period culminated in the Franco-Prussian War of 1871, after which the German Empire was consolidated. Germany also expanded its territory at the time, annexing the French territory of Alsace-Lorraine. At the end of 1871 there were 13,113 route miles in the German railway system. This conflict as well as the later expansion of strategic railways is covered in more detail in chapter 20.

It was only natural that Berlin, Germany's capital and greatest city, should become its largest rail center. In the north of the city, the Nord Bahnhof became the terminus for the Stralsund Rwy. Later, after the Stettner Bahnhof was built closer to Berlin's center, it handled the Stralsund traffic, as well as that of the Stettner line. The Custrin and Breslau railways came in from the east and met before terminating at the Ostbahnhof. Later, during the 1870s, Berlin's Stadtbahn was built, running from east to west through the city. While serving mass transit needs, it was also designed to handle long-distance passengers. Thus the Custrin and Breslau traffic was fed directly into the eastern end of the Stadtbahn, where a number of stations were available for entry and exit, and service to the Ostbahnhof was discontinued. Both Berlin's Stadtbahn and the subsequently constructed Ringbahn are discussed in chapter 28.

The Görlitz Rwy. came in from the east-southeast and terminated at the Görlitzer Bahnhof. Entering from the south were lines of the Anhalt and Dresden railways, running closely parallel to each other. The Dresden line, probably built first, originally terminated at the Dresdener Bahnhof. Later, this traffic was serviced at the Anhalter Bahnhof. A little to their west was the Potsdam Rwy., which terminated at the Potsdamer Bahnhof. Another line entered Berlin from the west-southwest. It carried traffic of both the Nordhausen-Frankfurt and the Potsdam-Magdeburg lines, and fed directly into the western end of the Stadtbahn. Also connecting at this point was a line from Hamburg.

Coming into Berlin from the west-northwest were two other lines running closely parallel, before joining and shortly terminating, at the Lehrter Bahnhof. These were the Hamburg and Lehrte railways. The Hamburg line, probably the earlier of the two, originally terminated at the Hamburger Bahnhof. Its traffic was later serviced at the Lehrter Bahnhof, which besides being a little closer to the city's center, had the added advantage of connecting to the Stadtbahn.

Circa 1875 Germany entered into negotiations with Austria, Hungary, France, Italy, Switzerland, Belgium, and Holland for the regulation of international railway transportation. The primary focus of these talks was the establishment of uniform principles regarding the rights of shippers and carriers.

In 1875, German companies constructed 1,517 miles of new railway, the most ever in a single year to that point. About the same time, Chancellor Bismarck attempted to nationalize Germany's railway system. However, his efforts to do so were frustrated by the actions of Bavaria, Saxony, and Württemberg. As a result the government had to content itself with expanding its power in Prussia. Between 1879 and 1885 the Prussian State system grew in size from 3,921 to 13,106 route miles. During the same period the aggregate of private lines was reduced from 8,463 to 1,548 miles. By this time the Prussian State Rwy. was well known for the strict style of military discipline it imposed on its employees. This practice was soon adopted by a number of the other railways.

The German government also made efforts to standardize signaling practices and other railway procedures throughout the entire nation. This official desire, if not obsession, for uniformity had existed for a long time, and it was also directed toward the manufacture of locomotives and other rail equipment. In its March 2, 1872 issue, *Railroad Gazette* reported about a speech on this subject, given by Herr Kaselowsky before the Berlin Association for the Improvement of Railways. "Without going so far as to maintain that the engines might be of uniform construction for all railways, yet he was perfectly convinced of the possibility of making for the passenger trains as well as the goods traffic an engine suited to the requirements of all the German railways, with a few exceptions in which there were abnormally sharp curves or steep gradients, and these might all be included in three or four different models. The only alteration, perhaps, that might be found necessary was in the construction of the grates, on account of the different qualities and properties of the various sorts of coal used as fuel."

While Kaselowsky made some good points in favor of his proposal, mainly regarding the economical savings, another individual pointed out "That it was just in consequence of the continual alterations and improvements in the construction of rails, engines, carriages and wagons that they had all been brought to their present degree of perfection . . . that it would be going too far, and have the effect of shutting out further progress were they to introduce the principle of normal uniformity." Also supporting this view was the locomotive manufacturer, August Borsig, a member of the association.

Narrow-gauge mania was received with varying degrees of enthusiasm in Germany. Early on, Saxony built a number of 2′5½″ gauge lines. Other lines were built in Hesse and some of the other smaller states. However, Bavaria had no desire to build them, and Prussian interest was only minimal. A few lines were built in a 4′7″ gauge. These included the 94-mile Offenburg & Singen line running through the Black Forest, which opened in 1866, and the 47-mile Weizen & Immendingen Rwy., completed in 1891, which ran through Buchberg Pass. At the end of 1890 German narrow-gauge lines totaled 542 route miles.

One interesting narrow-gauge innovation, introduced circa 1887, was Dolberg's portable light railway. Probably inspired by France's Decauville's system, it was unique in the ease with which track could be assembled and disassembled. Each section of track consisted of two light steel rails fastened at one end to a sleeper. About one-third of the sleeper was left free to act as a support for the next section of rails, which hooked into the previous set in a manner not unlike that used to assemble toy track. The rolling stock was of the double bogie type. Intended for use on farms and estates, it was demonstrated at agriculture shows, and gained some popularity in Germany and Austria.

In 1880 Germany's entire railway system comprised 20,891 route miles. At the end of 1881 the Bavarian State system consisted of 73 lines having a combined total of 2,668 route miles. Of this sum only 178 miles were double-tracked in the English fashion. Operating over this network were 1,013 locomotives, 2,469 passenger carriages, and 15,795 goods wagons. Total route miles for Germany and Luxembourg combined were 25,608 in 1890 and 27,100 in 1893.

In the early 1890s Chancellor Bismarck established the Imperial Railway Office in Berlin, locating it near the headquarters of the Prussian State Rwys. In 1896 the Prussian system purchased Germany's largest remaining private line, the Hessian Ludwigsbahn, and shortly incorporated other Hessian railways. At the turn of the century the German railway system was the world's second largest.

Important names in German railway history include Paul von Denis, August von Borries, and Dr. Wilhelm Schmidt. Born in Bavaria in 1804, von Denis attended France's l'Ecole Polytechnique. During his subsequent career he directed construction of the Nuremberg-Fürth and Munich-Augsburg lines, as well as a railway from Frankfurt to Mainz and Wiesbaden. Later he assumed responsibility for operations of the Bavarian State Rwys. Von Borries (1852–1906) served the Prussian State Rwys. from 1875 to 1902. In 1880, while chief mechanical engineer of the Hannover division, he built his first two-cylinder compound engine. Later, in 1884 he invented the intercepting valve, a device for improving starts. The von Borries compound system became widely used by many railways, and it inspired later work by Austria's Dr. Karl Gölsdorf. Near the end of the century Dr. Wilhelm Schmidt (1858–1924) perfected the old idea of the superheater, finally making it practical. This was an important breakthrough, as it further economized fuel use.

The first railway in the Austro-Hungarian Empire was a 3′7½″ gauge tramway, built by Franz Anton von Gerstner, which connected Linz (Austria) with Budweiss (now České Budějovice in the Czech Republic). It was constructed to provide a link between the Danube and a river draining into the North Sea. The first section of this route, between České Budějovice and Trojanov, extended for 33 miles. At the time of its inauguration on September 7, 1827, it was the longest in the world. Operated as a public toll road, this line was primarily used for the transport of salt, but other goods were also carried. Passenger service began over this route in 1832, the same year that it was completed to Linz. However, the introduction of steam engines only came in 1854, and horses continued to be used as late as 1872. Another early tramway was the Linz-Gmunden route, which opened on May 1, 1836. Chief engineer of this line was Franz Zola, father of Emile Zola, who later became famous as a novelist.

Much like Germany's Friedrich List, the Vienna professor Franz Reipl became an early advocate for steam railways in his country. He envisioned a network radiating outward from Vienna to the far reaches of the Austro-Hungarian Empire. Such a system, he thought, would both hold the empire together and contribute greatly to its general prosperity. Emperor Franz I rejected his ideas, but Emperor Ferdinand I was more receptive to these arguments. He was persuaded by Prince Metternich to build a 280-mile line from Vienna to Krakow, Poland. Saloman de Rothschild, the financier, was brought into this project, and later on he became a dominant figure in Austrian rail expansion.

This railway became known as the Kaiser Ferdinands Nordbahn. The initial section of this route, from Florisdorf to Wagram, was inaugurated on November 17, 1837. Robert Stephenson built the first locomotives for this line: the *Vindobona, Austria,* and *Moravia.* Although the Imperial government took a more active part in financing its own railways in the next decade, construction proceeded very slowly. By 1848 the empire's rail system only totaled 669 route miles. This included small portions of what would later become the Vienna-Trieste, Vienna-Prague, and Vienna-Budapest lines.

A trans-Alpine route from Vienna to Trieste, through Semmering Pass, was first promoted in 1838 by Archduke Johann, the general engineer director of the Imperial Army. In 1842, Karl von Ghega, an engineer who had helped construct the Nordbahn, was sent to America to observe the progress of the Baltimore and Ohio R.R., as it built through the Allegheny Mountains. Upon his return, surveys were commenced for what would become the Sudbahn, and a route was selected. It was decided to follow the B & O precedent and build at a continuous gradient of 1 in 45½ through the pass. However, construction was first started south of the pass, from Murzzuschlag toward Graz and Trieste. Much of this route went through present-day Slovenia.

Later, amid the political turmoil of 1848, the government decided to begin construction over the pass itself. This move was bold, considering the fact that there was some doubt that any existing adhesion locomotive could pull a train up the long grade. To resolve this problem, the Sudbahn held a competition in the manner of the Rainhill trials. It invited locomotive builders to submit heavy, powerful engines that would be tested for their ability to work the pass. What resulted was an event known as the Semmering trials, which began in July 1851. Although a winner was selected in this competition, none of the entries were really adequate for the job. Only in late 1853 did the Sudbahn find a suitable engine, one built by the Austrian Wilhelm Engerth. The route through the pass was completed in October 1853, and the entire line was inaugurated on July 17, 1854. By 1860 the Sudbahn had been extended to Udine, where it connected with the Lombardy-Venetian Rwy. The story of the Semmering trials receives greater attention in chapter 5.

In 1838 William Norris shipped the 4-2-0 *Philadelphia* to the Vienna-Raab Rwy. for operation over its Vienna-Gloggnitz route. A short time later the Vienna-Raab Co. and the Nordbahn received three similar engines from Norris. These locomotives quickly gained popularity in Austria because they had a front bogie, allowing better negotiation of curves. Thus they were better suited than English engines for operation in hilly terrain. Furthermore, they were simpler in design than the best English locomotives, and thus could be copied with less effort and expense.

By the end of May 1844, four other locomotives had been ordered from Norris. Later that year William and Octavius Norris moved from Philadelphia to set up an engine factory in the north part of Vienna. The works began production in 1846. However, by this time, other locomotive builders had established themselves, and were both copying the Norris design and making improvements on it. Thus the Norris's own venture soon failed.

John Haswell, a Scotch engineer, built the *Wien,* the first Austrian-made locomotive in 1841. This 4-2-0 engine, based on Norris's *Philadelphia,* was constructed at the Vienna-Raab shops, located near Vienna's Sudbahnhof. This shortly became the first Austrian locomotive factory. In 1842, Wenzel Guenther, a former employee of this shop, began building locomotives at Wiener Neustadt. His first locomotives were six 4-2-0s of the Norris style, however, these were larger than the American prototype. In 1844 he constructed the 2-4-0 *Kolos,* a very large engine for its time. By 1845 he had built 11 more 4-2-0 engines. George Sigl started building locomotives in Vienna in 1842. He also used the Norris engine as his model. Later he took over the unsuccessful Norris works, and in 1847 moved to Wiener Neustadt, where he assumed control of the Guenther operation. Sometime later Wiener Lokomotivfabrik AG was established at Florisdorf. In 1880 Germany's Krauss and Co. opened a locomotive works at Linz.

In the aftermath of the revolution of 1848, Kaiser Ferdinand abdicated in favor of his son Franz Josef. The events of that period threatened the cohesiveness of the empire, and seemed to point to an ever-greater need for having an extensive rail network. However, the imperial government was in a weakened position both politically and financially. Following completion of the Sudbahn, the imperial treasury was almost empty. Thus the private sector was invited to undertake the burden of building railways. The French Rothschild and Pereire banking families assumed the primary role in this endeavor. The Pereires' Credit Mobilier bought many of the state-owned railways, and established the privately owned Austrian State Railway Company, based in Paris.

In 1851 a railway was completed between Prague and Dresden. As the private sector became involved, railway construction intensified. Between 1856 and 1860 Rothschild constructed the Kaiserin Elisabeth Bahn, a line connecting Vienna and Salzburg. In 1858 von Ghega completed the Inn Valley Rwy., extending from Kufstein, where it connected with a Bavarian line, to Innsbruck. This became a separate part of the Sudbahn system. Later, in 1867 a southern extension from Innsbruck was opened through Brenner Pass to Ala, at the Italian frontier near Verona. This provided the empire with a second rail connection to Italy, and also opened a route for German-Italian traffic.

During this period many short lines were built as well. Among these was the Jassenova, Oravicza & Steyerdorf Rwy., a 10-mile Hungarian line having 1 in 50 grades and curves of 114-yard radius. Wilhelm Engerth and Pius Fink designed the famous locomotive *Steyerdorf* as a prototype for this route. It was exhibited in London in 1862.

In 1870 Rothschild completed the Kaiser Franz Josef Bahn. This route extended from Vienna to Prague and Pilsen. In 1872 the Nord-West Rwy. was inaugurated. It ran from Vienna into Moravia and provided a more direct route to Dresden and Berlin. Also opening that year was Rothchild's Kronprinz Rudolf Bahn, which extended from St. Valentin to Selzthal. Unfortunately, its namesake, the crown prince, committed suicide in 1889.

During the 1870s the Kaiserin Elisabeth Bahn constructed an extension from Salzburg to Worgl, where it connected with the Sudbahn's Inn Valley line. This provided an all-Austrian route from Vienna to Innsbruck. Previously part of this journey involved travel through Bavaria.

While the Austro-Hungarian rail system only totaled 1,483 route miles in 1858, this figure had reached 7,437 by 1872. Three years later, at the end of 1874, Austria alone had 5,963 miles of track. Operating over this system were 2,237 locomotives, 4,841 passenger carriages, 51,811 goods wagons, and 227 snowplows. By the end of 1885 the Austro-Hungarian system had grown to 13,957 miles. At the time the network in Hungary totaled 5,610 route miles. Working on this latter system were 1,503 locomotives, 2,587 passenger coaches, 31,697 goods wagons, and 138 snowplows.

The 84-mile Vorarlberg Rwy., between Innsbruck and Bludenz, was completed circa 1885 and opened a direct rail connection between Vienna and eastern Switzerland. Work on this project, begun circa 1880, included excavation and construction of the 6.37-mile Arlberg Tunnel. This tunnel, the world's third longest at the time of its completion, was separately inaugurated on November 19, 1883. Overseeing this project was Julius Lott, who had also directed construction of the route through Brenner Pass. Another line finished about this time was the Ivan Pass Rwy. in Bosnia-Herzogovina. In actuality, it was a separate 36-mile portion of the 113-mile Sarajevo & Konjica Rwy. It had a 2.43′ gauge, and included a rack section using the Abt system.

In 1867 the Kingdom of Hungary was reestablished, and the MAV, or Royal Hungarian State Rwys., soon gained control of privately owned lines. The Austro-Hungarian government formed an increasingly close relationship with Germany following the Franco-Prussian War. Thus it became increasingly influenced by that country's rail policies. During the 1870s, the government took a more active role in constructing new railways, and in the 1880s absorbed most private lines into a state-controlled system. Fearing Russian hegemony, both empires became active in building strategic railways to bolster their border defense systems. More information on this buildup can be found in chapter 20.

Some military railways were constructed in narrow gauge, and in 1890 Austria alone reported 60 route miles of such track. The empire's total route mileage in all gauges that year was 15,877. By 1893 this figure had increased to 17,609.

The only important rail competition that developed in Austria was for the Vienna-Berlin traffic. In the latter part of the century three rail lines contended for this business: the Nordbahn, the Imperial Royal Austrian State Rwy., and the Austro-Hungarian State Rwy. The Nordbahn had the central position, while the Imperial line ran to its west and the Austro-Hungarian route to its east. All of these lines terminated at Bidenbach, where they connected with the Saxon State system.

Important Austrian railway figures include John Haswell and Dr. Karl Gölsdorf. Haswell (1812–97) was born near Glasgow and gained early training working for the locomotive manufacturer William Fairbairn in Manchester. He was recruited to build engines for the Vienna-Gloggnitzer line and decided to remain in the country. Aside from building locomotives on the English and Norris patterns, he made many important innovations of his own. In 1851 he built Europe's first 0-8-0 type engine, submitting it as a contestant in the Semmering trials. A subsequent 0-8-0 goods engine, introduced in 1855, greatly influenced later European design. In 1861 Haswell built the *Duplex,* an experimental 4-2-0 engine having four cylinders.

Dr. Gölsdorf (1861–1916) originated very distinctive locomotive designs. In the 1890s these usually featured twin steam domes connected by a large pipe, and the von Borries two-cylinder compound system. In the mid-1890s he developed a starting device for compound engines, and after the turn of the century made other important design contributions.

TOP LEFT: Stephenson's 2-2-2 Patentee class *Der Adler*, Nuremburg-Fürth Rwy., 1835. TOP RIGHT: First German-built engine, 0-4-2 *Saxonia*, Leipzig-Dresden Rwy., 1839. BOTTOM: View of Bollerswand Viaduct and tunnel at Semmering Pass on the Vienna-Trieste Sudbahn, 1860.

PLATE 113

PLATE 114

Restaurant scene in a German train station, 1876.

Manufacture of locomotive wheels at Borsig's ironworks in Moabit, near Berlin. From a painting by Paul Meyerheim displayed at the Borsig Gartenpavillon in Moabit, 1878.

PLATE 116

Railway scene at Sigmundskron on the Bozen-Merance Rwy., 1883. Then a part of the Austro-Hungarian province of Tyrol, this picturesque area was absorbed into Italy following World War I. The terminus points of this line are now known as Bolzano and Merano.

PLATE 117

Arrival of a train carrying German Kaiser Wilhelm to meet with Austrian Emperor Franz Josef, Innsbruck, 1889. Located in western Austria, Innsbruck became an important rail center, handling east-west traffic between Switzerland and Austria, and north-south traffic between Italy and Germany.

Passengers boarding a train in the Christmas holiday season, circa 1892.

PLATE 118

PLATE 119

Horses frightened by a passing train, 1893.

Train guard checking passenger tickets on a rainy day, 1893.

PLATE
120

PLATE 121

Station scene at a Bavarian resort, circa 1896.

CHAPTER

(8)

Railways of the Russian Empire

In the early nineteenth century, transportation in Russia was mainly limited to movement over a vast network of perhaps 50,000 miles of waterways. These waterways, however, suffered the significant disadvantage of being frozen for almost half of each year. Roads were virtually nonexistent in most parts of the country. In 1817 Czar Alexander I began construction of an overland road between St. Petersburg and Moscow. Upon its completion in 1834, travel time between these cities was reduced to a mere 10 days. Transportation was only slightly improved in the Ural region. There, beginning at the Zmeinogorsk Mine in 1763, a few tramways had been laid for moving ore or serving other industrial purposes. Yet, like similar lines elsewhere in Europe, these were very small in scale and operated with animal traction. On one of these tramways in 1833, the engineer Tcherepanov built and tested Russia's first locomotive. Yet this development was of little importance, for at the time, the Russian government was somewhat resistant to the idea of railways.

Inertia and xenophobia aside, some were more perceptive regarding the strategic and commercial possibilities that railways apparently offered. In 1835, an Austrian engineer, Franz Anton von Gerstner, builder of the Linz–České Budějovice line, made a proposal to Czar Nicholas I for construction of an extensive railway system. The czar summoned von Gerstner for consultation in 1836. Soon construction began on a line of 6′ gauge between St. Petersburg and the imperial summer residence at Tsarskoe Selo. The formal inauguration of the line commenced on October 30, 1837, with a 2-2-2 Hackworth engine drawing a ceremonial train to the czar's palace. When the line was extended to Pavlovsk the following summer, it reached a total of 30 miles.

Initially, animal traction also was used over the route, being preferred when passenger loads were light. In addition, locomotive-drawn trains only stopped at the main stations. Horse-drawn cars brought passengers to and from smaller stops. Locomotives, under imperial command, were required to carry 11 horns and a trombone to warn the unwary. The early success of this line encouraged the czar, in 1839, to authorize construction of the 4′8½″ gauge Warsaw-Granitza Rwy. This line was to connect with an Austrian railway that would extend to Vienna. Although construction money ran out in 1842, this route was finally completed in June 1845.

In 1839 the czar sent a mission to both Europe and the United States to study different railway systems. This practice had been established a century and a half earlier by Czar Peter the Great to bring new technological knowledge into the country. Typically, on such trips, skilled foreign personnel were recruited to help implement the desired changes. In this instance the czar's officers returned with Major George Washington Whistler, builder of the Boston & Albany line over the Berkshires and father of the later famous artist. He had been enlisted to oversee construction of a line between St. Petersburg and Moscow, which would be called the Great Nicholas Rwy. Other Americans recruited to assist in this project included the Philadelphia engine builders Eastwick and Harrison, and Thomas and William Winans, sons of the Baltimore builder Ross Winans. By 1845, these individuals had set up a factory for the manufacture of locomotives at Alexandrovsk, near St. Petersburg. Soon they were manufacturing 4-4-0 locomotives for passenger service and the world's first 2-6-0 Mogul type engines for freight.

Construction began on the line in 1842 and it was inaugurated on May 7, 1847. A 5′ gauge was chosen and this was to become the standard for the country. The building of this 400-mile line was difficult. When the weather was favorable as many as 50,000 serfs might be employed at any given time;

however, little work could be done in the severe winter months. Five thousand workers may have died in this effort, primarily from recurring typhoid epidemics. Thomas Winans returned to Baltimore a millionaire in 1851, constructing for his Russian-French wife the beautiful mansion Alexandroffsky. When the American direction of the Alexandrovsk facility ceased in 1867, it had delivered 192 locomotives, 253 coaches, and 2,608 goods wagons for the line. It also had built 8 engines for the Volga-Don Rwy.

In the early 1850s, the manufacture of locomotives began at Kolomna, southeast of Moscow. When the Crimean War began in 1854, the total distance of Russian railways was only about 650 route miles. As a result, the movement of troops and supplies southward was largely limited to the traditional water routes. It was felt afterward that the resulting transportation delays contributed substantially to Russia's defeat in this conflict.

Czar Alexander II, who came to power in 1855, proved to be a greater patron of railways than his father. In the late 1850s construction began on five different lines in the area between the Black Sea and the Baltic. One primary purpose of these routes was to facilitate the shipment of grain. In the winter of 1861–62 unique experiments were made in Russia using two engines, designed by Nathaniel Grew, which drew trains of cars, mounted on skids, over ice between St. Petersburg and Kronstadt. These engines, with the front carrying wheels replaced by skids, were powered by a single set of studded drivers. In 1862 a line was finally completed between St. Petersburg and Warsaw. That same year a Warsaw-Bromberg Rwy. was inaugurated to connect Russia and Prussia. Also, in Finland, a line was opened between Helsinki and Hameenlinna. In 1864 the Kiev-Odessa Rwy. began operation with luxury-class rolling stock imported from Germany. By 1866 the country had 2,065 route miles of track.

By the end of 1870 this figure was approaching 7,000 route miles. At the time much of the existing network centered around Moscow. To the city's north-northeast ran a line passing through Yaroslavl to Vologda. It eventually would reach the Arctic port of Archangel. Another railway extended eastward to Nijni Novgorod (Gorky). This was intended to continue on to Kazan and thence to Tobolsk in Siberia. Extending southeast from Moscow was a line to Ryazan. This railway was intended to terminate at Orenburg. From there it was anticipated that connections would be established to the Caspian region, Siberia, and elsewhere. Another line ran from Moscow to the fertile provinces of the south. Extending from this route was a southeastern branch to the Sea of Azov and a southwestern branch to Kiev and Odessa. Running west-southwest from Moscow was a railway to Smolensk and Warsaw. A line to Riga branched north from this route. Completing this picture was the St. Petersburg line, which ran northwest from Moscow.

By this time a route had opened between St. Petersburg and Helsinki servicing the northern shore of the Gulf of Finland. Serving the Gulf's southern shore was a line extending from St. Petersburg to Port Baltic. At the end of 1871 Russia had 8,719 route miles of track. In 1874 1,168 more miles were added. This included the 323-mile Morschausk-Sysran route and the 122-mile Homel-Bachmatsch section of the Landovarovo-Romny Rwy. By the end of that year the country's network totaled 11,600 route miles.

At the time Russia was producing limited quantities of small wood-burning locomotives and importing other wood-burners from Germany, France, and Belgium. The Odessa Rwy., a company that did burn coal, imported it first from England and then later from Silesia. *Railroad Gazette*'s September 25, 1875 issue, quoting a consular report, mentions that these policies changed when "an agent of the Philadelphia firm (Baldwin) had brought the attention of the consuming departments to the fact that anthracite coal of best quality can be found around the town of Worenesk (Voronezh), in southern Russia, extending over 180 miles, and that whole regions were abounding with it in the southern half of the empire, sufficient to feed the locomotives of the whole continent."

This subsequently led to a development of coal mining in the Voronezh district, and the importation of some coal-burning locomotives from America. In 1872–73 Baldwin shipped 30 new engines to the country, and in 1874 an order for 50 locomotives was placed with the Grant Co. The Baldwin engines were guaranteed to haul 1,075 tons on a level. This was a significant increase in power over the existing Russian locomotives, which typically hauled 650 to 700 tons.

The coalfields of southern Russia turned out to be immense, occupying an area known as the Donets Basin, which lay partly in the Rostov Province of Russia and partly in the Luhansk and Donetsk provinces of the Ukraine. The completion of the Ekaterininsky Rwy. in 1884 connected the Donets coal district with the iron mines of Krivoi Rog, at the bend of the Dneiper River. Soon industrialization began to take place in or near cities like Yekaterinoslav, Rostov, and Kharkov. This ultimately led to a reduction in the use of charcoal for smelting and a decline for the ironworks of the Ural region.

In 1867, Russia began building a railway into the area between the Black and Caspian seas. This line started at Rostov and by 1875 extended to Vladikavkaz. A second railway, from the Black Sea port of Poti to the Georgian capital of Tiflis, was completed in 1872. Following a war with Turkey in 1877, Russia gained complete control of the territory between the two seas.

About the same time an oil boom had begun in the area around the Caspian port of Baku. Thus considerable support existed for building a railway through the Caucasus, connecting both seas and making Baku's oil available to Europe. Such a line also offered the additional possibility of opening up trade with the central Asian region on the Caspian's opposite shore. Rail construction began for connecting Tiflis and Baku, and was completed in 1882. The previously built route between Poti and Tiflis was improved, especially in the Suram Pass area, and a new branch was extended from Samtredi to Batum. Batum, a port on the Black Sea south of Poti, had previously been controlled by the Turks, and subsequently was chosen over Poti to handle the oil traffic. English-built locomotives were operated over this line, with 0-6+6-0 Fairlie engines used in the more mountainous areas.

Because the Russian conquests in central Asia were essentially a contiguous territorial annexation, the story of the Trans-Caspian or Central Asian Rwy. perhaps belongs here rather than in the chapter on Asia. This line, initiated primarily for military and strategic purposes, ventured into regions that had formerly been a part of the Persian Empire. Thus it played a significant part in helping the Russian government to establish a more permanent presence in the area. Built under the direction of Gen. Annenkoff, it was originally conceived as a temporary military railway, but was constructed more substantially for fear of the erosive effects of nature in the harsh desert. Raw materials like fuel, wood, iron, and water were scarce and had to be transported in with some difficulty.

The first portion of the line originally started at Mikhailovsk. It was built with Decauville narrow-gauge equipment previously used in a Balkan military campaign. Later this railhead was moved to Uzun Ada, also on the Caspian. Workmen and other personnel were housed in a long double-decker work train. Early on the line provided logistical support for a military assault on the Turkoman stronghold of Geok Tebe, but otherwise rendered no assistance in Russia's ongoing war efforts.

Once the 144-mile Uzun Ada-Kizil Arvat route was completed in 1881, it was seen that extending it farther might give Russia a strategic advantage in its regional competition with the British. This competition was then known as the "Great Game." It was also soon observed that real commercial possibilities existed for this railway. The moment the line reached Kizil Arvat, it began to carry shipments of merchandise. Traditional caravan routes that had existed for centuries were quickly altered to accommodate this new development. For instance, camel journeys from Khiva which previously had been made to Orenburg in 43 days could now be made instead to Kizil Arvat in 15 to 16 days. Trade routes from Afghanistan and Persia were soon diverted as well.

After various alternatives were considered, railway construction resumed in 1885. By July 14, 1886, the line extended 355 miles farther to Merv, located 30 miles north of the Persian border. This portion was constructed on the standard Russian 5′ gauge, and locomotives from the Kolomna works were used. During the same period the western terminus at Uzun Ada was diverted to Krasnovodsk, a military facility with a deeper harbor. By 1888 the eastern end of the line was extended to Samarkand, and by 1898 to Tashkent. A Merv-Kushk branch line opened in 1889. In 1892 large-scale cotton farming was introduced into the trans-Caspian region. In 1899 a branch off the Samarkand-Tashkent line was extended to Andijan near the Chinese border.

In the early 1870s Ferdinand de Lesseps, builder of the Suez Canal, had promoted a rail line from Russia through Turkestan into India. However, in 1874 the Russian government dropped its support for this plan. Later de Lesseps advocated a line reaching all the way from Europe to China, a part of which included the Trans-Caspian Rwy. From Samarkand he envisioned a line extending into Chinese Turkestan, touching at Tang-Kissar, Kashgar, and Yarkand, and continuing on, ultimately terminating in either Wuchang or Kankow.

In Russian-controlled Finland, the Wasa-Uleaborg Rwy. and, later, the Lulea-Ofoten line became the northernmost railways in the world. The latter ran from the head of the Gulf of Bothnia across Lapland to the Atlantic. Comparably speaking, these lines were approximately 1,000 miles farther north than the Canadian Pacific.

In 1887 there were 62 Russian railway companies, 10 of which were government-owned. On October 29, 1888, a special Imperial train was derailed at Borki on the Kursk-Kiev line. Twenty-one passengers lost their lives and 37 were injured. Among those hurt were Czar Alexander III and the czarina. In a strange aftermath that put human lives at risk, Russian engineers restaged the accident in order to gain a better understanding of it. In this sequel six train attendants were injured and the greater part of a train destroyed. In 1889 the 245-mile Pskov-Riga line was completed. By the end of that year the Russian Empire had a total of 19,489 route miles of track. Of that figure European Russia had 17,693 miles, Finland 1,135, and the trans-Caspian region 661.

Train speeds in Russia were slow throughout the nineteenth century. The April 27, 1888 issue of *Railroad Gazette* provides this report:

> The Russian Ministry of Railways is now engaged in considering whether or not it is possible to increase the speed of mail trains in the Russian Empire. At present they run at an average speed of 20 m.p.h., including stoppages, and it is desired to increase this to an average of 23 m.p.h. It is said that on some lines the speed of the mail trains does not average more than 16 m.p.h. The ordinary passenger train does not exceed 12 m.p.h., while on the best line in Russia, between St. Petersburg and Moscow, the ordinary passenger train speed is 15 m.p.h. The fastest train in Russia is the courier train between St. Petersburg and Moscow, which makes including stops, 28 m.p.h. This train carries no mails and consists only of a few first class carriages. Examples can be found on one or two branch lines of a train (scheduled to complete its journey) at 10 m.p.h., and, even then regularly arriving late.

On January 1, 1896, there were 8,123 locomotives in Russia, of which 4,020 were built in the country. Of the whole number, about 45% were constructed before 1880. Coal was burned on 40%, oil on 32%, and wood on 28%. Late-nineteenth-century Russian locomotive builders include Putilov in St. Petersburg, Maltzev in Raditza, Bryansk Engineering Works, Kharkov Locomotive Works, and Lugansk in Voroshilovgrad. In 1897 the Ostrolenka-Pilawa Rwy. was completed. This line was important because it provided a connecting link between three lines: the Warsaw–St. Petersburg, the Warsaw-Terespol, and the Vistula. Its most notable feature was a bridge over the River Bug.

In 1888 the first small part of what was to become the 5,777-mile Trans-Siberian Rwy., between Moscow and Vladivostok, opened. This initial section connected Samara and Ufa. In 1891 Czar Alexander III issued a decree to extend the line from Ufa to Vladivostok, an immense effort that would not be completed until 1904. Shortly after the czar's decree a branch was built north from Vladivostok to Khabarovsk. Next the railway was extended from Ufa to Tcheliabinsk, crossing the Caucasus. Czarevitch Nicholas attended the inauguration of this section in 1892. By 1894 the line had reached Tomsk. In December 1899 the line inaugurated its trans-Siberian express over the completed portion of the route. The section between Chita and Vladivostok ran through Manchuria and became known as the Chinese Eastern Rwy.

The trans-Siberian line opened up vast territories in Siberia, where the government handed out free land to settlers. During the 1890s there was a massive migration of Russian peasants eastward, especially into the Altai region. Soon its lowlands were given over to farming, while cattle were raised in the highlands. For a while the railway operated "church carriages," which carried all accessories necessary for conducting Orthodox services. These would make two stops a day at remote settlements along the line.

This railway also served important strategic needs by providing support for a stronger Russian presence in the Far East. At the time it was gaining increasing influence over northern China. However Japan, in the war of 1905, successfully checked Russian expansion. At the end of the nineteenth century, Russia's rail system was extensive, being the fourth largest in the world, but coordination among the many different lines was often lacking.

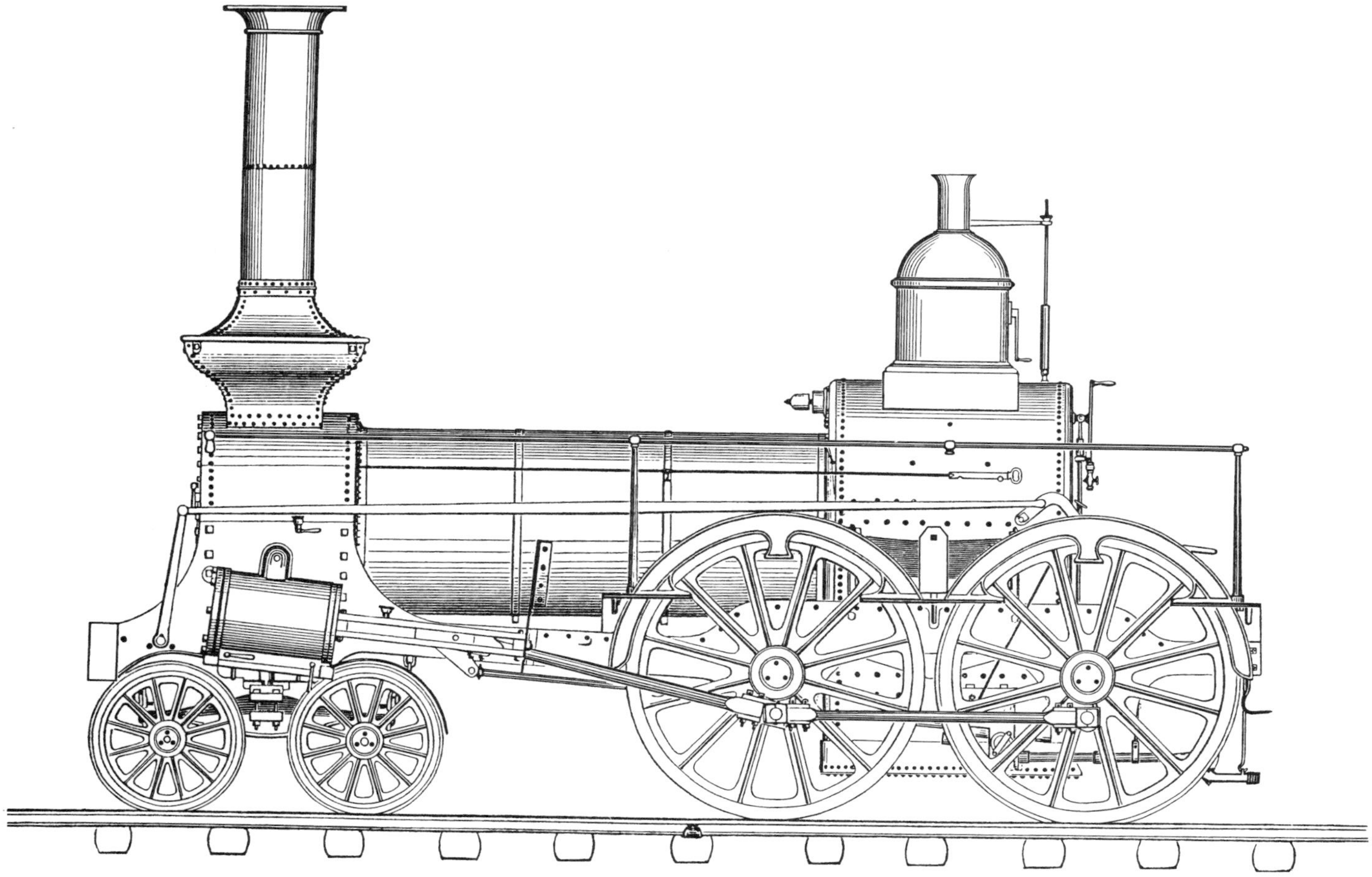

Locomotives for the St. Petersburg–Moscow Rwy. built at the Alexandrovsk Locomotive Works, St. Petersburg, by the Americans Eastwick, Harrison, and Winans, 1844: TOP: 4-4-0 passenger engine. BOTTOM: World's first locomotive of the 2-6-0 type, later called "Mogul." This engine was used for goods service.

PLATE 122

Top: Locomotive built for use on icy surfaces in Russia by Neilson & Co., Glasgow, 1861. **Bottom:** Battle between Russian troops and Polish insurgents following derailment of train near Czyzewe, Poland, 1863.

PLATE 123

Transportation of oil in Russia's Baku region, circa 1886: TOP: Oil train. BOTTOM: Oil wagons at the station in Baku.

The Trans-Caspian Rwy.: Clearing away sand from the path of a train, 1888.

PLATE 126

The Trans-Caspian Rwy.: Top: View of line in Kopet-Dagh region. Bottom: Supply train carrying water, provisions, and supplies for railway stations and guard posts along the route, 1888.

The Trans-Caspian Rwy.: Camel is struck by a passing train, 1889.

PLATE

127

PLATE 128

A scene at the Baltic Railway Station in St. Petersburg: expelled Jews being loaded onto a train, 1891.

Construction scene on the Trans-Siberian Rwy., 1897.

PLATE 130

Russian locomotives: Top left: 0-8-0 goods engine constructed at Bryansk Engineering Works, Moscow-Kazan Rwy., 1899. Top right: 4-4-0 tandem-compound express engine, Putilov Co., St. Petersburg, for St. Petersburg–Varsovie line, 1898. Bottom: 0-8-0 goods engine designed by Thomas Urquhardt and built by Maschinenfabrik Esslingen, Grazi-Tsaritsin Rwy., 1891.

CHAPTER

9

Other European Railways

Italy's first railway was the Naples-Portici line, which opened on October 4, 1839. The *Bayard*, an English-built 2-2-2 Patentee class engine, pulled the inaugural train. The railway was built by the French engineer Armand Bayard de la Vingtrie, and was part of a concession, granted by the Kingdom of the Two Sicilies, for construction of a network of lines extending from Naples to Avelino, Salerno, Caserta, and Capua. The Naples-Caserte line was completed in 1843, and was extended to Capua the following year.

At the time, Italy was a fragmented country of independent states, not unlike Germany. The Kingdoms of Piedmont and Sardinia were independent. Venice and Lombardy were under Austrian control, and in between these and the Kingdom of the Two Sicilies was another group, the papal states, under Vatican authority. As a result, states largely operated independently of each other in building their early railways. In 1840, the 8-mile Milan-Monza line was inaugurated. In 1842 the first portion of the Milan-Venice (Lombardy-Venetian) Rwy. was opened. This line was first proposed in 1837; however, construction was postponed for a time, due to Austrian military activities in the region. By 1860, this railway had been extended northeast from Venice to Udine, where it met a branch of the Vienna-Trieste Sudbahn. In 1847 the Pisa-Livorno Rwy. was completed in the Grand Duchy of Tuscany. In 1848, the Turin-Moncalieri line was inaugurated in Piedmont.

While Italy's earliest engines were of English manufacture, some Norris-style engines were later ordered. In 1846 two 4-4-0 engines were shipped from Philadelphia to the Romana Rwy. Between 1845 and 1859, the Cockerill firm of Belgium supplied others to the Pistoia-Lucca, Turin-Cuneo, and Royal Sardinian State lines. Italian railways adopted the American practice of having cabs on their engines early on. The first locomotive builder in the country was the Ansaldo company of Sampierdarena. Later manufacturers included the Pietrarsa firm of Naples and the Cerimedo company of Milan. Engines were also later constructed at company shops. Among lines building locomotives were the Southern Italian Rwy. at Florence and the Upper Italian line at its Turin facility.

The first Sardinian railway, the Turin-Genoa line, was built under the direction of I. K. Brunel. It opened on December 6, 1853, and went through Alessandria and the steep Giovi Pass. The problem of having sufficient adhesion for ascending this 1 in 27 grade was solved when it was discovered that two tank locomotives could be coupled back to back. In effect this created a single 0-4+4-0T engine, requiring two drivers but only one stoker. Railway construction in Italy could be difficult, not only in the mountain regions, but also in the south and Sicily. In the latter areas engineers encountered large formations of shifting sand and clay, continual landslips, crevasses, and soil disintegrated by volcanic action and earthquakes.

Papal state railways included the Rome-Frascati line; the Rome–Civita Vecchia route, which opened in 1859; and the Rome-Velletri Rwy., inaugurated on January 27, 1862. A special passenger car was built for use by Pope Pius IX. He was an early enthusiast for a national railway system, feeling that such a network would bring many pilgrims to Rome.

At the time tremendous political changes were occurring in the country. In 1859, France sent troops into Italy and fought a war with Austria. Following this the Kingdom of Sardinia and Piedmont annexed Lombardy, while giving up Nice and Savoy to the French in return for their support. In 1861 a further development took place with the crowning of Victor Emanuel as king of Italy. His kingdom now included all of the Italian peninsula except for the papal states and Venetia. Venetia was incorporated in 1866. During this period French troops were sent to Rome to prop up the papal government. However, once the Franco-Prussian war began in 1870, these forces had to be withdrawn. At this point Garibaldi's troops seized Rome, which was then declared to be the new capital.

Italy's unification in 1861 greatly facilitated railway growth. The new government wanted to expand the existing railway system to help integrate the country. Thus it offered to refund construction costs of new lines, once these had opened. By 1866 most of Italy's major cities had some kind of rail service. The country's second rail connection with Austria was established in 1867, when an Austrian line was extended through the 4,775-foot Brenner Pass to Italy's border. This railway continued north to Innsbruck and provided an important link to Bavaria.

Circa 1870 narrow-gauge mania came to Italy. One short line constructed was the 7½-mile Turin-Rivoli Rwy., which opened in 1871. At the end of May in that year, Italy had 3,919 route miles of track. One year later this figure had grown to 4,149.

With the opening of the Mt. Cenis tunnel in 1871, there was an increased flow of traffic between Italy and France. Carried over this route was the British government's East Indian Mail, which went on to Bologna and the port of Brindisi. Between the tunnel and Turin all traffic was carried over the Victor Emmanuel Rwy.

The St. Gothard Tunnel, which opened in 1882, provided a route for traffic between Italy on one side and Germany and Switzerland on the other. The Novaro-Pino Rwy., beginning at the tunnel's Italian entrance, became part of a main line to Genoa and the west coast. This 40½-mile route had 18 tunnels, 262 small bridges, and 22 large bridges.

At the end of 1881 Italy had 5,523 route miles of track. Of this, 2,452 miles were state-owned. By this time most lines had been organized into four main systems: the Upper Italian, the Roman, the Calabrian, and the Southern. These were later consolidated into two systems, the Mediterranean and the Adriatic. After its amalgamation, the Mediterranean system had about 40 different types of locomotives. The November 30, 1888 issue of *Railroad Gazette* reported that the company had set a policy to gradually reduce this number to six, two each for passenger, freight, and switch engines. Of these, one category was for normal operation and the other for lines having steep grades. In addition, the company also adopted a new type of express engine and two new locomotives for combined freight and passenger service. Thus it planned to operate with a total of nine different engine types.

Among many Italian lines built from 1861 to 1890 were the following: (1) the Aquila, Rieti & Terni Rwy., a 64-mile route in the Apennines; (2) the 105½-mile Benevento, Campobasso & Termoli Rwy., which, branching from the Naples-Foggia line, traveled through higher elevations of the Apennines to the Adriatic coast; (3) the Leopoldina Rwy., which ran from Livorno and Pisa to Florence; (4) the 105-mile Rome-Sulmona line, which crossed two distinct chains of the Apennines; on this route was the 11,484-foot Cocullo Tunnel; (5) the Santa Caterina & Rocca-Palumba Rwy. of Sicily, which ran through the 3¾-mile Marianopoli Tunnel.

In 1891 Italy had 8,117 route miles of track. By the end of the century, the Italian rail system had become extensive. However, there were still a few areas, primarily in the south, where the network remained incomplete. This system was nationalized in 1905.

Until 1883, Greece had only one railway. This was the 6-mile standard-gauge line from Athens to Piraeus, its port. By 1890, however, an additional 372 miles of meter-gauge lines had been built, and 204 more miles were under construction. The longest of the completed lines ran 190 miles from Athens to the Peloponnesus. A short time later, a standard-gauge line was constructed into Serbia to provide a connection to the greater European rail system.

After gaining its independence from the Ottomans in 1878, the new kingdom of Serbia was not friendly to the economic hegemony of its powerful neighbor, the Austro-Hungarian Empire. Thus the latter looked for a rail route to Constantinople bypassing Serbia, while the Serbs turned to the French for help in building railways. However, this relationship soon soured as well. After the scandal-ridden collapse in 1883 of its primary lender, the Bank Union Generale, Serbia nationalized its small railway system. By the end of 1892 the country had 337 route miles of railway.

Romania was formed in 1859 from the union of Moldavia and Walachia. However, until it declared its independence in 1877 it was under Turkish suzerainty. Its railway development during that time is covered in chapter 11. Soon after gaining independence Romania made an agreement with the Germans for assistance in building more railways. Shortly, however, many in the country began to resent Germany's growing presence there. This led to unrest that received encouragement from the popular press. It did not help that the German railway promoter Bethel Henry Strousberg was apparently diverting railway money into various shady activities. An international incident finally resulted when Germans were attacked by a Romanian mob. Having the choice of supporting either Strousberg or the Romanians, Germany's Chancellor Bismarck uttered one of his more famous quotes: "Such degenerate people cannot be held in check through good dinners but through a few strong battalions." The result was that Romania was forced in 1880 to comply with the original agreement. However, Austrians were brought in to complete construction. At the end of 1892 the country had 1,590 route miles of track.

Railways came late to the Iberian Peninsula. Compared to countries like England and France, both Spain and Portugal were poor and thinly populated. Furthermore, mountainous terrain in both countries made railway building difficult and expensive. Spain's first line was the 17-mile Barcelona-Mataro route, which opened on October 28, 1848. Perhaps unwisely, the country selected a gauge of 5′6″, which became its standard. Later, after lines were extended to the border with France, the change in gauges required costly and time-consuming transshipment. The influx of French capital, beginning in the mid-1850s, helped to stimulate Spanish rail growth. The Madrid-Alicante line was opened in 1858, and the Madrid-Barcelona Rwy. in 1860. Early Spanish locomotives were built by such firms as Kitson of Leeds, Cail of Paris, and the Sharp Stewart Co.

During the Carlist insurrection in the early 1870s, Spanish trains were robbed on numerous occasions. In *Our Iron Roads,* F. S. Williams reported: "Spain, again, is a land where travelers are occasionally attacked by brigands; by whom, during the Carlist times, expresses were run off the line, railway men were murdered, and against whom the passengers and their escorts of soldiers had to barricade themselves in houses till assistance arrived."

In *Wonders and Curiosities of the Railway,* William Sloane Kennedy reported on Spanish rail travel circa 1880: "There are long and irritating delays, and at about every stopping place you are amused to see the leisurely fashion in which engineer, fireman, and conductors will roll up cigarettes and never start until the last puff has been drawn. But on every Spanish train there is a *wagon reservado para Senoras;* and at the stations the refreshing cry, *Quien quiere agua?* mingles with the voice of the vendors of cool, delicious grapes, oranges, and lemons."

Spain later became dominated by four large railway systems: the Madrid-Zaragoza-Alicante company; the Madrid, Caceres & Portugal network; the Andalusian Rwy.; and the Norte. The latter company had main lines between Madrid and Irun, on the French frontier, and from Barcelona to Zaragoza and Pamplona. In 1891 it was reported that 1,674 locomotives, 4,868 passenger carriages, and 32,921 goods wagons were operated over the entire Spanish system.

In 1893 the San Cebrian Rwy. was opened from Madrid to the San Cebrian mines. During the last quarter of the century, a number of meter-gauge lines were built in Spain's mountainous regions. By 1900, these railways had a combined 1,450 route miles of track. The total for standard-gauge lines was 7,100 miles.

Portugal's first line was the Lisbon-Carregado Rwy., which was inaugurated on October 28, 1856. The country adopted the same 5′6″ gauge used in Spain. By 1866, a railway had been built to the Spanish border, and one could ride by rail from Lisbon to many points in Europe. At the end of 1871, both countries had a combined total of 3,818 route miles of track. By the end of October 1887, Portugal's total alone was 1,125 route miles, of which 455 were state-controlled. Total route mileage increased to 1,440 by 1893. Portugal, like Spain, also constructed a number of meter-gauge lines.

The first railway crossing the border of Switzerland was the Strasbourg-Basle line, inaugurated on September 19, 1841. The first line built inside the country was the Zürich-Baden Rwy., which opened on August 9, 1847. The first engine to operate over this route was the *Limmat,* a Norris-style locomotive built by Kessler of Esslingen. During the 1850s Basle became a converging point for several other lines. Among these was the Swiss Central Rwy., which included among its engineering achievements the 2,800-foot Hauenstein Tunnel. A contemporary line was the St. Gall & Appenzell Rwy., which opened in April 1856.

Beginning about the late 1850s, locomotives were constructed at the Swiss Central shop in Olten. Niklaus Riggenbach, the company's mechanical engineer, was in charge of this shop. In the next decade he became involved in designing a rack railway system for use on steeper grades. The first vertical boiler rack engines, for the Rigi line, were built at Olten. The whole development of rack systems, as well as construction of the St. Gotthard and Simplon tunnels, is covered separately in chapter 19. In 1875 Riggenbach founded the International Company for Mountain Railways and established a shop at Aarau. The Societe Suisse, otherwise known as the Swiss Locomotive Works of Winterthur, began production in 1873. This company soon became a major builder of European engines.

As locomotive technology improved, adhesion could be

used for steeper gradients. On the 5½-mile standard-gauge Mt. Uetliberg Rwy., built in 1875, adhesion engines operated on grades of up to 7%. The 27½-ton 0-6-0T locomotives used on this line were built by Kraus & Co. of Munich. Typically, an engine pulled a train of four passenger cars, each having a carrying capacity of 40. Speeds over this route ranged from 8 to 10½ m.p.h. The downhill descent was made with compressed air, employing an apparatus pioneered on the engines of the Rigi Rwy. The passenger cars for this line were built in the American style, with a central passageway. This was the customary practice in Switzerland.

At the end of 1871 there were 920 route miles of track in the country. By the end of 1884 this figure had reached 1,790 miles, a total including 59 miles of rack and funicular lines. Operating over this system were 619 locomotives, 1,807 passenger carriages, and 9,031 goods wagons. The network included 189 tunnels, 1,963 bridges, and 656 stations. There was an average of one station for every 2.73 miles of track. Total route mileage increased to 2,082 by 1893.

Following a referendum in 1898, Switzerland began nationalizing its major railway companies. Electrification began on some lines about the turn of the century. This resulted in considerable cost savings, as water-generated electricity replaced the use of coal, which had to be imported.

The pioneer line into Luxembourg was the Great Luxembourg Rwy. built from Brussels into the capital. By 1892 the Grand Duchy of Luxembourg had more miles of railway in relation to its population than any other country on earth. In all this totaled 276 miles, with 224 in standard gauge and 52 in meter gauge. The capital itself is situated in the south-central part of the country. It was served at the time by lines of the William Luxembourg Rwy.: one going north toward Liège, Belgium; another running west to Trier, Germany; one pointing south toward Metz, in German-controlled Alsace-Lorraine; and another, branching from the latter, cutting back northwest toward Brussels.

Luxembourg was blessed with considerable iron deposits that were located along its southwestern border opposite France and Alsace-Lorraine. This area was served by the Prince Henry Rwy., which from there ran northward along Luxembourg's western frontier. After intersecting the Brussels line, it continued to about midway up, at which point it made a diagonal northeast cut across the country. At the eastern frontier with Germany it turned southward, running along that border until it made a connection with the Trier line. Luxembourg did not maintain operational control over its entire system. The portion connecting to Alsace-Lorraine was controlled by the German railway authorities in Strasbourg. Similarly, Belgium's State Railway Administration operated the line from Brussels into the capital. It was there that the William Luxembourg company had its shops. The Prince Henry shops were located in Petrange.

The idea of railways was initially resisted in Holland. Because the country had an excellent waterway system, it took time before either the government or industry saw any need for this new technology. Furthermore, most Dutch capital was tied up in shipping. However, Holland's King William was an enthusiastic advocate of railways, and he wanted to build one between Amsterdam and Rotterdam. In order to attract capital, he gave a personal guarantee that dividends would always be paid.

The flat, boggy ground on which the road bed was to be laid was unexpectedly spongy and precarious. Unable to solve this quandary, the original construction engineer was replaced by F. W. Conrad. He finally created a firm foundation by using layers of faggots. Sometimes a number of layers were required, and in more difficult areas these were sometimes alternated with layers of rubble. Stakes and wattle were used for added support, and over this was laid sand and turf.

This line was constructed in a 6′4½″ gauge, but was later converted to the standard 4′8½″ width. The first section of this route was inaugurated on September 20, 1839. It reached from Amsterdam to Haarlem, a distance of 10½ miles. The first engines to operate on this line, *De Arendt* and *De Snelheid*, were built by Stephenson, and were of the 2-2-2 Patentee class.

By the time this line had extended another 30 miles to The Hague, it had crossed a total of 58 waterways. Initially, the construction of bridges over these canals had posed a real dilemma. Not only were there many to build, but they would have to be constructed on the low, flat land, while at the same time be of adequate height for vessels to pass underneath. Thus an immense amount of soil would have to be brought in. This predicament was circumvented by using bridges with removable sections that allowed boat traffic to pass. Variable conditions required different designs, but most were of the swing or lift types.

Following completion of this difficult and expensive line, railway construction proceeded both slowly and intermittently, with England and France providing most of the capital. It was typical that the financing country would also send in technicians and provide the rail equipment. Railways were built primarily to provide passenger service. There was little incentive to move goods by rail, when they could be moved cheaply by water. It was only after the British constructed the Dutch Rhenish Rwy., connecting Amsterdam and Cologne, that goods traffic increased somewhat. Unfortunately, however, the Dutch had insisted that this line be built in a 2 meter gauge; thus transshipment was required at the break of gauge point. This line only became profitable after it was converted to the standard width.

Circa 1882 the Backer & Rueb firm (later Maschinenfabriek Breda) began manufacturing tramway locomotives at Breda. At the beginning of 1883 there were 1,249 miles of track in Holland. The government controlled 653 miles, but leased all of these out to corporations, which operated them. The other railways were owned by 18 different companies. The longest of these lines extended for 167 miles. By this time, three companies were dominant: the Netherlands Rwy., which operated primarily between Amsterdam and Rotterdam; the State Railway Operating Company; and the Dutch Rhenish line.

During the 1880s, railway competition was intense, and adversely affected both the industry and the country. Thus, in 1890, the state consolidated existing lines into two systems, the State Rwy. and the Netherlands Rwy. The former operated mainly in the southern part of Holland, and the latter in the north. Operation practices on these systems were somewhat different. Because English engines were used on the State Rwy., drivers operated on the left side, and, accordingly, signals were placed to the left. However, the company adhered to the continental practice of running on the right-hand track. By 1891 the combined length of these two systems was 1,887 route miles.

The first railway in Denmark was a 19-mile route built by the Zealand Rwy. Co. It opened between Copenhagen and Roskilde on June 26, 1847. The Zealand company continued to expand, and Crampton-style engines were among those operated. Railway building in the country was slow throughout the century. Denmark, which consists of the Jutland Peninsula along with many islands, is not ideally suited for railway development. Jutland itself is relatively small, and most of its population is concentrated in coastal towns frequently separated by inlets and bays. Copenhagen, on the Isle of Zealand, is the major city of the country as well as its political and commercial center. Thus, most early rail activity took place there.

Railways were later built on the islands of Funen and Falster. Transportation between these islands and the peninsula was provided by steamboat. However, in 1872, experimental ferry service began for carrying rail cars across the Little Belt between Frederica and Strib. This proved successful, and beginning in 1883, a number of other ferry routes were added.

The first state railway was opened in 1862, on Jutland, between Aarhus and Randers. At the end of 1871, Denmark had 548 route miles of line. In 1885 the Danish State Rwys. (DSB) was established to consolidate all the state-owned roads. Denmark's first narrow-gauge railway was completed in 1891. Built primarily for transporting fish, it connected a fishing village with a fish market.

In 1867 Denmark's state railways owned 20 locomotives, 54 passenger carriages, 6 mail carriages, and 231 goods wagons. By 1893 these figures were 263 engines, 775 passenger coaches, 62 mail carriages, 4,240 goods wagons, and 22 steamboats. That same year the country's rail system had grown to a total of 1,289 route miles.

Norway's first railway was the Christiana (Oslo) to Eidsvoll line built by Robert Stephenson. It was 42¼ miles long, built in the standard 4′8½″ gauge, and inaugurated on September 1, 1854. In 1861 a branch was constructed from this line to the town of Kongsvinger. It was extended farther, in 1865, to the Swedish border. Perhaps inspired by this effort of private enterprise, the government subsequently undertook an extensive building campaign. From Oslo, railways were constructed in three directions to coastal ports: southwest to Stavanger, west to Bergen, and north to Trondheim.

Because these state railways were built through mountainous regions, their construction required several decades. In the meantime, Carl Phil, a Norwegian engineer, brought in narrow-gauge technology. Soon a number of 3′6″ gauge lines were constructed. These included the 56-mile Drammen–Lake Randsfjorden Rwy., the 24-mile Hamar-Elverum line, and the 30½-mile Trondheim-Storen route. In some cases portions of the major state railways were constructed in narrow gauge. The total length of Norway's rail network in 1891 was 970 route miles.

Railway construction began in Sweden in 1854. The first line inaugurated was the 11-mile Nora-Ervalla Rwy., which opened in 1856. The same year, on December 1, two lines were inaugurated by the Swedish State Rwys. These were the Göteborg-Jonsered and Malmö-Lund routes. In 1862 the country's first major railway was completed between Göteborg and Stockholm. Besides connecting Sweden's two largest cities, this 277-mile route also provided a link between the Baltic and North seas.

Malmö, an important southern port across from Denmark, was connected to this line by completion of the 231-mile Malmö-Falkoping Rwy. a few years later. By 1867 1,000 route miles of track had been laid in the country. Other lines completed by the late 1860s included the 40-mile Stockholm-Upsala Rwy., the 36-mile Laxa-Kristinehamn route, the 29-mile Kratrineholm-Norköping line, and the 15-mile Hallsberg-Örebro Rwy.

At some point, probably in the 1870s, the Swedish State Railways Co. began building locomotives at its shop in Motala. Perhaps a typical line of this period was the narrow-gauge Vadstena-Fagelsta Rwy., which opened in 1874. This line was absorbed into the Fagelsta-Vadstena-Odeshög Rwy. in 1886, which two years later, finally reached Odeshög. The first line to penetrate the Arctic circle was the Sweden & Norway Rwy., which opened in 1888. The inaugural train consisted of the engine and 41 cars, carrying a cargo of 1,000 tons of iron ore. It ran from the Gellivara Mountains to the company's quay at Lulea, in the Gulf of Bothnia. At the end of 1885 Sweden had 84 rail lines, 36 of which were narrow-gauge. Total route mileage in the country was 4,281, with narrow-gauge lines accounting for 858 miles. State-owned railways operated 1,483 miles. In 1890 Sweden had 4,970 route miles of track. That year the Uddevalla-Vänersborg-Herrljunga Rwy. became the first in the country to adopt the zone system for traffic control.

Railway inaugurations: Top: Scene at Eidsvoll Station on Norway's Christiana (Oslo)-Eidsvoll Rwy., 1854. Bottom: Religious blessing at San Sebastian on Spain's Norte Rwy., August 15, 1864.

PLATE 131

PLATE 132

Inauguration of the Rome-Velletri line, January 27, 1862, with both the railway and special papal train receiving a blessing. Pope Pius IX was a great fan of railway development and hoped that trains would bring many new pilgrims to Rome.

PLATE 133

British-built locomotives for the Iberian Peninsula: Top left: 2-2-0 engine, Sharp, Stewart & Co., Spanish State Rwys., 1859. Top right: Sharp 0-4-2 engine, Barcelona & Saragossa Rwy., 1860. Bottom left: Sharp 2-2-2 passenger engine, Seville & Jerez Rwy., 1859. Bottom right: 2-2-2 passenger engine of Beyer, Peacock & Co., for Portugal's South Eastern Rwy., 1862.

Top: Train crossing the River Po on the Mezzana-Corti Bridge, Voghera-Pavia Rwy., Italy, 1866. **Bottom:** Rear view of station in Naples, Rome-Naples Rwy., 1863.

PLATE

134

PLATE 135

Attack by insurgents on passenger train between Barcelona and Saragosa, Spain, 1870.

Waiting for a train at a country station, Denmark, 1881.

Funeral train en route to The Hague, carrying Holland's King William III, after his death on November 23, 1890.

CHAPTER

((10))

Canadian and Latin American Railways

THE first railway in Canada was that of the Albion Coal Mine, Nova Scotia, built in 1827 by Peter Crerar. Originally a tramway for carrying coal to Pictou Harbor, it was later extended, and opened as a steam railway in 1838. Its first engines were the *Samson, Hercules,* and *John Buddle,* all built in England by Timothy Hackworth. The Champlain & St. Lawrence Rwy., connecting St. John to La Prairie on the St. Lawrence River, was Canada's second line, but first to operate with steam. Construction began on this 14½-mile portage route in January 1835. On July 21, 1836, its inaugural train was pulled by the Stephenson-built 0-4-0 *Dorchester.* That same year, Robert Stephenson made a trip across the Atlantic to visit this line. The company's second locomotive was the 4-2-0 *Jason C. Pierce,* named after one of its promoters. Built by William Norris of Philadelphia, it was delivered in 1837. This line was an immediate financial success, and by 1846 had been extended 20 miles south to Rouse's Point on the American border. It was later absorbed into the Grand Trunk Rwy. system.

The Erie & Ontario line was inaugurated in 1839, but operated only with horse traction until 1854. In 1846 the St. Lawrence & Atlantic Rwy., an international line that would later connect Montreal and Portland, Maine, began construction in America. It, too, was later absorbed by the Grand Trunk. Other early Canadian railways included the Longueil & St. Hyacinthe line, which opened in 1848, and the Montreal & Lachine Rwy., which began construction in 1847. The latter line purchased a 4-4-0 locomotive from Philadelphia's Norris firm that same year. In 1850, the year this line was opened, the combined total of Canada's railways amounted to less than 100 route miles.

Beginning circa 1853 there was a lot of rail activity in the country. That year, the Simcoe & Union Rwy. was inaugurated. Also in 1853 the Norris firm shipped six 4-4-0 locomotives to the Great Western Rwy. This new line, connecting Toronto and London, was inaugurated on December 15, of that year. In 1853 the first Canadian locomotive builder, James Good, began production in Toronto. The Kingston Locomotive Works became a later manufacturer.

Commencing operations on August 25, 1854, was a small portage line on the Ottawa River, the Carillon & Grenville Rwy. Unlike other Canadian lines, which used the standard gauge of 4′8½″, this company chose a width of 5′6″. On October 1, 1856, the London & Port Stanley Rwy., a 24-mile line connecting London and Lake Erie, was opened. That same year the 5′6″ gauge Grand Trunk Rwy., between Montreal and Toronto, was inaugurated. By 1860 its total length was over 800 miles, extending from Port Huron at the Michigan border through Toronto and Montreal, to Riviere du Loup on the St. Lawrence River. Between Toronto and Detroit, its lines ran parallel to those of the Great Western, which were to the south. The Grand Trunk also connected with American lines at the Victory Bridge, built across the St. Lawrence at Montreal. By 1873, the year it converted to standard gauge, this system was over 2,000 miles in length.

Another major Canadian line was the Intercolonial Rwy., begun in 1867. Its purpose was to connect all the component provinces of the newly established Dominion of Canada: Quebec, Ontario, Nova Scotia, New Brunswick, and Manitoba. This work was completed in 1876. At Quebec, the Intercolonial line made connections with the North Shore Rwy. The latter line, 160 miles long, followed the north shore of the St. Lawrence between Montreal and Quebec. From its terminus on the river at Quebec it operated a steamer that ferried rail cars across to the Intercolonial.

Other Canadian lines also operated ferries as well as

freighters. The June 29, 1872 issue of *Railroad Gazette* reported on the operation of one such ferry, owned by the Great Western Rwy. In 1871, it made 10,008 trips between Windsor and Detroit, carrying a total of 111,680 freight cars, a small number of passenger cars, and a few locomotives. Navigation on the Great Lakes and canals largely shut down in the winter months, however, and more traffic went by rail. Some experiments were made in laying railways across ice, among these, one line across the frozen St. Lawrence.

Perhaps circa 1871, the Great Western and Grand Trunk lines, because of their Detroit-Buffalo connections, became involved in hauling the competitive Chicago–New York freight traffic. Their lines provided an alternative route, north of Lake Erie, while previously this tonnage had only moved to the south. At Detroit, traffic was exchanged with the Michigan Central R.R. In Buffalo it was exchanged with either the New York Central or Erie lines.

The Toronto & Nipissing Rwy., built circa 1871, became the first narrow-gauge line to open in North America. This 3′6″ gauge, 87-mile railway provided access from Toronto to its north and northeast, a thinly populated area with abundant timber resources. Later on, the 3′6″ gauge became the preferred gauge in Newfoundland and on Prince Edward's Island. The Newfoundland Railway Co. was incorporated in 1881 to build a 340-mile line from St. Johns to Harbor Grace. When the colonial government took over operations in 1896, only 96 miles had been completed.

In 1871 the Dominion of Canada purchased British Columbia from the Hudson Bay Co. Having the immediate intention of connecting this far western province with the rest of Canada, the government committed itself to building a Pacific railway. However, as Canada was still building the Intercolonial line, financing this line proved difficult. In hopes of gaining funds, the government turned to Philadelphia banker Jay Cook, financier of America's Northern Pacific R.R. Shady transactions soon took place, contributing to the collapse of Cook's bank and a subsequent U.S. stock market crash in 1873. This led to the "Pacific Scandal," resulting in the forced resignation of the Canadian government.

Construction on the Canadian Pacific line was begun in 1875, although the company itself was not chartered until February 17, 1881. By June 1883 the rail line extended for a distance of 1,704 miles. Large crews worked from both ends toward the middle, building through such formidable obstacles as Rogers Pass and Kicking Horse Pass. The line was finally completed on November 7, 1885, with a golden spike ceremony at Craigellachie, British Columbia. In 1888 the Canadian Pacific was expanded eastward to Mattawamkeag, in north Maine, where it connected with the Maine Central R.R. In the west, the railway decided to promote the scenic beauty of the Rocky Mountains to draw tourists, and built several amazing resort hotels such as the one at Banff, near Lake Louise. These hotels soon became playgrounds for the rich and famous, but also gained popularity with ordinary people.

Circa 1890 other railways were being built in western Canada. These included the Regina & Long Lake and Manitoba & North-Western lines. Later in the decade the amalgamation of a number of railways east and west of the Great Lakes resulted in the creation of another important system, the Canadian Northern. Canada's total system grew from 1,880 route miles in 1860 to 2,670 in 1870, to 6,886 in 1880, and, by 1893, to the impressive figure of 15,330.

Railways came to the Caribbean Sea early. In Cuba rail construction began in 1834, and by 1855, 375 miles of track had been laid. Among the early lines were the Havana & Guines Rwy. and the Cardenas & Bemba line. In 1838 the Norris firm shipped the 4-2-0 *Herrera* to the Havana & Guines Rwy., and subsequently sold other locomotives to Cuban lines. In 1880 Cuba had a total of 860 route miles of track. By 1893 this figure had increased to 1,000. By comparison, Puerto Rico reported only 11 miles.

On November 21, 1845, a 14-mile railway between Kingston and Angels Station was inaugurated in Jamaica. Many local notables, including the governor, the Earl of Elgin, attended this opening. The first train of 10 carriages was pulled by *The Projector,* built by the Sharp Brothers of Manchester. By 1851 this line was using six locomotives. In 1869 it was extended 11 miles from Angels Station to Old Harbour. In 1879 the railway was sold by its private owners to the Jamaican government. In 1885, two new extensions were inaugurated; a 24-mile line from Old Harbour to Porus and a 24-mile line from Spanish Town to Ewarton. In 1889 an American company bought the line and began more rail construction. It built a 66-mile extension through the mountains to Montego Bay that opened in 1894, and a 54-mile line from San Antonio to Bog Walk, which began service in 1896. Perhaps the most notable event on this railway occurred in October 1892, when a derailment caused a locomotive to fall 60 feet into the Bog Walk Gorge, killing one person.

In the early nineteenth century the only route for shipping between the Atlantic and Pacific oceans was through the stormy seas around Cape Horn, at the bottom tip of South America. Unable to find water passages anywhere else, various governments began to consider the alternative possibility of using an overland route through one of the several narrow isthmus areas of Central America and Mexico. At such a location, passengers, mail, and freight could be disembarked at one port, be transported a short distance overland, and then be put aboard another ship. This idea became increasingly realistic as railways began to develop and show their potential.

In 1835, the American president Andrew Jackson sent an emissary to study the feasibility of constructing a railway or canal in Panama, which was then a part of New Grenada (Colombia). A concession for building a railway was granted, but no construction was forthcoming. In 1847 a French company was given a concession but was unable to raise sufficient capital. Following the discovery of gold in California, Americans took a new interest in building the line. A company was organized that included John Lloyd Stephens, William H. Aspinwall, Henry Chauncey, and others. Work began on this 5′ gauge, 47-mile railway in May 1850, at Manzanillo Island on the Atlantic. Massive amounts of earth were moved to connect it to the mainland. Docking facilities, shops, warehouses, and saloons were soon built, and in time this became the port of Aspinwall, later renamed Colon. Working conditions were very difficult from the beginning due to the heat, flies, and mosquitoes; and many workers died from yellow fever, malaria, cholera, and dysentery.

To avoid potential adhesion problems on steep grades, a third center-rail system was devised to improve traction. This was the innovation of George E. Sellers, who built several 4-4-0 locomotives in Cincinnati. Shortly, however, plans were revised, and the center-rail idea was discarded. Sellers' engines were shipped minus the center-rail driving apparatus, the first arriving in 1852, and were used in construction of the line. The railway was inaugurated on January 28, 1855, connecting Aspinwall with Panama City, on the Pacific. Its completion totally revolutionized commerce between the two oceans.

Later, in 1869, the Panama line lost much of its business due to the completion of America's first transcontinental railroad. Its stock fell from $360 to $80 a share in the first week following the news. In 1879 the railway was purchased by a French company that had gained a concession to construct a ship canal. This attempt was a failure, however. Just after the turn of the century, the United States purchased the railway, and used it to facilitate construction of the Panama Canal, which finally opened in 1914.

On August 22, 1837, a concession for building a Mexico City–Vera Cruz rail line was granted by President Anastasio Bustamante to Francisco Arrillaga. This concession was forfeited after two years, however, after Arrillaga failed to lay any track. The first Mexican line to be completed was an 8.08-mile railway between Vera Cruz and El Molino. It was built in the 4′8½″ gauge that became the standard for the country. President Jose Joaquin de Herrera inaugurated the line on September 16, 1850.

Nothing further happened on the Mexico City–Vera Cruz route until 1855, when construction began following President Santa Anna's grant of a concession to the Mosso brothers. The first section of what became the Mexican Rwy., or La Compania del Ferrocaril Mexicano, was opened on July 4, 1857, connecting Mexico City and Guadalupe. Robert B. Gorsuch of New York was the builder. This line changed hands several times afterward, and suffered through the political turmoil of the 1860s, which included the reign of Emperor Maximilian and the chaotic times following. During this period English engineers took over the work. Lines constructed from both ends were finally joined near Maltrata in late 1872. On January 1, 1873, President Lerdo de Tejada inaugurated the completed line. The railway was costly and arduous to build, ascending to an altitude of 7,600 feet from the sea-level port of Vera Cruz. One portion of this ascent was a rise of 4,700 feet in a distance of 25 miles. The construction of the Metlac Viaduct in a section of the route known as the Infernillo proved especially difficult. It consisted of 9 spans of 51 feet each, having a curve of a 325-foot radius and a 4% grade. The English-built Fairlie-style engines operating on this line gained a good reputation for their power, flexibility, and excellent adhesion.

In the narrowest part of southern Mexico, between the Pacific and Gulf of Mexico, was the Tehuantepec Isthmus. Like Panama, and other narrow stretches of land in Central America, it received early attention as another possible commercial route between the two major oceans. Thus planning began for construction of either a railway or a canal across the isthmus. A concession for this purpose was first granted in 1842, but little was accomplished in the period following, nor after a number of succeeding concessions were given. It was only after Porfirio Diaz rose to power in 1876 that the political situation changed sufficiently for there to be new interest in a project across the isthmus.

In 1881, James B. Eads, builder of the Mississippi River bridge at St. Louis, began promotion of an idea to build a level isthmus railway. It would have a number of parallel rails, where entire ships (of up to 7,000 tons) could be drawn by locomotives from one ocean to the other. It was argued that such a railway had much to offer. It would be much less costly and require considerably less time than the effort of building a canal. Its maintenance costs would also be inexpensive in comparison. It would eliminate the high costs of transshipment required by the Panama Rwy., and attract the bulk commerce between the two oceans, the greater portion of which still went around Cape Horn. Because its location was substantially farther north than Panama, additional savings of cost and time could be made on most interoceanic shipments. There also would be a strategic advantage for the United States, in that the closer proximity of this location made it potentially easier to defend against acts of foreign aggression.

In one version of this plan, a cradle car sufficient for carrying a ship was designed that had 1,380 wheels, 115 for each of 12 parallel rails. Another version of the plan had 6 rails. These were intended to run straight toward a few critical points where changes in direction were necessary. At these points would be giant floating turntables. A liberal concession for the line was granted by Mexico on May 28, 1881. Despite this, however, there were problems in raising funds. Lengthy negotiations with the Mexican government failed, and Eads turned to the U.S. Congress for help. Unfortunately, he died in 1887, before any assistance was forthcoming, and his fantastic dream was never realized. Earlier, in 1879, a few miles of an ordinary railway over this route had been laid. Further progress was made in 1894, but the line was not complete until 1906. It connected the ports of Coatzacoalcos on the Gulf with Salina Cruz on the Pacific. Although it initially took business from the Panama Rwy., its success was temporary. The opening of the Panama Canal in 1914 made it useless, except for hauling local traffic.

Eads' proposal, no doubt, owed something of its inspiration to James Brunlees, a prominent English engineer who in 1872 was promoting a similar ship railway through Honduras. Brunlees' plan was more modest, however. It called for a line having 6 parallel rails with a total gauge of 25 feet. Each special cradle car would weigh 700 tons, have 240 wheels, and have a load capacity of 2,000 tons. Hydraulic lifts would be employed to move the ships from the water to the cars and off again. It was thought that 5 to 10 locomotives would be required to move each ship, with a rack system being required on the grades. The estimated cost of building such a railway at the time was $75 million.

Although the politically turbulent era of the 1860s was not good for railways in Mexico, the Las Cumbres Rwy. was opened in 1868. Beginning with the 1872–76 presidency of Tejada, railway building was promoted, with English engineering firms being favored over those of America. The president's antipathy for the United States also extended to the idea of connecting Mexican railways with their American counterparts at the northern border. He assisted the British-owned Mexican Rwy. when its financial status became shaky, by permitting it to raise its rates, build a wharf in Vera Cruz, and extend a line from that city through Jalapa to Puebla. During this period, a short railway was constructed in the Yucatan, from Merida to the port of Progreso. On December 5, 1874, Tejada granted a concession to a company headed by Sebastian Camacho for constructing a line from Mexico City to Leon. However, after the successful revolt by General Pofirio Diaz in 1876, this concession was revoked.

Mexico quickly became friendlier to American interests. By 1880 American companies had received concessions for building five rail lines having a total route mileage of 2,500 miles. One of these was for the Mexico City–Leon line, which became the Mexican Central Rwy., or the Ferrocaril Mexicano Central. The concession also called for the extension of this line from Leon to Paso del Norte. On April 10, 1884, this 1,224-mile railway was inaugurated. In July 1891, a branch line through San Luis Potosi to the port of Tampico was completed. This route improved conditions for trade between interior Mexico and Europe and the eastern United States.

Another American-built railway was the Guaymas-Nogales line, or Ferrocaril de Sonora. It was begun in 1879 and completed in 1882. On June 1, 1880, a first General Railway Law was passed, limiting companies to a grant of 99 years, after which time all their properties would revert to national control. On September 13, 1880, a concession was granted to an American company for construction of a narrow-gauge line from Mexico City to Laredo, and for another line from Mexico City to Manzanillo. Both lines became parts of the Ferrocaril Nacional Mexicano, or Mexican National Rwy. Later, financial difficulties led to the sale of this line to an English company. The Mexico City–Laredo portion opened on September 28, 1888. It was converted to the standard gauge just after the turn of the century.

By 1883, 2,779 route miles of track had been laid in the country, and much more construction was underway. Legislation passed on July 1, 1883, provided comprehensive regulations regarding the building, maintenance, and operation of lines. Many concessions were given out at the time. In its August 24, 1883 issue *Railroad Gazette* gave this brief description of the government plan: "Three transcontinental or east and west trunk lines, with two north and south trunk lines. This for the country northward of the city of Mexico, which is the bulk of the republic. For the rich triangle southward from the city of Mexico to Guatemala, a little system of its own was laid out, giving it outlets to Atlantic and Pacific ports and down the Isthmus, to connect in time with the South American system."

The *Gazette* article also described the American-Mexican connection, perhaps somewhat arrogantly, as follows:

> This is the railway field of Mexico, but in considering it, it must not be thought of as a distinct and separate system of itself, like the railways of England or of India, for instance. It is simply an extension of the enormous railway system of the United States, which at four different points already has projected itself across the border and into the interior of Mexico. The whole southwestern system of our country converges at San Antonio, and from thence crosses the Rio Grande at Laredo, whence in its onward march it has already passed Monterey. The Southern Pacific throws an arm across the border at Eagle Pass, and another at Nogales, while at El Paso the Texas Pacific already has 400 miles of Mexican connection.

In 1892 the Mexican International Rwy. was completed. It extended from Piedras Negras to Durango, and provided access to coal deposits in Coahuila. At the time, built under Federal concessions, were 89 railways in the country running a total of 6,546 route miles. Reported separately from these were 219.5 miles of city tramways, 114.7 miles of suburban railway, 44 miles of private lines, and 142.6 miles built on the Decauville system. By 1900 Mexico's rail network was substantial. Its continued growth through the early part of the twentieth century allowed for improved travel, commerce, and the extensive extraction of its many natural resources.

Following the Panama Rwy. other lines were constructed elsewhere in Central America. However, due to the general poverty and lack of commerce in the region, they were few in number and took a while to appear. The following account from Costa Rica, published in *Railroad Gazette*'s August 17, 1872 issue, gives some idea of the problems faced.

> In the first days of last May, a costly expedition was inaugurated to search for the location of a new railway line between El Limon and La Angostura, passing through Tuis and Chirripo, under the immediate orders of Senor Don Alfred F. Sears, a most able engineer who has figured prominently in some of the most important railroads of the United States. This expedition was accompanied by Don Jesus Bonnilla, who suggested the expedition, and by Don Jose Maria Figueroa, and other influential persons of Cartago. The only result obtained has been the loss of three men in the mountains, and the full conviction that at present, at least, a railway is absolutely impractical between Limon and Angostura, passing through the valley of Tuis and Chirripo. One of the bridges of this line would need to be three miles long and 3,000 ft. high, costing $20,000,000.

In spite of these obstacles, however, railways soon came to the country. The *Railroad Gazette*'s April 17, 1875 issue reported a fire in the machine shop of the Costa Rica Rwy. at Port Limon that destroyed the building and three locomotives. Later, the publication's September 28, 1888 edition reported three operating railways: a 70-mile line from Port Limon, on the Atlantic, to Corrillo; a 25-mile railway between Alajuela and Cartago, passing through the capital, San Jose; and a line of 12 miles from Puntarenas, on the Pacific, to Esparta. At the time, travel and commerce in the country primarily depended on cart roads. These provided essential connections between points on the rail routes. By 1892 two of the lines had been connected, creating a 113-mile route between Alajuela and Port Limon.

Guatemala's first line, completed in 1880, had a 3′ gauge, and ran 13 miles from the Pacific port of San Jose to Esquintla. A 30-mile line connecting the coastal town of Champerico to Ritalhulca was opened in 1883. Later, a 72-mile road between Guatemala City and San Jose was inaugurated. This route was built with English rails but used American rolling stock. In 1897 the Iztapa Rwy. was completed. Branching from the Central Rwy. at Obero, it ran to the port of Iztapa. In 1893 Guatemala reported a total rail network of 118 miles. That same year Costa Rica's total mileage was 231, while Nicaragua had 99, El Salvador 62, and Honduras 37.

In 1816, Richard Trevithick, builder of the world's first locomotive, made a visit to Peru. While there he proposed the construction of a 9-mile line to connect Lima and the port of Callao. Political instability at the time made this impossible. However, years later, in 1851, the line was built, becoming one of the earliest in South America. The first railway on the continent was the Demarara line in British Guiana (Guyana), inaugurated on November 3, 1848. Another early railway was the 31-mile Caldera-Copiapo line in Chile, which was begun in 1850 and completed in 1852. At the time Copiapo was a center for the mining of copper and silver. The railway's first locomotive was a 4-4-0 Norris, but it subsequently bought British engines. The American William Wheelwright built this line, and later in 1852, began construction of the 114-mile Valparaiso-Santiago Rwy.

The first Brazilian line was the Mana Rwy., a short route connecting Rio de Janeiro and Petropolis. This line began service on April 30, 1854. Its 5′3″ gauge became the country's standard. Shortly, construction began on a larger railway, the Dom Pedro Segundo, named for Brazil's ruler. Its first section, one of 25 miles from Rio de Janeiro to Queimadas, was inaugurated on March 29, 1858. Branch lines were later extended to Minas Geraes and São Paulo. Other Brazilian railways began construction about the same time. The Tijuca Rwy., a line near Rio de Janeiro, opened in 1862. The Paulista Rwy. was inaugurated some years later. It extended 50 miles from the port of Santos to São Paulo, and rose to an elevation of 2,500 feet. There were difficult engineering problems on this line, especially in the construction of the Serra Viaduct.

Over one half of Brazil's coffee crop was grown in the São Paulo region, and this railway provided the only route for its export. In the March 27, 1872 issue of *Railroad Gazette,* it was reported that a landslide had occurred on February 22, of that year, causing severe damage to the line: "The road for miles was moved from its foundation and buried under earth and rocks, so that, it is reported, two months' work will be required to put it into working order again." The article went on to suggest that this disaster would have "far reaching effects" on the world coffee market.

By the end of 1874, Brazil had 836 route miles of line in service, plus 1,034 under construction. In addition to this, concessions had been granted for 2,392 more miles of railway. A later Brazilian line of interest was the Prince of Grand Para Rwy. This meter-gauge line, through mountainous terrain, employed a Riggenbach rack system to overcome grades of up to 15%. In 1886 the Dom Pedro Segundo Rwy. reported having a total of 118 locomotives. Of these, 98 were of American manufacture, 18 English, one French, and one Belgian. This line later became the Central Rwy. of Brazil.

The first line in Argentina opened August 30, 1857. Construction on the South Argentine Rwy. began on March 7, 1861. Work was initiated on a line between Rosario and Cordoba on April 20, 1863. It was built by Wheelwright and Thomas Brassey, the famous English engineer. On July 9, 1866, the Gualeguay Rwy. was inaugurated. On September 11, 1866, the 100-mile Ferro-Carril Del Oeste began service. It ran from Buenos Aires to Chivilcoy, and was built by Manuel G. Gonzales. Another Wheelwright-built line, the Central Argentine Rwy., connected Rosario with Cordoba, and opened in 1870. Argentina's national gauge became 5′6″. Its railways were almost entirely British-owned. By 1888, the country had 4,535 route miles of track, 700 locomotives, 2,000 passenger coaches, and 15,000 to 20,000 goods wagons.

Paraguay, which occupies a central position between Brazil and Argentina, had only one railway in the nineteenth century, the Paraguay Central line, which ran from Asuncion to Perapo. Uruguay, also located between Argentina and Brazil, but farther to the south, had several railways. One of these, the Central Uruguay line, completed in 1892, was 352 miles long. It began at the port of Montevideo and ran northward through the middle of the country, terminating at Rivera on the Brazilian border. This railway opened up a vast fertile area for agricultural purposes.

Extensive rail building in Peru did not begin until 1868. Soon lines were constructed from Pisco to Yca, Mejia to Arequipa, and at a coal mine near Pasco. In 1870 work was initiated on railways connecting Arequipa to Puno and Callao to La Oroya, both to gain access to the mineral-rich Andes. The June 1, 1872 issue of *Railroad Gazette* reported a total of 28 railways in the country, most of which were still under construction. The total projected route mileage for these lines was 2,310. Among these were a 301-mile railway from Tacna to Puno and a 209-mile line connecting Juliaca and Cuzco.

Circa 1873, the Callao, Lima & Oroya Rwy. had reached completion. In its 136 miles it had begun at the Pacific port of Callao, ascended the Andes, crossing the continental divide at an elevation of 15,848 feet, and then dropped down to La Oroya at an altitude of 12,000 feet. This standard-gauge line became noteworthy, as it reached the highest elevation of any railway in the world for any gauge without the aid of a rack or center-rail adhesion system. It was built by the American Henry Meiggs, and was famous for its Agua de Verrugas Viaduct, which at the time had the highest iron piers in the world.

The *Railroad Gazette*'s August 24, 1872 issue reported that construction on the line had been fraught with unexpected and appalling difficulties. "For example, the section of road between Chosica and Rio Seco has been built with such a frightful sacrifice of human life that at one time it was feared that it would be impossible to get laborers to work there at any price. A strange disease, known as the 'Oroya fever,' attacked the men all along this section, and, baffling every effort of the medical board to subdue or even arrest its progress, slew them by the hundreds." At the time the railway had six American-built engines, five Rogers and one Danforth. When it reached final completion, the line had a total of 30 tunnels, plus 30 bridges and viaducts. In March 1889 the Verrugas Viaduct was destroyed, either by a landslide or by flood. The central pier was carried away, and the two spans resting on it collapsed. In 1890 this line became the Central Rwy. of Peru.

The 232-mile Arequipa & Puno Rwy., also built by Meiggs, reached its highest elevation at about 14,600 feet. Elsewhere in the Andes other countries, like Bolivia and Chile, constructed high-altitude railways in their efforts to reach natural resources. Most of these lines were built in narrow-gauge.

About the same time rail construction was also beginning, if not already in progress, at other places on the continent. On June 27, 1883, there was an inaugural ceremony for the 23-mile 3′ gauge railway from the port of La Guaira to Caracas in Venezuela. In attendance were the nation's president, various government officials, and thousands of spectators, many of whom had never before seen a railway. In its short distance this line ascended from sea level to 3,200 feet before descending 250 feet into Caracas. Most of the goods traffic went uphill while downhill shipping largely depended on the size of the coffee crop. The line had many sharp curves and went through deep cuts and tunnels and alongside the edges of precipices.

In its October 7, 1892 issue, the *Railroad Gazette* described two unique dangers of this line:

> The land slips are very frequent during the rainy season, which lasts about six months. As many as 202 have been known to occur during one night, and this is a constant source of danger. Sometimes the trains have been imprisoned for three days between two huge land slips. On one occasion, the slips were so numerous, that on the whole track there were not 400 yards of clear line remaining unencumbered, and the traffic was stopped for a fortnight. On another, two bridges were washed away . . . The quantity of locusts that infest the line is so great that they frequently stop the trains. Even when only a few are crushed, the rails and wheels become so greasy that adhesion is destroyed, and slip occurs to such an extent that the trains often run back down the incline.

Another Venezuelan railway was the Puerto Cabello & Valencia line. Having a 3′6″ gauge, it opened in 1887 and included the Trincheras Incline, a section requiring rack locomotives. The Lange & Livesey rack system was employed for this grade, with 0-4-2T rack engines supplied by England's Beyer-Peacock Co.

Valencia, next in commercial importance to Caracas, was connected to the capital on January 20, 1894, by completion of the Great Venezuela Rwy. This 112-mile, 3′6″ gauge project, known earlier as the "Krupp concession," was undertaken by the Grosse Venezuela Eisenbahn Gesellschaft of Berlin and Hamburg. Construction began at both ends, with the two crews joining somewhere near the midpoint of the route. This line ran through a rich countryside that produced sugar, coffee, various food crops, and livestock. The company's engines were mostly of German manufacture, with the American-style passenger coaches provided by Charlier, Van der Zyphen & Co. of Deutz.

The most important of South America's nineteenth-century lines was the Transandino Rwy. A trans-Andes route to connect Chile with Argentina had first been proposed by William Wheelwright in 1860. On January 26, 1874, the Argentine government granted a concession to Mateo and Juan Clark to build a line from Buenos Aires, by way of Mendoza, to the Chilean border. Construction did not begin, however, until 1883. A first section, one of 232 miles from Villa Mercedes to Mendoza, opened in 1886. The 429-mile portion across the pampas, between Buenos Aires and Villa Mercedes, opened in 1888.

Because only a comparatively short distance of track was needed on the Chilean side, construction was held up until substantial progress had been made in Argentina. The Chilean government had granted a concession for this work to Clark & Co. on November 13, 1874. The distance for the line was estimated at under 50 miles. Construction would start from the terminus of the Central Rwy., at Santa Rosa de los Andes, and proceed through the Uspallata Pass to connect with the Argentine line at the mouth of a tunnel. Work on this section finally began on April 5, 1889. Shortly, in December 1890, labors on both sides ceased due to lack of funds. Efforts resumed in 1898, and the railway was completed early in the twentieth century. Connecting the port cities of Valparaiso and Buenos Aires, it was the only South American line to cross the Andes and link the two oceans.

In 1880 South America had a total of 6,444 route miles of track in service. Four years later this figure had increased to 9,515. By 1893, the continent's network had doubled in size, reaching 19,514 miles. Of this total, Argentina's system led with 8,023 miles, followed by 6,651 for Brazil, 1,735 for Chili, 974 for Uruguay, 882 for Peru, 500 for Bolivia, 287 for Venezuela, 218 for Colombia, 158 for Paraguay, 63 for Ecuador, and 23 for British Guiana.

MR. F. W. GATES, ONE OF THE DIRECTORS. MR. T. REYNOLDS, FINANCIAL DIRECTOR. MR. BRYDGES, MANAGING DIRECTOR. MR. G. H. MINGAYE, PAYMASTER. MR. R. EATON, LOCOMOTIVE SUPERINTENDENT. MR. W. C. STEPHENS, SECRETARY IN CANADA. MR. G. FORSYTH.

PLATE 138

Locomotive built at Hamilton shops of Canada's Great Western Rwy. Designed by G. Forsyth, this 0-6-0 goods engine was built during a business downturn (1858–60) when there was little else to do.

Top: 4-4-0T locomotive, Hunslet Engine Co., Leeds, for 3′6″ gauge Prince Edward Island Rwy., 1872. **Bottom:** Use of a snowplow on the Grand Trunk Rwy., 1862.

PLATE
139

PLATE 140

Interior of a "colonial" class sleeping car on the Canadian Pacific Rwy., 1888. The word *colonial* referred to "colonists" or settlers, thus it roughly corresponded to American immigrant coaches, although this car seems to be luxurious by comparison.

TOP: 4-6-0 "ten-wheeler" type engine built at Canadian Pacific's Montreal shops, circa 1893. MIDDLE: 2-6-6T Rogers "double ender" style engine, Matanzas Rwy., Cuba, circa late 1870s. BOTTOM: 2-4-4T Rogers "double ender," Matanzas Rwy., circa late 1870s.

Top: Baldwin 0-4-2T Forney style engine, Mexico's North Central Rwy., circa 1887. **Bottom:** Fairlie articulated engine pulling work train over Metlac Viaduct, Mexican Rwy., circa 1881.

Artist's conception of a ship in transit on the proposed Interoceanic Rwy., circa 1885. This line, promoted by James B. Eads, was intended to cross Mexico's Isthmus of Tehuantepec.

PLATE
143

Top: 4-6-0 passenger engine, Neilson & Co., Glasgow, Mexican Rwy., 1895. Middle: 0-8-0 goods engine, Neilson & Co., Mexican Rwy., 1895. Bottom: Baldwin 4-6-0 compound engine, Mexican National Rwy., 1893.

2-6+6-2T Fairlie-style engine pulling a train on the Mexican Central Rwy., 1893. This giant locomotive was designed by F. W. Johnstone, and built at the Rhode Island Locomotive Works, Providence.

PLATE
145

The Panama Rwy.: Scenes at the Atlantic terminus of Aspinwall (later renamed Colon): Top: Train awaiting departure for Panama City, 1865. Bottom: A view of tracks along city's main thoroughfare, circa 1877.

PLATE 146

PLATE 147

Decauville railway equipment used in excavation of Panama Canal: Scene at Emperador, 1885.

PLATE 148

Agua de Verrugas Viaduct, Callao, Lima & Oroya Rwy., Peru, 1872. Built by the American contractor Henry Meiggs, this structure had, at the time, the highest iron piers in the world. In March 1889 it was destroyed, either through the action of a landslide or a flood.

PLATE 149

Rack railway ascending Mt. Corcovado with a view of Rio de Janeiro, Brazil, circa 1890. This line was inaugurated in 1884.

PLATE 150

Baldwin locomotives built for South America: Top: 2-4-2 Columbia type passenger engine for Antofogasta Rwy., Chile, circa 1887. Bottom: 2-10-0 Decapod type goods engine for Dom Pedro Segundo Rwy., Brazil, 1885.

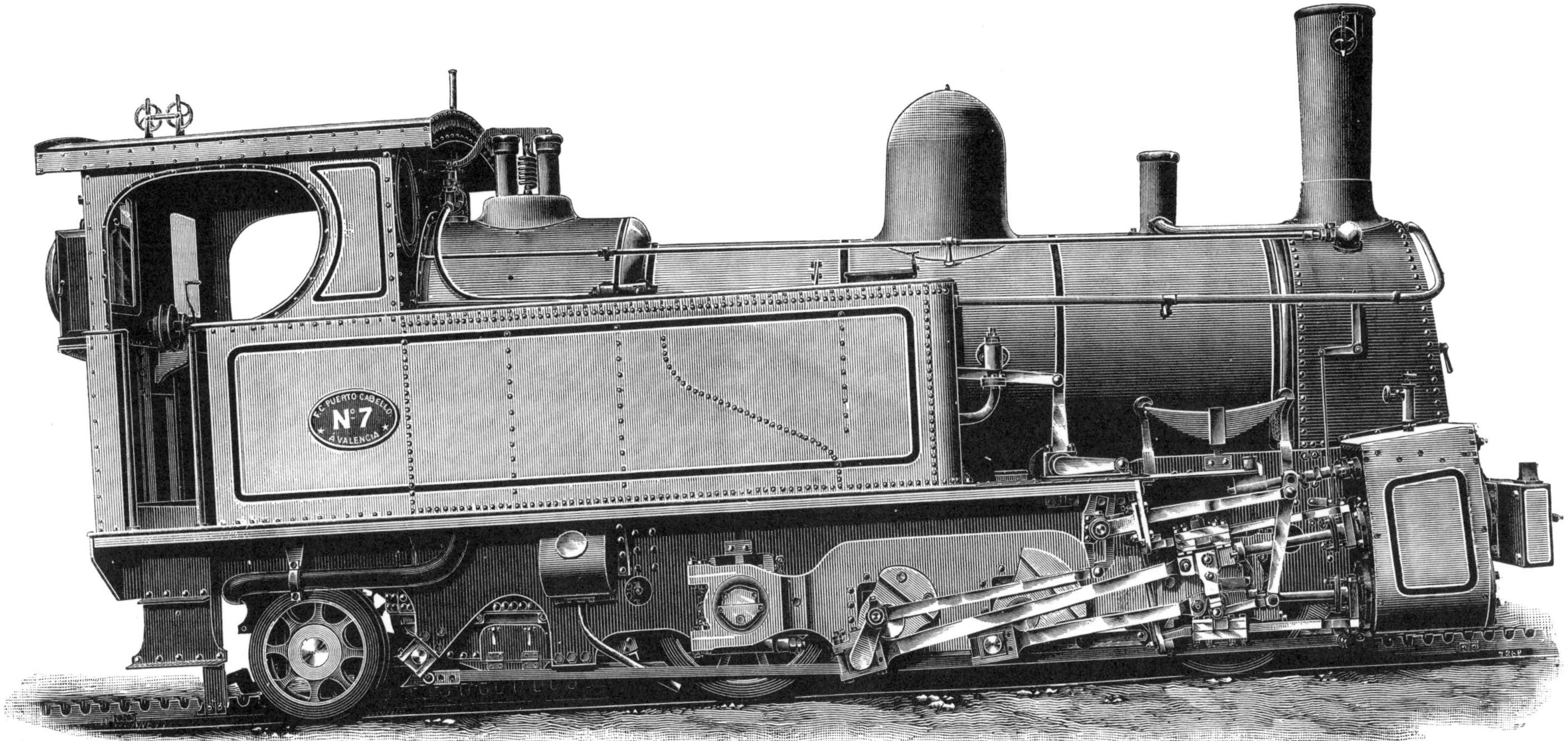

PLATE **151**

TOP: Baldwin 2-8-0 Consolidation type freight locomotive, Paulista Rwy., Brazil, 1895. BOTTOM: 0-4-2 rack engine, Lange & Livesey system, Beyer, Peacock & Co., Manchester, 1887. This locomotive was built for use on the Trincheras Incline of Venezuela's Puerto Cabello & Valencia Rwy.

PLATE 152

Brooks locomotives built for Latin America, 1896: Top: 4-6-0 "ten-wheeler" for Cuba's Cienfuegos–Santa Clara Rwy. Bottom: 4-8-0 "twelve-wheeler" for the Central Rwy. of Brazil (formerly the Dom Pedro Segundo line).

CHAPTER

((11))

Asian, Australian, and Pacific Railways

THE story of railways in Asia perhaps began in colonial India, on New Year's Day 1848, when Lord Dalhousie replaced Lord Ellenborough as Britain's governor-general. Dalhousie, an ardent advocate of railway building, had previously served on England's Board of Trade. He saw the great potential railways offered in furthering Britain's colonial ambitions in India. Beginning in 1843, rail lines had been proposed for the country, but Lord Ellenborough opposed them. Under Dalhousie's oversight, a well thought out and excellently constructed railway system began to emerge.

India's initial line was built by John Chapman, and it later became the first part of the Great Indian Peninsular Rwy. Extending 21 miles from Bombay to Thana, it opened on April 18, 1853. The line's first locomotives came from the Vulcan Foundry in England. Robert Stephenson was retained as a consulting engineer and served until his death in 1859. By 1858, under the supervision of James Berkley, this railway had been extended across the Western Ghat Mountains. The most difficult heights were overcome by employing a system of switchbacks.

Another early line was the East Indian Rwy. Begun in 1851 by the East Indian Co., it was built under the supervision of Rowland MacDonald Stephenson and opened in 1855. This 120-mile line connected Calcutta to the Raneegunge coal mines. It was later extended to Cawnpore, and in 1857, was used to transport troops during the Indian Mutiny. This insurrection alerted the British to their vulnerability in the country. From then on railway building took on a greater urgency.

Westward construction continued on the East Indian Rwy. until it met the Great Indian Peninsular Rwy., on March 8, 1870; creating the first line across the subcontinent. The Great Madras Rwy., which opened in 1856, was also to expand and attain importance. The Indian standard gauge became 5′6″. The early lines, financed by private capital, were well built in accordance with Dalhousie's rigorous specifications. However, while stimulating the flow of commerce, they remained unprofitable due to the enormity of their initial construction costs. Also contributing to railway expenses was the fact that for some years, most fuel coal had to be imported from Wales around the Cape of Good Hope.

One interesting development occurred in 1863, when an Indian prince, the Guicowar of Baroda, had built at his own expense a 2′8″ gauge tramway line between Meagaum and Dauboi. Both passengers and goods were carried, with traction being provided by oxen. In 1873 probably the first locomotives ever built in Asia were constructed at the shops of the Bombay, Baroda & Central India Rwy. in Parel. The three engines were constructed for goods service. This line had begun in 1860.

In 1870 India's system totaled 5,000 route miles. In 1871, the government began building a number of less expensive narrow-gauge railways, either to serve as secondary lines or for mining and military purposes. The French, who occupied Pondicherry, built a small line from Pondicherry to Villapuram that connected with the Great Madras Line. In 1882, among the country's major lines, the East Indian Rwy. was operating 536 locomotives; the Great Indian Peninsular Rwy., 447; the Said, Punjaub & Delhi line, 159; the Indus Valley & Kandahar, 156; the Great Madras Rwy., 150; the Oudh & Rohilcund, 125; the Bombay, Baroda & Central India line, 100; and the Eastern Bengal Rwy., 48. That same year the East Indian Rwy. had a total of 1,507 miles of line, of which 469 miles were double-tracked. By this time railways were increasingly substituting native for European employees, even for positions as high as engine driver and station master.

During this period, a number of impressive railway stations were constructed in major cities. The terminus of the Madras Rwy. was completed in 1873. Its architect, George Hardinge, designed it in the Romanesque style. Victoria Terminus, the Great Indian Peninsular station in Bombay, was begun in 1877 and completed in 1887. While basically a Gothic Revival structure, it boasted many fantastic features and contained over 40 large offices for the railway's administrative staff. More details on this imposing station can be found in chapter 14.

As Russian influence expanded ever further into Central Asia, the British became increasingly concerned about the vulnerability of India. The northwest frontier region had been a traditional invasion route into the country for thousands of years, and for strategic purposes the British decided to build railways into this region. The two traditional gateways through this immense mountain barrier were the Bolan and Khyber passes, but the Hurnai and Kurram passes also posed a possible threat. Earlier, the engineer William Andrew had promoted a railway to be built through the Turkish Empire, Persia, and Afghanistan to link Europe with India. He succeeded only in creating, by 1855, the Scinde Railway Co. This company was initially organized to build a 108-mile line from Karachi to extend up the Indus River. At the beginning of the Second Afghan War in 1878, the line terminated at Sukkar, still some distance from the frontier mountains. Thus it was not much help in Gen. Stewart's advance to Kandahar, which cost the lives of perhaps 40,000 camels.

Late in 1878 a light 5′6″ gauge railway line was constructed from Sukkar to Sibi at the base of the Bolan Pass, a distance of 133 miles. An attempt to build a line through the nearby Hurnai Pass was halted after a military reversal. When Gladstone defeated Disraeli in 1880, Britain's foreign policy became less actively aggressive, and railway projects on the frontier were put on hold. However, in 1883 construction commenced on rail routes through both the Hurnai and Bolan passes, with that of the Hurnai using the standard Indian gauge and the more difficult Bolan line a narrow gauge.

Work on the Hurnai line turned into an immense effort, with British and Sikh soldiers, Afghans, Baluchis, and Pathans all engaged in the work under Major James "Buster" Browne. Construction continued for several years, with multiple engineering problems consuming 100 tons of dynamite. Periodic epidemics of cholera and malaria as well as dysentery and scurvy plagued the workers. Local tribes were hostile to this effort, and sometimes murdered or robbed soldiers and other railway personnel. In 1885 severe floods washed out bridges and a substantial portion of trackage.

In the meantime the Bolan Rwy. reached Quetta by August 1886, and was extended farther to Bostan, where it connected with the Hurnai line. In 1888 construction resumed, taking the line to Chaman near the Afghan border, an effort that required the digging of the longest tunnel in India. Afghanistan refused to allow penetration of the railway past this point, seeing it, no doubt, as an agent of British hegemony. It is perhaps worth noting that up to our present day, no permanent railway has ever been built in that remote land. By 1883, the line along the Indus River was extended from Sukkar to Peshawar. Just after the turn of the century, other railways were constructed in the area, but no line was ever built through the Khyber Pass.

In its August 21, 1891 issue *The Engineer* reported that there were 2,317 railway stations in India. Employed in its entire system were 4,607 Europeans, 5,673 "East Indians," and 252,413 "natives." By 1893 India's rail network had reached the impressive total of 18,042 route miles. This system was substantially superior to any other in Asia and the sixth largest in the world. While it advanced the purposes of the British, it also served the people of India by connecting the previously isolated portions of the country, improving commerce, helping in the creation of a middle class, and awakening a new awareness of national identity. Not all of these lines were under British control. In 1895 it was reported that native rulers owned 10 railways combining to a total of 1,400 route miles. The following year construction began on the Ahmedabad-Parantij Rwy., the first line begun by private enterprise and financed with rupee capital.

Railways, perhaps, also influenced the country in another subtle way. Quoting the London publication, *Iron,* the July 10, 1875 issue of *Railroad Gazette* gave the following report.

> In a lecture last week, before the National Indian Association, Mr. Framjee R. Vicajee, of Bombay, stated some curious facts regarding the social and political effects of the Indian railway system. As regards the first, the lecturer remarked that, as was shown by official returns, fifteen-sixteenths of the traveling public went third class, and that this promiscuous mixing of wholesales and retails was likely to have the humanizing result of breaking down altogether the strict line at present dividing one caste from another. In England, the only caste which the railway class system protects from contamination is that which is based on wealth, and from this caste chiefly proceeded the opposition to the Midland reform scheme; but

> in India, which is a poor country, really high class people travel third class to save their money, as Mr. Vicajee quaintly observed, while in England they travel first class to save or assert their caste.

In Ceylon (Sri Lanka), also under English control, railway development lagged behind that of India, partly because much of the large island was mountainous and required a considerable amount of tunneling. The 74½-mile Colombo-Kandy line was opened in 1867. This was followed by a 17-mile branch to Navalapitiya, inaugurated in 1874. The 28-mile Kalutara Rwy. was opened in 1879. In 1888 there were 182 miles of track in the country, with the Colombo-Nanuoya route being the primary line. By 1896 Ceylon's network had increased to 297 route miles.

The next area in Asia to feel the impact of railways was the Ottoman Empire. Circa 1850 this empire occupied a considerable area, extending from the European provinces of Bosnia, Serbia, Wallachia, and Moldavia, southward through Albania, Macedonia, and Thessaly; eastward to the Black Sea; and then across the Anatolian Peninsula into Central Asia; upward between the Black and Caspian seas; and then southward as far as Baghdad and Basra. Ottoman rule also extended over Syria, Lebanon, and Palestine. Although its size was vast, the empire was slowly disintegrating, and it had acquired the reputation of being "the sick man of Europe." This deterioration of Ottoman power fed Russian expansion in the region, and soon led to the loss of the territory between the Black and Caspian seas. The successful struggles leading to independence for Romania (Moldavia and Wallachia) in 1877, and Serbia and Bosnia in 1878, also had strong Russian backing.

It was against this tragic historical backdrop that railways were introduced to the empire. Shortly after the Crimean War ended in 1856, a British company constructed the first Ottoman railway between the Black Sea port of Constanta and Cernavoda, on the Danube, in what would later become Romania. Between 1857 and 1859 a second line was built by the English in Anatolia (modern-day Turkey). It connected the Aegean port of Izmir with Aydin, and used 10 Stephenson-built locomotives.

During the 1860s a few other railways were constructed. One of these, in Bulgaria, extended from the port of Varna to Ruse, on the Danube, opposite the Romanian city of Giurgiu. When a line was built from Bucharest to Giurgiu in 1869, traffic between the two railways was ferried across the river.

About this time in France, the Seguin brothers vigorously promoted the idea of an Ottoman rail line that would extend all the way to Basra, and later, to India. While they had the enthusiastic support of Emperor Napoleon III, this dream never materialized.

During this period the Ottoman government had little money for building rail lines, and primarily granted concessions only to companies that had raised their own funds. It did offer, however, certain guarantees of profits. When, in the late 1860s and early 1870s, these failed to materialize, the government reneged on its commitment. This had the obvious result of driving away investors.

In 1872, the prominent German engineer Wilhelm von Pressel was invited by the Ottoman government to devise a master plan for the construction of railways in the empire. He came up with a 10-year plan suggesting a 1,677.7-mile trunk line connecting Constantinople with the Persian Gulf. In addition, an extensive branch and feeder system of 1,118.47 miles was proposed. Sultan Abdul Hamid's government accepted these suggestions, but decided to give strategic and military concerns priority over economic interests in construction of the system. Also at the time, a new method was adopted for providing security to investors. It guaranteed railways a certain amount of income for each mile of line in operation.

Shortly, a number of new rail projects began. Lines were built from Nis to Vidin and from Kasaba to Alasehir. In 1873 construction for the Oriental Rwy., the most important line in the European provinces, was initiated. Completed in 1888, it extended from Constantinople through Edirne to Sophia and Belgrade, finally opening an all-land route for the famed Orient Express passenger train. From Edirne a branch extended to the Greek port of Salonica. The inauguration of this railway was briefly described in the August 17, 1888 issue of *Railroad Gazette*, as follows: "After the final spike had been driven, the railway was consecrated by pouring upon the virgin rails, so to speak, the blood of four properly sacrificed sheep, accompanied by a loud prayer delivered in chorus by the ulemas."

Another important Ottoman railway was the Anatolian line. Construction on this route, between the port of Izmit and Ankara, began in 1873, but was soon halted. Only in 1890 were efforts resumed. In two years the line had reached Ankara, and by 1895 was extended south to Konya. While most Ottoman railways were slow to stimulate commercial activities along their routes, the Anatolian line was a significant exception. After 10 years of operation, grain production in its service area had increased by 400,000 tons annually. Other railways in Anatolia included the Izmir-Cassaba, Mersina-Adana, and Moudania-Broussa lines. Additional extensions had also been made to the Izmir-Aydin Rwy., which by 1896 totaled 318 route miles. That same year the entire Ottoman Empire had 2,423 miles of track, with Asian railways accounting for 1,211 miles.

For the Arab provinces, railway construction came late. The first inaugurated was a 53-mile French-built meter-gauge line, completed in 1892. It connected Jerusalem with the port of Jaffa. The first locomotives, five in all, were built by Baldwin, while the carriages and wagons came from France. The railway had four iron bridges, and besides the termini, five stations. The uphill trip from Jaffa to Jerusalem required 3½ hours, ascending 2,476 feet in elevation. However, the downhill trip only took half that time. Later Arab lines included the Hamidie Rwy., named after Sultan Abdul Hamid, which ran between the port of Acre and Damascus, and the Beirut & Damascus Rwy. This latter route, also built by French capital, was first extended in 1894 from Damascus to the wheat-growing Hawran region, and then later into Beirut.

In 1897 the Ottoman government used its railways successfully in support of a war with Greece. About the end of the century, the Ottoman government embarked on its most ambitious railway project to date, beginning construction of a line from Damascus to Medina, primarily for the use of religious pilgrims. It would be some years into the twentieth century before railways reached Baghdad. Authorization for a Konya-Baghdad Rwy. had only been granted in 1903. However, in neighboring Persia, the Belgians had constructed that nation's first railway years earlier: a 10-mile line between Teheran and Shag-Abdul Azima, was inaugurated on June 20, 1888.

When Commodore Perry's American fleet visited Japan in 1854 a small locomotive and cars were brought for purposes of demonstration. The Japanese did not initially favor the idea of railways. However, in 1868, after the Meiji Restoration ended a long policy of excluding foreigners, planning began for a line between Tokyo and Yokahama. Helping to bring about this change were Marquis Ito and Count Okuma. Viscount Inouye was chosen to direct railway matters. The first railway had a 3′6″ gauge and was 18 miles in length. Built by the English, it was begun in 1870 and had its inauguration on October 14, 1872. The Vulcan Foundry built the first 2-4-0 locomotives used on the line. Emperor Mutsuhito, shunning the ancient custom of seclusion, was among the dignitaries at the opening and rode on the first train. Within a year this line had carried 1,223,000 passengers.

In its November 9, 1883 issue *Railroad Gazette* reported on this line. Its Tokyo terminus was built in stone and had two large entrances in front. Waiting rooms were provided, offering both foreign and domestic newspapers for sale. Passenger trains ran between this station and its Yokohama counterpart every 1¼ hours, leaving at both ends simultaneously. Waiting passengers were alerted for boarding by a bell rung five minutes prior to departure. At a second bell further access was barred and the train departed. On the route was Kanagawa, a place of religious pilgrimage. The 57-minute train ride followed the coast of the bay, passing through rice fields. The coaches were of three classes, with the first class having three compartments. Second-class carriages were open, in the American style. However, passengers sat at the sides, on two long seats facing one another. Third-class coaches were similar, but only offered benches for seating.

In 1873 construction on a 27-mile line connecting Kyoto and Osaka was begun. On May 11, 1874, a 20¼-mile line was inaugurated connecting Osaka and Kobe. *Railroad Gazette,* in its September 11, 1875 issue, offered this report: "Most of the subordinate officials are Japanese. The engine drivers, but not the firemen and foremen plate-layers, are foreigners. The works of the line, which include large river bridges, three short tunnels and some heavy earthwork, appear to have been constructed with care and solidity. They are reported to have withstood the action of heavy floods last year without injury or interruption of the traffic."

At some point Japan initiated a policy whereby the English constructed railways on the main Island of Honshu, the Germans on the southern Island of Kyushu, and the Americans on the northern Island of Hokkaido. Between 1881 and 1891 a 374-mile trunk route between Tokyo and Kobe was constructed. In 1893 the 150-mile Tokyo-Naoetsu Rwy. was opened. Construction involved penetrating through Japan's greatest natural barrier, Usui Pass. The Usui Toge Rwy., which operated through this section, had a 3′5″ gauge, and employed 0-6-0T engines using the Abt rack system. That same year the nation's network totaled 1,877 route miles. By 1896, 2,290 miles were being operated and 1,368 more were under construction.

Among those providing locomotives to Japan at the time were the British firms of Dubs, Neilson, Kitson, and Sharp Stewart. On April 26, 1893, the first engine built in Japan was successfully tested, and was soon followed by others. A two-cylinder compound type, it was constructed at the Kobe shops of the Imperial Government Rwys. In 1896 440 locomotives were operating in Japan. In 1897 Baldwin began building 2-8-2 Mikado type engines for the country. By this time a number of companies were manufacturing rail carriages and wagons. These included the Japan Railway Company, the Viscount Inouye Co. of Osaka, Nippon Shario Kabushiki Kaisha of Nagoya, and the Imperial Government Rwys. at shops in Tokyo and Kobe.

Railway development in China came somewhat later and encountered a nation more resistant to new ideas. The first line, between Shanghai and Woosung, was built by Jardine Matheson, a British firm, and opened on December 1, 1876. It was about 12 miles long and there were four stations, at Shanghai, Kangwan, Woosung Bar, and Woosung Creek. The first two locomotives, supplied by the Ransomes & Rapier Co. of Ipswich, were named *Pioneer* and *Celestial Empire*. Unfortunately, a local man was fatally struck on the railway shortly after its opening. The populace, being somewhat frightened of the trains and perhaps stirred up by higher local authorities, protested and demanded a halt to the line's operation. After the Chinese government reimbursed the British for the cost of the line in November 1878, it was dismantled. The scrap was removed to Taiwan and left to rust on a beach. A later railway was built in 1897, reconnecting these two cities.

In 1877 Marquis Li Hung Chang petitioned the Imperial Government on behalf of the China Merchants' Steam Navigation Company of Shanghai to allow the British-owned Kaiping Mining Administration to build a 7-mile line. It was to run from the Tangshan mining area near Tientsin (Tianjin) to the river so as to more easily supply the company with coal. This petition gained approval, but due to conservative opposition, construction was postponed. After a revision of the proposal four years later, this line was built using light rail and animal traction. It became the first railway built by the Chinese.

Shortly, a locomotive, secretly assembled from scrap materials, was put on the line. This immediately provoked strong opposition from the court officials who previously had blocked construction. They argued that the noise of the locomotive disturbed the tranquility of the ancestral spirits and that its smoke might damage the rice crop. This time, however, the more progressive side prevailed, and the engine received official approval. On June 9, 1881, it was christened "The Rocket of China." Two tank-engines were acquired for the line the following year. In August 1888 an extension was completed into Tientsin. By 1894 this line had been extended to a length of 175 miles, becoming the first part of the Peking-Mukden (Beijing-Shenyang) Rwy. In 1896 Marquis Li Hung Chang paid a visit to Great Britain. He traveled in a special train and was received as a guest of honor at the Neilson Company's Locomotive Works in Glasgow.

Besides Marquis Li Hung Chang, another early advocate of Chinese railways was Marquis Tso Tsung-Tang. He was interested in their construction for purposes of national defense. For this reason, he proposed the construction of a line along the Grand Canal from Peking to the Yangtze River. He died shortly after, and the Imperial Court passed over his proposal in favor of one that would extend a line from Peking to Hankow (Wuhan). Marquis Chang Chih-Tung, the leading proponent of this plan, went to Hankow, and soon built up an industrial complex for the manufacture of steel rails. This facility was the Hanyang Steel Works, which had the first Bessemer furnace in China. He also set up a mining company for the extraction of iron at Taiyeh and coal at Pingsiang. By 1894 a 20-mile railway had been set up to facilitate shipment of the coal to the furnace.

The disastrous Sino-Japanese War of 1894–95 left China in a reduced state. The world's major powers all soon made demands on China for concessions and territorial areas of influence. In 1896 Russia was able to get permission for extending its trans-Siberian Rwy. straight east through Manchuria to Vladivostok. This saved it from the enormous expenses of taking a more northerly route through its own territory. This section, between Vladivostok and Manchouli, became the Chinese Eastern Rwy., and was built between 1897 and 1904. Russia also soon gained the Liaotung Peninsula and built the ports of Dairen (Luta) and Port Arthur (Lushun). At the same time it was granted permission to build two railways, one from Linyu (Shanhaikuan) to Mukden and Kirin (Chilin) and the other from Linyu to Port Arthur and Dairen. In 1898 it received authorization to build a line from Port Arthur and Dairen northward to connect with the planned Chinese Eastern Line.

Great Britain obtained Weihai and Hong Kong, and extracted from the Imperial Government a nonalienation agreement covering the Yangtze Valley. It gained a number of important railway concessions totaling 3,000 miles, but construction was delayed as official attention was first turned to the African Boer War, and then to the Boxer Revolt of 1899. After the murder of two German missionaries in Shantung Province, Germany was able to use this as leverage, in 1897, for gaining access to Tsingtao (Ch'ingtao). Using similar pretexts in 1898, it gained permission for a line from Tsingtao to Tsinan (Chinan), and a second line from Tientsin to Ihsien.

France carved out its own territory of influence in South China, and got nonalienation promises covering the provinces of Guangxi (Kwangsi), Hunan (Yunnan), and Guangdong (Kwangtung). Its intentions seemed primarily directed toward establishing trade links and constructing railways in the southern provinces that could be connected to lines built up from Indo-China. Before the end of the century, France was able to gain permission for building a number of lines in this region. Previous murders of French missionaries and army officers helped to provide timely leverage. Japan, also, was able to gain authorization for building lines.

As early as 1896, Americans had been enlisted to survey the Peking-Hankow route. However, America's war with Spain distracted it from completing negotiations with China to build the line. Thus the concession went to the Belgians. The British, in 1896, had already completed an 80-mile section connecting Lukouchiao with Paoting. However it, too, was turned over to the Belgians. Few, if any, of these newly authorized lines were completed by the end of the century. However, most were under construction and, by 1903, China had 2,708 route miles of rail line.

In Southeast Asia rail development was late, sporadic, and primarily limited to coastal areas. In French-controlled Indo-China the earliest railways were the Phu Sang Thuong–Lang Son line, a narrow-gauge military railway in the Tonkin Area, and the Saigon-Mytho line, which ran 44 miles through Cholon. In 1896, when Paul Doumer became governor-general, planning began for a major railway system that would extend from Saigon to Tonkin. Construction began in 1898, but the first section of the line did not open until 1906. An 1898 plan proposed five railways, including one from Hai Phong to Yunan in China and another from Hanoi to Vinh.

In British-controlled Burma the Mandalay Rwy. in upper Burma was inaugurated circa 1888. The earliest railway in Thailand was a 20-mile line from Bangkok to the port of Paknan at the mouth of the Menam River. It was begun in July 1891 and inaugurated on April 11, 1893. By 1896 a line between Bangkok and Korat was under construction.

The Dutch, who controlled much of what later became Indonesia, also introduced railways. All of these were of a 3′6″ gauge. The Sumatra Rwy. had both standard 0-6-0 tank locomotives and rack locomotives of the Riggenbach type for steep grades. Maschinenfabrik Esslingen built the latter. Both coal wagons and passenger carriages were built in the American style, with the latter being made of teak wood at the railway workshops at Padang and built in two classes.

The stations and other facilities were described in the May 10, 1895 issue of *Engineering* as follows:

> At Emmahaven, Padang, and Padang Panjang having a platform and one track roofed over; at Fort de Kock and Solok only the platform is covered; at the roadside stations no covering is provided. Principal stations contain a post and telegraph office, the offices necessary for the service, and a refreshment room. Roadside stations are all of uniform type, and comprise a room which serves as station master's office, ticket office, etc., and a covered shed for the shelter of passengers. At some of the country stations, goods sheds have been erected of brick, with corrugated iron roofs. No carriage or coal sheds have been built. The coal is stacked near the water cranes, and placed in readiness in rattan baskets on small wooden platforms, and is taken on the engine whilst watering. At the principal stations, and at those roadside stations where rack-rail sections commence, running sheds for six or nine locomotives have been built. They each contain the usual hydrants, ashpits, a small work shop for light repairs, and an office for the superintendent.

Australia's history of railways began in 1827 with the building of an inclined tramway at Newcastle. It was constructed to move coal from a mine to a nearby waterfront and used horse traction. In 1836 in Tasmania a 5-mile wooden tramway was constructed at the Port Arthur prison settlement. The small wooden cars that carried visitors across the Tasman Peninsula were pulled with convict labor. On October 4, 1849, the Sydney Rwy. Co. in New South Wales was formed. In 1850, it began rail construction under the Irish engineer F. W. Shields, who later quit his position over a salary dispute. Shields had begun the line using the standard Irish gauge of 5′3″. However, James Wallace, the Scotsman who replaced him, insisted on changing the previous work and building the line in the standard English 4′8½″ width. The first section of this line, one of 13½ miles connecting Sydney and Parramatta, was inaugurated on September 26, 1855. Stephenson-built locomotives were the first used on the line.

In the meantime, other rail lines had made their appearance. On May 18, 1854, the Port Elliot & Goolwa Rwy. in South Australia was inaugurated. This line began passenger service over its 7-mile length using horse traction. On September 12, 1854, the 2½-mile Melbourne & Hobson's Bay Rwy. in Victoria was inaugurated, offering the first steam service in the country. Unfortunately, two Stephenson locomotives ordered for the line were delayed in shipment and did not arrive in time for the inauguration. Informed in advance of this, the company hastily constructed a primitive locomotive for the event. Mechanical failures proved an embarrassment, however, and the line had to be closed until the other engines arrived. This railway used the 5′3″ gauge, having made changes in its original construction plans to conform with Shields' work on the Sydney line. This led to a gauge dispute that was never resolved, with Victoria and New South Wales holding fast to their chosen widths and Queensland, Tasmania, and Western Australia all later deciding for a gauge of 3′6″. Other early Australian railways included the Adelaide-Gowler Bridge line, which opened in 1859, and the 95-mile Melbourne & Ballarat Rwy., which opened on April 10, 1862.

There were few navigable rivers in Australia and railways were very much needed. However, the country lacked capital and financing was difficult. Beginning in 1855 with the Sydney line, the government came to the aid of a number of struggling railways. Between 1850 and 1890, the country's population increased from 400,000 to 3,250,000. Australia had about 1,000 route miles of railway in 1870, but as population and capital increased, rail construction exploded. Much as in America, but on a reduced scale, rail lines helped to bring in new immigrants and new prosperity for industrial cities, agricultural areas, and mining communities. The first locomotives built in Australia were completed in 1870, one by Railway Works, Sydney, and another by Mort & Company's Dry Dock Engineering Works, Balmain. Later engine builders include Victoria Foundry, Ballarat; Phoenix Foundry, Ballarat; Adelaide Locomotive Works; and Vale & Lacy, Sydney.

After construction of a bridge over the Murray River, the rail systems of New South Wales and Victoria met at Albury on June 14, 1883. That same year lines from New South Wales and Queensland were connected at Jennings. On May 1, 1889, following completion of a bridge over the Hawkesbury River, direct rail service was inaugurated between Sydney and Newcastle. This provided a final link in a through rail connection that extended from Farina in South Australia, through Adelaide, Melbourne, and Sydney, to Charleville in Queensland. By 1897 Australia had 13,888 route miles of line.

In New Zealand there was an even greater problem concerning population and adequate capital. The first line in the country was the horse-drawn Dun Mountain Rwy. on South Island, which opened on February 3, 1862. The first steam line in New Zealand was the Lyttleton & Christchurch Rwy. on South Island, inaugurated on December 1, 1863. It was 4¼ miles long and had a gauge of 5′3″. The *Railroad Gazette* in its October 26, 1872 issue reported that an English firm had contracted with the government to build 125 miles of line. These included a 42-mile railway from Invercargill to Mataura, a 41-mile line from Auckland to Waikato, and a 34-mile route from Dunedin to Clutha. At the time it was projected that a 400-mile trunk line would be built through the provinces of Southland, Otago, Canterbury, and Nelson. Other early lines in the country were the Otago G.N. Rwy. and the Rimutaka Incline line. This 3-mile railway, beginning service in 1878, used the Fell center-rail adhesion system. By 1877 all of New Zealand's railways had been converted over to a national gauge of 3′6″. Total route mileage in the country was reported to be 44 in 1870, 542 in 1875, 1,253 in 1880, 1,654 in 1885, and 2,036 in 1893.

Rail technology found its way to even more remote destinations in the Pacific. Among these places was the island of Oahu in Hawaii. The April 13, 1888 issue of *Railroad Gazette* mentions the 3′ gauge Hawaiian Rwy., a 20-mile line skirting the coast. This railway had both British- and American-made rolling stock. In its September 24, 1897 edition *The Engineer* briefly reported on the Oahu Railway & Land Co. The main line of this company was from Honolulu to Waianae, a distance of 33.8 miles. However, branches and sidings increased this figure to 38.5 miles. The company also operated three horse and cattle ranches, and in addition, owned scows, wharves, and coal elevators at Honolulu. There were plans at the time to extend the line to Kabuku. Also on the island were private plantation lines built for hauling sugar and pineapples. By the early part of the twentieth century Oahu's various small railways totaled over 1,000 route miles. However, today, none remain in operation.

Top: Scene on East Indian Rwy., 1863. Bottom: Oxen pull mixed train on 2′6″ gauge Meagaum-Dauboi tramway constructed by the Guicowar of Baroda (Gujerat), 1863.

PLATE 153

Top: Departure of Duke of Edinburgh from Parell Station, Great Indian Peninsular Rwy., 1870. **Bottom**: Terminus of the Madras Rwy., designed in the Romanesque style by George Hardinge, Architect, 1873.

Tigers move to escape an approaching train, 1895.

PLATE

155

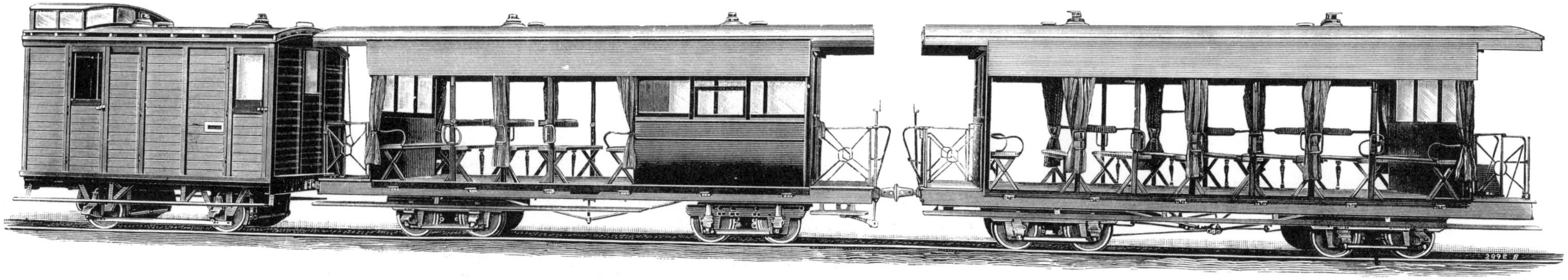

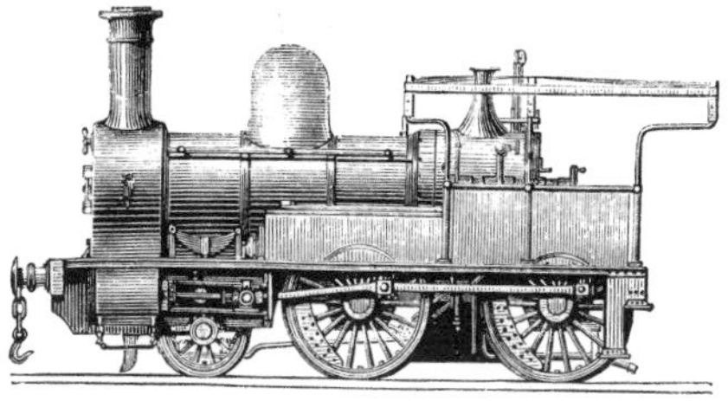

PLATE 156

Rolling stock for Indian railways: TOP: Passenger carriages and brake van for 30″ gauge line, Alexander Penney & Co., London, 1895. MIDDLE LEFT: High-sided open goods wagon for same line. MIDDLE RIGHT: Sharp 0-8-0 goods engine, Indian Branch Rwy., 1865. LOWER MIDDLE CENTER: 2-4-0 engine, Maschinenfabrik Esslingen, East Indian Rwy., 1867. BOTTOM LEFT: Sharp 0-4-0T, Darjeeling & Himalayan Rwy., 1895. BOTTOM RIGHT: 2-4-0 passenger engine, Dubs & Co., Glasgow, Oude & Rohilcund Rwy., 1886.

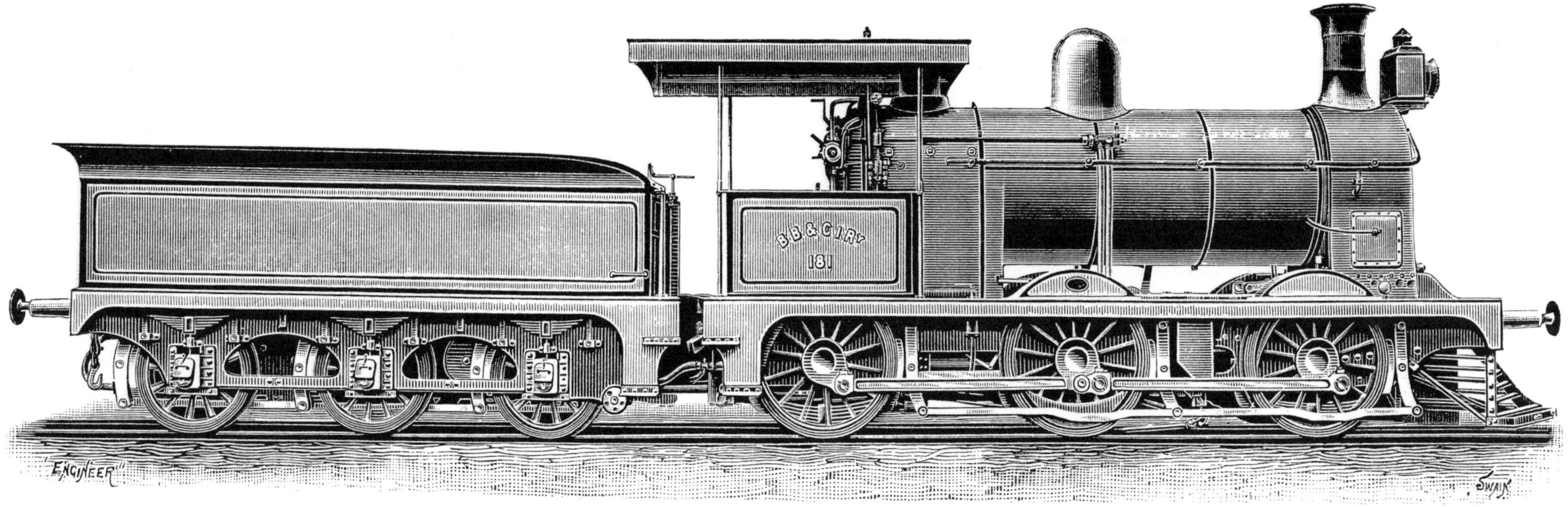

Top: 0-6-0 compound goods engine, Sharp, Stewart & Co., Bombay, Baroda & Central India Rwy., 1895. Bottom: 0-8-0T engine for use on the Bhore Ghaut Incline, Neilson & Co., Great Indian Peninsular Rwy., 1873.

PLATE 157

Top: Special carriage built for sultan of Turkey to coincide with inauguration of Istanbul-Andrinople Rwy. **Bottom:** "Railway travel in India."

PLATE
158

Railways in the Turkish Empire: Top: Station and facilities at Philippopel, 1876. Bottom: The first rails laid in Syria, the Tripoli Tramway, 1880.

Japanese railway scene in Shinagawa, a suburb of Tokyo, 1887.

PLATE 160

Inauguration of China's Tientsin-Tschingyang Rwy., November 20, 1886. This line used portable Decauville equipment.

PLATE 161

PLATE 162

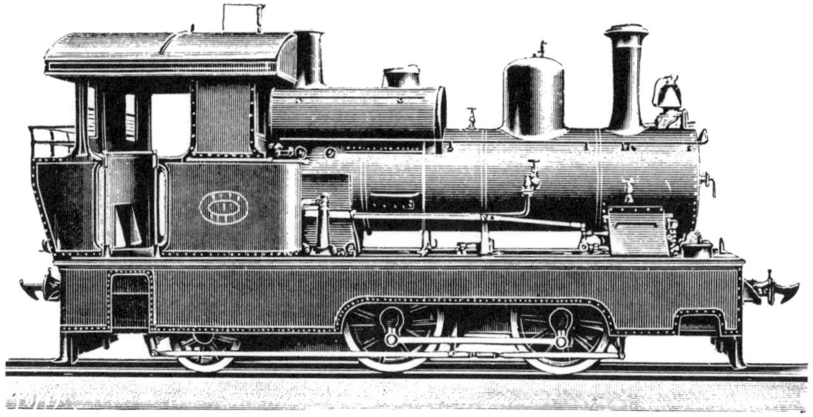

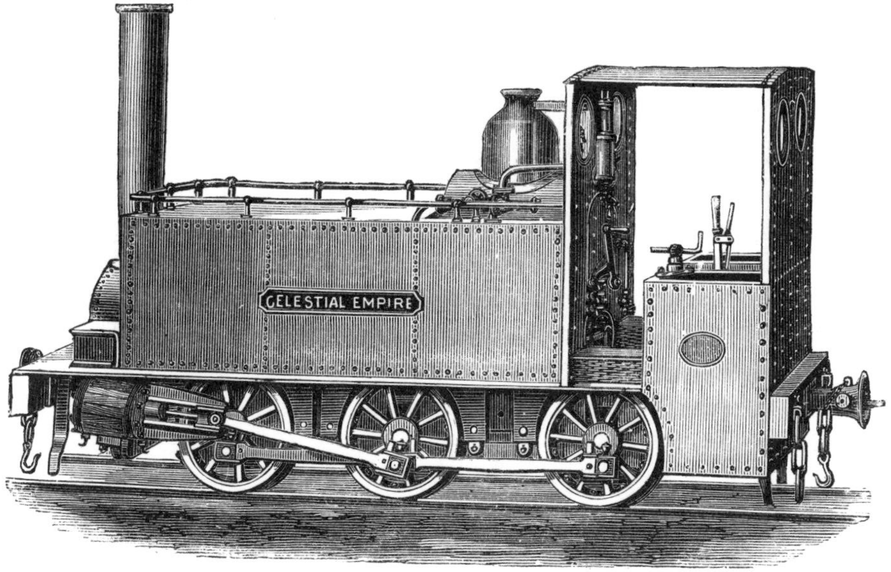

TOP: 4-4-0 passenger engine built for Imperial Chinese Rwys., Dubs & Co., Glasgow, 1887. MIDDLE LEFT: 0-4-2T oil-burning compound engine for use in Java, Sachsische Maschinenfabrik, 1900. MIDDLE RIGHT: China's first locomotive, Ransomes & Rapier Co., Ipswich, England, Shanghai-Woosung Rwy., 1876. BOTTOM LEFT: Application of Decauville railway technology at a sugar cane plantation in Java, circa 1880s. BOTTOM RIGHT: China's second engine, Ransomes & Rapier Co., Shanghai-Woosung Rwy., 1876.

PLATE 163

Moorabool Viaduct, Melbourne & Ballarat Rwy., Australia. This line of 95 miles opened April 10, 1862, and served a gold mining area. The bridge was 1,450 feet in length and 110 feet high.

TOP: Baldwin 2-6-2 Prairie type passenger engine, New Zealand Rwys., 3′6″ gauge, 1885. BOTTOM: Rogers 2-4-2 passenger engine, New Zealand Rwys., circa 1878.

PLATE 164

CHAPTER

12

African Railways

DURING the nineteenth century Africa was in the process of being colonialized by European powers. Thus it was countries like England, France, Belgium, and Germany who first introduced railways to the continent. Virtually all of these lines were built either for strategic reasons or to extract resources. Strategically, they could be used to either expand or reinforce the colonial presence in the region. Railways provided the means for rapid movement of troops and supplies, and on occasion, were used effectively to support military campaigns, against both native forces and competing colonial powers. This strictly military dimension of African railway history, including the Boer War, is covered separately in chapter 20. Otherwise, railways established important communication and trade links, and as in all other countries, made possible the greater flow of people, goods, and ideas.

The idea for an Egyptian railway was first proposed by Thomas Waghorn. Waghorn, a river pilot, was an early promoter of a Mediterranean–Red Sea route between England and India. At the time, all transport between Britain and its Indian colonies moved by sea around the bottom of Africa. In the 1830s the English engineer Thomas Galloway conducted surveys in Egypt for a railway. However, this project was dropped because of economic constraints and strong foreign protest.

Later, in 1852, construction of an Egyptian railway was begun under the direction of Robert Stephenson. A line was first built from Alexandria to the Nile, and then upriver to Cairo. This 120-mile route was inaugurated on January 1, 1856. When a railway between Cairo and Suez was completed in 1858, Waghorn's dream became a reality. Soon most of the passenger and mail traffic, which had previously gone around the Cape of Good Hope, traveled over this railway. However, train speeds were slow, only averaging about 10 m.p.h. During this period Stephenson had a special passenger train built for use by Khedive Abbas I, the Egyptian ruler.

Unfortunately, Abbas died in 1854, and never got to ride on his elegant new train. He was succeeded by Muhammad Said, a friend of Ferdinand de Lesseps, a Frenchman who was promoting a Suez canal scheme. Having already built their railway, the British strenuously opposed this plan. Later de Lesseps made a trip to England, and to some extent, overcame these objections. Finally, in 1866, a concession was given to de Lesseps for his canal. By November 1869 it was sufficiently navigable for a festive inauguration to be held. Once the canal was opened for regular traffic, competing railway business came to a standstill. As a result the Cairo-Suez line was dismantled. However, rail construction continued up the Nile, eventually reaching Aswan.

Circa 1860, the French began building railways in Algeria. Early lines included the Algiers-Blida Rwy., inaugurated on August 15, 1862, and the Oran-Relizane line, which was later extended to Algiers. A general plan for the country's railway development was drawn up in 1879, taking into consideration both commercial and strategic needs. Of primary commercial importance was a central trunk route from Tunis in the east to Morocco in the west. Built in the standard gauge, it was to run through the most important cities of Algeria's interior. Subsidiary narrow-gauge lines were to be built from the seacoast to connect with the trunk route at various points. This central line was completed in 1884, and by 1888 Algeria's rail system totaled 1,683 route miles.

At the time connections had been established from all of Algeria's port cities to the trunk line. There were six railways in the country that year, with the East Algerian Co. operating 492 miles; the Franco-Algerian Co. 415 miles; the Paris, Lyon, Mediterranean Co. 318¾ miles; the Bone Guelma Co.

271½ miles; the West Algerian Co. 164½ miles; and the Mokta-El Hadid Co. 20½ miles. In 1888, a line was constructed southward to Beskra, at the edge of the Sahara. Earlier, in 1880, a survey had been made for a trans-Sahara railway. The idea for such a line was to connect the French possessions of Algeria, Senegal, and Gaboon-Congo. Although a route south of El Golea seemed feasible, the line was never built.

Circa 1871, the French began building railways in Tunisia. These were mostly of the meter gauge. The Tunis-Bardo and Tunis-La Goulette and La Marsa lines were both authorized that year. A line from Tunis to Dacha Djanbot was completed in 1876. By 1878 this route had been extended to the Algerian border. Later Tunisian railways included the Tunis-Bizerta and Tunis-Sousse lines, both completed in the mid-1890s.

The railways of North Africa were similar to those of Europe in that they were largely built between cities having significant levels of population and long histories of commerce. In the rest of Africa, where more primitive social and economic patterns prevailed, this was not the case. There, much like in America, railways had a pioneering role. They helped to establish commerce and centers of population, where none previously had existed. South Africa had become such a pioneer region, with Dutch settlers originally occupying the area. However, as Britain expanded its colonial interests in India, South Africa became important, and Britain established a presence there as well. Later a war was fought and the Dutch, or Boers, as they were called, lost. They retreated into the interior and established the separate republics of Transvaal and the Orange Free State.

The first railway in South Africa was a 2-mile line, opened on June 26, 1860, between Durban and The Point. The Cape Town-Wellington Rwy. was completed about the end of 1863. In 1870 South Africa had a total of 69 route miles of track, all but six being in the Cape Colony. In 1872 the Cape government took over the existing railways and established a standard gauge of 3′6″. About this time rail construction was greatly stimulated by the discovery of diamonds near Kimberly, and later, in 1886, by the gold rush in the Witwatersrand area of Transvaal.

The terrain of South Africa consisted mainly of vast plains, separated by ridges of hills, but with few mountains. Initially this seemed ideal for railway building. However, once construction began, problems revealed themselves. The Cape was very dry, and had been largely denuded by bush fires and casual tree cutting. This intensified the action of the sun and desert winds, leaving a hard surface of baked clay on much of the land. Rainfall in the Cape was minimal, and that of the Orange Free State only somewhat better, except near Table Mountain. Thus there was a general scarcity of both water and wood, excluding a few areas like the Anatola Forest. Natal and the east coast were better provided with both trees and moisture, although rainfall was mainly limited to the summer. The major portions of the Transvaal and Bechuanaland also were dry.

Commenting on this problem, the October 27, 1893 issue of *Engineering* said: "On the whole, the effect of the climate of South Africa is such that the seasons consist of long protracted droughts, interspersed by brief periods of serious storms and floods. The rain, being all bottled up, so to speak, for long periods, falls, as it were, all of a heap. No ordinary water openings along the railways can possibly provide for the escape of these avalanches of water, so little of which can possibly percolate below the barren and baked surface of the ground, and most of which flows rapidly off as surface water; hence the railways in South Africa are perpetually exposed to being carried away, and new waterways have constantly to be added at great expense."

After diamonds were discovered near Kimberly, a line was built there from Cape Town, opening in 1885. Later, after Cecil Rhodes gained power, this railway was extended to Mafeking and finally to Bulawayo in southern Rhodesia (Zimbabwe). Following the discovery of gold in the Transvaal, railway construction began toward that area. In 1886 an Anglo-American company received permission from Portugal to build through its East African territory (Mozambique) in order to reach the Transvaal goldfields. Beginning from the port of Lourenco Marques, on Delagoa Bay, 54 miles of the line had been laid by 1888. However, 4 uncompleted miles remained within Portuguese territory, and because of difficult terrain, construction was not finished within the time limits specified by the charter. At this point the Portuguese government declared the concession forfeited, and attempted to seize the railway. Not pleased to have its interests threatened this way, England sent a gunboat to the scene and the Portuguese backed off.

Later, the Delagoa Bay & Transvaal Rwy. ran into financial difficulties. There were construction delays and the line was only rescued with the aid of Dutch financing. Finally reaching Pretoria, it was opened on July 8, 1895. However, the British-built Cape Elizabeth-Pretoria Rwy. had arrived there first. It was inaugurated on January 1, 1893. This line also had stops at Bloemfontein and Johannesburg. Other South African railways included the Durban-Johannesburg line, the Cape Central Rwy., which connected the Cape

Town–Bulawayo and Cape Elizabeth–Pretoria routes, and the Grahamstown–Port Alfred line.

The entire African continent had only 1,083 route miles of track at the end of 1871. In 1880 this figure had grown to 2,842 route miles. By 1884 route mileage had climbed to 4,075. Of this 40% was in the British colonies of South Africa, 29% in the French colony of Algeria, and 23% in Egypt. In 1893 total African route mileage was 7,208. By 1897 this figure had increased to 8,169.

Elsewhere in Africa, during the nineteenth century, fragmentary railway systems were built at points largely isolated from each other. The Dakar–St. Louis Rwy., in French West Africa, was inaugurated in 1885. Built through a rich peanut-producing area, it soon became very successful. This line connected St. Louis, now the capital of Senegal, with the port of Dakar. A little farther to the east, in the French Soudan (present-day Mali), the French constructed a military railway between Medine and Bafoulabe, a distance of 69.59 miles. In 1897 there were plans to extend this line eastward to Kona, as it would command the whole of the British Niger Territory.

During the 1880s the 2′ gauge Beira Rwy. was completed in Portuguese East Africa. The 200-mile Harar-Djibouti line, begun in 1894 by the French, connected the port of Djibouti to the Ethiopian city of Harar. This line was difficult to build, as it was begun at sea level and then ascended to an elevation of 6,000 feet.

The Kayes Niger Rwy. was first planned in 1863, but construction only began in 1881. The builders of this route encountered many obstacles, including difficult terrain, tribal conflicts, yellow fever epidemics, and severe logistical problems. This route was completed in 1906. In German East Africa (Tanzania) the first portion of the meter-gauge Usambara Rwy. was opened in 1895. It connected Tanga with Pongwe, an island town on the Pangani River.

By 1897, 117 miles of the Congo Rwy. was under operation in the Belgian Congo (now Zaire). Having a 2′7½″ gauge, it ran from Matadi to Tumba. However, construction continued toward Stanley Pool (present-day Kinshasa), which was reached March 18, 1898. Although this 240-mile line was under Belgian control, many of its construction workers were British subjects from Sierra Leone. In 1900 the Conakry-Niger Rwy. was inaugurated. It ran from the port of Conakry (Guinea) into the interior. Also at the time railways were being built in Uganda and elsewhere.

Cecil Rhodes' great dream was to build a railway from the Cape to Cairo. Although such an effort would have encountered many obstacles, it probably could have succeeded, given the colonial realities of the time. Another railway dream never realized was a transcontinental route across the middle of Africa connecting Lagos with Berberah. In the earlier part of the twentieth century a number of other railways were built, yet the African network remained fragmentary, a sort of ghostly skeleton for the potentials that were never realized. In two World Wars, Europe squandered its wealth and manhood, and lost these colonial possessions. While this is probably just as well, something was lost as a result, perhaps a guiding vision and a certain discipline. While the colonial political systems were no doubt harsh and brutal for those forced to serve them, still they could mobilize individuals in efforts that transcended local or tribal realities. The infrastructure of railways remaining in Africa today perhaps represents the best aspect of the colonial legacy.

TOP: Inaugural train of Algiers-Blida Rwy., Algeria, August 15, 1862. View showing station of Hussein-Dey. BOTTOM: Special "Summer Saloon" carriage constructed at Saltley Works, Birmingham, for the viceroy of Egypt, 1858.

PLATE
165

North African railway inaugurations: Top: Ceremony in Algiers for French-built line to Blida, August 15, 1862. Bottom: Inaugural train at Tunis Station on Tunis–La Goulette Rwy., August 31, 1872.

PLATE

166

PLATE 167

During a famine in Algeria, starving people rush to luxury Wagons-Lits dining car in hopes of begging food from the halted train.

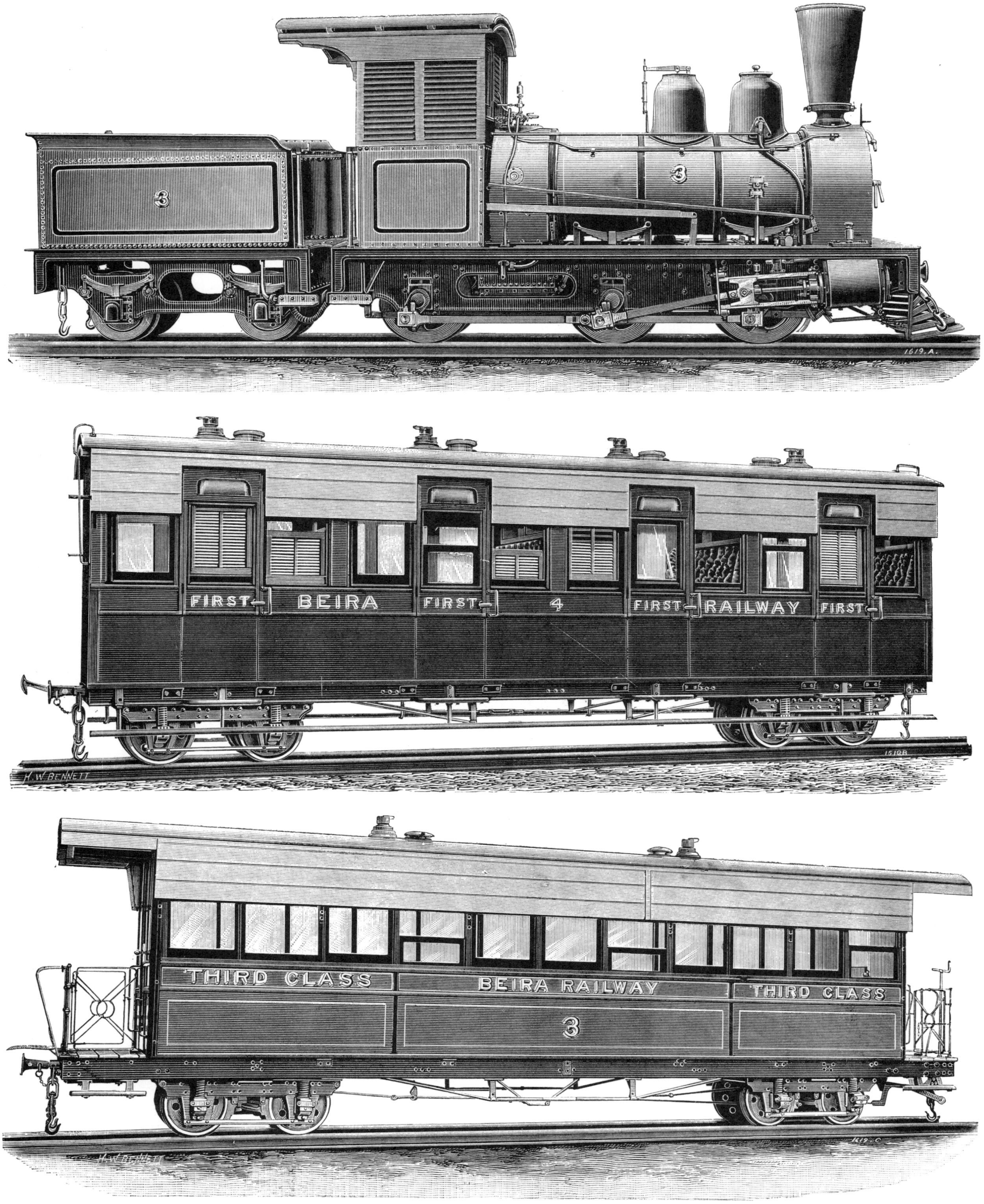

Rolling stock for the Beira Rwy., Portuguese East Africa (Mozambique), Brush Co. Falcon Works, Loughborough, 1893: Top: 0-6-0 engine. Middle: First-class coach. Bottom: Third-class coach.

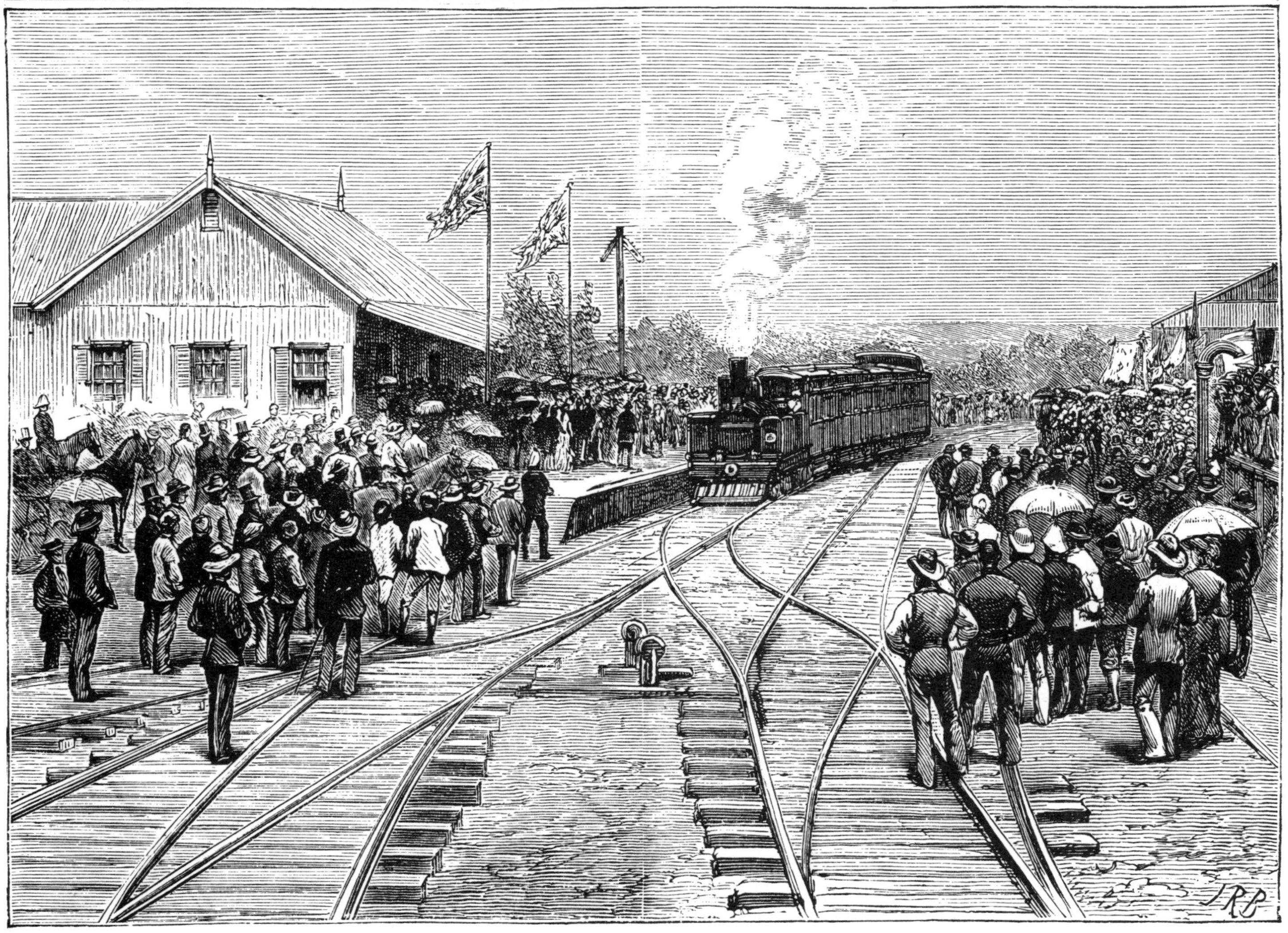

Top: Sheep crossing track in front of a train, Durban-Johannesburg Rwy., South Africa, circa 1890s. **Bottom:** Inauguration of the 60-mile Durban-Pietermaritzburg line, Natal Province, South Africa, December 1, 1880.

Construction on the Congo Rwy., 1898: Top: Tank engine pulling living quarters of European personnel. Bottom: Work train. The Belgian company for this 240-mile railway was formed in 1887. Construction began in 1890, and the terminus of Stanley Pool (now Kinshasa) was reached March 18, 1898.

PLATE 170

CHAPTER

(13)

Passenger Service

PASSENGER carriage design was slow to develop in Britain. The Stockton & Darlington's *Experiment* of 1825 was essentially a stagecoach with flanged wheels. Beginning circa 1830, and for some decades afterward, first-class coaches were normally built with three compartments. In essence these consisted of three stagecoach bodies joined to each other and mounted on a four-wheel sprung chassis. Each compartment was separate, and had doors, windows, and bench seats. The earliest second-class carriages were often open, but usually had a canopy overhead. These were soon enclosed, but compartments were made smaller so that extra ones could be added. First- and second-class coaches typically had oil lighting and spring buffers. Initially, luggage was stored on the roof, but on some coaches, this space was used for extra passengers. However, as speeds increased this practice became more dangerous, and it was soon discontinued. The Great Western line introduced the first six-wheel coaches. These longer carriages had a third axle providing support in the middle with lateral movement allowed in the journals so turns could be made.

Into the 1840s third-class coaches remained open, leaving passengers vulnerable to the weather. In some cases, there were no benches to sit on, and riders were expected to stand. Usually, also, these carriages were attached to slower goods trains. In 1844 Gladstone passed an act requiring all future carriages of this class to be enclosed. These new third-class coaches came to be called the "parliamentary" type. Gladstone's legislation also stipulated that railways provide third-class passengers at least one train a day, traveling at a minimum speed of 12 m.p.h., and that the fare for such service not exceed a penny per mile. While railways complied with this new law, some kept the open carriages, adding a fourth-class option.

A brief critique of the new "parliamentary" passenger carriages was provided by the *Pictorial Times* in its September 13, 1845 issue: "Bristol & Gloucester: No night lamp, one door only on each side. A sight of the country is confined to the passengers who are fortunate enough to get near the door. No provision is made for the admission of air in bad weather, when the doors and windows are closed. This badly lighted and ventilated carriage carries 54 passengers. Eastern Counties: No night lamp; when the shutters are closed there is no look-out, and no air but that which steals its way through four Venetian blinds. Great Western: No night lamp. No look-out for seated passengers. When the shutters are closed the carriage is badly lighted by Venetians." The nicest of these coaches was that of the Midland line. Besides having a stylish design, it provided decent lighting, comfortable seating, and a window on the door of each of its three compartments. It might be mentioned here that no effort was made to heat any British passenger carriages until 1852.

Initially American passenger coaches followed the English precedent, evolving from the earliest stagecoach variety to the type in which carriage compartments were joined together. With the addition of bogies, the length of these cars was often extended, and some were built with a dropped frame. However, a third stage in car design quickly followed, in which individual compartments were eliminated in favor of a longer open style. Initially long benches, running the width of the car, were employed for seating.

An early coach of this type, the *Columbia,* built by Ross Winans in 1831 for the Baltimore & Ohio R.R., had open platforms at each end and an entrance in the middle. Most early cars of this style had no springs, but these were soon added. Center aisles and backs for the seats also shortly appeared. Having more space than the compartmentalized type, these coaches were also safe to heat. This was done with coal stoves. A minimal amount of lighting was provided by placement of a glass-covered lamp at each end. Ventilation for these coaches was a problem, however, and one often had the choice between breathing fresh air tinged with engine soot or a stale, and often unpleasantly warm, "atmosphere."

This new American coach, described at the time as "a democratic palace instead of a nest of aristocratic closets," represented a radical departure from the British design. There, the privacy offered by separate compartments, not to mention the rigid division of coaches into different classes, was considered essential, as it prevented the sort of democratic mixing of classes that many Britons found distasteful. Thus while the new American cars represented an advanced stage of development in terms of design efficiency and economy, their acceptance in England, and on the Continent, would be delayed for many years.

One important car builder of this period was Richard Imlay of Philadelphia. He made an early attempt to improve ventilation by raising the center part of the roof the length of the coach. This became known as the clerestory roof. However, this innovation was quickly forgotten until the 1860s. Imlay also put water closets in a number of his coaches, and made some early sleeping cars. Unfortunately, his business failed in the aftermath of the 1837 panic. Beginning circa 1840 a new arched style of roof appeared, replacing the previously popular flat roof. This style in turn gave way to the clerestory type in the 1860s.

In the early years of railroads food and drink were either carried onboard by the traveler or purchased from local people at train stops. Soon stations and trackside restaurants began offering food and refreshment, but this was often of poor quality. The first dining cars were built by the B & O line in 1848, and offered simple meals. However, for many years to come most railroads continued to schedule meal stops.

On the Continent early passenger coaches were greatly influenced by the English model. In many instances these were built by carriage builders who converted or expanded their operations to serve the new railway industry. French coaches typically came in three classes: enclosed carriages built on the stagecoach model, open carriages, and wagons. In 1839, a coach of each class was ordered by the Koechlin firm at Mulhouse. Using these as prototypes, the company began manufacturing its own carriages. The Montpellier-Sète Rwy. offered five classes of passenger service. Its first-class coaches were built by S. Chaumont of Paris. Its second-class carriages were constructed in Montpellier. They had three compartments and could carry 30 passengers if the upper deck seats were used. In 1840 the St. Etienne–Lyon line introduced longer bogie cars of the American type. Entrances and exits were made from platforms at each end. This style did not catch on with other railways, and these carriages were later replaced with more conventional designs. Enclosed double-deck passenger carriages appeared in the 1850s. They were first used for suburban traffic on the Paris-Lyon and Ouest lines, and quickly became popular.

While trains initially made all stops along routes, a new practice emerged where certain trains omitted stops where traffic was minimal. This led to the concept of the express, which, by eliminating all but the most important stops, provided much faster service. Other trains continued to make all stops and came to be known as "locals."

Beginning circa 1838, in Britain, railways constructed special luxury coaches for state use. In 1843 Queen Victoria traveled in a lavish carriage over the London & Birmingham line. The following year the South Western company built a similar coach for use by the royal family when traveling to the Isle of Wight. Joseph Beattie, superintendent of the company's carriage department, designed it and oversaw its construction. In its October 26, 1844 issue the *Illustrated London News* gave this description:

> The carriage, externally, is 17 ft. in length by 7 ft. 4 in. wide. It is placed on wooden wheels which have been successfully used on this railway, and, from their principle of construction, afford comfort and perfect security. The ground color of the body is the same as that of her Majesty's private carriages, dark maroon; in the center panel are emblazoned the royal arms; the commode and door handles are massive silver, with the royal arms elaborately chased; and the windows are of superbly embossed and ground glass. A crown and cushion surmount the center of the body, and along its entire length is an elegant scroll cornice; the carriage itself being similarly decorated.
>
> The interior is divided into two compartments, ⅔ of the length being assigned to that occupied by her Majesty and Prince Albert; the other compartment being for the royal children and suite, and accessible by an almost imperceptible door, opening into either compartment, and hung on a patent self-adjusting hinge. The lining is of light drab flowered silk damask, trimmed with crimson and white silk, and embossed pile figures. The roof or ceiling, is of white damask, embroidered with crimson; and at each angle are embroidered the national emblems in silver, encircling the imperial crown of gold. In the center of each ceiling is a crown, surrounded by the rose, shamrock, and thistle, carved, and painted after nature. Immediately over each of these crowns is placed one of the improved lamps . . .

> The draperies throughout are of crimson and white figured satin damask, richly lined and trimmed, and surmounted by a white and gold cornice. The blinds are of peach-blossom silk, with crimson silk and silver tassels. Beneath the center window, and near the seats for her Majesty and the Prince, are placed white marble slabs upon white and gold consoles. The carpet is of the finest Axminster manufacture; beneath it, and between the framework and body of the carriage, is a layer of the patent Kamptulicon (composition of cork and Indian rubber) which entirely prevents the unpleasant vibration incident to railway travelling.

Perhaps most artistic in its decoration was a carriage built for Pope Pius IX, and used on the Rome-Naples route. It had three compartments: an anteroom, a saloon containing a chair of state, and sleeping quarters. The February 12, 1859 edition of *Frank Leslie's Illustrated Newspaper* offered this description:

> Externally the car is ornamented with bas-reliefs in bronze gilt, representing the Apostles, with figures of Faith, Martyrdom, etc., and other Roman Catholic emblems. The Papal arms are conspicuously displayed upon the side of the car, surmounted by a cross, while a large tiara crowns the front compartment. On the panels are crosses, and the letters P.P.IX., together with heads of cherubim . . . The interior of the carriage is ornamented in a manner still more luxurious than the outside. The sleeping compartment is an anodyne in itself; the waiting room is sufficiently splendid to reconcile any one to an hour or two's delay of audience; and the central saloon is decorated with every variety of sanctified ornament. It contains no other chair than that provided for the Pope; but a small table is placed in one corner.

It is perhaps only natural that sleeping cars would develop in America rather than in Europe. As opposed to the European traveler, the American had fewer options. Trains were slower, travel distances often longer, and many rail routes traversed sparsely settled regions offering few accommodations for the weary traveler. The initial idea for a sleeping car apparently goes back to a railroad surveyor, R. F. Morgan, who displayed a design for such a car at Faneuil Hall, Boston, in 1829. In 1836 the Cumberland Valley R.R. built the first sleeper, operating it between Baltimore and Philadelphia.

It was not long before other companies copied this idea. It was reported that triple-decked sleeping berths were used in 1848 on a line between Atlanta and Griffin, Georgia. Later, Theodore T. Woodruff developed an idea for a hinged, pull-out berth, located above the car windows, and secured a patent in 1856. The man who really capitalized on this idea, however, was George M. Pullman. After taking his first ride on a sleeper in 1858, he immediately became interested in their design. Beginning in 1859, he tested a number of ideas on the Chicago & Alton Railroad. Shortly, however, he left for Colorado and made a small fortune selling supplies to miners.

Returning to Chicago in 1864, he organized the Pullman Palace Car Co., and introduced his luxury sleeper *Pioneer* the following year. In a story disputed by some railroad historians it is said that after President Abraham Lincoln's assassination, the *Pioneer* carried his body on the funeral train. Whatever the truth of this story, the *Pioneer* received considerable publicity and soon similar Pullman cars were operating on a number of railroads. In 1867 Pullman introduced a combination sleeper-diner type called the "hotel" car. In 1868, the *Delmonico,* his first all-dining car, began operation on the Chicago & Alton line. These developments, coming just before the completion of the Pacific Railroad, could not have taken place at a more timely moment.

The distance from New York to San Francisco by rail was approximately 3,000 miles. The rail journey over this route, moving both day and night, took six days. Thus for those who could afford it, and many could, riding in a Pullman car helped to make the ordeal of the journey bearable. The following description of a Pullman train comes from the March 20, 1875 issue of *The Illustrated London News:*

> The train consists of a variety of cars—the drawing room sleeping car, which can be quickly converted into a day car, without leaving any evidence of the purpose for which it has been used at night; the parlor car, for short lines and day travel only; and the hotel car, which is the same as the drawing room sleeping car, but also contains kitchen and buffet, with adjustable tables to accommodate a full complement of passengers for long journeys and excursion parties. The internal decorations and fittings are magnificent. The woodwork is of walnut, with ornamentation of gold and ebony, silver-gilt mountings, or medallions in bronze. Above are fresco ceilings, from which descend elegantly designed lamps, which give forth a brilliant flame. Handsome carpets cover the floors; the seats are of cushioned velvet; highly polished mirrors in gilt frames decorate the walls, those at the end giving an appearance of great length to the car. In fact, so far as convenience, comfort, and luxury are concerned, Pullman's Palace Cars are, to all intents and purposes, first class hotels, of one story, mounted on wheels.

Pullman's cars were later exported to a number of other countries. In addition to leasing his cars to railroads, he supplied the porters and other staff as well, giving them special job training. Pullman also built luxury cars for the private use of the wealthy. These could be coupled onto trains and taken to any location. Circa 1886 he introduced the Vestibuled Limited Train composed of handsomely designed coaches connected by flexible vestibule passageways. These provided safer and easier movement for passengers between cars. By enclosing spaces between coaches, vestibules also reduced wind resistance and kept cars more rigidly in line.

A short article from the December 7, 1888 issue of *Railroad Gazette* described how "state of the art" Pullman service had become:

> There are now in the yards of Pullman's Car Company, Pullman, IL, two complete vestibuled trains ready for delivery to the Union Pacific. These trains will compose what is to be known as "The Golden Gate Special," a limited overland express, to be started on its first trip Dec. 5. It will run once a week from Council Bluffs to San Francisco as a solid train. The cars are the finest the Pullman Company can produce, and the style of the train is very much after the fashion of the "New York and Chicago Limited." It will be heated by steam, lighted by electricity and furnished with hot and cold water throughout. The car next to the engine will be a combination baggage room and barber shop with bath rooms; the next is the dining car, which is followed by two sleeping cars, and bringing up the rear is a composite car (parlor and sleeper) equipped with ladies' bath rooms and an observation room. The rear platform of this car is not vestibuled, and the windows extend to the floor, allowing every advantage to sightseeing travelers. A large reading room is also a feature of this car, though both the sleeping cars on each train are equipped with well selected libraries.

Pullman did have a number of competitors, the most prominent being Webster Wagner. Wagner began building and operating Woodruff-designed sleepers for the New York Central in 1858. When Cornelius Vanderbilt gained control of this line in 1866, he imposed terms, forcing Wagner to reorganize his company. It was first called the New York Central Sleeping Car Co., and later the Wagner Palace Car Co. On January 13, 1882, Wagner burned to death in one of his own coaches in a train wreck at Spuyten Duyvil, New York. The company was then operated under the Vanderbilts until they sold it to Pullman in 1900.

Georges Nagelmackers, born in Liège in 1845, came from a prestigious banking family. His father was involved in railway financing and had connections to Belgium's King Leopold I. After a failed romance, the young Nagelmackers traveled to the United States, arriving just after the Civil War. During this year in America he traveled a lot and observed railroad practices. He saw George Pullman's *Pioneer* while on display, and as a distinguished foreign visitor was invited to ride on it. Later he visited Pullman and toured his workshops. He was especially interested in the structural aspects of Pullman's heavy cars and the quality of service provided on them.

Prior to this visit, Nagelmackers had dreamed of creating a transcontinental European express train. At the time most trains terminated at national borders. The first international through train, operating between Basle and Rotterdam, had only been inaugurated in 1863. This development occurred after major banking interests united behind the project and exerted considerable political pressure on the governments involved.

Nagelmackers' American trip gave him a more expanded vision of what might really be possible. On returning to Belgium he enlisted the help of his father, who in turn sought the aid of the new king Leopold II. Shortly Nagelmackers formed a company and began plans for an international express train to operate between Paris and Berlin. Five special carriages were ordered from a firm in Vienna. These were received around the beginning of July 1870. The Franco-Prussian War, which began on July 19, thwarted this effort.

Undiscouraged, Nagelmackers next conceived the idea of a special express train to operate from the Belgian port of Ostend, through France, to the port of Brindisi in Italy. Once put in operation this train became an immediate success. E. H. Cookridge in his book, *Orient Express,* explains why: "And Britain was soon to prove the best customer for the new train. Scores of government officials, diplomats, and army personnel, as well as quantities of urgent mail, had to be carried quickly and safely between the mother country and India, the pearl in the Imperial Crown of the British Empire, and to Egypt and Africa. If an express train existed running from a Channel port . . . to southern Italy, up to ten days' traveling time could be saved as compared with the stormy and uncomfortable sea route through the Bay of Biscay and the western Mediterranean." When the Mt. Cenis tunnel was opened on November 17, 1871, this cut the time of this journey by an additional 18 hours. Unfortunately for Nagelmackers, French railways refused to grant him running powers through the tunnel, and the British government shortly terminated his contract.

Almost at the brink of going out of business, Nagelmackers formed a partnership with Colonel William Mann. The colonel, a flamboyant Civil War veteran with a shady past, had patented a sleeping car design in the United States. Perhaps from necessity he had left America and thus began promoting his "Mann Boudoir Car" abroad. After he and Nagelmackers met in London, their partnership was formed.

In 1873 some new boudoir carriages of a six-wheel design were built. Hearing that England's prince of Wales would be attending a wedding in St. Petersburg, Mann pulled off a promotional coup, persuading the prince to ride in one from the English Channel through Berlin to the Russian border. In the meantime Nagelmackers was taking advantage of this publicity, arranging contracts for supplying coaches. By 1875 51 boudoir carriages were being operated on a dozen different lines. Nagelmackers was not totally happy with Mann's style of business, however, and in 1876 bought him out. At this point the Compagnie Internationale des Wagons-Lits et des Grands Express Europeens was formed. Initially it continued to lease former Mann coaches. These were attached to various international express trains, such as ones operating between Paris and Vienna, and Ostend and Berlin.

In 1880 Nagelmackers introduced his first restaurant car for operation between Weimar and Bebra. At the same time he was still continuing his efforts to create the transcontinental express he had long dreamed of. Partly through the help of Leopold II, who had many royal relatives and contacts, he overcame the resistance of railway companies and various governments. Finally, he had contracts in hand to bring his long labors to fruition.

Nagelmackers' dream train, which became known as the Orient Express, was intended for operation between Paris and Constantinople (now Istanbul). However, as there were missing links in the rail route between these points, improvisation was necessary. Beginning in October 1882, trial runs of this train began between Paris and Vienna. The official inauguration took place at Paris's Gare de l'Est on October 4, 1883. Considerable publicity was generated beforehand and a select group of distinguished individuals, including aristocrats, state officials, diplomats, bankers, railway officials, and journalists, was invited for the initial run. From Paris the route ran through Munich, Salzburg, Vienna, Budapest, and Bucharest to Giurgiu. From there the Danube was crossed by ferry to Ruse, Bulgaria. Passengers then continued by rail to the Black Sea port of Varna. The rest of the trip was completed by steamer.

While the train was relatively short it was unsurpassed in its luxury. The two passenger carriages were of the bogie style. Each had three double and two single compartments, accommodating four and two people, respectively. During the day compartments functioned in the usual first-class manner, but at night they were converted for sleeping. Ahead of the coaches was a guard's van with space for mail, and at the rear another for carrying passenger luggage as well as extra stores of food and drink. The bogie-style dining carriage was well appointed, and included at its opposite ends, a gentlemen's smoking lounge with a library and a ladies' drawing room.

During the course of this inaugural run there was lavish feasting, and formal receptions were held at stops along the way. After this festive beginning, service was continued twice weekly in both directions. The popularity of this train was greater than anticipated, and in 1885, daily service was begun over the Paris-Vienna portion of the route. In 1888 the overland rail line was finally completed to Constantinople, shortening the time of the journey to 4½ days. The new part of the route ran through Nisch, Zaribrod, Slivnitza, Sofia, Ichtiman, Bazardjik, Philippopolis, Mustapha Pasha, and Adrianople.

After inauguration of the Orient Express, the company's business rapidly expanded in Europe. Soon, its sleepers and dining carriages were operated on many rail routes, and the company also initiated new luxury trains. In Italy it took over a number of Pullman concessions. Beginning in 1889 Wagons-Lits service was inaugurated in Algeria. In 1898 Nagelmackers built a special state train for the French government. Immediately used for receiving the czar of Russia, it made a positive impression. Soon arrangements were made for a St. Petersburg–Vienna–Nice–Cannes luxury train, and also for Wagons-Lits service on the Trans-Siberian Express, inaugurated in December 1899.

In 1873 France's Nord Rwy. introduced a new type of coach having an outside arcade. This allowed passengers free movement between compartments on the same car. As this design evolved, the outside corridor became enclosed. This style eventually became favored in many parts of the world, replacing the old type, which had separate outside doors for each compartment. In 1874 Giffard introduced a carriage with a unique suspension system. Springs were mounted in the usual places to suspend its four-wheel chassis. The coach body was not mounted to the chassis, but instead, was suspended over it by large springs positioned at each end. This reduced oscillations and provided a smoother ride for passengers.

At the opposite end of the social ladder from Nagelmackers' trains were those operated by the Compagnie Generale Transatlantique. Specializing in the transportation of emigrants bound for America, this company operated an international train between Basle and Le Havre. It had coaches of two types, one with 40 seats and a buffet, and one with 80 seats. In the latter, which was of the open type with a center aisle, the backs of seats continued upward as a high partition that had an expanded area at the top. Here children's cots could be fitted or luggage stored. Emigrants from Austria, southern Germany, Italy, Switzerland, and France rode on this train, and in 1883 the company carried 35,000 passengers.

A correspondence of sorts existed on the other side of the Atlantic. There an inexpensive immigrant service was provided on some lines. Often immigrant trains were run separately at the lowest level of priority. In other cases immigrant cars were attached to other trains. Providing nothing but basic transportation, these cars were built cheaply, often simply converted from older rolling stock.

In the 1870s train speeds began to increase. On June 1, 1872, a new express began operation between Paris and Rome. The train left Paris at 8:00 a.m. and, averaging 40 m.p.h., arrived in Rome at 6:00 p.m. the following day. In its March 27, 1875 issue, the *Railroad Gazette* reported that throughout the German Empire (excluding Bavaria) express and fast trains averaged 28 m.p.h., including stops. Ordinary passenger trains averaged 21 m.p.h. In the former category the fastest trains were on the Berlin-Potsdam-Magdeburg line, averaging 34 m.p.h. In the latter category the Maerchen-Posen line fared the best, with trains averaging 25 m.p.h.

A report in the March 9, 1883 issue of the *Railroad Gazette* gave figures for average speeds (including stops) of fast trains on some major routes in France. Among these we will mention three: Paris to Bordeaux—39½ m.p.h., Paris to Calais—34⅓ m.p.h., and Paris to Marseilles—31⅙ m.p.h. The fastest portion of the Paris-Marseilles run was over the 77-mile stretch from Valence to Avignon, where trains averaged 43 m.p.h. The fastest average in the country was reported over a 29.2-mile distance between Ruffec and Angouleme, where a 44¾ m.p.h. speed was achieved. At the opposite end of the scale was an omnibus train running on the Girdle Rwy. from Paris to St. Lazare. The average speed over this 21¾-mile line, including its 28 stops, was 12½ m.p.h.

An important French train at this time was the Train de Luxe. Although operated by the PLM Co. between Paris and the Riviera, service began out of Paris's Gare du Nord terminus in order to accommodate the considerable traffic between Britain and southern France. Passengers arriving from Calais would board this train, which departed every evening at 7:50. Its journey took it over the Nord tracks to the Ceinture line, and then proceeding southward, it was soon on the main line of the PLM.

In the late 1880s the fastest American train speeds were in the competitive areas along the Atlantic seaboard. Passenger trains on the Philadelphia & Reading's line between Bound Brook and West Trenton averaged 54.4 m.p.h. Baltimore & Ohio trains between Washington and Baltimore averaged 53.3 m.p.h. To the west, train speeds progressively declined. Circa 1893 the New York Central's *Empire State Express* was averaging 51 m.p.h. over its 440-mile run between New York and Buffalo. The company's *Exposition Flyer,* running that summer between New York and Chicago, a distance of 980 miles, averaged a speed of 49 m.p.h., typically completing the journey in 20 hours.

The Engineer, in its January 3, 1896 issue reported that the average speed for the New York Central's *Empire State Express* was 53⅓ m.p.h. At the time express speeds in England averaged 51¾ m.p.h. For Germany the average was 51¼, France 49.88, Belgium 45.04, Holland 44.73, Italy 42.34, and Austria-Hungary 41¾ m.p.h.

In its May 27 and November 4, 1892 issues the *Railroad Gazette* reported on the introduction of vestibuled corridor trains in Germany. Initially these trains were operated on the Prussian State Rwys. between Berlin and Cologne. The coaches had four-wheel bogies and compartments with side corridors. Each train consisted of five passenger carriages plus a combination mail and baggage coach. The latter was coupled directly behind the engine, and was followed by a carriage having four second-class compartments and a porter's compound. Each of the remaining coaches had two first-class and four second-class compartments. The porter's area functioned as a buffet from which hot and cold dishes and drinks were served to passengers in their compartments. In Germany vestibuled coaches came to be called *Durchgangswagen,* and trains of this type "D-trains."

Beginning in 1872, under James Allport, the Midland Rwy. inaugurated some unprecedented changes in Britain's passenger business. That year the company made all of its trains available for third-class passengers. Two years later it abolished second-class service, reducing the first-class fare to that previously paid for second-class. In 1875 the company decided to upholster all third-class passenger coaches. These costly and revolutionary changes were not welcomed by competing railways, but a precedent was set that other lines were eventually forced to follow.

In 1873 the North British Rwy., using a six-wheel chassis, built Britain's first sleeping coach. On June 1, 1874, the Midland line introduced Pullman service to the United Kingdom, operating both sleepers and parlor carriages. In *The Railways of Britain,* Jack Simmons makes this observation: "Although the Midland Pullmans were not an unqualified success, they helped to popularize a number of most desirable improvements in British carriage design. They were well ventilated; they were heated. The only heat ordinarily provided in British railway carriages hitherto had come from 'footwarmers,' tins filled with hot water, or later with acetate of soda." And they also had bogies. The Midland soon built more bogie coaches, including the first of the six-wheel type in 1875.

On November 1, 1879, the Great Northern inaugurated Britain's first dining coach service between London and Leeds. The following month, Britain's first all-Pullman train began operation on the London, Brighton & South Coast Rwy. In 1881 the same company made the first experiments with electric lighting on passenger carriages. In 1882 the Great Northern Rwy. introduced the first coach of the side corridor type. In 1892 the Great Western line initiated the first vestibuled corridor service in Britain. During the mid-1890s dining carriage service was introduced to Asia on Ceylon's Rohilkhand-Kumaon Rwy. By 1897 similar coaches were operating on India's Bombay-Calcutta "Limited Mail" train.

After the Midland's changes of the mid-1870s, its competitive position and revenues improved. In 1876 the company completed its central route between London and Scotland. Although this course over the Pennines was more arduous and time-consuming than the rival east and west coast routes, it did offer competition that the Great Northern and London & Northwestern companies could not ignore. However, responses on their part were not immediate. In 1883 the London & North Western cut 10 minutes off the scheduled time of its London-Glasgow express. However, its socially conservative management was resistant to changing its third-class policies.

During this period the east coast route opened all of its trains to third-class passengers except for the prestigious Flying Scotsman. This famous express departed Kings Cross each day at 10:00 a.m. for Edinburgh. Finally, in 1887, the Great Northern relented and declared the Scotsman available for third-class travel. As this train was already an hour faster than the competing London & North Western train, departing from Euston, war was declared. In 1888 a speed contest erupted between the two companies. At the onset the eastern route was covered in 9 hours while the western run took 10 hours. During the course of competition these times were reduced to 7 hours 26¾ minutes and 7 hours 38 minutes, respectively.

At the time Britain's trains were the world's fastest. Each day trains there traveled 63,000 miles at an average speed of 40 m.p.h. or greater. In comparison only 14,000 miles were traveled daily at such speeds in America. The American press took a great deal of interest in the British competition, and the headline for the August 7, 1888 issue of the *New York Times* read "Flying Over the Rails." The accompanying article provided an account of the final London-Edinburgh run ending the competition. This press coverage also stimulated interest in American train speeds. The New York Central quickly responded, promising to better the British speeds in its run between New York and Buffalo.

Speeds on the North Western's route were improved by a technical innovation allowing trains to take on water at full speed. Long, shallow troughs, containing water, were laid between rails. As a train passed over a trough, a scoop was lowered from the tender, quickly drawing water in. This system, which reduced time required for water stops, was also tried by France's Ouest line and by the Pennsylvania R.R.

At the end of the competition in Britain, the two lines agreed to a truce, with the east coast route scheduling its run for 7¾ hours and the west coast for 8. In 1895 this rivalry was renewed, only this time by trains running at night between London and Aberdeen, a distance of over 500 miles. Aside from company honor, the primary goal of the competitions in 1888 and 1895 was to capture the business of sportsmen traveling to Scotland for the annual grouse shooting season. These individuals typically traveled in style, usually being accompanied by wives or female companions, numerous servants, dogs, luggage, and guns. This was known in the railway business as the "The Grouse Traffic."

In an essay entitled "Racing to Scotland," published in *The Railway-Lover's Companion,* Roger Fulford writes: "One of the difficulties in arranging these races was that the trains had to travel over the lines of several different companies. From Euston you ran as far as Carlisle on the London & North Western. From Carlisle you went to Aberdeen by the Caledonian Railway . . . The King's Cross route was more complicated. The Great Northern took you to York. There the North Eastern took over to Edinburgh, and from Edinburgh the North British continued on to Aberdeen, each change of company meant a change of engine. The advantage to the west route of having only a single change of company at Carlisle was considerable. To make up for this the advantage of geography lay slightly with the east route."

Although both routes ran separately for most of their distance, they converged at Kinnaber before proceeding to Aberdeen. Thus whichever train arrived there first won the race. On August 15, the west train reached Kinnaber one minute ahead of the east one. The next day both arrived at exactly the same time, but the Caledonian signalman, in a gesture of sportsmanship, gave the track to the east train. For the greater part of the competition the west had the edge. This was mainly because stops were made without adherence to timetables. The Great Northern had more difficulty in persuading its partners to likewise comply. The fastest times were set by the east on August 21 at 8 hours, 40 minutes, and by the west on August 22 at 8 hours, 32 minutes. Both times broke a world speed record previously set by the New York Central on its New York–Buffalo run.

From its beginnings in 1825 to the end of the century the passenger coach went through many interesting design changes. Over the course of this period coaches became increasingly longer and heavier.

Lighting evolved from oil and kerosene lanterns to gas lamps and finally to electric lights. Heating, if it existed, began with coal stoves or footwarmers, but later was done with steam. Carriages were originally built of wood and this continued to be the norm for most of the century. However, wooden carriages were extremely vulnerable when involved in wrecks, and as a result, many lives were lost. Safety improved with the introduction of compressed air and vacuum brake systems and the use of metal in coach construction.

By the end of the century carriages had become relatively sophisticated in design. European railways had begun incorporating many design features originally pioneered in America, including bogies and clerestory roofs. However, there was a general reluctance to give up the compartment system for the open type with the middle aisle. Instead, the side-corridor style became increasingly favored. In some countries like Russia and Switzerland the open type of carriage was preferred, and elsewhere it was often used for third-class coaches.

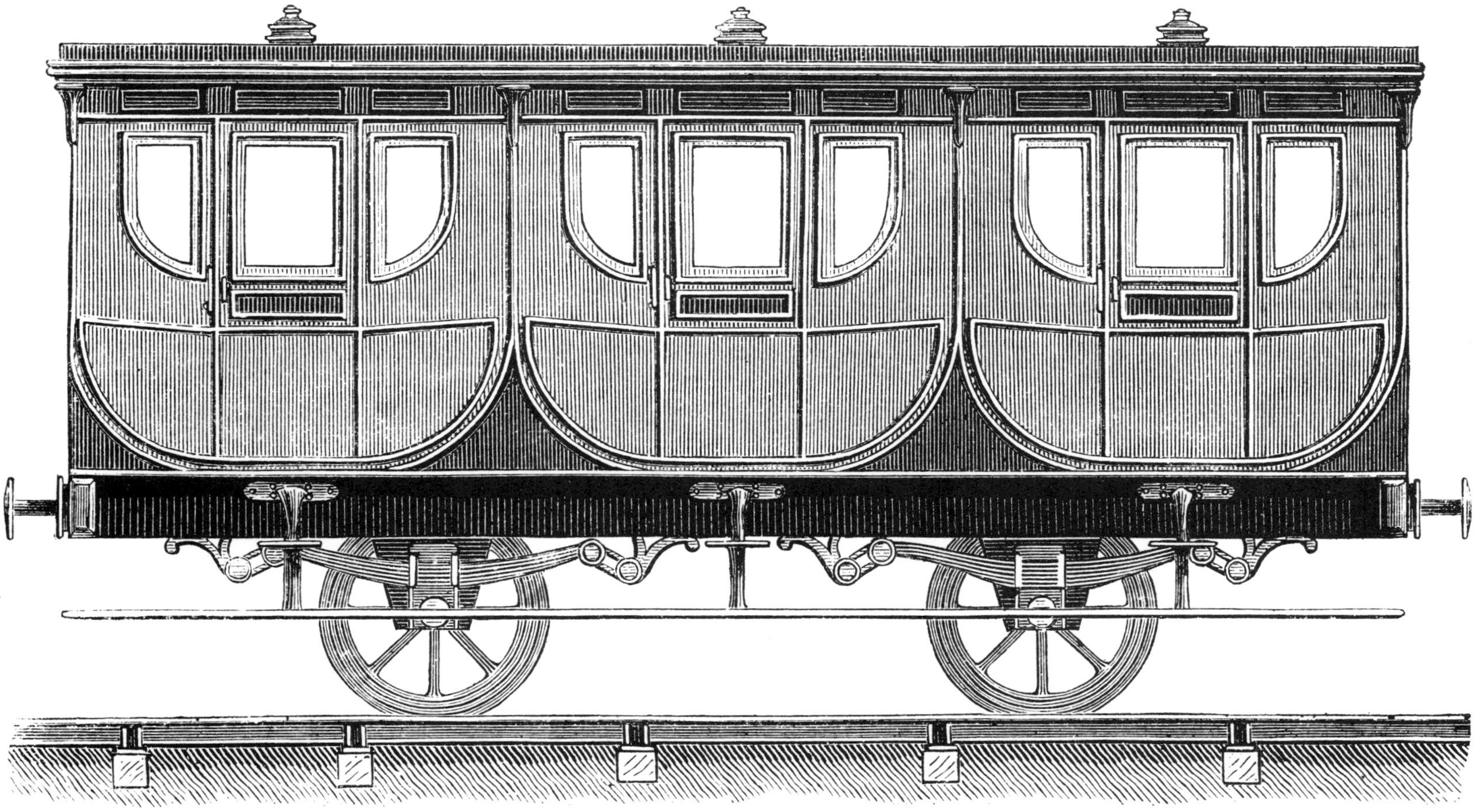

TOP: French first-class passenger carriage, circa 1850s. BOTTOM: Belgian first-class carriage with attached smoking saloon, 1862.

PLATE 171

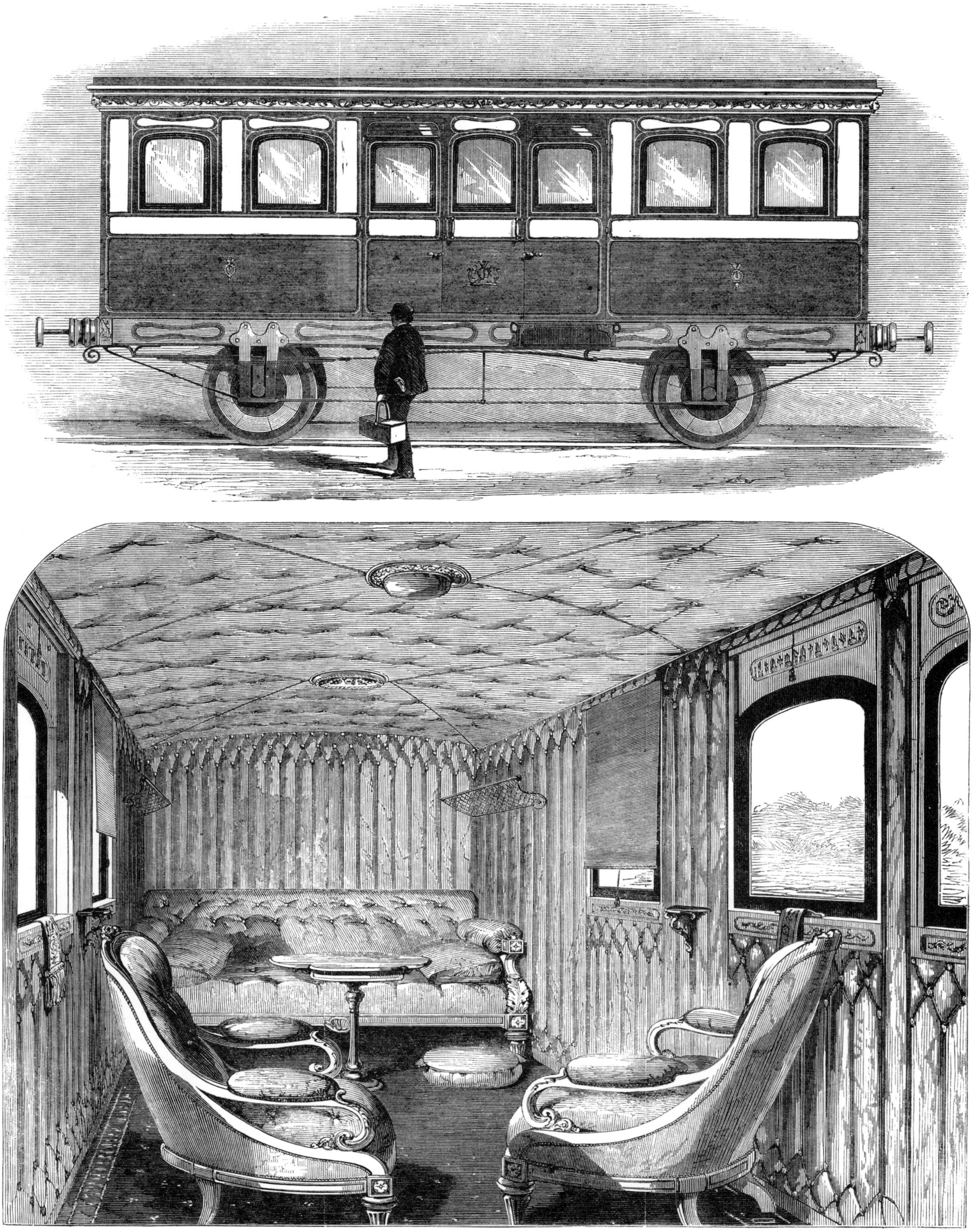

Exterior and interior of royal coach built for England's Queen Victoria, London & North Western Rwy., 1861.

PLATE

172

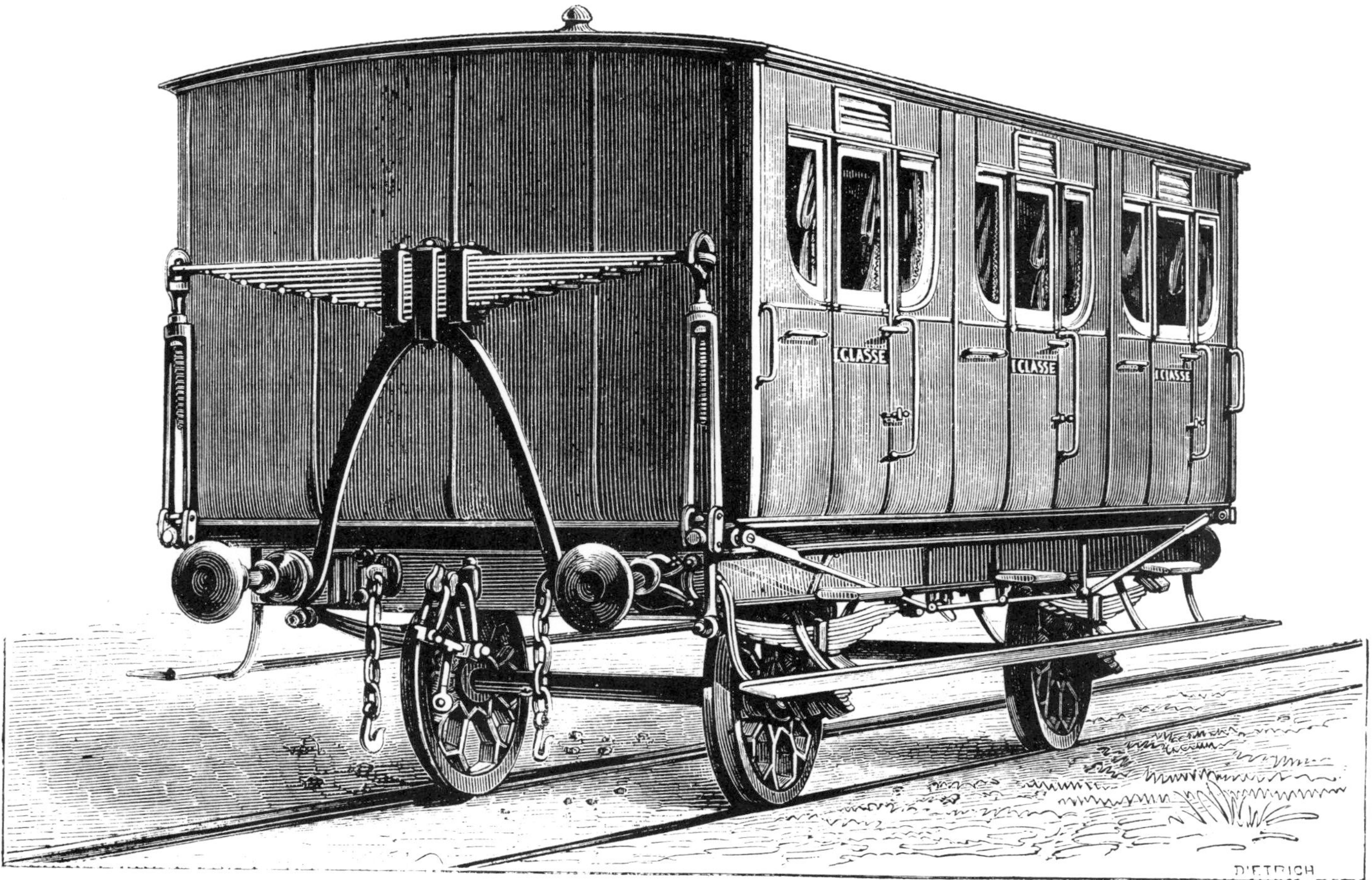

Top: Iron passenger car as advertised by La Mothe Co. of America, 1855. **Middle:** French passenger carriage with exterior corridor for communication between compartments, Nord Rwy., 1872. **Bottom:** First-class passenger carriage designed by Henri Giffard, Nord Rwy., 1874.

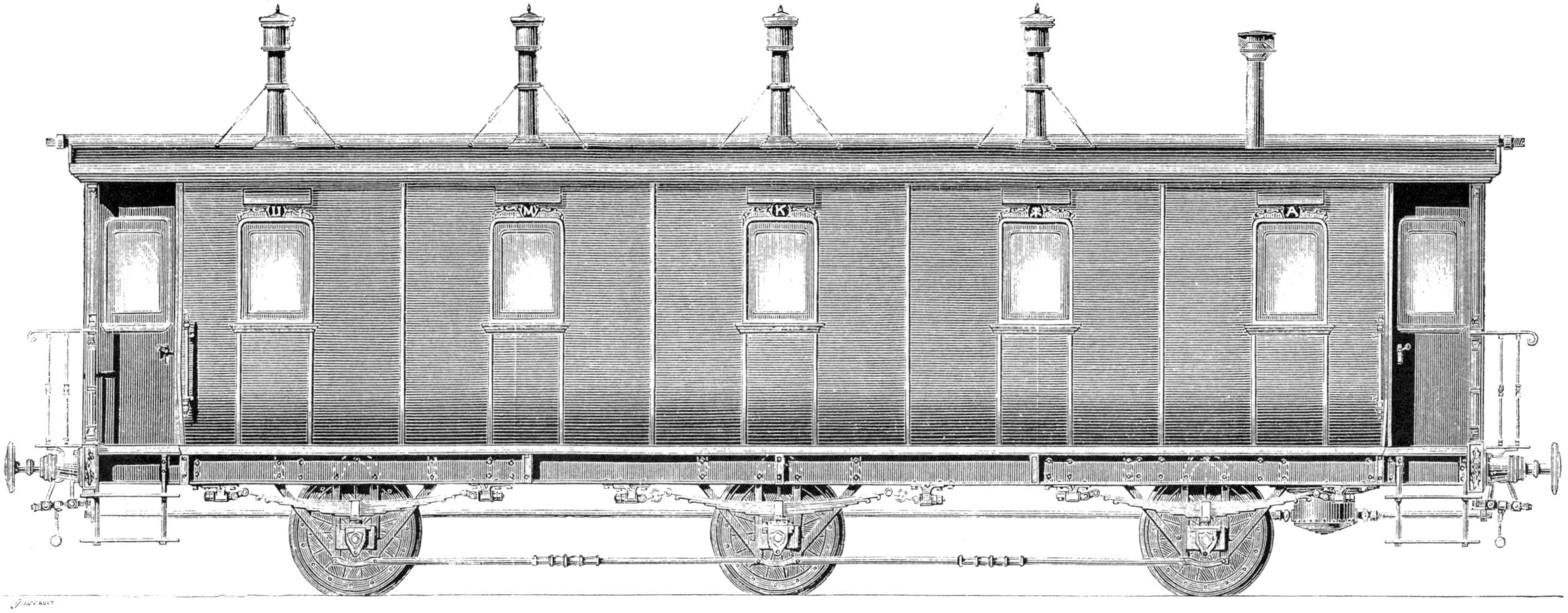

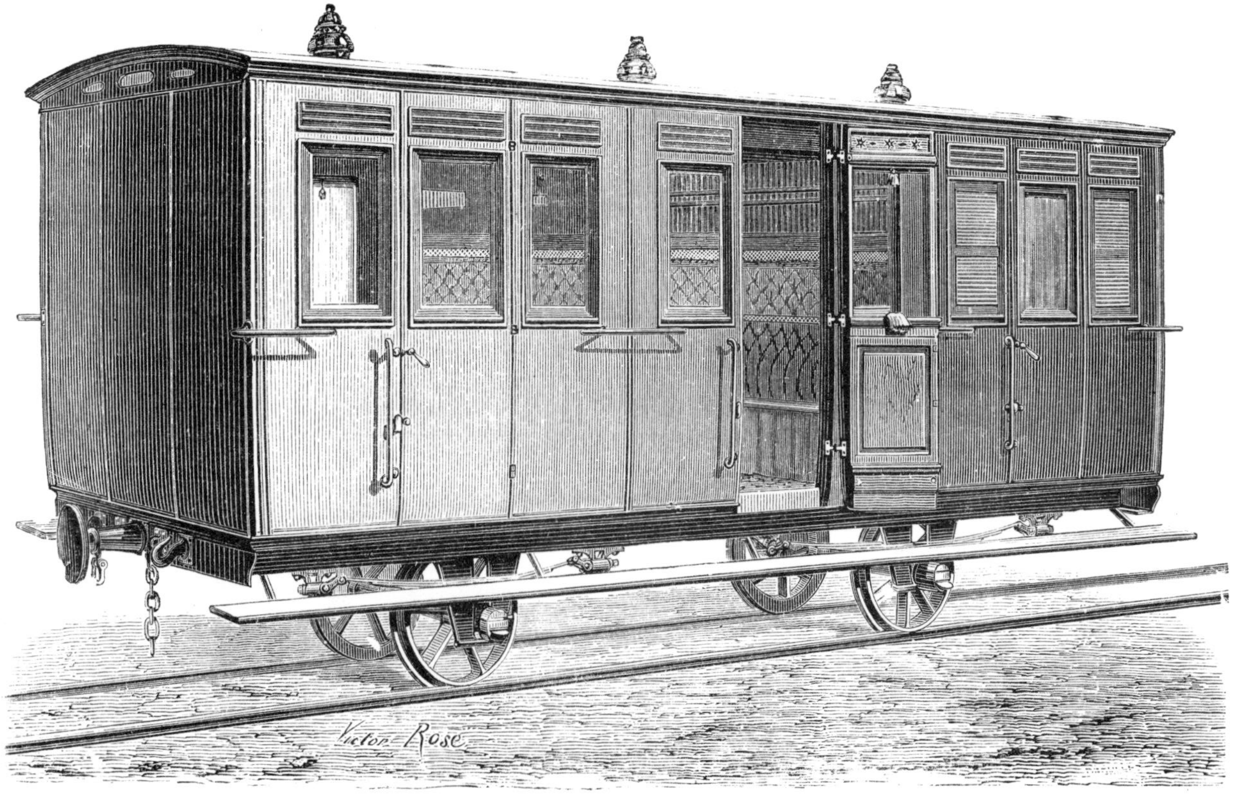

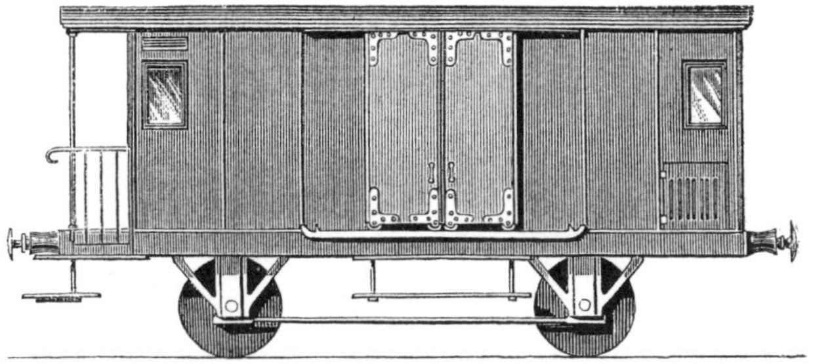

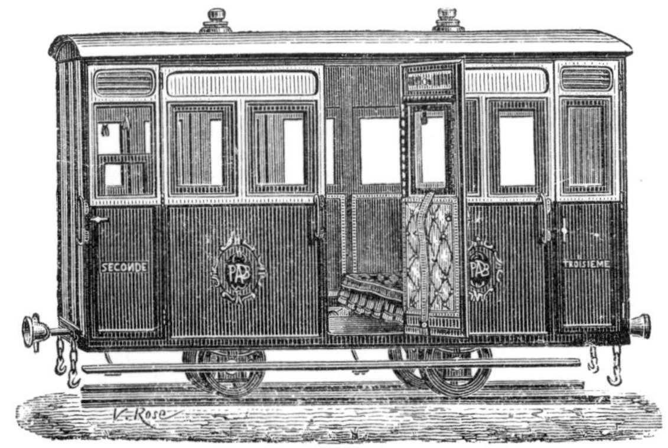

PLATE 174

Top: First-class carriage, Moscow-Kursk Rwy., built at company shops, Moscow, 1872. Middle right: Baggage van design for 2′9″ gauge line by C. E. Spooner, engineer, 1870. Bottom left: First-class coach for export to Peru, Compagnie Francaise de Materiel de Chemins de Fer, circa 1875. Bottom right: Narrow-gauge coach, Decauville Co., circa 1880s.

Top left: Interior of passenger car, Pennsylvania R.R., 1875. Top right: Interior of Pullman parlor car, 1875. Bottom: Interior of American sleeping car, Erie R.R., 1867.

PLATE 175

PLATE 176

American narrow-gauge passenger cars: TOP: Car built by Jackson & Sharp Co., Wilmington, Delaware, for Denver & Rio Grande R.R., 1871. BOTTOM: Car built by Billmeyer & Smalls Co., York, Pennsylvania, for Nevada's Eureka & Palisade R.R., 1875.

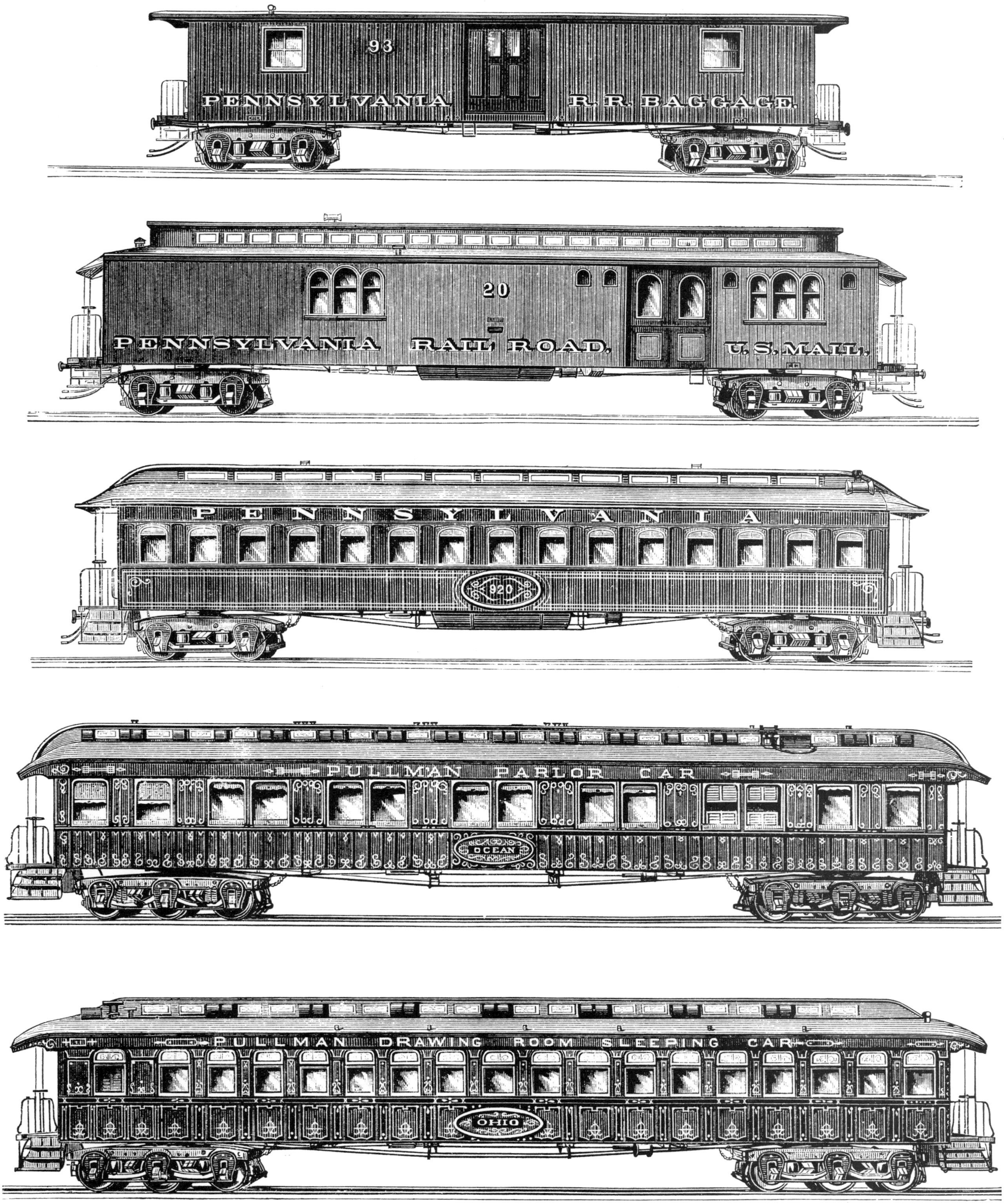

Typical rolling stock pulled on Pennsylvania R.R. passenger trains, 1875: Top: Baggage car. Upper middle: U.S. postal car. Middle: Standard passenger car. Lower middle: Pullman parlor car. Bottom: Pullman sleeping car.

PLATE 177

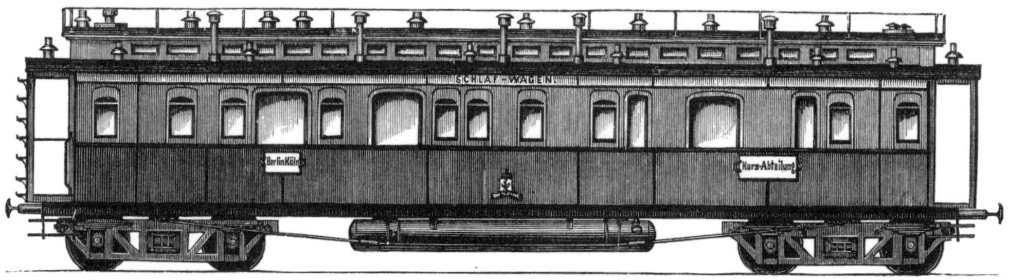

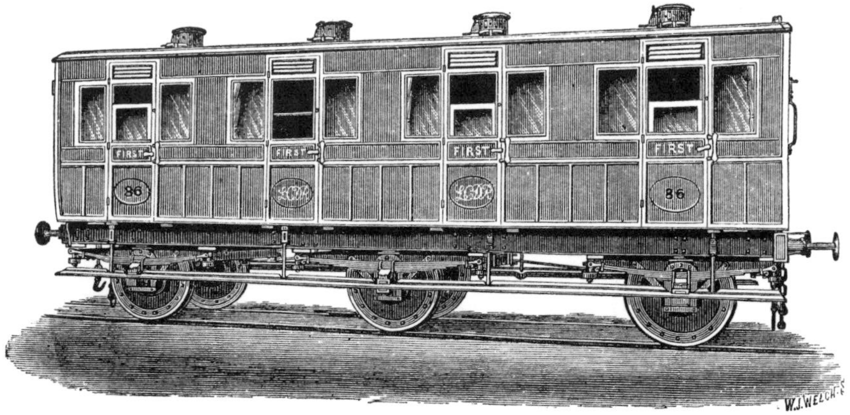

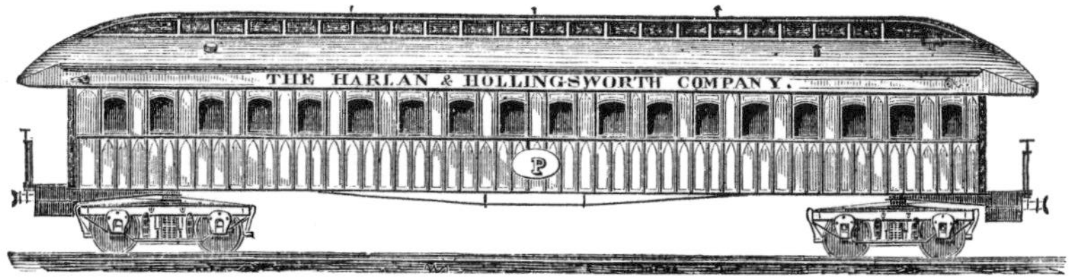

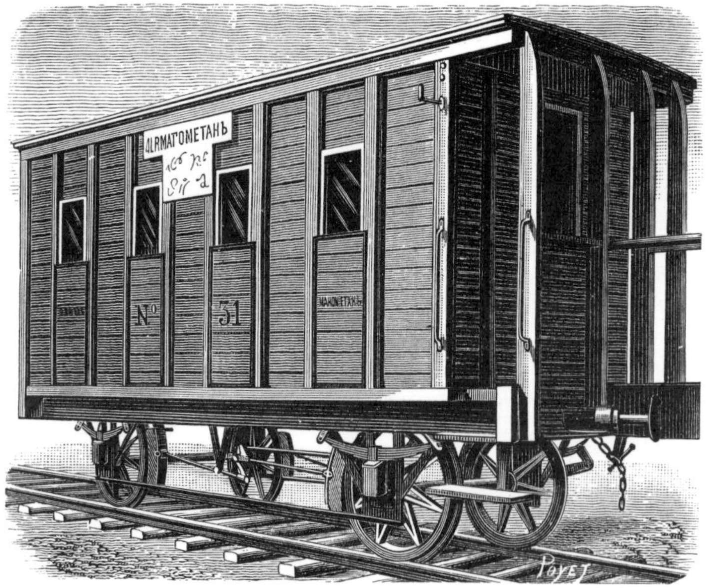

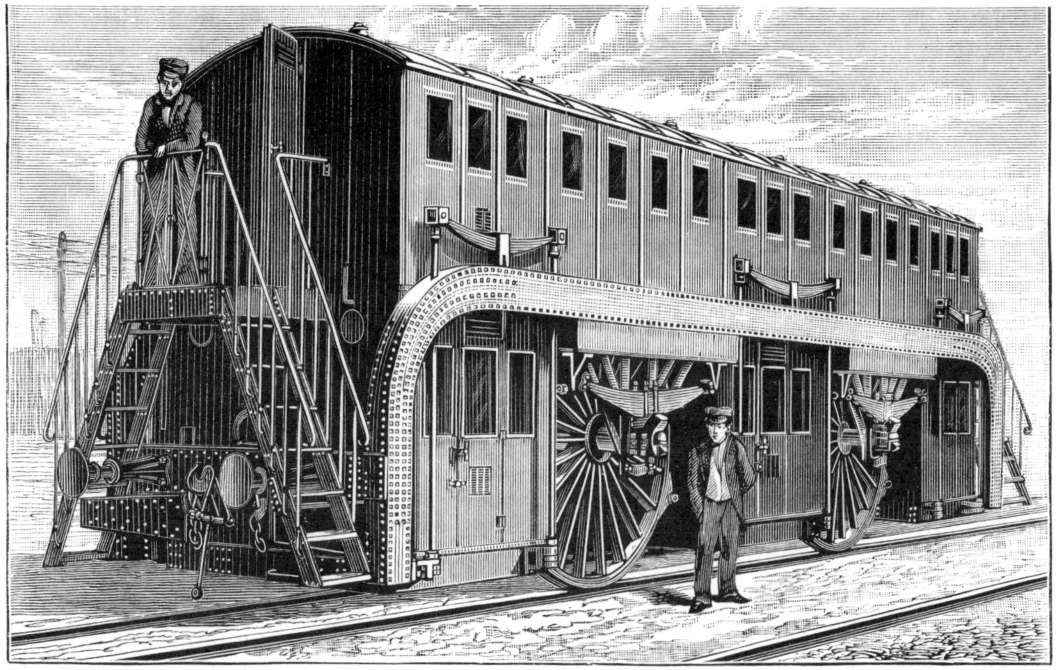

PLATE 178

Top left: Third-class dining saloon, Great Northern Rwy., 1894. Top right: Decauville baggage wagon, 1893. Upper middle left: Sleeping carriage, Prussian State Rwys., circa 1890. Middle right: First-class carriage, London, Chatham & Dover Rwy., circa 1875. Lower middle left: American passenger car, circa 1872. Bottom left: Passenger carriage, Trans-Caspian Rwy., 1888. Bottom right: Estrada's experimental large-wheeled double-decker carriage, France, 1886.

Dining on American railroads: Top: Meal stop for a Lackawanna passenger train, 1877. Bottom: Dining service on the Union Pacific–Central Pacific transcontinental line, 1870.

PLATE 179

LEFT: American baggage room, circa 1888. RIGHT: Buying tickets in Amsterdam, circa 1880s.

PLATE 180

PLATE 181

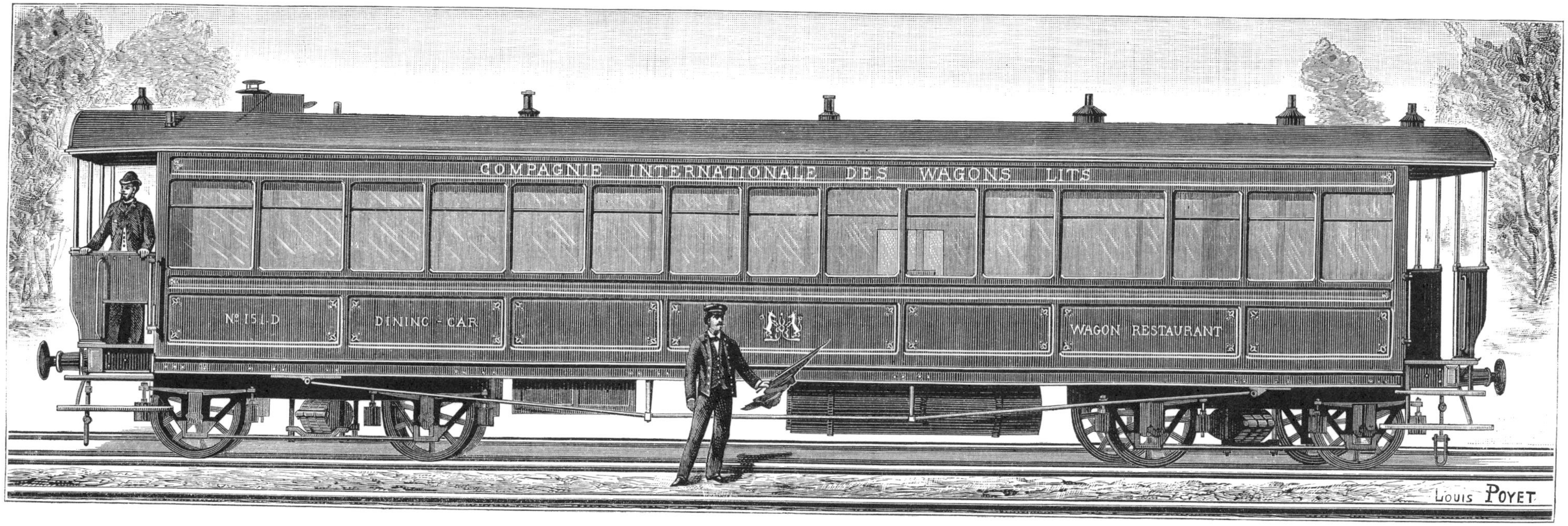

George Nagelmackers' original Orient Express, 1884: Top: View of entire train. The Est Rwy. provided the "501" class 2-4-0 express engines used by the company. Bottom: Luxury Wagons-Lits dining carriage.

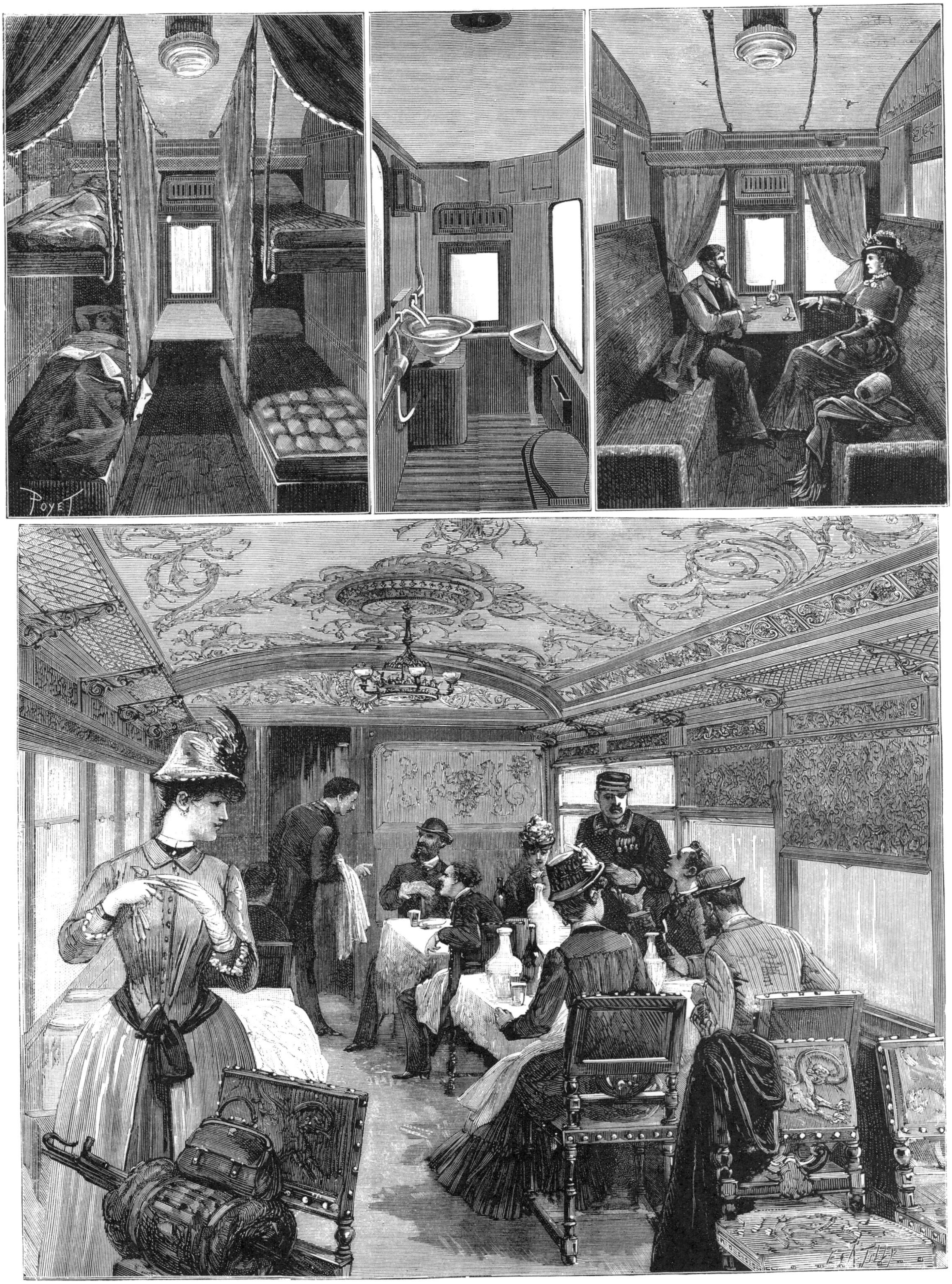

The Orient Express, 1884: TOP LEFT: Interior of passenger compartment as arranged for sleeping. TOP CENTER: Toilet. TOP RIGHT: Passenger compartment arranged for daytime use. BOTTOM: Interior of dining carriage.

PLATE 183

In 1888 a line was opened through Bulgaria making possible a direct land route for the Orient Express, all the way from Paris to Constantinople. In this picture the inaugural train receives a salute from Bulgaria's Prince Ferdinand.

Top: New vestibule passenger cars, Pennsylvania R.R., 1888. Bottom: French passenger train entering a tunnel, 1899. Note brake van at end of train in which train guards rode: This practice, initiated by the British, was followed in most parts of the world.

PLATE
184

CHAPTER

((14))

Passenger Stations

Railway passenger stations were typically built at origin and destination points of lines, and sometimes at intermediate locations along the route. They initially came into being to serve two primary functions: selling tickets and providing shelter from bad weather. As railways evolved the function of stations became increasingly sophisticated; thus they grew in size and also began to incorporate other elements. Platforms were added to provide greater convenience for entering and exiting coaches, but they also demarcated areas of safety for passengers. Train sheds were soon built as well. Usually constructed over more than one set of tracks, they provided protection from the weather both for passengers and for rolling stock not in use. Also built at stations were various facilities for train servicing.

Perhaps from the outset, station size was primarily determined by an estimation of the future traffic that would need to be serviced. Thus early city stations might require at least a few rooms, while a single room would suffice for a country stop. As time passed the size of all of these structures increased to incorporate such features as waiting rooms (sometimes separated for men and women, or by class of service), water closets, toilets, ticket booths, baggage rooms, express offices, stationmaster offices, and, later on, telegraph offices. Larger city stations often housed railway company offices, and also began to provide amenities such as restaurants, book stalls, smoking rooms, barber shops, reading rooms, and hotel accommodations.

Railway stations could be differentiated in various ways. In the *Encyclopedia of North American Railroading,* Freeman Hubbard lists five types: "(1) Through station, built beside tracks or under them in such a way that trains pass through it directly in either direction, a setup which eliminates a lot of switching. (2) Way station, usually built to straddle the tracks, often consisting of buildings on both sides of the tracks and connected by a tunnel or an overpass . . . (3) Stub station, one in which the tracks stop abruptly, so that trains must head in, back in, or depart in reverse . . . (4) Loop station, one having an oval-shaped track connecting with the mainline so that trains proceeding via the loop head directly into their tracks beside the depot platform. (5) Terminal station, built at the end of a railroad line, usually at the end of the main line." It might be noted here that these categories are not mutually exclusive. For example, most railway terminals of the nineteenth century could also be classified as stub stations.

Another way of differentiating stations was by their style of layout, a method discussed by Carroll Meeks in *The Railroad Station: A Social History.* While a number of approaches to station design were tried in the early years, three primary types emerged: the one-sided combination, two-sided, and head. The *one-sided combination* type was essentially the fundamental prototype from which all other layouts emerged. Its distinguishing feature was that all station facilities were on one side of the track(s). Many early stations of this type had no platform at all. Thus if passengers had to cross tracks during boarding or disembarking, their lives were at risk. While a single platform became typical for this type, additional ones might be constructed as traffic required. Footbridges, subway passages, cross-platforms, and even drawbridges were used to provide a means of transfer.

The two-sided type. This style came into use at termini as a means to separate the arriving and departing traffic. Meeks states: "This meant that incoming and outgoing passengers were kept apart, so that arriving trains did not hold up departing ones and both could wait longer at the platforms." Stations of this type had facilities on both sides of the track.

However, the principal building housing the ticket booth, waiting rooms, and baggage rooms was on the departure side. If facilities on both sides were identical, it was called a "twin" type. By 1850 this style became favored for large terminal use. Initially these stations had two platforms; intermediate platforms were often added later, creating a disruption in the easy flow of passengers and baggage. It was later learned that elevated versions of the two-sided type made good through stations.

The head type. Essentially the same as Hubbard's stub station, this became the favored layout for termini. It solved traffic flow problems of the other types by placing tracks and platforms in the rear, all parallel to each other. The "head" or front part of the station was used as a concourse, allowing passengers easy access to all platforms. The head type also provided flexibility; any track could be diverted for inbound or outbound service, depending on the needs of the moment. The only disadvantage of this type was that passengers often had to walk farther.

The new phenomenon of the railway station provided architects with interesting new challenges. In the course of the century many traditional styles were resurrected and incorporated into station design. Although this period was very eclectic, there was perhaps a consistent and natural sequence of developments. Carroll Meeks identified four distinct periods that characterized this era: the emulation of one style (1830–50); the synthesis of many (1850–60); the takeoff of creativity (1860–90); megalomania (1890–1914). Meeks's observations might be remembered as we proceed further.

The earliest practices established by English railways for the handling of passengers had their precedent in those used by stagecoach lines. Thus in many cases arrangements were made with nearby inns to both sell tickets and provide accommodations. A converted warehouse in Darlington served as the Stockton & Darlington's first terminus in that town. In Stockton a wooden coach shed sufficed for the same purpose. The first railway station built in England was the Liverpool Road Station, opened in Manchester on September 15, 1830. Next to open was the Crown Street Station in Liverpool. Both structures were relatively simple in design and had a single platform. The Crown St. terminus introduced the world's first train shed. Built of wood and 30 feet wide, it also provided protection for a second track used for storage.

The London & Birmingham's Euston Station, built in 1838, was a more ambitious structure. It had two platforms, one for each direction of travel. It also had the world's first iron train shed, designed by Robert Stephenson and Charles Fox. It was built in two spans and was 200 feet long. This station's most famous feature was a classic Greek Revival arch supported by four Doric columns. Designed by Philip Hardwick and placed in front of the station, it suggested the true potential of railways: providing a gateway into a much larger world. Shortly later, the new head type of station appeared. Early examples of this style include depots built in 1841 by David Mocatta at Brighton and George Townsend Andrews at York.

After midcentury a number of important stations were built. King's Cross in London, terminus for the Great Northern Rwy., was completed in 1852. Designed by the architect Lewis Cubitt, it was divided into two train sheds, each 800 feet long, 105 feet wide, and 91 feet high. One served arriving trains and the other departures. Each was covered by an arched glass roof, receiving support from laminated wood ribs. These rested on simple corbels, and were reinforced with iron. The front facade, built of brick and faced with stone, was original in design and almost modern in its simplicity. Two great semicircular windows marked the termination of the sheds. In between them was a squared clock tower.

Cubitt's arched glass roofs indicated a new trend in the air, largely possible due to new techniques, developed by Joseph Paxton, for using iron and glass. These included iron vaulting, ridge and furrow roof glazing, and prefabrication. Paxton first achieved his breakthrough in 1840, constructing a giant greenhouse at Chatsworth House. Using the same techniques, he built a structure to house palm trees at Kew Gardens in 1848. His masterpiece, however, was the much larger Crystal Palace built for the London Exhibition of 1851. It was to have a seminal influence on future developments in railway architecture.

Paxton's work influenced construction at Great Western's London Paddington Station, which opened on January 16, 1854. This structure resulted from a collaboration between I. K. Brunel and architect Matthew Digby Wyatt. The train shed, providing space for 10 tracks and four platforms, was 700 feet long by 240½ feet wide. Covering it were three sections of arched roof extending longitudinally. Each was covered with glass along its top portion, while farther down corrugated iron was used. Wrought iron ribs provided reinforcement, resting on girders where the sections joined. These, in turn, received support from iron columns placed beneath every third rib. At the far walls the ribs received similar support from cast iron pilasters.

Also following this trend was Birmingham's Grand Central Station, opened by the London & Northwestern on June 1, 1854. A massive structure for its time, it was situated on New St. and also served the Midland, Stour Valley, and North Staffordshire lines. Having both booking and company offices for the various railways, it featured a large refreshment room, divided by rows of massive pillars into three areas. The train shed's single arched glass roof, which reached a height of 80 feet, received support from iron trussing, and was 1,100 feet long and 205 feet wide. Laid out below were 10 lines of track and four platforms.

The *Encyclopedia of Railroads,* edited by O. S. Nock, provides this observation of subsequent developments: "The desire to cover ever wider spaces in one span gave impetus to development of the arched iron roof, the dramatic impact of which, more than any other, epitomizes nineteenth century structural engineering. It was an impact that was fully exploited by the railways as one company after another sought to build bigger and better train sheds from the 1860's onwards, particularly in America. As older stations became obsolete or outgrown they were replaced by larger, more grandiose structures which, without a great arched shed would have been considered incomplete."

During this period construction also began on combination station-hotels. The South Eastern's Charing Cross terminus in London, designed by E.W. Barry, was one such structure. Built near the Thames embankment, it opened January 11, 1864, with railway offices occupying the lower portion of the building and the hotel the upper part. White brick, with terra cotta ornaments and cement moldings, was used for facing the exterior. The roof owed inspiration to the Louvre of Paris.

While the hotel could be entered from the refreshment room of the station, its main entrance was on the West Strand, close to the corner of Villiers St. A grand staircase led to the rooms on the upper floors. Also available was a general coffee room, a ladies' coffee room, a ladies' retiring room, a reading room, a smoking room with balcony overlooking the station, a billiard room, and a large meeting room. The train shed was covered in the new fashion by a single arched roof. The shed extended almost out to the river, at which point it joined a viaduct. This four-tracked viaduct, extending south across the Thames, provided the means of entry and exit for the trains.

On September 1, 1865, the South Eastern Rwy. opened up a second London terminus, this one at Cannon St. In *British Railway History,* Hamilton Ellis gives this impression of both of the company's termini:

> Very imposing stations were built at both termini, with large hotels fronting the Strand and Cannon St. respectively, in all the pompous grandeur of Eighteen-Sixty-Gothic. Both stations had great semi-circular iron roofs, and the outer end of Cannon St. was further embellished with a pair of rectangular towers with leaden domes and spires, which might have belonged to some somber Victorian Escorial. Charing Cross, though plainer, had a large Gothic cross in the forecourt, perpetuating the memory of Queen Eleanor's ancient great cross at Charing. While Cannon St. was used principally by business passengers to and from the City, Charing Cross became a general main-line station . . . Public opinion was very favourably impressed by these magnificent stations.

In building large stations, English railways tended to separate assignments, giving their engineers the duty of constructing train sheds, while commissioning fashionable architects for designing the front buildings. By this time there was a real competition in station construction. Station architecture made a statement about the railway, its ownership, its management, its city of location, and its architect. Ideally, perhaps, a large station might embody an image of power, prosperity, glamour, stability, success, confidence, progressiveness, and longevity—in short, qualities desired by most of humanity.

Various architectural styles provided inspiration for building stations, with one becoming the fashion for a time only to be replaced by another. Thus over a period of time styles such as Classical Greek, Italianate, Gothic, Renaissance, Romanesque, and Byzantine were all explored for their potential and were sometimes combined with each other. The Great Western's Temple Meads Station of Bristol, built in 1841, was an early example in the Gothic Revival style. However, the ornate wooden roof of its train shed was borrowed from the Tudor era.

Another Gothic structure was the Midland Company's St. Pancras terminus in London. Sir George Gilbert Scott designed the front portion and W. H. Barlow its 243-foot-wide shed. Latticed arches, rising directly from the platforms, provided support for the roof, culminating in a slightly Gothic point. Scott's work in front was equally impressive, making this truly a beautiful building. At its opening on October 1, 1868, this station was considered to be the finest in the world. An adjoining station hotel, also designed by Scott, was completed in 1873.

While the discussion so far has mainly focused on large termini, the architecture of smaller stations should receive some mention. Among East Anglia's better designed stations were Newmarket, Needham Market, and Stowmarket. The latter two were designed by Frederick Barnes in the Tudor style. Hamilton Ellis suggests that Barnes "would adapt anything from a cottage to a large manor house to the needs of the railway." H. A. Hunt was another important architect of the time. Turning for inspiration to the Jacobean style, he built stations at Stone in 1848 and Stoke-on-Trent in 1852, both in Staffordshire. Other attractive stations included Sandon, Staffordshire; Fenny Stratford, Burkinghamshire, in the half-timbered style; Battle, Sussex, built in 1852 and designed by William Tress in the Gothic style; and at Eastbourne, Sussex, an eclectic design with central clock tower built in 1884.

Beginning in the 1840s the Great Western Co. adopted policies resulting in good station design. The Midland also acquired a reputation for good depots. This company came to be known for a characteristic glass gabled awning, while small Great Western stations were remembered for cantilevered canopies. In Scotland the Highland Rwy. gained recognition for its neat stations, built of native stone, with corbie-stepped gables. Other railways, like the London & South Western and Great Northern, gained opposite reputations, their names often associated with stations of inadequate quality. During the 1870s, after the turf wars had largely been settled, companies focused on consolidating their existing territories. This often resulted in greater standardization of design practices.

Circa 1888 Clapham Junction, located in the southwestern part of London, handled more trains than any other station in the city. Its ticket offices were located in a large well-lighted tunnel, which ran at a right angle to the nine parallel platforms. Situated on these were waiting rooms, urinals, water closets, book and newspaper stalls, and refreshment rooms. This through station was an important transfer point for passengers. An average of over 1,500 passenger trains made stops each week. Of this figure, about 900 were operated by the London & South Western line and 500 by the London, Brighton & South Coast Co. Also stopping at the station were trains of the London & Northwestern; London, Chatham & Dover; Great Western; and Great Northern lines. All the empty London & South Western trains were cleaned and made up here, and then sent to Waterloo Station.

Because the scarce capital of early American railroads was virtually all invested in rolling stock, track, and bridges, little was left for stations. Much as in England, railroads often turned to nearby inns, taverns, or general stores to handle the business of selling tickets or holding small consignments of baggage and freight. In some locations railroads began to build unmanned platforms to protect passengers from the weather. These usually had signaling devices for stopping trains. The earliest manned stations rarely had platforms, but were built somewhat like a house, usually of wood. Often they had extended overhanging roofs for weather protection. But their main purpose was to provide facilities for the stationmaster and storage space. In later decades many of these early structures were torn down and replaced by more substantial buildings.

The oldest American station remaining today is a wood and stone structure dating from 1831. It was built by the Baltimore & Ohio R.R. at Endicott's Mills, Maryland. Although the building was originally intended for freight use, a two-stall engine house was soon added, and then passenger facilities moved in. Prior to that time passenger tickets had been sold at a nearby hotel.

When the Baltimore & Ohio built an extension into Washington, D.C., in 1833, a three-story structure at Pennsylvania Ave. and 2nd St. N.W. was selected to be its terminus. This building had previously been a boarding house. For some time, however, the company's tracks could not be extended to the site because an adjacent tailor shop blocked the path. Thus tickets were sold behind this shop until a generous settlement with its owner cleared the way for its demolition. Tracks were then extended to the terminus, and later, a belfry and car shed were added.

In 1835 a creatively designed station was opened by the Boston & Lowell R.R. in Lowell, Massachusetts. It was constructed in the Classic Greek style, and included a track laid through a colonnade. Boston's Old Colony Station, which opened in 1847, was more conservative in appearance, but was one of the first to seriously cater to the comforts of travelers. Besides having a general waiting room, it offered a barbershop, newsstand, smoking room, lady's salon, boot black's stand, and telegraph office.

At New Haven, in 1848, an imposing station in the Italian Villa style, designed by Henry Austin, was opened. While generally keeping to this style it also incorporated elements of Indian and Chinese architecture. Another important depot completed that year was the Boston & Providence terminal in Providence, Rhode Island. Designed by Thomas A. Tefft, it was built in the Romanesque style.

By the 1850s larger and better designed stations were built in increasing numbers. Railroads were beginning to enjoy some prosperity and had more resources to allocate. Furthermore, business activity in many places had increased to the point where larger facilities were badly needed. In 1852 a new B & O terminal was opened in Washington. Designed by the architects Niernsee and Neilson in the Italian Villa style, it had an exterior of brick, brownstone, and stucco. Its main features, however, were a 70-foot tower and 340-foot iron-roofed train shed. During this same decade, larger stations were opened in such cities as New York, Philadelphia, Baltimore, Harrisburg, and Chicago. These larger buildings provided dining facilities, and sometimes, overnight hotel accommodations.

Despite incorporating certain Romanesque details, Cleveland's Union Passenger Station, which opened in 1866, was undistinguished in appearance. It provides a good example of an indigenous American style one railroad historian has called the "train barn." Giving it further interest is the name of its designer, Amasa Stone Jr. Stone was then president of the Cleveland & Erie R.R. Later, as president of the Lake Shore line, he was disgraced by the Ashtabula tragedy.

This structure was of the one-sided combination type. It was built jointly by the Cleveland & Erie, Cleveland & Toledo (later Lake Shore), Cleveland & Columbus, and Cleveland & Pittsburgh rail lines. Situated at the foot of Water St. next to Lake Erie, it was 603 feet long by 180 feet wide. The station's walls were of sandstone. Its slate roof received support from 48 Howe trusses, built of iron. At its summit was a tin-covered deck extending the length of the building. Six tracks were laid longitudinally through the station on its north side, while two others were terminated midway in.

Rising from the midpoint of the south side was a relatively small tower. At its base was a door, which served as the station's entrance. Facilities for reception, dining, baggage, and express were located on the first floor of the south side. Company offices and a reading room were located upstairs. This building had many windows for providing light. While lacking beauty, it was probably well designed for withstanding Cleveland's severe winters.

As lines expanded and traffic increased following the Civil War, station facilities, especially in the areas of more established commerce, were improved. Throughout America, as perhaps nowhere else, the train station provided a vital connecting link with the rest of the nation. In small towns especially, depots played an important role, bringing relief from an otherwise isolated and often monotonous existence. The arrival of a train became a daily social ritual for the community, bringing excitement and activity, news from the outside world, and a chance to encounter people, both friends and strangers, both travelers and the merely curious. Here one might receive letters or newspapers, exchange gossip, pick up an express shipment, send a telegram, meet an attractive member of the opposite sex, observe the general commotion, admire the locomotive and cars, or board the train.

In time hotels, livery stables, cafes, and other business establishments often built up around stations, providing the nucleus for a larger community. Small-town stations might typically have a waiting room, a ticket and telegraph office, a baggage room, and separate restrooms for men and women. In keeping with the spirit of the age, ornamental devices such as clock towers, bay windows, stained glass, dormer roofs, turrets, elaborate chimneys, finials, gables, and gingerbread were employed by architects to make stations look more attractive and stylish. This inclusive style became known in America as "railroad Gothic." Standardized station design also came into vogue in this period.

The first passenger station in New York City was established in the early 1830s by the New York & Harlem R.R. While traction on this line was originally supplied by horses, it added a locomotive in 1837. The company's charter forbade it from operating steam engines south of Fourteenth St. Even north of this barrier conditions were crowded, and locomotives gained a reputation for being noisy, dirty, frightening to horses, and prone to accidents with other traffic. The most notable occurrence was a boiler explosion at Union Square in 1837, which killed the engineer and wounded 20 others. In response to public pressure, an 1844 law raised this barrier up to Thirty-second St. It was raised again in 1858 to Forty-second St. So it was against the backdrop of this and similar legislation that railroads conducted business in the city and planned their stations.

Circa 1851, the New York and Hudson River R.R. built a small station at Thirty-first St., on the far west side. Later the company built a terminus at Chambers St. and West Broadway. However, its charter forbade steam operations south of Thirtieth St., so horses were used to convey cars between the two points. The city's first substantial station was jointly built by the New York & Harlem and New York & New Haven railroads. Inaugurated in 1857, it was located on Fourth Ave. between Twenty-sixth St. and Twenty-seventh St., and became known as the Madison Square Station. Horses provided intermediate conveyance here as well.

In 1861 the New York & Hudson River Co. opened a new terminus at Thirtieth St. and Tenth Ave. The first train to arrive was one carrying newly elected President Lincoln, on his way from Springfield, Illinois, to his inauguration in Washington. Following his assassination, Lincoln's funeral train departed from this site on April 25, 1865.

During the 1860s shipping magnate Cornelius "Commodore" Vanderbilt decided to invest in railroads. He acquired the New York & Harlem in 1863, then the New York Hudson River R.R. in 1864, and the New York Central in 1867, combining the latter two into the New York & Hudson River R. R. His Grand Central Station, completed in 1869, served these lines, as well as the New York & New Haven. This imposing structure was situated at Forty-second St. and Fourth Ave. (now Park Ave. South), and was designed by the architect John B. Snook. It was built of brick with iron trimmings in the Renaissance style, and featured, in the head area, Mansard-style slate roofs. It was the most expensive railroad station built up to that time in North America.

Having a shape like the letter *L*, the main building abutted the train shed on two sides. In its basement were four restaurants: two for both sexes, and two for gentlemen only. Also on this floor was a police station, a billiard room, and four stores. In the first story were waiting rooms, ticket booths, telegraph offices, dressing rooms, and newspaper stands for each company. The New Haven R.R. occupied the front part of the building, with the other lines on the New St. side. The train shed extended for a length of 650 feet, with its arched roof built in the St. Pancras manner, only omitting the Gothic point. It was covered with corrugated iron and glass. The northbound tracks leading away from the station were originally at ground level; however, after many accidents with crossing traffic, an open cut was made for train use with bridges constructed overhead for street traffic. This structure was expanded in 1898, under the direction of architect Bradford Lee Gilbert. Later, however, it was completely rebuilt and reopened in 1913.

During the nineteenth century New York City could only be accessed by rail from the north. No bridge or tunnel provided rail connection to either Long Island or New Jersey. Thus this traffic required the intermediate use of ferries. Beginning with the Camden & Amboy line in the 1830s, various railroads established New Jersey terminals within some proximity of Manhattan. By 1874 a number of these stations existed. The Hoboken Terminal served both the Delaware, Lackawanna & Western and Morris & Essex lines. From it two ferries operated across the Hudson River; one serving Christopher St. and the other Barclay St.

To its south was the Erie Railroad's Jersey City terminus. Ferries ran from it to Twenty-third St., where the company's main office was located, to its Chambers St. depot, and also to Castle Garden to pick up immigrant traffic. Below the Erie another terminal served the New Jersey and Midland railroads. From it ferries operated to Desbrosses St. and Courtlandt St. This terminus later housed the Pennsylvania line. Farther south at Communipaw was a terminal housing the Central R.R. of New Jersey and later the Philadelphia & Reading line. Its connection to Manhattan, the Communipaw Ferry, stopped at Pier 15. Southernmost was the Camden & Amboy's terminus in Amboy, served by a ferry from Battery Park Place.

At this same time the Long Island Railroad had a terminal at Hunter's Point, Long Island City, on the East River. It could be reached from Thirty-fourth St. via the Hunter's Point Ferry. Another ferry ran upriver to the terminal from James St., at Pier 31. The Long Island R.R. also serviced the main part of Brooklyn, stopping at Prospect Park Depot and Bedford Station. In building a terminal farther north at Weehawken, New Jersey, in 1891, the North Hudson County R.R., a commuter line, had to contend with the high cliffs of the Palisades. It solved this problem by extending a partial viaduct out over the Hudson and building a small station at its end. Beneath this a ferry dock was constructed with a steam-powered elevator conveying passengers in between.

An important American station of the 1870s was the Boston & Lowell terminus in Boston. Situated on Causeway St., it was built in the Renaissance style and had a principal tower 148 feet high, a central dome of 113 feet, and an east tower of 104 feet. Designed by L. Newcomb & Son, it was planned in a way not only to handle the average traffic needs of the line, but to facilitate a seasonal rush of passengers between points farther south and the mountain resorts of New Hampshire.

This summer traffic largely arrived at 6:00 and 6:30 a.m. on overnight trains from New York. An express train for the mountain resorts left at 8:00 a.m. Thus the line's intent was to create a station so complete in its conveniences and appointments that travelers would not have to go to a hotel. A ladies' grand reception room occupied the central front of the building. To its right and left, respectively, were waiting rooms for men and women. A dining room was situated to the left of the main tower. Also on the same floor were a barber shop, baggage rooms, toilets, and drinking fountains.

The main court, 92 feet long and 52 feet wide, extended from floor to ceiling, beyond three stories, and gave visitors an ample impression of the magnitude of the structure. The station had a single enclosed train shed 565 feet in length. The terminus itself was situated within 50 feet of the Eastern Railroad Depot and was also in close proximity to the Fitchburg and Boston & Maine stations.

The Pennsylvania Railroad's West Philadelphia Depot, situated at Thirty-second St. and Market St., was opened on

May 6, 1876, just in time to handle a mass influx of visitors for America's Centennial Exposition. Facilities included a restaurant, ladies' and gentlemen's waiting rooms, dressing rooms, restrooms, a telegraph office, a counter for the sale of newspapers, storerooms, and various offices used by the Pennsylvania, Pullman, and Western Union companies. Adjoining the station on the east was a two-story building for handling outgoing baggage. Situated on the west was a room for incoming baggage. Twelve tracks were laid in the yard: four for arrivals, four for departures, and four for storage or washing of cars. Covering both tracks and platforms were two train sheds, each 800 feet long by 94 feet wide. These had curved tin roofs, which were supported by iron columns.

On December 5, 1881, the Pennsylvania R.R. opened its Broad Street Station in Philadelphia. Train access to it was made over a viaduct that became known as the "Chinese Wall." The station originally had two arched sheds. However, in a subsequent expansion program, these were demolished and replaced by a single arched structure, the largest of its type ever built. It was 595 feet long, 300 feet wide, and 108 feet high. While eight tracks were originally laid in the station, four others were later added to accommodate increased traffic. In the early 1890s the railroad completed an impressive terminal building adjacent to the shed. Located across the street from City Hall, it was designed by the architect Frank Furness. In addition to providing the amenities typical of a large city station, it housed the central offices of the company. The completion of this facility finally gave Philadelphia a station befitting its status as a ranking American city.

Chicago's Grand Central Station, built in 1890 by the Baltimore & Ohio R.R., had stained glass windows, marble floors, a huge fireplace, various other ornamented features, and an immigrants' room. The Union Station in St. Louis was America's largest when it opened in 1894. Its style has been variously described as Romanesque, Norman Revival, and Chateau. Built from Missouri granite, it was designed by Theodore C. Link and Edward B. Cameron. Its beautiful interior was decorated in the art nouveau style. South Boston Station, terminus for the Old Colony, Boston & Albany, and New Haven lines, opened its doors in 1899. Serving 50 million passengers in 1900, it was the world's largest railway station until 1910.

Much as in England, early French stations were usually of the one-sided combination type; however, in many cases these had no platforms. Paris's St. Germain station at Rue de Stockholm, built in 1838, was constructed on a bridge above the tracks. The Gare de Sceaux, built in 1846 at Paris's Place d'Enfert, was of the loop type, with the approaching train stopping at a platform in front of the station.

As passenger traffic increased, larger stations were built. The Gare de l'Ouest (now Gare Montparnasse) had two platforms, one for incoming traffic and the other for outgoing. Paris's Gare St. Lazare, was built from 1843 to 1847 as terminus for the St. Germain, Rouen, and right-bank Versailles lines. Its triangular trussed shed roof, designed by Flachat and constructed of iron and glass, anticipated Paxton's Crystal Palace. This facility had administrative offices, a waiting room, a concourse, and two platforms. Between the platforms were a number of other tracks. Later absorbed into the Ouest system, the station remained unchanged for many years, inspiring in 1877, a series of 10 paintings by the artist Claude Monet. Finally, in 1886 Gare St. Lazare went through a major renovation and was greatly expanded.

Flachat built other train sheds in Paris, including one of 113 feet width for the renovated Gare de l'Ouest in 1852. François Reynaud designed the original Gare d'Orléans (now Gare d'Austerlitz), which was built from 1843 to 1847. Polonceau later designed a train shed for this terminus. While planning other sheds for the P.O. system, he devised a new system of triangular truss construction where the stiffeners or struts were placed perpendicular to the rafters.

Paris's Gare de l'Est was designed by François Duquesney, and opened in 1852. It gained worldwide fame for its aesthetic merits, and was considered by many to be the most beautiful station of the century. Its front facade derived inspiration from the projection of the train-shed roof, precipitating at this juncture a huge arched window. The shed itself was a tied arch structure, 97 feet wide, having latticed ribs springing up from the far walls. There was an almost cathedral-like quality in the building's exterior appearance, unfortunately destroyed when it was later enlarged. As expansion was only possible on the right, the original structure was duplicated there. Then, as Carroll Meeks observes: "The old and new units became end pavilions connected by a long and uninteresting block, a composition characteristic of the late nineteenth century."

The first French station to employ the head type layout was Paris's Gare du Nord, built in 1847. It was rebuilt in the early 1860s and again in the 1880s. The architect François Reynaud designed the original structure. In its second incarnation, M. Hittorff was its architect. The April 18, 1863 issue of *Illustrated London News* provides us with a glimpse of this building:

> The architecture of this new *embarcadere* is as unpretending as it is graceful; a simple cornice crowns the center of the facade, which is decorated with eight pilasters, rising two by two, and indicating the form and proportions of the nave. The central arch has an elevation of rather more than 100 ft., and is surmounted by a statue representing the city of Paris, supported on the right and left by eight other statues personifying the principal cities to which the Northern line serves as a means of connection—London, Vienna, Berlin, Cologne, Brussels, St. Petersburg, Amsterdam, and Frankfurt . . . It includes five principal parts, as indicated by the facade, namely, a grand hall on the right, a salles de depart, followed by what we may call the station proper where the tickets are distributed; and on the left, the salles d'arrivee, and covered accommodation for the vehicles in waiting. In the interior are nine lines of rail, separated by four large and six smaller platforms. The building composing the facade comprises the Salon Imperial with its accessory chambers, the post and telegraph offices, the bureau of the stationmasters, and waiting rooms.

In general, French railways acquired a reputation for the architectural unity of their stations. This practice required a close integration of the front buildings and train shed. Victor Laloux's Gare de Tours, which opened in 1898, perhaps carried this idea to its ultimate expression. Its twin train sheds were projected forward, culminating in a beautiful art nouveau facade. In a way similar to Gare de l'Est, large semicircular windows were used to demarcate the juncture of the shed roofs. Both were flanked by giant columns, each having elaborate decoration and surmounted by an enthroned female figure.

Laloux's much larger terminus at Paris's Quai d'Orsay, received a different treatment. As engines at this station were all electrified, a train shed of the traditional type was no longer required. Tracks were routed into the station via a tunnel underneath the Seine, and the terminal built above them. The concourse, ticket booths, and waiting rooms were all placed at street level. These surrounded an open area, providing a view of the trains below. Staircases connected the two levels, while above was an elegant vaulted ceiling with skylights. This Beaux-Arts style building, completed just in time for the Paris Exhibition of 1900, had a profound impact on subsequent station design. Also completed that year was Marius Toudoire's rebuilding of the Gare de Lyon. This structure, noted for its enormous clock tower, also influenced later designs.

Early German and Austrian stations were built as economically as possible. They were usually of simple construction, and often had no platforms. They were often built in a pleasing rustic style, which blended well with the environment. Some stations in larger cities were built with greater architectural ambitions. Among these was Potsdam Station, opened in 1838, and the Thuringer Bahnhof of Leipzig, designed by Eduard Potsch and completed in 1844. The train shed of this latter structure was fronted in an Italian style, with an arcade having four large doors flanked by two towers. The large doors allowed locomotives and rolling stock out into a courtyard, where from a turntable they were turned around or routed to another track. During an expansion of this structure under Brandt in 1855–56, the towers were raised and wings attached to each side.

While England provided service for three separate classes of passengers, Germany had four. Thus waiting rooms were separated accordingly, and in the case of stations serving a number of railways, these facilities were multiplied by the number of railways. As this was not only expensive, but wasteful of space, consolidation of facilities began. At the time the cross-platform or midway, adjacent to the train gates, was widened and turned into an auxiliary waiting space. Also during this period German architects explored the full potentials of the one- and two-sided type stations. An example of the former was a multiplatform station in Hannover having one set of tunnels for passenger traffic and another set connecting the baggage platforms. Berlin's Stettiner Bahnhof, which opened in 1876, was designed by its architect, Theodore Stein, to have the appearance of a head type building. However, in reality it was of the two-sided type, having the main entrances and exits in the wings. Vienna's Nord-West Bahnhof was also built in this manner.

It was at this time that the two-sided type reached another stage of evolution. Meeks provides this observation: "Just as the one-sided type had been made more efficient by the use of two levels, so the two-sided type was given a renewed value by raising it on stilts. Elevated two-sided stations were no longer limited to terminal stations, but were extremely practical as through stations. The name 'island' is appropriate to such stations because they are isolated on viaducts completely surrounded by streets." Johann Eduard Jacobsthal was considered by many to have developed this type. In his Alexanderplatz Bahnhof in Berlin, completed for the Stadtbahn in 1885, the elevated shed became the station. Subsidiary rooms were incorporated either within the shed or below. This profoundly simple style became very popular in Germany. Later structures of this type include stations built at Hamburg and Dresden about the turn of the century.

Berlin's Anhalter Station, built from 1872 to 1880, was considered important in terms of its design integration.

Drawing from Germany's Romanesque heritage, the architect Franz Schwechten gave this station an imposing, if not monumental appearance through a succession of graduated elevations. Beneath the curved roof at the front of the shed, nine arches were arranged to create an arch of arches. Within each was a large window. Flanking the shed at both ends were truncated towers. Crowning its summit was a heroic winged figure, seated on a throne.

During the 1870s Frankfurt am Main had three railway stations, each serving a separate region. It was decided to consolidate these into a single facility, and Georg Eggert was selected as architect. To maintain the traditions of these railways, Eggert designed three identical arched train sheds placed side by side. Each shed had the following dimensions: 610 feet long, 183 feet wide, and 93 feet tall. The combined width of all three was 549 feet, and through them 18 tracks were laid. Entrance to the station was through a majestic domed hall, with a cross-platform providing access to the train sheds. Meeks comments: "No medieval cathedral was more spectacularly calculated to exploit the properties of its materials. The five luminous volumes—three sheds, cross-platform, and domed vestibule—formed a graduated sequence of ascending heights and crossing axis. Externally these glass cages, rising above low wings framed by dome pavilions and slender towers, commanded attention. The rather elaborate ornament of the masonry portions served to emphasize the simplicity of the vaults as a boxer's fancy shorts set off the lines of his torso." Construction of this huge station began in 1879 and was only completed in 1888.

Following the unification of Italy, many new stations were built. The largest of these were in Genoa, Milan, Trieste, Turin, Rome, and Naples. Italian architects at the time tended to ignore the latest international trends and built in a traditional way. However, Rome's Stazioni Termini, designed by Salvatore Bianchi and opened in 1874, owed some inspiration to the Gare de l'Est. It was of the pseudo head variety, however, being in fact a two-sided type. Turin's Stazione di Porta Nuova, completed in 1868, was also built in the two-sided style. Another two-sided type was Milan's Stazione Centrale, completed in 1873. Inspired by the Louvre, it was designed by the French architect J. L. Boudhot and featured a number of pavilions and Mansard-style roofs.

Russia's St. Petersburg Station, completed in 1837, was of the one-sided type. The Smolensk Station, following the early British practice, was built in the Italianate style and included a campanile. The *Railroad Gazette* in its June 15, 1888 issue provided this view of Russian country stations: "The wayside stations in Russia are more picturesque than those of Western Europe. They seem all to be built of one pattern. They are of wood, of course. Russia is the country of wood, as distinguished from the countries of stone; just as it is the country of the plain as distinguished from the countries of the mountain. The wayside railway stations are one-story buildings, neatly and simply built. They are roofed with thin sheet iron, painted red-brown. At first it is difficult to distinguish the iron from shingles. But when, as is sometimes the case, it is painted green to resemble verdigris, the iron is distinguishable. At every station there are three objects conspicuous in the front of the building: a clock, a bell and a large thermometer."

In 1894 Lisbon's Central Station was completed. Under the previous arrangement a terminus in the far eastern part of town served lines coming in from the north and east, while a station at the extreme western end served both suburban traffic and a line coming in from the west. This was an inefficient system, however, as both termini were too far removed from Lisbon's central area and inconvenient for both suburban and through traffic. The various lines were extended to the northern part of the city, where a junction was made. From there a tunnel 2,843 yards long brought traffic into the commodious new terminal, situated overlooking the principal boulevards and Praca de Dom Pedro.

To the end of the century small stations in a few European areas continued to be built in a rustic traditional style. These included Germany's Black Forest region, Scandinavia, and southern France. As the fin de siècle was bringing down the curtain on the nineteenth century, a number of architects and designers became strongly influenced by the similar art nouveau, Jugendstil, and Vienna Secession movements. Rebelling against traditional and classical mindsets, they turned for inspiration to nature. What resulted was a more colorful and decorative approach to building design, giving structures a more unified organic look. While this movement was especially strong in Belgium, France, central Europe, and Finland, its influences spread elsewhere. Stations reflecting this trend included Vienna's Karlsplatz Station, Central Station in Antwerp, and the Gare de Tours.

Elsewhere in the world station styles were largely derived from Europe. However, in a few areas like India, a strong indigenous architectural tradition existed. This led to the creation of composite-style stations. One structure of this type was the symmetrically designed "Mogul Gothic" station at Gwalior. The most imposing, however, was Bombay's Victoria Terminus, built for the Great Indian Peninsular Rwy.

between 1877 and 1887. Designed by Frederick William Stevens in what was called the "Gothic Saracenic" style, it became a remarkable fusion of the Occident and Orient. In *The Railway Station: A Social History,* Jeffrey Richards and John M. Mackenzie give this description: "It is an extraordinary hybrid, combining domes with spires, fretted and decorated round-headed arches with lancet and rose windows all in a warm red color. Middle English, Venetian, Romanesque, and Orientalist features jostle with each other in an architectural epic of commercial and religious power. Marbles and granites were imported and the building was embellished with appropriate statuary and ornamentation . . . To this day, it signifies almost better than any other building the Victorians' love of spectacle and sentiment and their confidence that a new world order could be created out of a massive syncretism."

Top: King's Cross Station, London Terminus of the Great Northern Rwy., 1852. **Bottom**: Gare de l'Est, Paris, 1852.

PLATE 186

Train shed view of Grand Central Railway Station, Birmingham, England, 1854.

PLATE 187

Front facade of Gare du Nord, Paris, 1863.

PLATE 188

View of London's Charing Cross Station & Hotel, South Eastern Rwy., 1864. This is an example of a terminal where a hotel functions as the head house or front facade for the rest of the station.

PLATE 189

View into train shed of London's Cannon Street Station, 1866. Like Charing Cross Station, this terminus also served the South Eastern Rwy. However, its primary patrons were passengers commuting in and out of London's business district. Charing Cross, by contrast, handled the bulk of the main line traffic.

Train shed view of London's St. Pancras Station, 1869. This Gothic-style structure, built by the Midland Rwy., was considered at the time to be the world's finest.

PLATE 190

PLATE 191

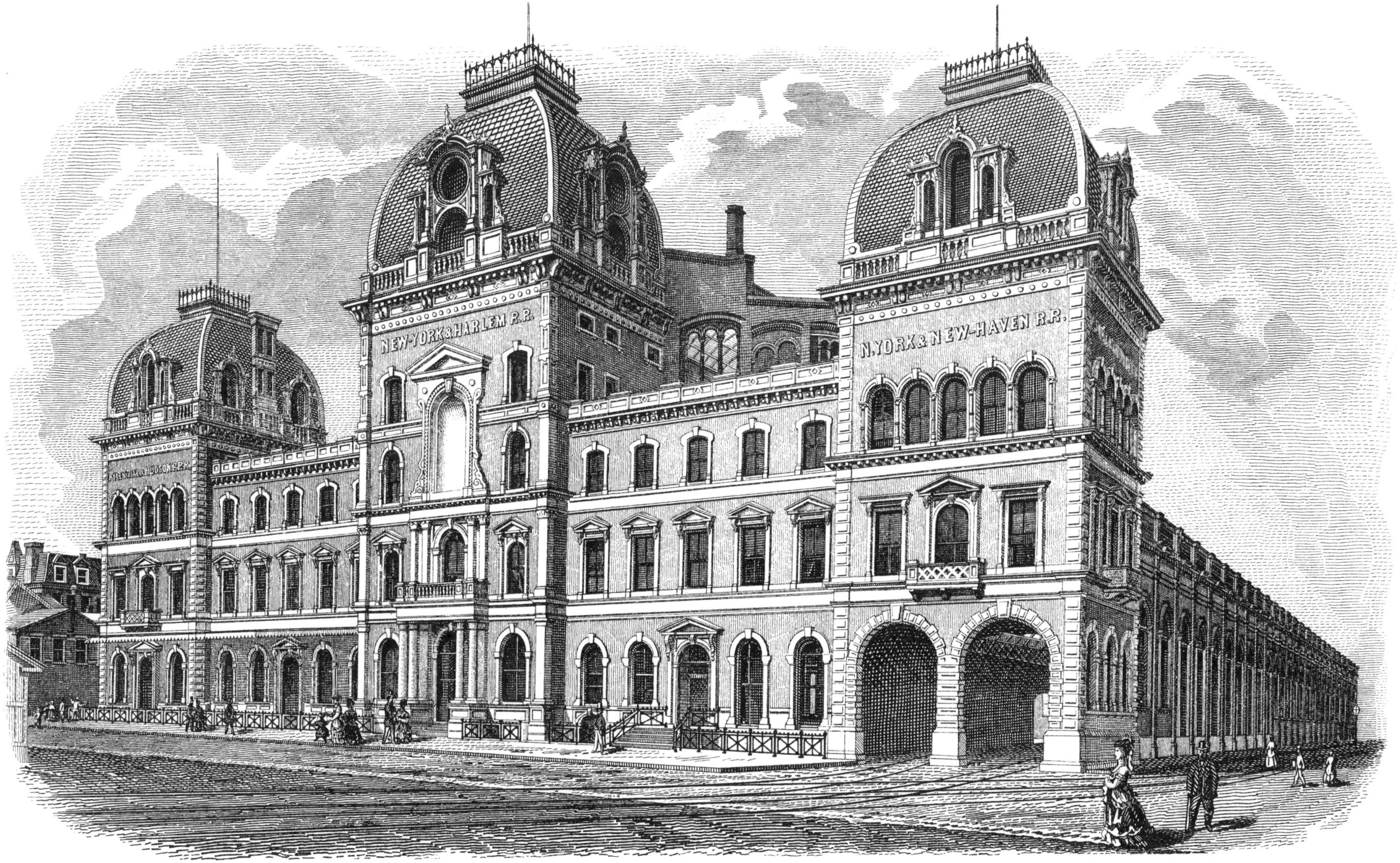

View of New York's Grand Central Station, 1870. This impressive terminus served the New York Central & Hudson River; New York & Harlem; and New York & New Haven railroads. The two arched entrances on the right provided a route for horse cars.

Top: West Philadelphia Station, Thirty-second and Market Sts., Pennsylvania R.R., 1876. Bottom: Way station at Ridley Park, Pennsylvania; Philadelphia, Wilmington & Baltimore R.R., constructed in 1872.

PLATE
192

PLATE 193

Union Railroad Station, Worcester, Massachusetts, 1875. Designed by William Ware and Henry Van Brunt, this station served the Boston & Albany; Providence & Worcester; Norwich & Worcester; Worcester & Nashua; and Boston, Barre & Gardner railroads. The tower was 212 feet high, and the entire building was faced in stone.

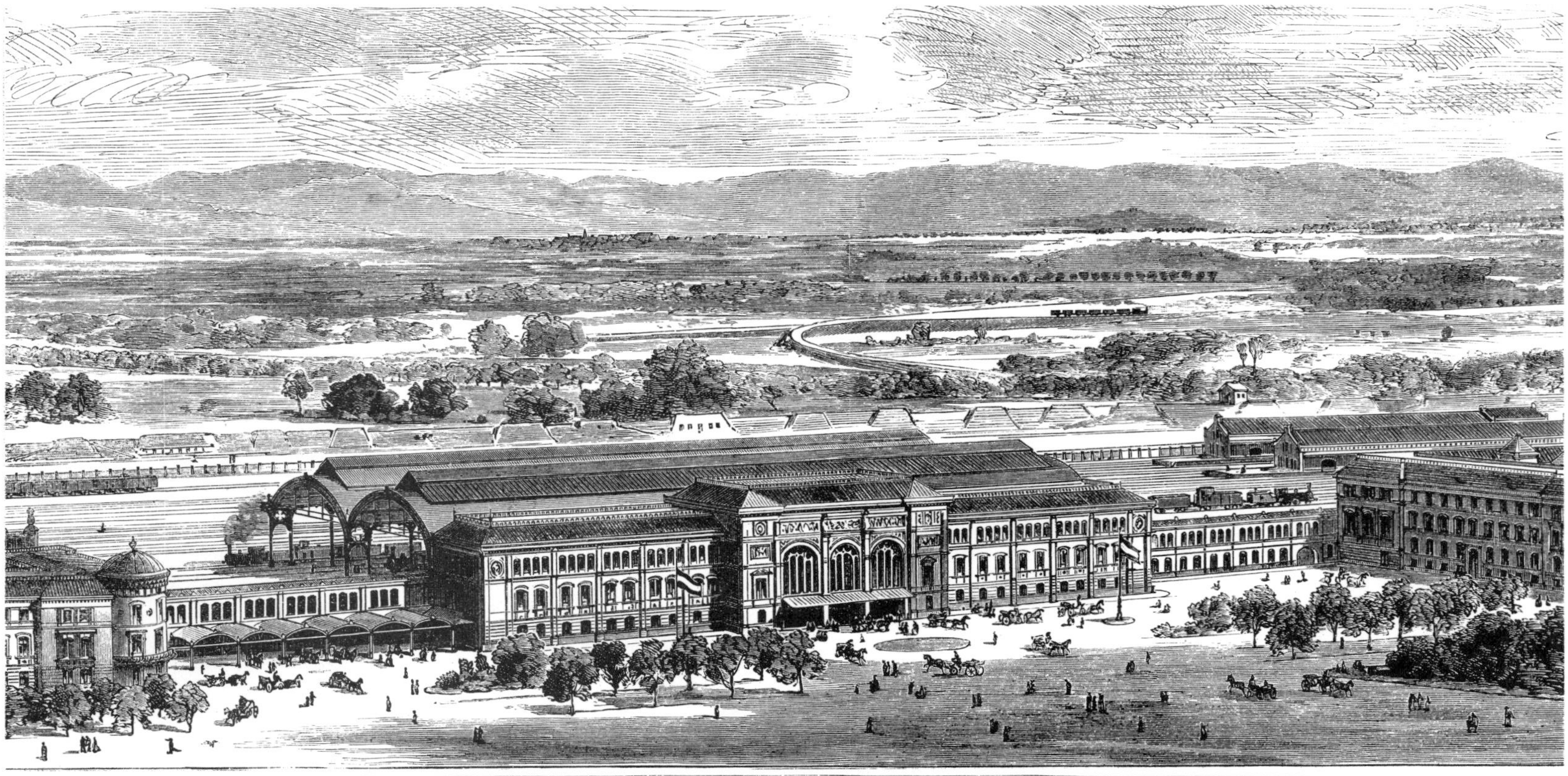

Top: Bird's-eye view of new station in Strasbourg, Germany (now France), 1885. This structure is a through station of the one-sided type.
Bottom: Stazione Termini, Rome, circa 1874.

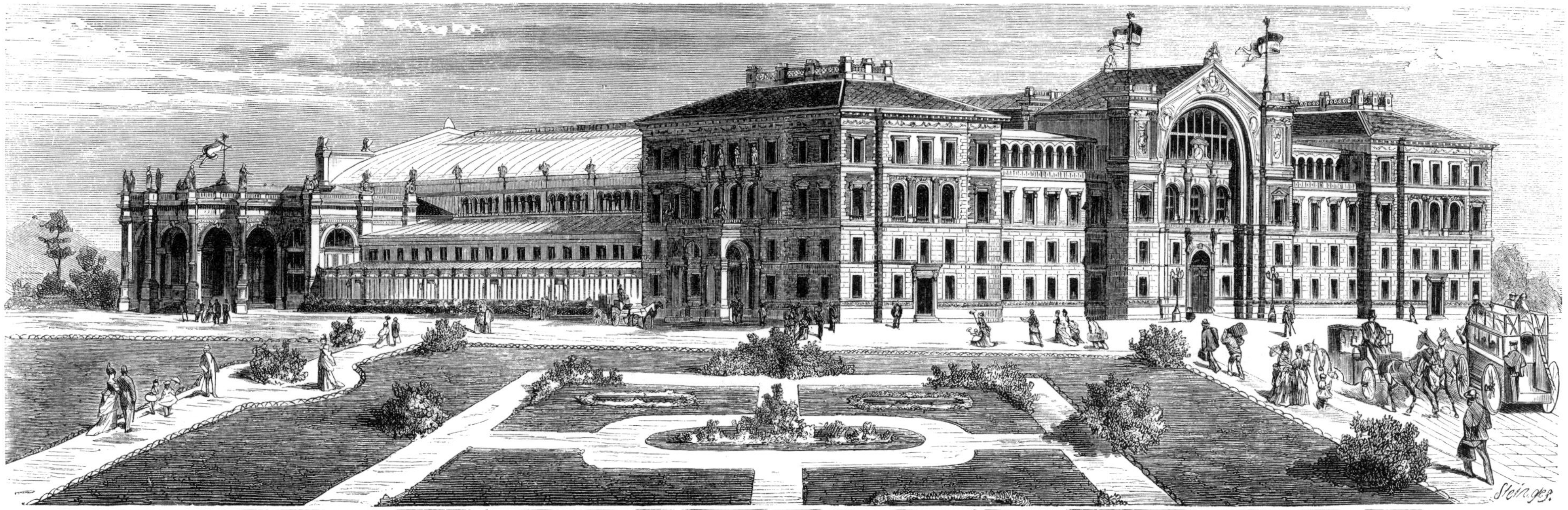

PLATE 195

TOP: Nordwestbahnhof, Vienna, circa 1880. BOTTOM: Station in Tournai, Belgium, 1880. The Vienna terminus was of the pseudo-head type, being in reality a two-sided style station. The Tournai structure appears to be a through station of the one-sided type.

Exterior and train shed views of Berlin's monumental Anhalter Bahnhof, 1880. This station, designed by the architect Franz Schwechten, served the Anhalt and Dresden railways, and took eight years to build.

PLATE
196

PLATE 197

Top: A common fixture at many stations was the footbridge that allowed passengers to safely move from one platform to another. This one was made by the Arrol Brothers of Glasgow. Bottom: View into train shed of Central Railway Station, Christiana (Oslo), Norway, 1881.

PLATE 198

Bird's-eye view of Gare St. Lazare, Paris, Ouest Rwy., circa 1886. Note double-decker passenger carriages used for commuter traffic. In the upper left corner is the Pl. de l'Europe.

New stations in Montreal, Quebec, 1889: Top: St. James Street Station, Grand Trunk Rwy. Bottom: Windsor Street Station, Canadian Pacific Rwy.

CHAPTER

15

Freight and Goods Service

THE essential service railways provide to the public is to move space. A space in a coach or wagon is filled, with either passengers or goods, and then moved to a different location. At this point the service is terminated, and the space is emptied. For this, the railway receives compensation. In order for it to make a profit, and ultimately to survive, its compensation will have to exceed the costs of rendering the service.

The freight rate practices of early English railways were largely inherited from the canal industry. Essentially, rates were determined by an equation involving weight, space, and distance, although other factors were taken into consideration. These included the size of the shipment, its value, whether it was fragile or perishable, and its vulnerability to weather. Another important factor was whether there was any competition from other carriers. When railways had competition, they tended to reduce rates to attract business, sometimes even below the cost of providing the service. Otherwise, they tried to charge "what the traffic would bear."

Heavyweight raw materials, like coal and iron ore, were at the bottom of the rate scale. These materials required little space in relation to their weight, and were invulnerable to damage or the weather. Thus they were carried in bulk, on cheap open wagons, at the lowest level of priority. Rates went up from there, with the highest typically assigned to small shipments of fragile, lightweight merchandise. Valuable shipments of gold, money, or jewels, although technically freight, fell into a totally different category. Coaches or wagons carrying these shipments were usually guarded and often attached to express trains.

The Surrey Iron Rwy., and other early English lines, adopted a practice commonly used on canals. This was to allow cartage contractors to operate over their routes. The contractors supplied their own wagons, as well as horses or locomotives, and paid the railways a toll based on the class (or classes) of goods that were being carried. This policy was initially followed on the Stockton & Darlington line. However, in the charter granted by Parliament in 1828 to the Liverpool & Manchester Rwy. there was a provision allowing the company to function as a common carrier in its own right. This practice, which gave the railway more control over its operations, was also found to be more profitable. Thus, soon, cartage contracting was largely discontinued.

At Liverpool Station hoists and jiggers were provided to help those loading or unloading goods. Soon the Liverpool & Manchester and other early English lines began building separate warehouses to serve this need. The Manchester & Leeds Rwy. had warehouses that could store 100,000 sacks of flour. In 1832 the Glasgow & Garnkirk line built a goods station that had platforms built to the same level as the floors of the railway wagons. This made the task of moving freight much easier. Tracks were also laid along the nearby river bank so that exchanges could be made between the wagons and boats. By 1860, in Paris, the Gare du Nord had 86 acres of space reserved for goods, while the Gare de Lyon had 62 acres.

Goods trains were typically longer and heavier, and operated at slower speeds than passenger trains. The earliest goods wagons were sturdy four-wheel carts with slanted sides used for hauling coal or ore on tramways. The coupling of these cars was originally done with chains. While coal wagons were the first to be used on the Stockton & Darlington line, these were found to be inadequate for most classes of freight. Beginning in the late 1820s, experiments were undertaken to improve wagon design. This ultimately led to the development in Britain of three primary types of wagons: the flat platform truck, the open wagon with sides, and the enclosed wagon with sliding door.

Other wagons were built for specialized purposes. This group included enclosed wagons for carrying different classes of livestock, all the way from chickens to horses. Wagons were initially built of wood. However, in 1844, Daniel Gooch, of the Great Western line, built an enclosed wagon with both the underframe and body made of iron. Experiments were also undertaken to build larger wagons having a greater carrying capacity. However, the ordinary 10-ton wagon proved adequate for most goods-hauling needs, and became the standard car of most railways.

In America, during the 1830s, primitive wagons began differentiating into prototype versions of boxcars, flatcars, gondolas, and hoppers. The latter two types were well suited for carrying coal, ore, and rock. Flatcars were found to be useful for hauling several types of freight, among them large, bulky items needing no protection from the weather. Animals, like cattle or pigs, soon rode in cars that were essentially cages on wheels. Covered hoppers were built to transport products like perishable grains. Boxcars provided weatherproof safekeeping for manufactured goods or other vulnerable items. Although these cars were initially built with four wheels, bogies were soon added. Appearing somewhat later was the caboose, which was attached at the rear of the train. In the middle of this car the roof was raised to provide a lookout point, from which the progress of the train could be observed.

The improved goods service that railways provided quickly accelerated business activity in England and elsewhere. As railway networks expanded, new industries arose and businessmen had more opportunities and options. Factories no longer had to be situated near sources of raw materials or near a navigable river or canal. They only needed to be near a rail line, and as time went on, that could be almost anywhere. While early railways had carried coal to water, where it was then transferred to barges, they later transported it directly from the mines to the cities. In London, the Northumberland & Durham Coal Company began using the North London Rwy. to carry coal from the West India Docks into the suburbs. Later, the Great Northern line, carrying coal directly from the mine, built into the city and took over much of the business. In general, goods revenues became increasingly important to British railways, and by 1852 they surpassed the income from passenger service.

The effect of railroad freight service in the United States was even more pronounced. Unlike England, which had a good network of roads and canals preceding the introduction of railways, America had almost nothing. At the time, the country's commerce largely depended on oceans and navigable rivers. There were few canals and roads, with the latter usually being of poor quality. As a result, the country's settled regions developed largely autonomous economies. The introduction of railroads soon brought this isolation to an end. As these areas became connected by rail, trade between them rapidly increased. Perishable agricultural products and other locally sold goods found their way to larger markets. Regions began to specialize more. Factories were located in places favorable for production, and their goods carried by rail all over the country. As time passed, America's separate regional economies evolved into an interdependent national economy.

As regions and localities specialized, railroads did likewise. Pennsylvania companies like the Philadelphia & Reading, Lackawanna, and Lehigh Valley gained a reputation for hauling coal. In the Midwest railroads like the Chicago & Northwestern and Chicago, Burlington & Quincy brought in bountiful grain harvests. This grain was moved from Chicago, by other lines, to milling cities like Buffalo and St. Paul. The Duluth, Missabe & Northern and Duluth & Iron Range railroads transported Minnesota iron ore to Duluth, where it was shipped out over the Great Lakes. In the South some railroads specialized in hauling cotton. Following the Civil War, the Atlantic & Great Western and Philadelphia & Erie lines carried out immense quantities of oil. America's largest freight carriers, the New York Central & Hudson River and Pennsylvania lines, operated between the East Coast and Midwest, and carried almost everything.

British lines also began to specialize in certain ways, adapting to the needs of shippers along their routes. For many carriers the transport of agricultural products became increasingly important. As railways expanded and rail equipment improved, these perishable products could be carried safely at ever greater distances. Milk producers were among those who took advantage of this opportunity. During the 1870s milk could be shipped at a halfpenny a gallon for a short distance, and at a penny for longer distances. This inexpensive rate encouraged many producers to expand the range of their business. Among these were producers in the rich meadowlands near Derby, who began shipping to London. Their milk was carried into Derby on both early morning and afternoon local trains. From there, these shipments were transported by mail train to London. At the time, 15,000 gallons of milk were consumed daily in the city, all of it brought in by rail.

The shipment of livestock also accounted for a sizeable amount of railway business. Wagons of livestock were usually grouped together and run separately as high-priority trains. This was done both to ensure the survival of the animals and to avoid unnecessary loss of weight. Weight loss was greatest where shipping distance was longest. As a result the practice of shipping "dead meat" began. This became a substantial business in the northern reaches of Scotland. The two railways serving this area, the Highland and Great North of Scotland, picked up shipments of slaughtered beef all along their routes. Carcasses destined for the London market were cut in half, with each side sewn up in canvas and closely packed into wagons. From 4 to 6 tons of beef were carried in each wagon. Two branches of the Great North of Scotland line brought these wagons into Aberdeen. In the winter months this averaged about 25 wagons daily.

These left Aberdeen at noon, attached to an express train of the Caledonian Rwy. This train, carrying London goods only, had two engines. At Perth, it picked up other meat wagons brought in by the Highland line. At this same stop, the train divided in two. One train took the east coast route, picking up additional meat all along the way, and the other went via the west coast, receiving new cars as far as Rugby. The first train was scheduled to arrive at Camden Station just before midnight, its total journey taking about 30 hours. The second train was scheduled to arrive at 2:40 a.m. Upon arrival these shipments were promptly unloaded and quickly dispatched to the Newgate and Leadenhall markets.

The availability of rail transport also revolutionized Britain's fish trade. This was a more irregular traffic, dependent upon the season and catch sizes at different parts of the coast. During the height of the mackerel season, special trains arrived in London with about 90 tons daily. During herring season 10 to 12 wagons arrived daily from points as remote as Banff, Peterhead, and the northeast coast of Scotland. Cod fish were sometimes carried live, in large tanks of water.

In *Our Iron Roads,* Frederick S. Williams provides a glimpse of fish activity elsewhere:

> The fish traffic at many country stations is very large. "At Nottingham," remarked an official, "it is almost the largest we have for money. It arrives by the mail due very early in the morning. This train brings fish from the east coast—from Hull, Berwick, and the whole of Scotland. Hull and Grimsby also send a train here by the Manchester, Sheffield & Lincolnshire, leaving Grimsby at five, and arriving at 10:30. Here it shunts off the trucks that are for Nottingham, and for the Erewash Valley and Mansfield lines; it then goes forward to Bristol. Yes it supplies Bristol with fish. The other night it came in with 48 wagons and two engines, and was late because the catch had been so large that they could not get it all loaded at the usual time. Grimsby is our best fishing port for winter and summer in England . . . The London market depends on Grimsby for its crimped cod, much of which comes through Nottingham."

It was the procedure at London goods terminals to receive outbound freight in the evening from local shippers. These consignments would be loaded into wagons and then sent out at night. Inbound shipments would arrive in the early morning and would usually be delivered to consignees before the beginning of the business day. A typical scene of the 1870s in Camden Station is described by Williams:

> Fancy 1,500 men nightly occupied in loading and unloading goods in the goods sheds of a single company; vans arriving from all parts of the metropolis, beginning at 6:30 p.m. and ending at 9:30; a little army of men struggling with the bulky packages, which they deposit on their respective platforms, from whence they are loaded into the railway wagons placed alongside, and dispatched at once, train by train, to the remotest parts of the kingdom. The scene appears at first one of inextricable confusion—men battling with bales, barrels, crates, and hampers, amidst the noise of voices and clamor of machinery. Yet the whole is proceeding with regularity and dispatch, and in the course of a few hours the last train outwards has left, and the station is wrapped in quiet until the time of the early morning arrivals.

Following America's Civil War, a number of odd-gauge lines, especially in the South, were converted to the standard. About the same time various railroads began to make car interchange agreements, which allowed the cars of one line to be operated over those of another, thus eliminating the practice of transshipment. Another development, after the war, was the formation of fast freight companies. These were usually created by cooperating railroads for the purpose of consolidating small shipments and providing expedited service. Among these companies were the Green Line, White Line, Red Line, Blue Line, Star Union, Merchants Dispatch, Great Western Dispatch, and Traders Dispatch.

The earliest American oil cars appeared circa 1865. The original arrangement was simple: two round wooden vats mounted on a flatcar. Developed a short time later was the familiar horizontal cylinder design. Another contemporary innovation was the refrigerator car, capable of carrying perishable products from coast to coast. The earliest of these was built for George Hammond in 1867. J. Tiffany introduced an improved design in 1868. However, it was not until circa 1872 that the problem of refrigerated air circulation was satisfactorily worked out. While these cars hauled a number of different products they were most effectively used by the meat industry of Chicago. Large firms such as Armour, Swift, Hammond, and Cudahy found that their mass-produced beef could be shipped to such markets as Boston, New York, and Philadelphia and sold at prices that undercut locally slaughtered meat.

During this period rail competition greatly intensified, and lines resorted to underhanded tactics, either to gain advantage or merely to survive. Most typically they would cut rates where the competition required it and raise them elsewhere. In 1870 the New York Central and Erie lines engaged in a fierce competition for hauling livestock from Buffalo to New York. After both sides made repeated rate reductions, the New York Central set its rate at one dollar a carload. This was far below what it cost the line to provide the service, as the normal rate had been $160 per carload. Officials from the Erie saw a good opportunity. They bought a substantial number of cattle and shipped them to New York over their competitor's line.

In the Midwest, railroads often owned local grain elevators, where they frequently charged excessive storage rates. They also often downgraded grain when buying it from farmers, and then upgraded it before selling it to the large milling companies. In Pennsylvania, competing lines published rates for shipping oil that only the small companies paid. They secretly granted generous rebates to the largest shipper, John D. Rockefeller's Standard Oil Company. In one six-month period in the late 1870s, Rockefeller received a rebate from four railroads of $10 million.

The reverse side of rate wars was pooling. This practice was a secret and illegal collusion among competing railroads where they came to an agreement about dividing the business. Sometimes a revenue pool was held in common and then divided up. In other cases each company might agree to carry a certain percentage of the traffic. Some lines also colluded in dividing up territory, determining where they might extend lines and where they could not.

In Kansas, between 1867 and 1870, over a million head of cattle, mostly driven from Texas, were shipped east from Abilene on the Kansas Pacific R.R. The Atchison, Topeka & Santa Fe, perhaps observing this success, built a parallel line to the south; first to Newton and then to Dodge City. It was successful in intercepting the Texas herds, and in doing so, captured most of the Kansas Pacific's business. Yet, this success was only temporary. Soon, the Missouri, Kansas & Texas R.R. built through Oklahoma into Texas. At the same time, Texas railroads were also building into the cattle region from the east.

The era of the free range was coming to an end. Ranchers, as well as farmers, were claiming territory and erecting fences. These blocked the old trails and effectively brought an end to the long cattle drives. In its issue of April 27, 1883, *Railroad Gazette* reported: "Owing to the cattle range of south and southwest Texas being greatly curtailed by settling up of lands and devoted to agriculture, many large cattle raisers have determined to remove their herds to northwest Texas, particularly in the panhandle region. As it is almost impossible to secure any well defined trail on account of the improvements, the transfers will be made almost entirely by rail. Ward & Rugley, of Kansas City, have inaugurated the move and have secured a rate of transportation for themselves and several friends to have 75,000 head of cattle shipped via the Gulf, Colorado & Santa Fe and the Fort Worth & Denver roads from Milheim to Wichita Falls at $35 per car."

For many animals, being shipped over rail brought an untimely end. One such accident occurred on May 30, 1890, when a broken bogie caused an accident on a train of double-decked cars. These were loaded with approximately 5,000 Merino sheep, en route from California to Chicago. The accident occurred on the Atlantic & Pacific line 2 miles west of its junction with the Santa Fe R.R. in New Mexico. All but two cars were destroyed and about 1,000 sheep were lost. Local Indians worked all night skinning their carcasses.

Elsewhere in the world, the rapid expansion of railway networks made possible unprecedented levels of commerce and trade between nations. However, complex situations occasionally arose. One interesting case involved the difficulties faced by certain Austro-Hungarian lines in gaining access to domestically produced grain and timber. These railways provided routes to the German border, from which point shipments could continue to North Sea ports and eventually to England.

However, there was an interesting mix of other markets and carriers that also competed for this yield, as well as contending suppliers. Steamboats on the Danube carried much of this traffic to the Black Sea. From there it was loaded into ships, as was Russian grain from the nearby port of Odessa, and carried to various European ports, primarily those in the Mediterranean. Domestic rail carriers with connections to the Adriatic ports of Trieste and Rijeka also wanted this tonnage. While the railway rates were higher than the river rates, these ports were much closer to Mediterranean and Western European grain markets. An additional factor was the availability of American grain, which could be shipped at a cheaper rate to the North Sea, than into the Mediterranean. The net result of this interplay of possibilities was that if the aforementioned Austro-Hungarian lines were to carry any tonnage, they would have to offer very low rates.

Unfortunately, their natural partner in this endeavor, the

Prussian State Rwys., had inflexible rate practices. So the Austro-Hungarian lines came up with a plan to carry shipments to ports along the Elbe River, where they could then be sent downstream by barge or steamer. The inevitable result of this maneuver was a serious dispute with the Prussian State Rwys. The *Railroad Gazette,* in its March 9, 1883 issue, gave the following report on subsequent developments:

> The great conflict between the German and Austrian railways with regard to through traffic between Austria and the North Sea or North German points has been settled by agreement according to which there shall be no secret rates, but all changes must be made public at least 14 days before they take effect; rates made by the railways in connection with vessels on the Elbe, the Oder, the Danube and the Rhine must also be submitted to this regulation, and the rail rate must be charged separately from the river rate . . . Any rebate that may be made in favor of any of the ports of Bremen, Hamburg, Lübeck, or Stettin must be made for them all alike; the Austrian railways must give notice of any secret rates on the Elbe boats connecting with them, of which they have knowledge, as early as two weeks before the opening of navigation and every two weeks afterward, in order that the German railways may the better compete with the boats.

About this time in Germany the average goods train consisted of 36 wagons, with a carrying capacity of 370 tons. However, the average load was only 180 tons.

Goods train speeds remained low throughout the nineteenth century. One of the reasons was a resistance, on the part of railways, to invest in improved braking systems. On the Continent goods were typically carried at speeds no faster than 10 m.p.h. Late in the century, some British lines installed vacuum brakes on wagons of their express goods trains. On the Great Western and London & North Western lines, these trains sometimes reached speeds of 30 m.p.h. The situation was much the same in America. In the 1880s a few railroads began to install airbrakes on their cars. However, it was only after Congress passed railroad safety legislation in the 1890s that a serious effort was made to install these systems.

Because of the laissez-faire attitude of the British government, railways were largely free to charge whatever rates they liked, especially in areas where there was no competition. Where rivalry existed, carriers sometimes colluded to keep rates high. As these practices increased, public dissatisfaction grew. Between 1866 and 1884, after receiving numerous complaints from shippers, the government made three separate inquiries into the rate practices of railways. However, in each case little was done, and the railways involved were largely exonerated. In 1873 a Railway Commission was established to settle disputes between shippers and carriers, but this had little effect on the existing situation. In 1885 the government made a more serious effort to investigate railway rates, but only in 1888 was the first legislation passed that began to change things. During the latter part of 1889 and throughout 1890, hearings were held before both the Board of Trade and a joint committee of both houses of Parliament. This resulted in acts that came into effect on August 1, 1892, amending the powers and classifications of nine of the country's leading lines.

In America, similar problems led to efforts by individual states to regulate shipping rates. However, in 1881, the Supreme Court ruled that while states could regulate rates within their borders, they had no power over interstate commerce. This setback eventually led Congress to pass the Interstate Commerce Act, signed into law on February 4, 1887. This important piece of legislation created the Interstate Commerce Commission, a ruling body destined in the twentieth century to cause considerable harm to the nation's rail carriers.

PLATE 200

Coal train at Le Creusot, France, PLM Rwy., 1870.

Use of overhead traveling cranes for transshipment at Middlesbrough Docks, North Eastern Rwy., 1873.

Hauling livestock: TOP: Loading cattle for shipment at Halleck, Nevada, Union Pacific R.R., 1877. BOTTOM: Unloading shipment of cattle and sheep for Christmas market at Tottenham Station, Eastern Counties Rwy., circa 1850s.

PLATE 202

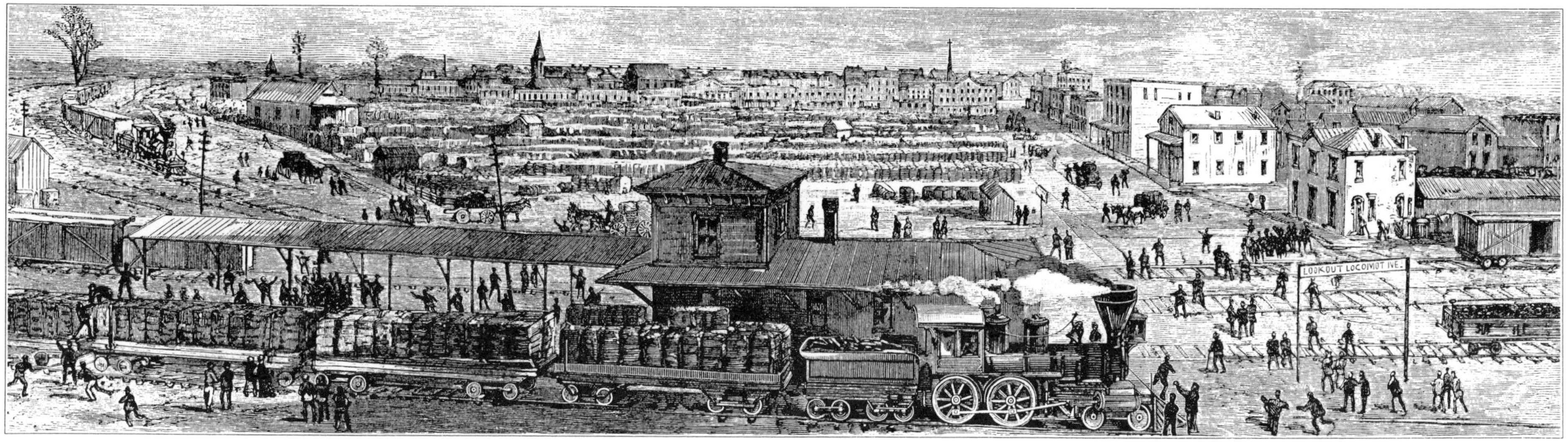

Top: Shipping cotton bales out of Meridian, Mississippi, 1879. Bottom: View in Oil City, Pennsylvania, Atlantic & Great Western R.R., 1871. The company's engine no. 71, shown here, was built by the Jersey City Locomotive Works in January 1864. Note also the early wooden car with three vertical tanks used for hauling petroleum.

Top: Busy activity at the gravity coal piers in Hoboken, New Jersey, on the Hudson River opposite New York City; Delaware, Lackawanna & Western R.R., 1882. **Bottom:** An English colliery scene at Talke-on-the-Hill, near Hanley, Staffordshire, 1873.

PLATE 205

View of New York Central & Hudson River R.R.'s grain elevator, Hudson River at Sixty-first St., New York City, 1877. The grain elevator was a uniquely North American invention. At the time these could be found in American and Canadian cities involved in the grain trade, but nowhere else.

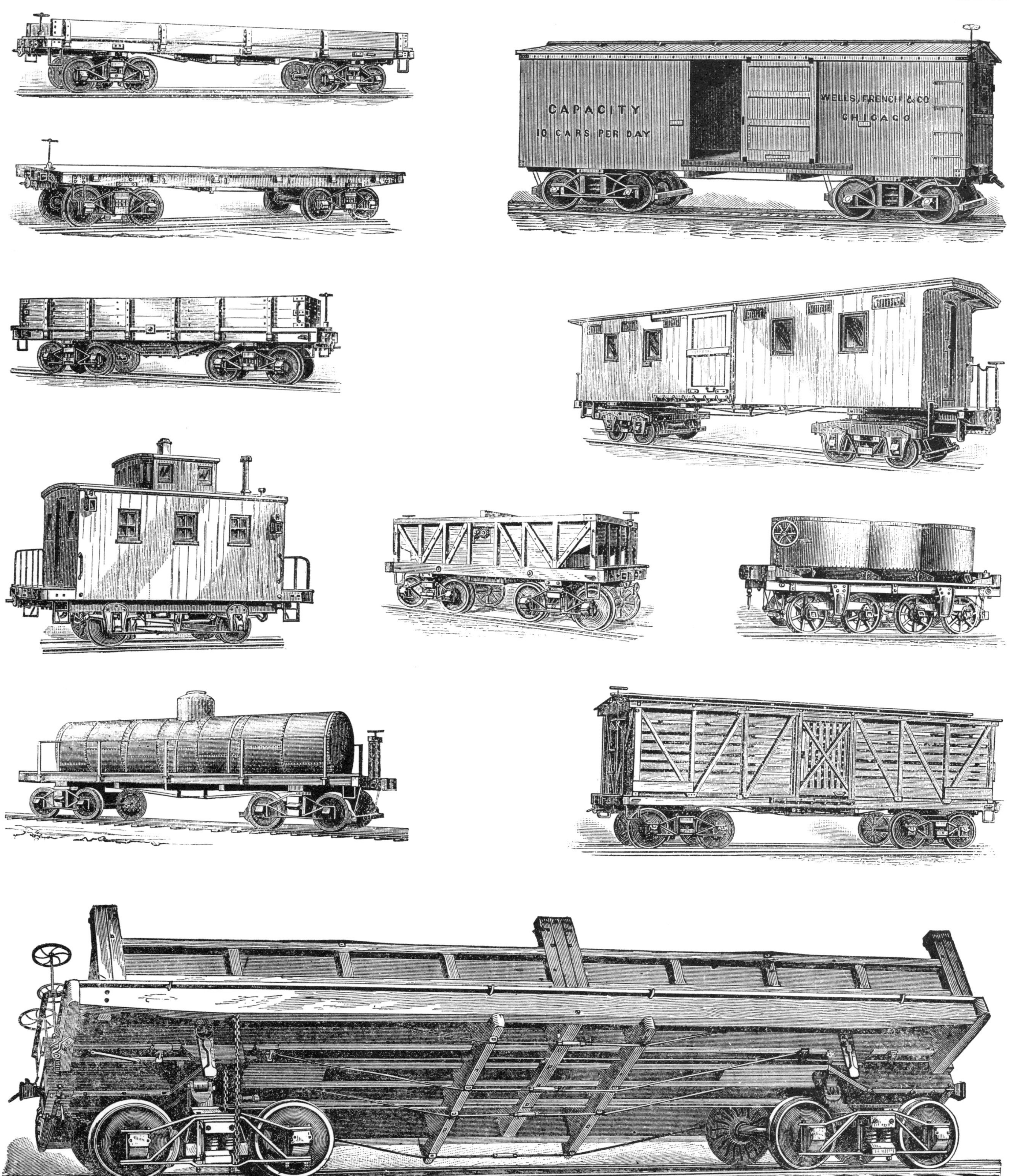

American freight cars built circa 1870s: Left, descending: Gondola, flat car, hopper bottom gondola, four-wheel caboose, tank car. Middle center: Hopper bottom coal car. Right, descending: Boxcar, milk car, iron hopper coal car, cattle car. Bottom: Screw lever dump & coal car, New England Car Co., 1881.

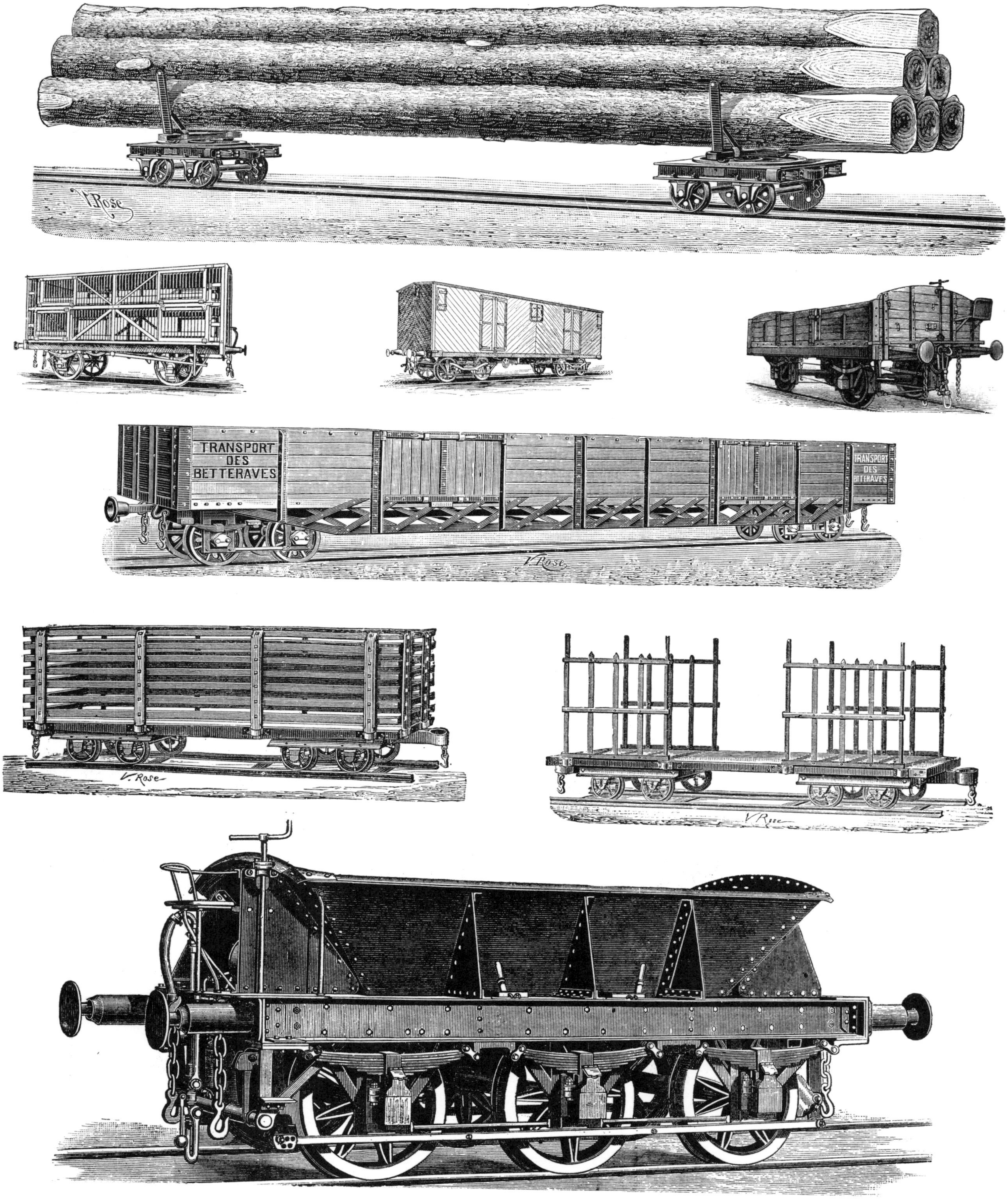

Top: Decauville lumber trucks, circa 1893. Upper middle left: Sheep wagon, Victorian Rwy., Australia, circa 1900. Upper middle center: Refrigerator wagon, New South Wales, circa 1900. Upper middle right: Open wagon, Germany, circa 1880. Middle: Open wagon for transporting beets, Decauville Co., circa 1893. Lower middle left, lower middle right: Decauville platform wagons with removable sides, circa 1893. Bottom: Iron ore wagon, Swedish & Norwegian Rwy., 1888.

PLATE
207

Top: Open goods wagon built by Leeds Forge Co., 1894. Middle left: Open van of Varsovie-Vienna Rwy. Middle center: Platform wagon of Württemberg State Rwys. Middle right: Württemberg wagon with movable sides and extremities. Bottom: Open wagon with sides. Note platform to provide weather protection for train guard.

PLATE 208

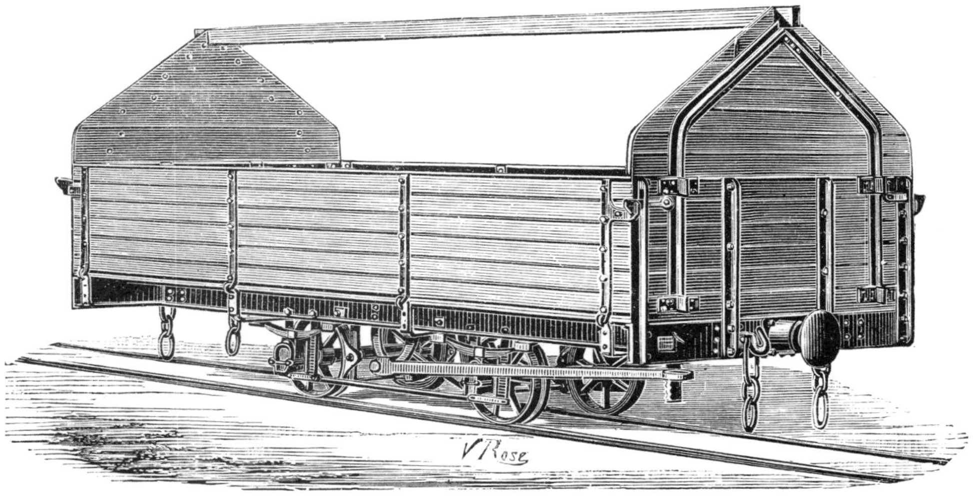

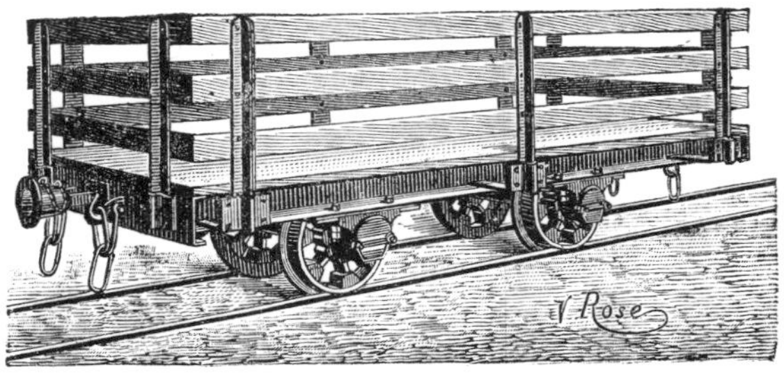

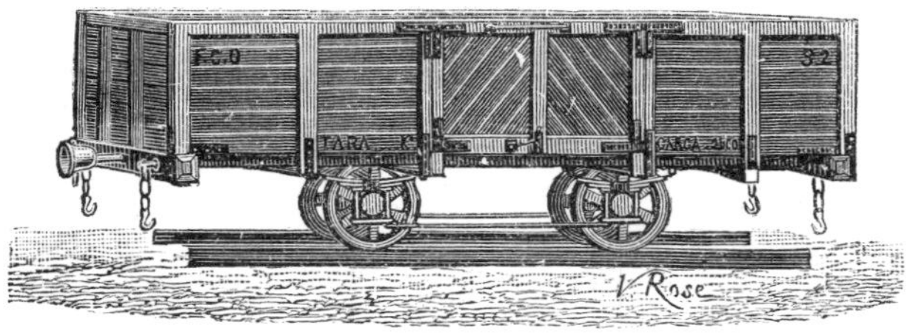

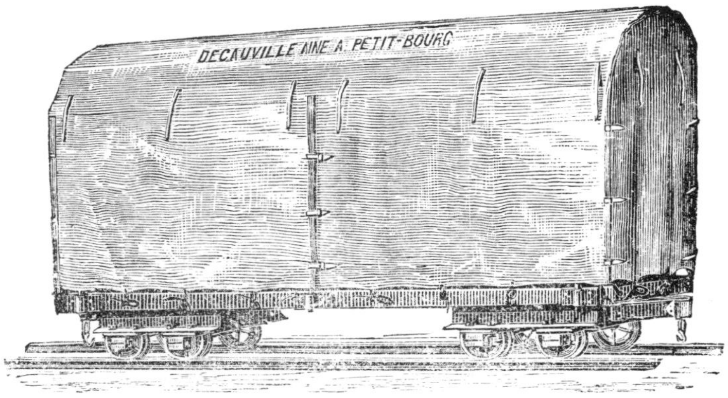

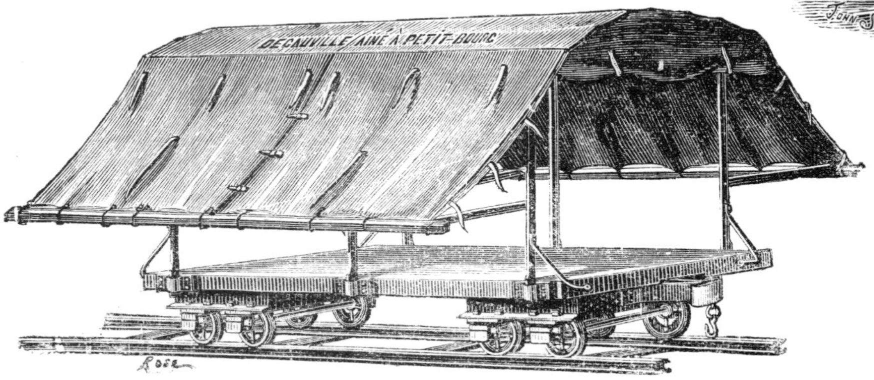

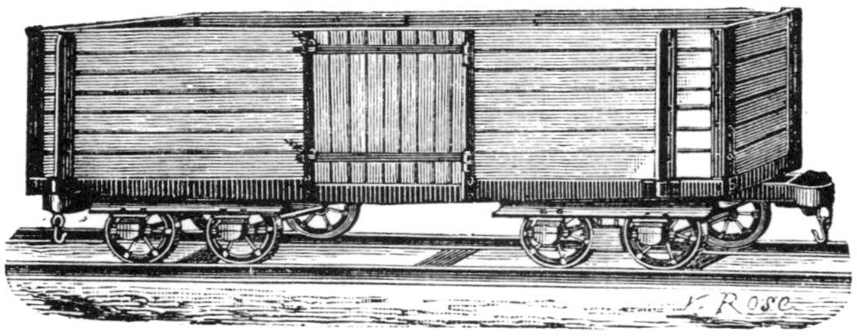

TOP LEFT: Half-box wagon. TOP RIGHT: Platform wagon with removable sides. UPPER MIDDLE LEFT: Wagon having adjustable canopy in closed position. UPPER MIDDLE RIGHT: Uncovered wagon. MIDDLE RIGHT: Enclosed British wagon with bogies. LOWER MIDDLE LEFT: Adjustable canopy wagon in open position. LOWER MIDDLE RIGHT: Uncovered wagon with bogies. BOTTOM: Enclosed wagon with bogies. Aside from the British van, all wagons are Decauville, circa 1893.

PLATE 209

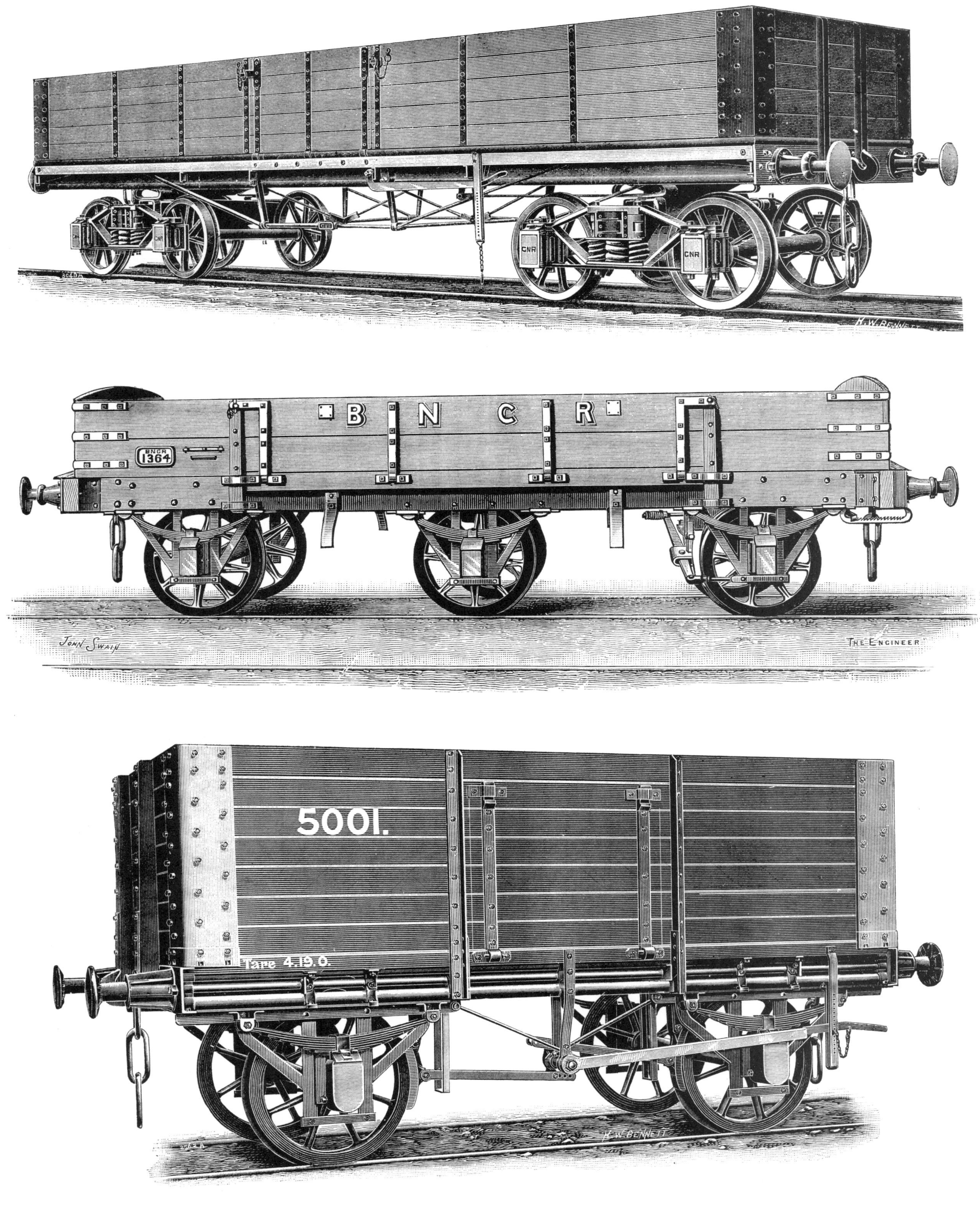

Open British wagons: Top: Bogie wagon with tubular framing, Lancaster Railway Carriage & Wagon Co., 1895. Middle: Three-axle wagon of Ireland's Belfast & Northern Counties Rwy., 1890. Bottom: Wagon with tubular frame, John Fowler & Co., Leeds, 1894.

PLATE
210

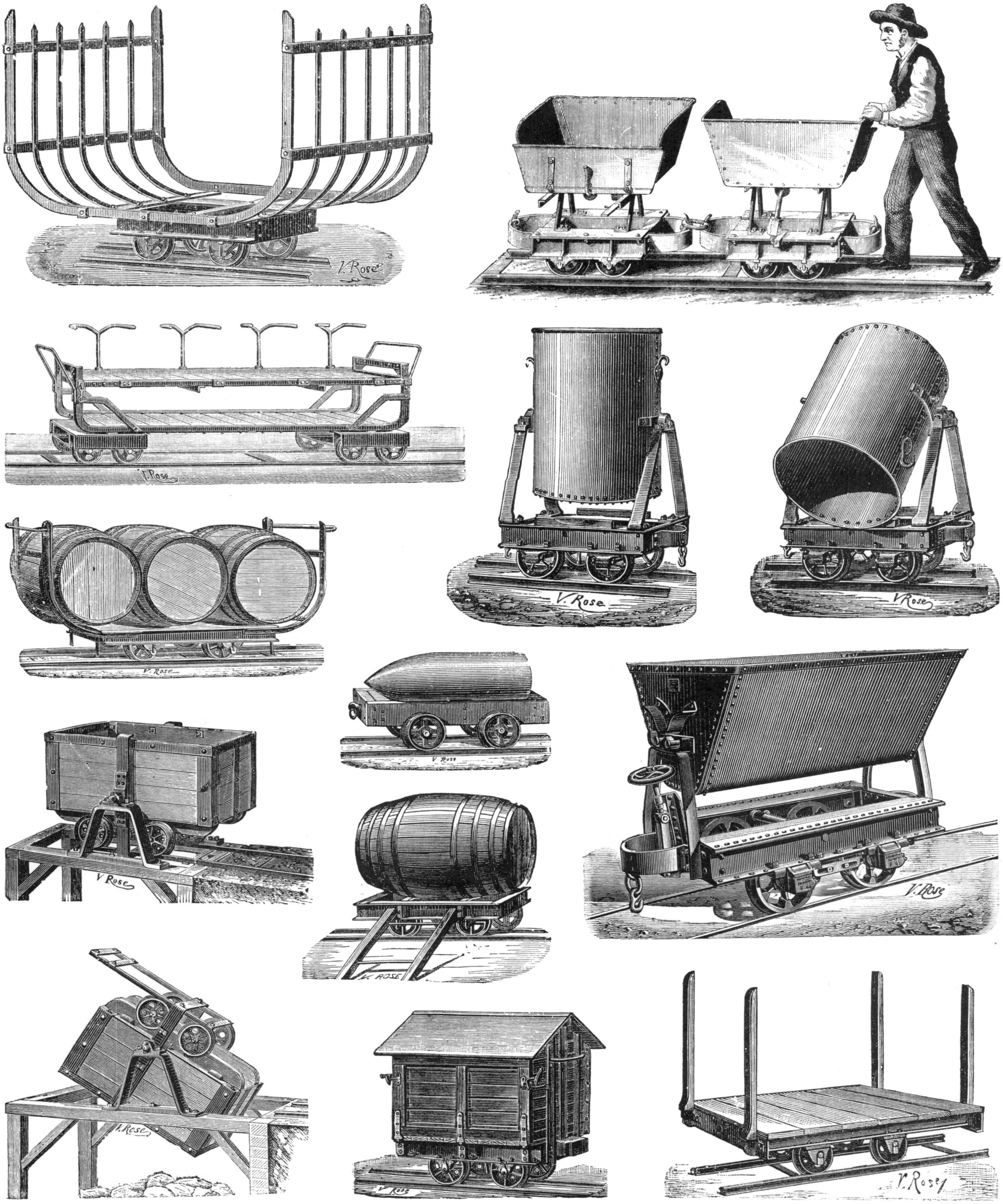

The Decauville Co. specialized in various applications to facilitate a more effective use of its portable narrow-gauge tramways. The small wagons or trucks on this page were used in mines, quarries, factories, wineries, in agriculture, by the military, in construction, and possibly for other purposes.

CHAPTER

(16)

Railway Signaling

Early in railway history signaling methods were introduced with the intent of improving efficiency and ensuring safety. The initial techniques involved the use of trumpets, flags, and lanterns. Perhaps the first *fixed* signal was used by the Stockton & Darlington line in 1827. It consisted of a triangular board, marked "danger," which pivoted on the top of a post. In 1834 the Liverpool & Manchester Rwy. began using rotating disc signals. That same year the company developed a night signal. A lamp that showed either a red or white light was affixed to the top of a post. Other early methods included the use of signboards or hand and arm gestures during the daytime and fires at night. The first signal used on the Stockton & Hartlepool line was a candle placed in a station window. When present it indicated that the train should stop, and when absent, that the train should continue.

In 1838 the Great Western Rwy. adopted large disc and crossbar signals, which were sometimes raised as high as 60 feet. They could be easily seen from a distance and gave engine drivers more time to brake if necessary. Large holes were drilled into the face of these signs to reduce wind pressure that might otherwise blow them down. The London & South Western line began using disc signals in 1840. That same year the "highball" signal was introduced in England. If the ball was raised to the top of a mast, it indicated that the line ahead was clear. This type of signal did not gain acceptance in Britain, but for a time became common in America.

Semaphore signals, controlled by levers connected to wires, made their debut on the London & Croydon Rwy. in 1841. These were set to three different positions, with the horizontal indicating "stop," the 45 degree suggesting "caution," and the vertical designating "all clear." All the levers controlling the switches and signals were placed together on a raised bridge. Mr. Gregory, the inventor of this system, later developed a stirrup-foot control for signals.

By the early 1840s, signals were grouped into three primary categories—line signals; station, tunnel, and gate signals; and junction signals—and rules were codified accordingly. However, this was done differently on each line. At the time English railway practices, including signaling methods, were beginning to be widely copied elsewhere.

In early English railway practice signalmen functioned much like policemen. They were positioned at intervals along the track where they were responsible for the safety of passing trains. Shacks were soon improvised to give them protection from the weather. Later, signal cabins were built having levers to control signals.

Intertwined with the function of signaling was that of switching. Switches, or "points," in English terminology, had been invented early and many different types were built. Their purpose was to allow a train to move from one track to another. In the earliest switches, the switchman (pointsman) had to hold the switch lever during the entire time of the train's passage to avoid derailment. Improved designs eliminated this defect.

Dispatching in the early years depended on a system of maintaining time intervals between trains. However, as both traffic and train speeds increased, this became increasingly dangerous. Thus some method had to be devised to improve communications, both between the dispatcher and signal cabins, and among the signalmen.

A possible way out of this dilemma was the electric telegraph. William F. Cooke and William Wheatstone patented the earliest system in 1837. On July 25, it was tested on the London & Birmingham line between Euston and Camdentown. This technology seemed to offer a number of possibilities for railway use. Shortly, other railways conducted similar experiments. Beginning in 1838 the Great Western tried it between Paddington and West Drayton. A major problem, the company found, was the general illiteracy of both stationmasters and signalmen, a situation that did not improve for at least another decade. For this or other reasons the Great Western discontinued use of the telegraph for a while. However, other lines used it, and in 1851 the South Eastern line developed a simpler method using bell code signals. Later a single-needle telegraph came into being. A method was devised whereby the movement of trains could be communicated by the position of the telegraph needle supplemented by bell code signals.

In 1842 Cooke introduced a new dispatching procedure. It was based on the idea that controlling actual distances between trains was safer than maintaining time intervals. In his "block" system, the line was divided into a number of territories, or "blocks." No train could enter into the next block until any ahead had departed. While this system was a great improvement, it would take some time before it gained widespread acceptance.

The first use of the "torpedo" or "fog" signal occurred circa 1843. These devices were essentially tins filled with gunpowder. In situations of danger, they were placed on tracks ahead of approaching trains. Exploding when struck by the wheels of the engine, they alerted the driver to the emergency ahead. Another safety procedure was to hang lamps at the front and rear of the train. For halted trains, a guard with flags or lantern was positioned 400 to 500 yards behind to stop any following trains. Lamps were sometimes mounted on locomotives to denote a train's special status. For example, in 1844, express mail locomotives on the South Eastern Rwy. were marked by three lanterns.

In comparison to England, America's signaling practices were more basic and slower to evolve. This was perhaps because in the early years American railroads had less traffic, less nighttime activity, and fewer intersections with other lines. Some stations maintained a sort of "crow's nest" from which the station agent would keep a lookout for locomotive smoke. Upon seeing this sign, he would scurry down and toll a bell to alert local citizens of the train's approach. At some point a system using locomotive whistles evolved. These signals provided instructions to the train crew and warned the unwary of the approach of the train. Perhaps the most important of these was the single short whistle indication for "down brakes." It alerted the brakemen to begin setting brakes as there was a stop ahead.

In 1844 Samuel F. B. Morse successfully tested his telegraph system. However, it was not until 1851 that it was first tried for dispatching, and this was due to an emergency situation. The extensive use of telegraph by the United States Military R.R. during the Civil War helped in spreading the practice, and it finally gained wide acceptance in the 1870s. Initially telegraph communications were received at depots along the route and manual station signals were used. Soon semaphore signal masts were built at stations, and later a more extensive network of signals based on the block system evolved.

The first attempt of this kind probably occurred in 1863, when the Camden & Amboy installed manually operated signals at certain points between Trenton and Kensington. Clear and red glass panels were used to indicate "go" or "stop." These were raised or lowered by rope, and were lighted by kerosene at night. This system was so successful that by 1870 it was extended all the way to Jersey City. A somewhat similar arrangement was reported in use in 1867 on a coal line up Broad Mountain near Ashland, Pennsylvania. Wooden signal towers were constructed at intervals along this heavily traveled single-tracked route. During the day trains were signaled by use of colored balls, and at night by colored lanterns.

On many American single-tracked lines a cheaper method was employed. A staff was given to an engineer upon beginning his run. It gave him right of way over a certain distance of the route ahead. Upon completion of that distance the staff was turned in, and if the journey was unfinished, another procured. In instances where trains met at a siding or a station, staffs had to be exchanged before the trains could continue.

The common procedure for switching in the early years was to first set the signal and then make the switch. Because this practice was dangerous, especially at junctions and terminals, efforts were made to link switches and signals in a way to eliminate any possible contradiction. The first attempt of this type was at East Retford Junction in 1852. In 1856 M. Vignier of France's Ouest Rwy. developed a mechanical interlocking device. That same year John Saxby patented a similar idea in Britain. Both versions were later replaced by much more sophisticated machinery. The Saxby and Farmer firm became Vignier's main competitor. It eventually manufactured much larger systems capable of controlling a great number of switches and signals. Safeguards were also developed to alert signalmen of problems, such as failures of switches to open or close properly. America's first interlocking machine was imported from England in 1870 by the United Railroads of New Jersey (formerly the Camden & Amboy) and installed near Trenton. However, American lines were slow to adopt these new devices. Even after domestic production began they were very expensive and thus were only installed where absolutely necessary.

Cost savings could result from this new technology, however. Work previously done by dispersed signalmen and switchmen was consolidated and could now be done by one person. Rapid adoption of signal boxes in Britain during the 1870s revolutionized the existing system, both increasing safety and eliminating many jobs. The Railway Regulations Act of 1889 finally codified changes that had largely taken place in Britain. Among its requirements was the adoption *for all passenger lines* of the block system and interlocking of all switches and signals.

In England the typical shift for a signalman was 12 hours. While work was normally routine and repetitive, it could, on occasion, become more interesting. One such incident is described by the *Railroad Gazette* in its July 20, 1883 issue:

> English papers relate how the signalman of Llandudno saved the Irish mail passengers from what might have proved a terrible fate. The signalman at the junction received a message from the signalman at Conway, the next station toward Holyhead, that a light engine was coming. The junction signalman, knowing that the Irish mail was due, decided to run the engine into a siding in order to permit the express to pass. He accordingly put up all the signals against the light engine, but to his extreme astonishment the engine came straight into the junction at full speed, swept round the corner, dashing past all the danger signals, and disappeared from view down the line toward Chester. A moment's reflection convinced the signalman that both driver and stoker must be asleep. He accordingly wired to the Colwyn Bay station signalman: "Engine coming; driver asleep; put fog signals on line." The man at Colwyn Bay was equally prompt, for, running out of his box, he had hardly time to lay a number of signals when the engine came thundering along and an explosion followed which effectually awoke the men. The engine was stopped and run back into a siding.

In America the Pennsylvania R.R. followed the example of the Camden & Amboy, and by 1876 had built an extensive block signaling system along its entire main route. Manned signal towers were situated every one to 2 miles apart, and each signalman had telegraphic contact with both dispatcher and other towers. While this new system was an improvement over older methods, it was still dependent on employees who at times were careless or inattentive, and accidents still occurred.

What was needed was a sensitive automatic system activated by the movement of trains. Thomas S. Hall and Dr. William Robinson saw this potential early and by the late 1860s had both begun to experiment with primitive electric devices. Hall's experiments resulted in the enclosed disc signal. It was first employed on a 16-mile stretch of the Eastern R.R. in 1871. These signals were later used by a number of railroads, including the Reading, Lackawanna, Illinois Central, Santa Fe, and Lehigh Valley. Hall died in 1880, but his company continued for another 45 years. Although Robinson received a patent in 1872 and formed the Union Electric Signal Co., his efforts were less successful. However, George Westinghouse recognized the value of his work, and in 1881 absorbed Robinson's firm into his own Union Switch & Signal Co. Railroads initially balked in adopting these systems because of their expense. In 1891 the Cincinnati, New Orleans, & Texas Pacific R.R. began installing automatic electric signals on its line, and had completed 80% of the task by 1900. Few other companies followed this example; by 1900 only 2,280 miles of track were worked by electric signals.

Following a collision on November 19, 1895, the New York & Brooklyn Bridge Cable Rwy. introduced what was probably the first use of telephone dispatching, which allowed direct communication between the dispatcher's office and the gripman or motorman. At the time, when traffic was at its maximum, cars were run at intervals of one to one and a half minutes apart. However, in dense fog, problems could occur. Electrical communication with the moving cars was made via an overhead wire strung along the tracks that normally provided current for car lighting at night.

At some point semaphore signals began to be lighted at night. In the British system red indicated "stop," green "caution," and white "clear." Later, during the 1890s, the three-position semaphore was discarded in favor of a simpler two-position type. The adoption of the two-position semaphore resulted from a problem of signal arms failing to set in the correct position, which created the potential for confusion and accidents. Red remained the night color for "stop," while white designated "clear." (This was later switched to green.)

Another problem of the semaphore remained unsolved, however. This was a tendency for signal arms to freeze in position during bad weather. This defect became obvious in 1876, when a wreck occurred during a snowstorm on the Great Northern line at Abbots Ripton. A signalman acted correctly in setting his signals to halt the approaching *Flying Scotsman.* Unfortunately, a signal arm stuck in a "clear" position, allowing the express to run into a goods train. This accident receives more attention in chapter 22.

Signal boxes at large stations often had so many levers that several signalmen were required for their operation. The movement of one train alone might require 10 to 20 lever changes, and working levers was exhausting, requiring great physical stamina. To improve this situation, M. Moutier, chief engineer of France's Nord Rwy., conceived in 1898 the idea of a single route lever system. Under his plan a single lever activated all the appropriate interlocking switches and signals along a particular route a train was to take. This system was eventually worked out to prevent any intrusion by other trains.

Electric signaling was first introduced in Britain on the Liverpool Overhead Rwy. in 1893. In this system extending rods attached to trains struck contact boxes. This activated semaphores to set at the "stop" position. At the same time signals for the previous block were cleared, thus allowing any halted train behind to proceed into the next block. From this beginning electric signaling slowly evolved in Britain. Beginning in 1898, near Crewe, signal boxes of the London & North Western Rwy. were converted to a new system where operation of the signals and switches was done electrically. Short levers, still with mechanical interlocking, replaced the longer levers of previous use. These activated electrical contacts that transmitted current to signal and switch operating equipment. Around the turn of the century automatic signals replaced signal boxes in some of London's newly constructed underground tube lines. Electric signaling technology continued to improve and eventually revolutionized railway operations on a worldwide scale.

Early English signals and signaling devices: Top left: Junction signal post, Great Western Rwy., 1844. Top right: Top of signal lamppost (Forsyth's system), Liverpool & Manchester Rwy., 1846. Middle left: Station signal indicating "all right," G.W. Rwy., 1844. Middle right: Line signal designating "all right," G.W. Rwy., 1844. Bottom left: Line signals showing "caution," G.W. Rwy., 1844. Bottom right: Line signals indicating "stop," G.W. Rwy., 1844.

PLATE

212

Locomotive lamp arrangements as tried by Mr. Forsyth on the Liverpool & Manchester Rwy., circa 1846. Approaching trains could be identified by their lamp configuration, and the engine driver, knowing his own configuration, would look for a corresponding signal.

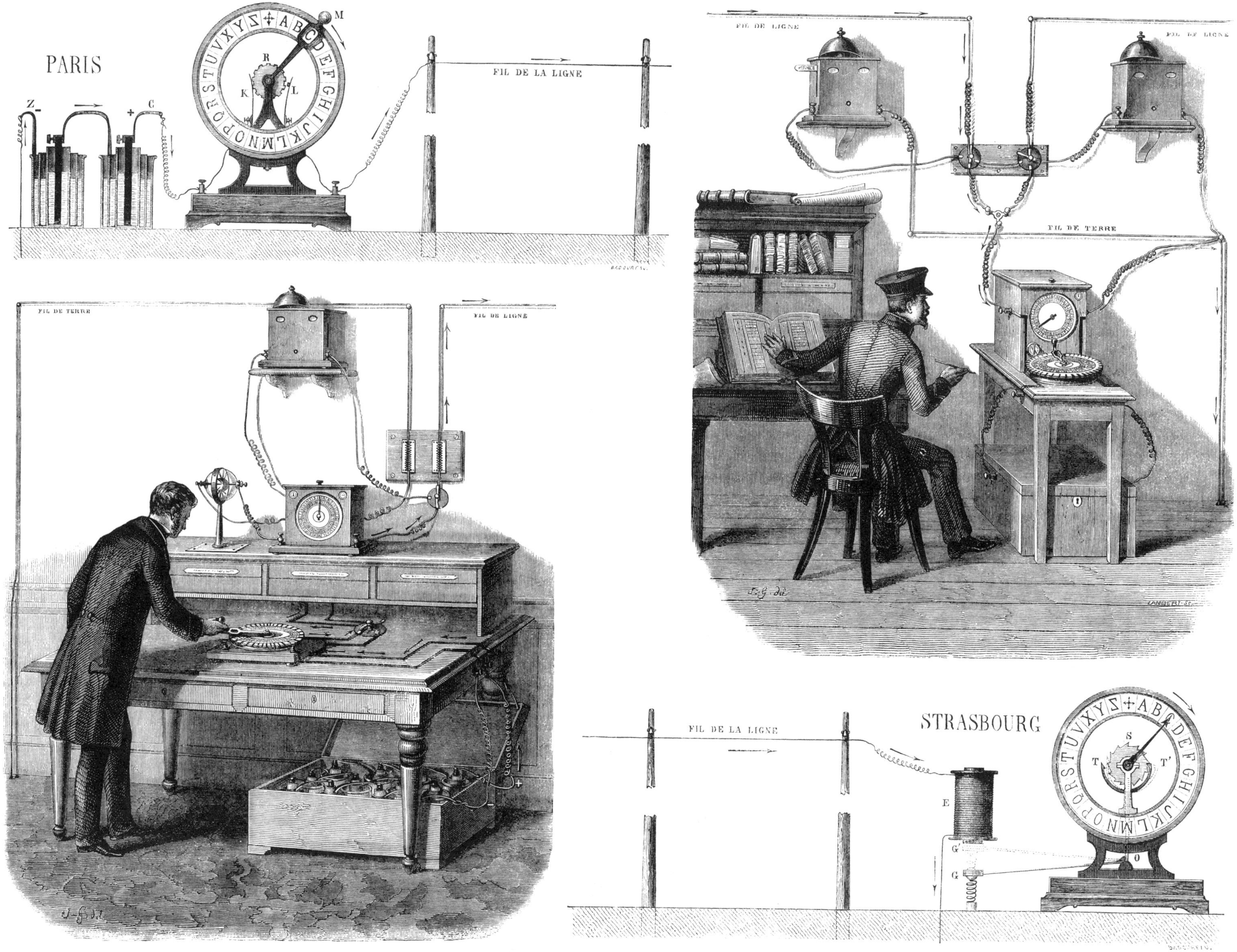

PLATE 214

Telegraph message being sent from Gare de l'Est in Paris to Strasbourg using the Breguet telegraph system, 1861.

Early American methods of signaling: TOP: Highball signal suspended from pole at Fort Plain, New York, New York Central R.R., 1856. BOTTOM: Dummy locomotive pulls coal train past a signal tower on Broad Mountain, near Ashland, Pennsylvania, 1867. This route was heavily traveled in both directions, requiring signal towers for traffic control. Colored balls were used in daytime and colored lanterns at night.

PLATE
215

Exterior and interior views of A.B. signal box at London Bridge Station, South Eastern Rwy., 1866.

American locomotive strikes torpedo signal, warning of danger ahead, 1874. In Britain these were called fog signals.

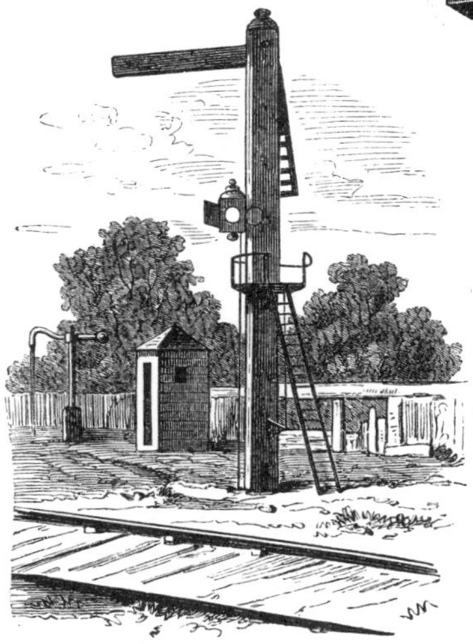

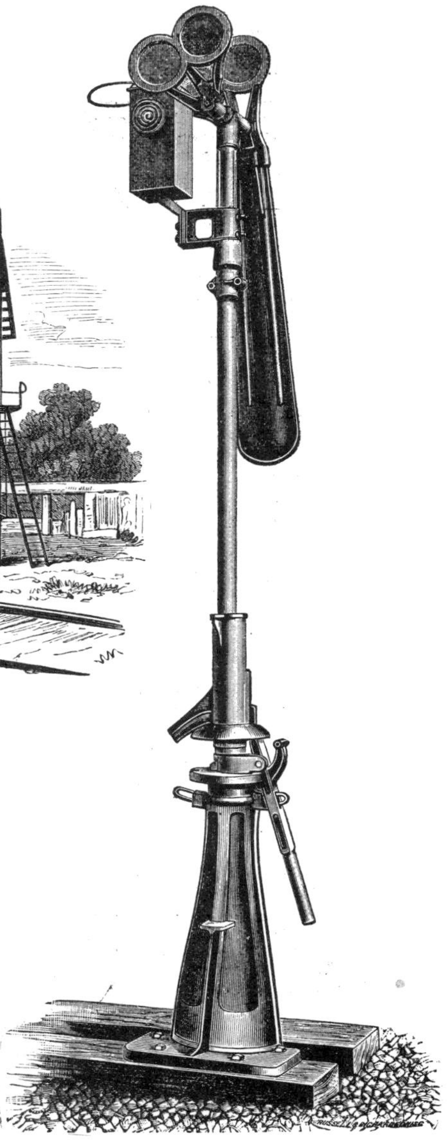

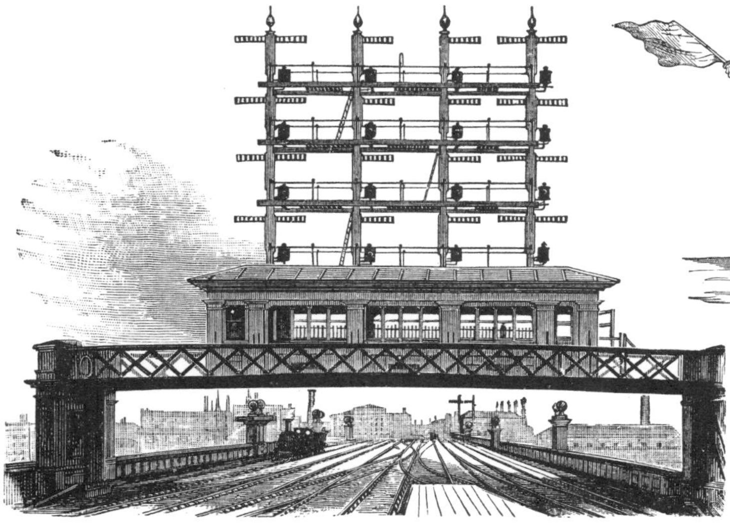

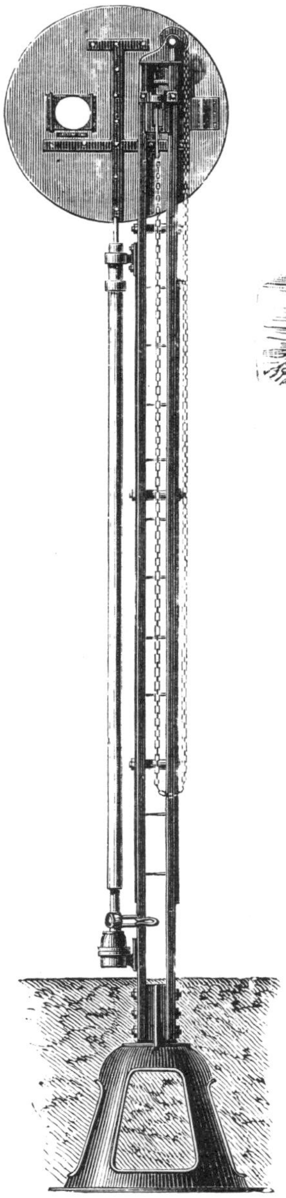

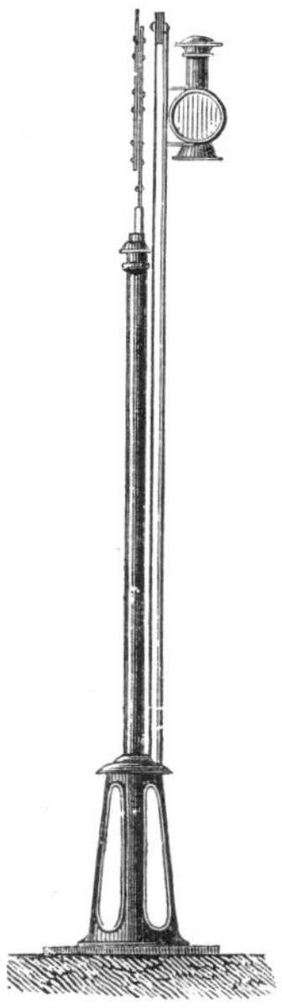

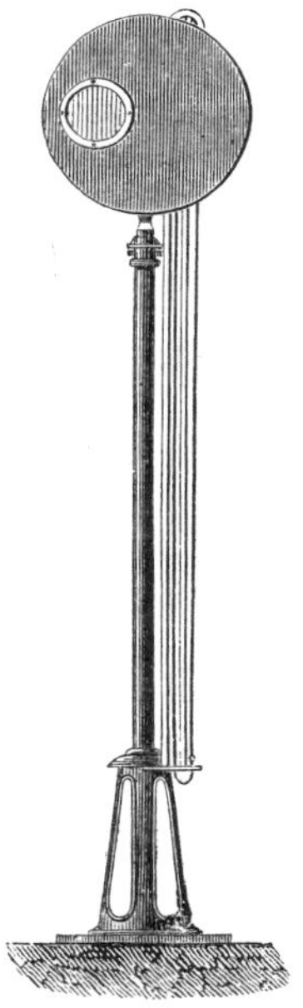

TOP LEFT: French signal box, circa 1880. TOP CENTER: German semaphore signal. TOP RIGHT: Cumming's semaphore switch stand (America), 1886. BOTTOM LEFT: Signal box at London's Cannon Street Station, South Eastern Rwy., circa 1880. BOTTOM CENTER LEFT: French signalman circa 1860. BOTTOM CENTER: Disc signal. BOTTOM CENTER RIGHT: Signal lantern raised on pole. BOTTOM RIGHT: Disc signal raised on pole.

Pointsman on France's Ouest Rwy. operating Vignier lever, which controlled both switches and signals, 1891.

PLATE
219

CHAPTER

(17)

Railway Construction and Operations

THE first step in the creation of a railway was determining the origin and destination points. Following the raising of some capital, a more specific route was determined. The relative merits of various intermediate cities and towns were considered, as well as locations of potential sources of revenue. Once a general plan was ascertained, surveys were made. From this information the chief engineer drew up final plans for the road. It would normally be the course having the fewest elevations or curves and requiring the least number of bridges or tunnels. Thus, generally speaking, an attempt would be made to avoid hills and valleys, and to cross rivers or streams as close to the source as possible. Railways also learned to build through areas where land was less expensive or in the hands of the less powerful. In countries like Britain, it was often necessary to cross the estates of wealthy aristocrats. This could result in parliamentary opposition or obstructive litigation, which in some instances was only overcome with exorbitant payoffs.

In planning a line it was important, in so far as possible, to achieve a balance between the cuttings and embankments. If the road was constructed at too low an average level, there would be an excessive amount of material left over, resulting from an uneconomic use of labor. If it was built at too high a level, then the required soil would have to be conveyed in at great expense from elsewhere.

Work of this kind could sometimes be very difficult. In *Our Iron Roads,* Frederick S. Williams describes one such undertaking:

> The Haslingden cutting, on the East Lancashire line, was probably one of the most difficult works of the kind ever undertaken. Nearly half a million yards of peat, gravel, and sand were removed from it, and it had to be cut through a bog-hole, the material of which, being saturated with water, sometimes came in faster than it could be taken out. The peat was twenty feet or more in thickness; and for three months, during the summer of 1848, all the earth that two locomotives and their trains could bring was carried away to form an embankment without obtaining a foundation. The difficulty was at length overcome by the company's engineer, Mr. Perring, who sunk large masses of stone at the required points, and these, forcing away the peat, provided a solid bed over which the line could be carried.

In laying out the roadbed at curves it became the practice to raise the elevation of the outer area somewhat over that of the inner area. This caused trains to lean slightly inward on curves, balancing the centrifugal force of the curvilinear motion. In Britain it became the practice to entrust construction to special contractors who assigned jobs to various subordinates. Vast armies of workmen, called "navvies," carried out the bulk of the work by hand. This term had come from the earlier era of canal building, when workers were first called "navigators," later shortened to "navvies." Track was laid as quickly as the completion of roadbed, bridges, and tunnels would allow. Switches were added whenever a connecting link was needed between tracks. Once other structures, like stations, water towers, and signaling devices, were in place, the line would be opened.

In America large companies extending for hundreds of miles were broken up into divisions. These were separated by towns or cities called division points. British railway towns like Swindon, Crewe, and Wolverton often served the same function. These locations, and often, terminal points, had repair shops, roundhouses, and yards, and became centers of activity for the company. Arriving non-through trains would be broken down in the yard with switch engines, and the cars shunted to a number of tracks having different destinations and priorities. As other cars were added to these tracks, new trains would be formed. The yards also stored maintenance equipment like cranes, snowplows, and work cars.

In these communities train crews and other employees normally lived close by so that they could be called for duty at any time. For the entire nineteenth century, American rail employees worked 10 to 12 hours a day, six days a week, which was typical for the industry worldwide. However, on occasion, they had to labor much longer, and sometimes had to return to work after only a few hours of sleep.

In America the chief officers of a railroad were usually the *president, secretary,* and *treasurer.* In charge of legal matters would be the *general counsel.* Running daily affairs would be the *general* manager. He would oversee the operations of the comptroller, purchasing agent, traffic manager, car accountant, division superintendents, superintendent of transportation, superintendent of machinery, and superintendent of roadway. Under the *comptroller* would be various auditors and clerks. The *purchasing agent* would work in coordination with the local storekeepers. The *traffic manager* would oversee the activities of the claim agent, general passenger agent, and general freight agent, with the two latter supervising traveling agents, local agents, and rate and division clerks. The *car accountant* would direct the work of lost car agents.

Under each *division superintendent* might be station agents, local storekeepers, a master mechanic, a train master, and a road master. Each *station agent* would direct the operation of his station and be in charge of various clerks and laborers, as well as the yard master. The *yard master* in turn would direct the local switch engine crews, switchmen, and brakemen. The local *storekeepers* would work in close coordination with the railroad's purchasing agent.

The *master mechanic,* who was jointly supervised by the *division superintendent* and the *superintendent of machinery,* oversaw the work of the repair shop supervisors and roundhouse foremen. Working below these people would be hostlers and cleaners, mechanics, laborers, car inspectors, and greasers. The *train master* would work in coordination with the *superintendent of transportation,* and under him might be the dispatcher, telegraph operators, conductors, trainmen, and personnel in the signal towers. The *dispatcher* in each division operated under a working schedule that fixed the position and time of all trains in operation. The *road master* would receive direction from both the *division superintendent* and the *superintendent of roadway.* Under him would be bridge and road supervisors. The former would oversee bridge foremen, watchmen, carpenter gangs, and mason gangs. The latter would direct wood and water tenders, floating gangs, construction trains, and section foremen. *Section foremen,* in turn, supervised section gangs and track walkers.

Trains were assigned priority at various levels in a hierarchy that favored passengers, mail, and perishable freight, like cattle or refrigerated foods. To accommodate scheduling, a slow-moving coal or work train would be moved to a siding, allowing an express or mail train from either direction to pass by. Double-tracked systems had existed as early as the 1840s in America, but were rare throughout the nineteenth century. In Europe, and especially Britain, double-tracked systems were much more common, with one track used for the "up" traffic and the other for the "down."

Delays and breakdowns were common, and usually tended to affect the movement of other trains. It was the dispatcher's job, despite these obstacles, to somehow orchestrate all the trains safely to their destinations with a minimum of inconvenience and lost time. The widespread adoption of telegraphic dispatching by the 1870s, and the somewhat synchronous development of manual semaphore signaling, followed later by automatic block signals, helped to make his job a little easier.

In America about 40% of all railroad employees were train crews, consisting of a conductor, engineer, fireman, and a number of brakemen. The conductor was in charge of the train and was under the supervision of the dispatcher, who provided him with written "train orders." Conductors were required to do all of the paperwork required on the trip. The paperwork for a freight train was considerable, listing every car on the train, the company, number, where it was picked up and dropped off, and the like. Passenger conductors dealt with the public, took the tickets, handled money, constantly checked the time to maintain the schedule, and signaled the engineer to resume the journey after making each stop. Freight conductors rode in the caboose at the rear of the train to watch for any sign of trouble. Once air brakes were adopted on freight trains, freight conductors monitored a gauge in the caboose for signs of any drop in air pressure. Typically a freight conductor had been promoted from being a brakeman, and could hope for a promotion to passenger conductor.

Although on an equal pay scale with conductors, engineers were subject to their direction. The engineer operated the locomotive, and had to know the route with all of its bends, curves, grades, signals, bridges, tunnels, blind spots, and stops. Because the time required for braking was so long, especially before air brakes, engineers had to be constantly alert for any signs of danger. Before attaining the position of engineer, a person first had to serve as a fireman. The primary job of the fireman was throwing wood, and later, shoveling coal, into the firebox in an effort to maintain the steam pressure at a desirable level. Firemen also were responsible for taking on water at water towers, lubricating the engine, watching its gauges, and, during the few decades when it was fashionable, polishing the locomotive and tender to a high shine.

The entry-level job for train crew service was that of brakeman, which, before the adoption of air brakes, was a very dangerous job. During those years, in making stops, the engineer might reverse the motion of his drivers, but the only brakes were on the individual cars, with each having to be applied separately by turning a wheel. At the whistle signal from the locomotive of "down brakes," the brakemen at various points on the train would immediately begin jumping from car to car and tightening the brake wheels until completing their sector. To release the brakes, another signal was given and the process was reversed.

While riding on the top of a boxcar, the brakeman would have to watch for tunnels and low bridges. In darkness or bad weather and on fast trains his job often became treacherous, and many brakemen fell off, frequently to their death. In the late 1880s train deaths resulting from the falls of crewmen averaged approximately 1,000 a year. Brakemen at the front and rear of trains had the added duties of functioning as switchmen and flagmen. When connecting or disconnecting cars with the link and pin coupling devices, they risked the additional danger of having fingers sheared off. Automatic couplers, first invented in 1867, were not widely used until the 1890s. Once trains were equipped with air brakes, the number of brakemen on freight crews was reduced, and those on passenger trains were assigned other duties.

In America's South, until the end of the Civil War, black slaves were often used as railway personnel. From that time on, however, free blacks were used only in the lowest menial positions. Things were only slightly better elsewhere for African Americans, but the Pullman Co. hired many for service as porters or dining car waiters. In its February 23, 1883 issue, the *Railroad Gazette* gave this report on a black engineer: "The only colored locomotive engineer that we know of in the north is employed on the switch engine in the yard of the New York & New England R.R. at Fishkill Landing. He is a competent and faithful workman, and is respected by his employers and his fellow workmen."

In the British system, which spread to Europe and elsewhere, railway operations were similar to those in America, but there were differences. The closest equivalent to the American conductor was the train guard. In charge of his train, he was to be in attendance at the station 30 minutes before its departure. While there, he was under the authority of the stationmaster, and it was his job to inspect his train, making sure that it was properly loaded, sheeted, coupled, greased, lamped, and organized in the correct manner, and that the brakes functioned properly. During the journey he rode in the brake (or guard's) van, a wagon functioning much like the American caboose. When the engine driver wished to slow the train, he would sound the appropriate whistle, and the guard would immediately apply brakes in his van. He would also function as a signalman, going back when necessary to signal any following trains. The guard also handled the paperwork of the trip, transcribing it into an official journal that would be sent to the superintendent's office.

The guard's job originally evolved from that of the brakeman. Frederick S. Williams, quoting an unnamed train guard, provides this description:

> In my early days we were called brakesmen; we had no brake vans; we had to ride on the top of the carriages or on the loaded vans, anywhere we could, and to get on and off anyhow we could. And on a frosty night it *was* getting off. Our limbs were often benumbed with cold; we were sometimes so stiff with cold that we had to be lifted off, and some, when they were lifted off, were found to be frozen to death. When trains were running down an incline, we had to scramble or jump from one wagon to another in order to put on perhaps five or six brakes. When you jumped upon a load of goods sheeted down, you could not possibly tell what you jumped upon, and, in consequence, many men lost their limbs or their lives. When a train stopped for anything, the brakesman had to go back a quarter of a mile; he was not to return till he was whistled in, and then he had to run in as quick as he could for fear he should be overtaken by a following train.

On British passenger trains there were normally two guards. The head guard rode in the van at the end of the train, while the junior guard rode in the van behind the tender, where he was responsible for luggage and parcels. Once these were disposed of at the end of a journey, he would search all second- and third-class carriages for leftover items and also check on the condition of the coaches. The head guard would make a similar search in the first-class section.

In the British system the engine driver and stoker were required to sign in for duty at least an hour ahead of the train's departure. They would then inspect the engine and put the distinguishing lamps in place. The driver was also required to inspect the "notice case" to see if there were any instructions regarding his train or the conditions of the route over which it would travel. Once under steam the driver's position was mainly on the footplate. Here he was positioned to control the regulator, and also to instantly grab the reversing lever, should an emergency stop be needed. The driver was not only required to keep a constant lookout, but also to listen to the sound of the engine, for any sign of irregularity.

For engines to operate at maximum efficiency, some skill was required on the part of the stoker. Coal could not just be thrown into the middle of the firebox, but had to be distributed evenly into every area. The best method for good firing was to stoke little, but often. Frederick S. Williams states:

> The grand aim of first rate stoking is to keep the steam at one pressure; that is to say, the needle of the pressure gauge should, as nearly as possible, point to a full boiler pressure, up hill and down dale. To accomplish this—and it is done on the crack engines every day—firing must be studied. Engines are not alike. Some are robust, others very delicate, but the generality of engines require the exercise of trained skill to jockey them. Some engines steam best with a low fire, and others may carry fuel up to the fire-hole. Nearly all engines are affected by cross winds. The firing should be done when the steam is just on the point of blowing off—a condition which generally happens while the engine is on a rising gradient, for the fierce blast causes a maximum supply of oxygen to pass through the fire and the tubes, which generates great heat and much steam.

On completion of a run the engine and tender were uncoupled and pulled over an ash pit. There the fireman disposed of the remaining fire and cleaned out the firebox. The engine was then returned to its shed, and before the crew signed out, checked for needed repairs.

An article from the July 13, 1883 issue of *Railroad Gazette* provides a uniquely different glimpse of locomotive driving. This piece, originally run in the *New Orleans Times-Democrat,* was entitled "The Engineer as a Musical Critic":

> "I was loafing around the streets last night," said Jim Nelson, one of the oldest locomotive engineers running into New Orleans, "and as I had nothing to do I stepped into a concert, and heard a slick-looking Frenchman play a piano in a way that made me feel all over in spots. As soon as he sat down on the stool I knew by the way he handled himself that he understood the machine he was running. He tapped the keys away up one end just as if they were gauges and he wanted to see if he had water enough. Then he looked up as if he wanted to know how much steam he was carrying, and the next moment he pulled open the throttle and sailed out on the main line as if he was a half hour late.
>
> You could hear her thunder over culverts and bridges, and getting faster and faster, until the fellow rocked about in his seat like a cradle. Somehow, I thought it was old '36' pulling a passenger train and getting out of the way of a special. The fellow worked his keys on the middle division like lightning, and then he flew along the north end of the line until the drivers went around like a buzz-saw, and I got excited. About the time I was fixing to tell him to cut her off a little he kicked the dampers under the machine wide open, pulled the throttle away back in the tender, and, Jerusalem Jumpers! How he did run! I couldn't stand it any longer, and yelled to him that she was pounding on the left side, and if he wasn't careful he'd drop his ashpan.
>
> But he didn't hear me, No one heard me. Everything was flying and whizzing. Telegraph poles on the side of the track looked like a row of cornstalks, the trees appeared to be a mud bank, and all the time the exhaust of the old machine sounded like the hum of a bumblebee. I tried to yell out, but my tongue wouldn't move. He went around the curves like a bullet, slipped an eccentric, blew out his soft plug, and went down grades fifty ft. to the mile, and not a confounded brake set. She went by the meeting point at a mile and a half a minute and calling for more steam. My hair stood up like a cat's tail, because I knew the game was up.
>
> Sure enough, dead ahead of us was the headlight of the special. In a daze I heard the crash as they struck, and I saw cars shivered into atoms, people mashed and mangled, and bleeding, and gasping for water. I heard another crash as the French professor struck the deep keys away down on the lower end of the southern division, and then I came to my senses. There he was at a dead standstill, with the door of the fire-box of the machine open, wiping the perspiration off his face, and bowing at the people before him. If I live to be a thousand years old I'll never forget the ride that Frenchman gave me on a piano."

When derailments or other types of mishaps occurred, special trains were sent out to restore the line to an operable condition and return damaged rolling stock to the shops. In Britain these were called breakdown trains, and they had priority over all other traffic when dispatched to the scene of an accident. Frederick S. Williams described a typical British breakdown train circa 1880. It consisted of about seven wagons: two tool vans, a riding van, a van carrying wood "packing," a breakdown crane, a wagon carrying "balance blocks," and one used for supporting the "jib." One tool van carried ropes of varying diameters and lengths, "snatch blocks," pulley blocks of various sizes, shovels, bars, screw-couplings, "clips" for securing broken axles, and other assorted tools. The second carried oil and naptha lamps, axes, crosscut and hand saws, traversing jacks and bottle jacks of various sizes, and hydraulic jacks. These vans also carried special appliances for righting overturned vehicles and putting them back on the track.

The function of the riding van was to carry up to a 40-man work crew. It also contained various types of lamps, flags, and fog signals, and had a stove for heating, with oven attached. The crew sat on "box-seats" that carried additional tools. The wood "packing" van carried about 3 tons of wooden planks of various sizes. These could be used for a number of different purposes, including the supporting of jacks. The crane, usually worked by five men, could lift 5 tons unsupported, but when secured, had considerably more strength.

To ensure safe operation tracks had to be regularly inspected and maintained. This task was assigned to the section master, who was given a territory consisting of a certain mileage of track. Under him were track walkers who made inspections and the section gang who made repairs. In severe rainstorms these men were put on constant patrol to see that ballasting was not washed out from under the tracks or over them.

The following article from the September 13, 1889 issue of *The Engineer* provides a glimpse of problems and dangers that section personnel could potentially face.

> William Dixon, a section foreman on the Oregon Short Line, on July 15th, found a rail in the track about six miles west of Soda Springs, Idaho, which was out of line on account of the pressure of the adjoining rails. With one man to help him, he undertook to replace it with a shorter rail. They had removed the spikes from the inner side of the rail, and Dixon took a claw bar to pry the rail out of its place. As soon as the rail was relieved from the pressure of the other rails, it sprung as if it was made of spring steel to a distance of 15 ft., striking Dixon and breaking both of his legs below the knees. There he lay, both legs broken, a rail out, a passenger train nearly due, and only one man to help him in his misery and warn trains in time to avert disaster. He ordered his man to get the hand car on the track and help him upon it, and with the broken bones protruding through the flesh, he started down grade, which at that point is heavy, and with nothing but a shovel to push himself along, he set out for the men on the next section, while he sent his man on foot in the other direction to meet the passenger train. Dixon had to go about three miles, and it was about three hours before he could have his injuries attended to. One leg was then amputated, and there was hope of his life and the other leg being saved. Later, however, it became necessary to amputate the other leg, and his death followed in three days.

If accidents resulted from faulty track, then in some countries, the section master was held legally responsible. The March 2, 1883 issue of *Railroad Gazette* relates this story from Germany:

> One car of a passenger train was thrown from the track near Trendelburg. After investigation by experts it appeared that the derailment must have been caused by the track being out of gauge at that place. The track-master whose duty it was to inspect and keep up that part of the road was tried; charged with neglect of duty. Evidence was given showing that the rails immediately after the accident were about ⅔ of an inch too far apart, and that the inspection required by the regulations should have shown it. The accused claimed that he had tested the gauge the day before the accident and found it correct. He argued that the gauge was widened by the derailment, and that the material used, which was condemned material of a more important road, made it probable.

In 1883 railways in the German Empire employed 171,763 individuals. Of these 1,318 were women. In France, the same year, the total number of employees was 193,893, of which 7% were women. In 1883 France's PLM Rwy., which operated 3,682 miles of line, reported 51,538 employees.

At the time there was an interesting contrast between French and British railways in the area of wages and benefits. French engine drivers were paid from $740 to $1,068 annually. The British driver was paid from $612 to $795. French stokers were paid between $398 and $503, while British stokers received from $351 to $519. Guards and road watchmen got between $260 and $370 in France, and $270 and $390 in Britain. However, French employees were paid when they did not show up for work while British employees were not.

If a worker was discharged in Britain he would receive from one to two weeks' notice of dismissal. In France he would receive two to six weeks of pay following his termination. In France the employee had a free pass for riding over company lines, and only had to pay one-quarter fare on other railways. His dependents could ride for one-tenth the normal fare on company lines. British railways granted no such privileges. The French railway system was also more generous with retirement and other social benefits. An employee could retire at age 55, provided he had served the company 25 years. He would draw half his regular pay for the average of the previous six years, provided the pension did not exceed $1,200 a year.

In Britain a worker could draw a pension provided he had contributed to the pension fund for 10 years. However, he could not retire until age 60. In calculating his pension the salary figures of his entire career were added and then divided by the number of years served. If he had 35 years of service he could expect to receive half this amount. If the British employee died, his widow and children would receive no benefits. In France a widow married three years to an employee who had at least 15 years of service would receive half his pension. In the event of her death the children would receive half of her benefit up to age 18. There were sick funds in both countries; in France there were also cooperative stores where employees could purchase living items at lower prices. While the French system provided more security for rail employees than did Britain's, the latter country at least made some effort to provide a minimal social safety net. Too often in many other countries, rail workers were exploited and then casually discarded.

French engraving showing details of construction on a double-tracked main line, 1861.

PLATE

220

French engraving giving bird's-eye view of a railway center showing ideal setup of railway station, engine house, shops, and other facilities, circa early 1860s.

PLATE

221

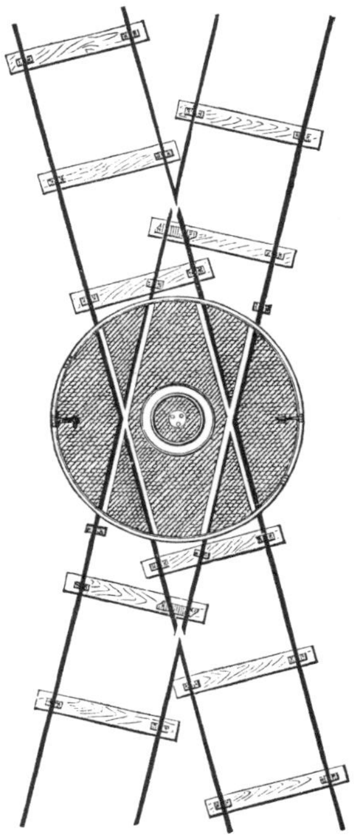

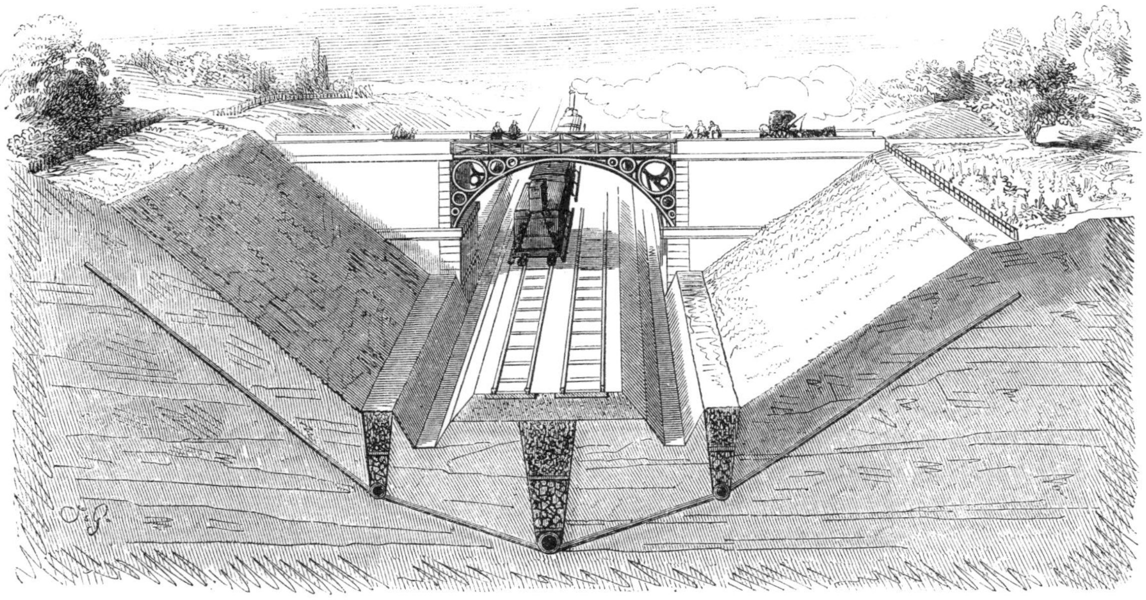

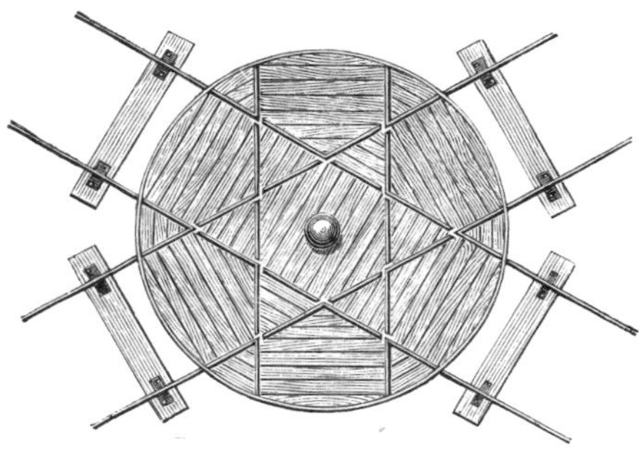

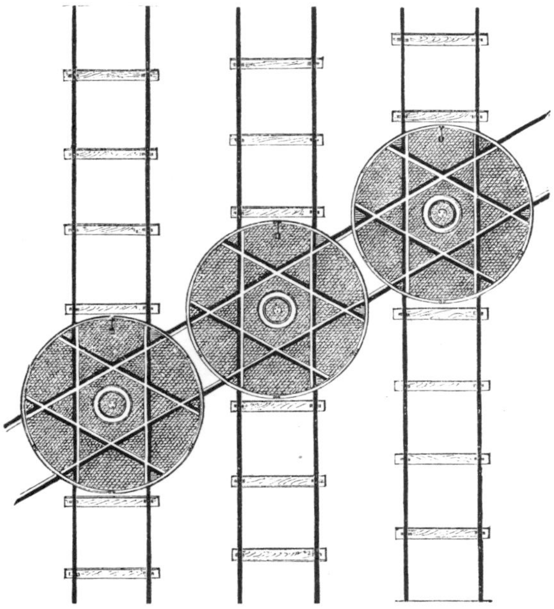

TOP LEFT: Use of turntables in France, 1865. TOP RIGHT, MIDDLE RIGHT, BOTTOM LEFT: Various configurations of turntables and tracks. MIDDLE LEFT: French engraving showing ideal drainage system for double-tracked line, circa early 1860s. BOTTOM RIGHT: Testing of a new type of "Frogless" switch, Chicago, St. Louis & Pittsburgh R.R., 1885.

Scottish engine driver and stoker during Christmas holiday season, 1890. At the time there were widespread rail strikes in the country.

PLATE 223

PLATE

224

Working conditions of the American freight train brakeman, circa 1888: Left: In good weather. Right: In difficult weather. Brakemen were responsible for operating the brakes on a number of cars, so they had to move from car to car, a risky undertaking in difficult weather.

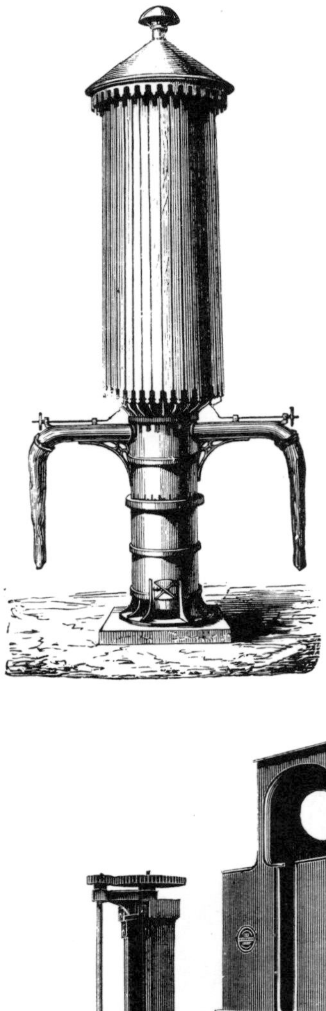

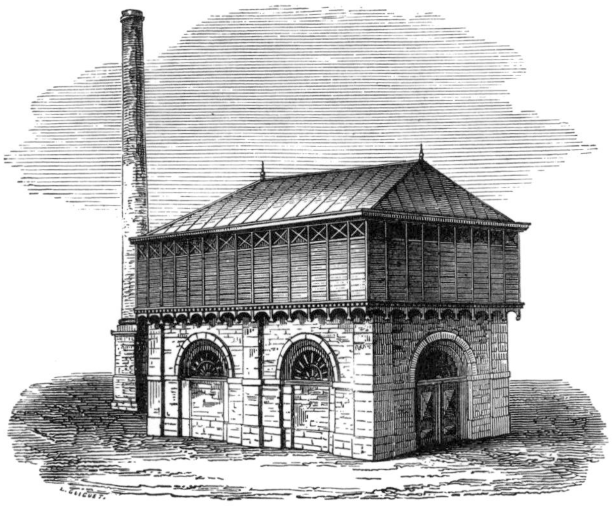

Top: Engine houses, Troyes, France, Est Rwy., 1893. **Middle left:** French water pump. **Middle center:** French water reservoir. **Middle right:** American water tank and windmill, circa 1880. **Bottom:** Locomotive lifting tackle built by Sharp, Stewart & Co., Barry Rwy., Wales, 1895. *Note:* The 0-6-2T was a popular engine in the region for hauling coal.

American devices for improving right of way construction and maintenance: Top: Ballasting apparatus, 1881. Middle: Andrew's leveling attachment for earth cars, 1880. Bottom: The McGrew ditching car, 1886.

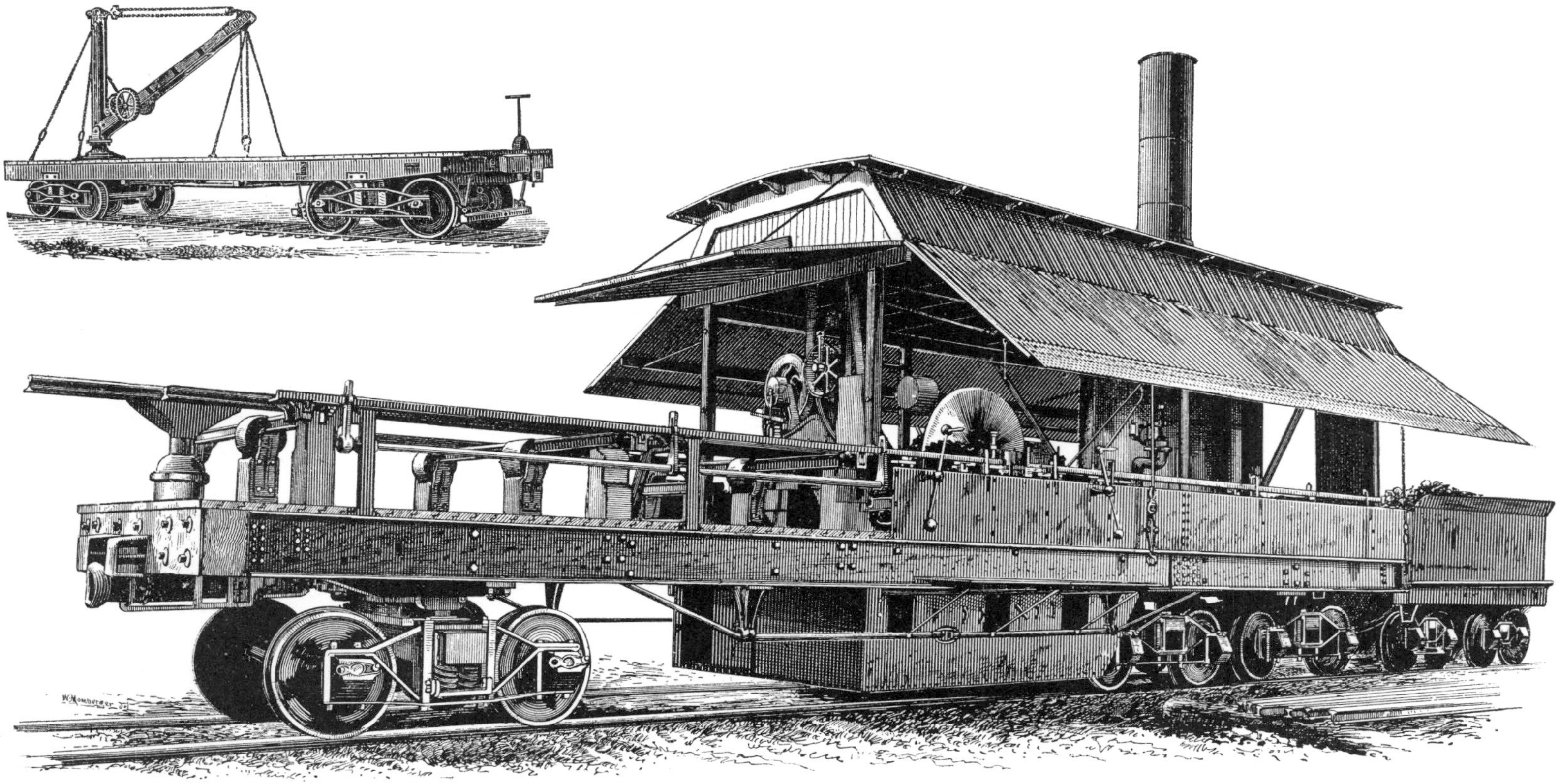

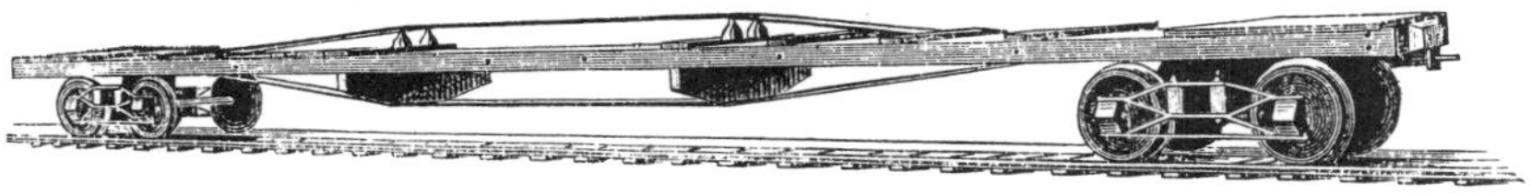

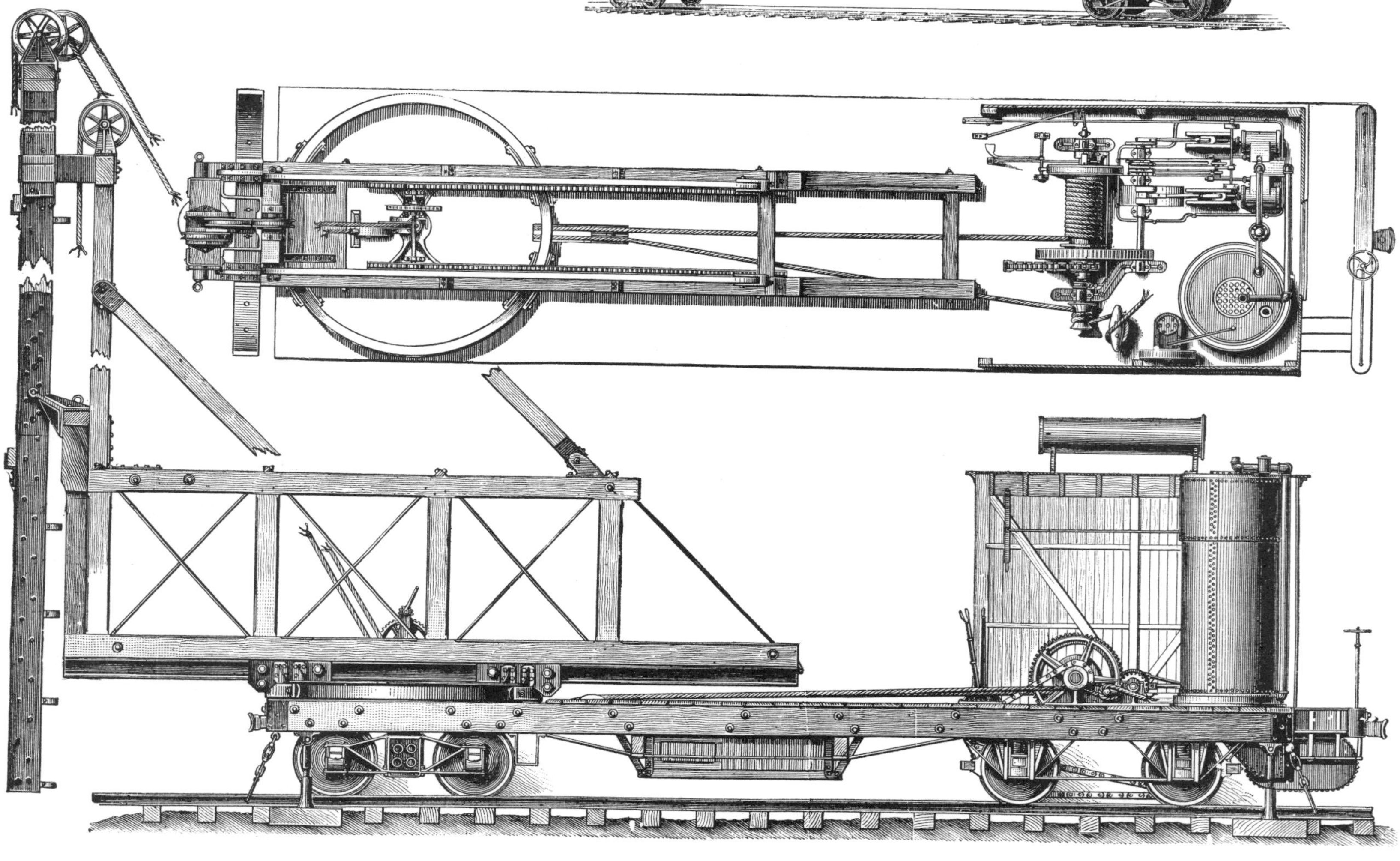

American work or construction cars: Top left: Derrick car, circa 1879. Upper middle: Rail saw car, Michigan Central R.R., 1886. Middle right: Ferry push car, circa 1879. Lower middle: Top view of steam pile-driver car, Industrial Works Co., Bay City, Michigan, 1885. Bottom: Side view of same car.

English locomotive cranes: Top: Built by R. & W. Hawthorne, Leslie & Co., Newcastle, for use in railway goods depots, shipbuilding yards, iron and steel works, docks, and quays, 1895. Bottom: Built by Black, Hawthorn & Co., Newcastle, 1876.

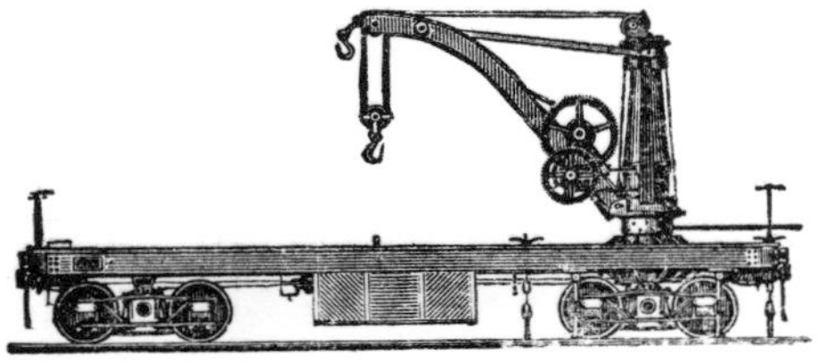

Top: Locomotive steam crane, built by Thomas Smith Co., Rodley, England, 1890. Upper middle center: British break-down train, circa 1880. Lower middle left: Wrecking car, Industrial Works Co., Bay City, Michigan, 1889. Bottom: 10-ton railway crane, R. C. Gibbins & Co., Birmingham, England, 1893.

PLATE 229

PLATE 230

Combined steam shovel and wrecking car, Bucyrus Foundry & Manufacturing Co., Bucyrus, Ohio, 1883.

TOP: Russell Wing Elevator Snow Plow, Ensign Co., Huntingdon, West Virginia, 1893. BOTTOM: Snow excavator patented by George Cox, Paterson, New Jersey, 1888.

PLATE 231

Top: Steam inspection car designed by Jay Noble, St. Louis, Missouri, 1876. **Bottom:** Steam inspection car built by Rogers Locomotive Works for Mejia & Arequipa Rwy., Peru, 1869.

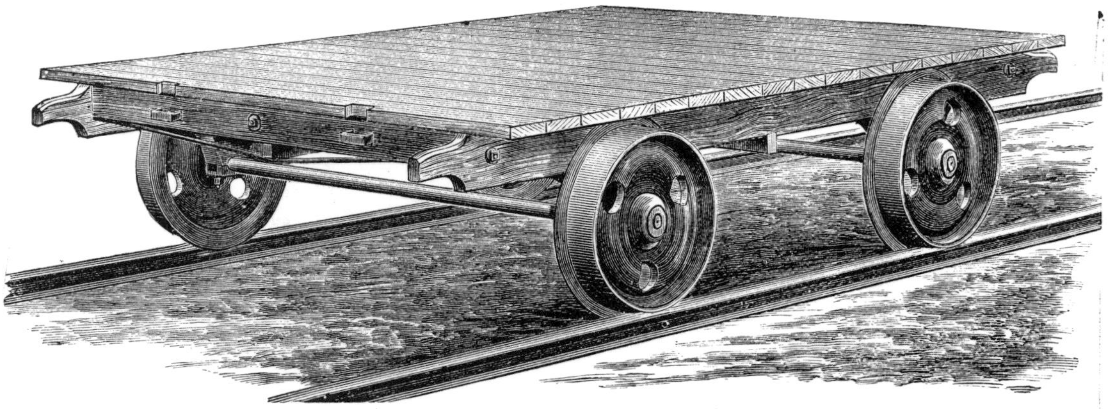

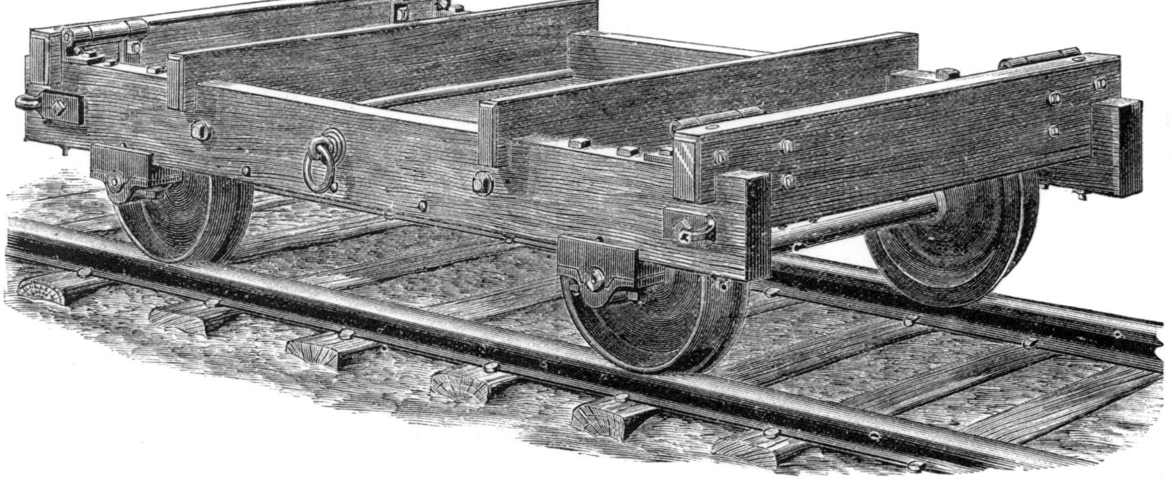

American cars for construction and line maintenance: Top left: Kalamazoo railroad velocipede, circa 1880s. Top right: Push car, Bucyrus Co., 1883. Upper middle center: Hand crank car, circa 1879. Upper middle right: Kalamazoo two-man velocipede. Middle left: Heavy-duty section hand car, Bucyrus Co., 1883. Lower middle center: Express hand car, circa 1879. Lower middle right: Section hand car. Bottom left: Two-man velocipede, Sheffield Co., Three Rivers, Michigan, 1881. Bottom right: Bucyrus iron or rail car for track laying, 1883.

CHAPTER

18

Railway Bridges

BRIDGE building has its origins in prehistory when fallen timber or stones were maneuvered into place to cross a small chasm or stream. From such humble beginnings a great technology arose, and an important part of this development occurred during the nineteenth-century construction of railways. During this period bridges were built from a variety of materials, including wood, stone, brick, iron, and steel. They were formed into a variety of styles, including girder (or beam), arch, trestle, pile, truss, cantilever, trussed arch, lattice, and suspension, and various combinations thereof.

The general practice in railway engineering, from its beginning, was to lay out the course of a line in such a manner that grades and curves would be as gentle as possible. Any deviation from dead level or from a straight line was uneconomical from the standpoint that there would be an additional cost in time, in consumption of fuel, and in wear and tear on the equipment. In situations involving sudden loss of elevation, such as occasioned by creeks, riverbeds, and canyons, a bridge would be required.

While a short distance between opposite banks can be connected by a single *span,* bridging longer distances requires the use of intermediate *piers* for supporting *multiple spans.* The foundations for these structures somehow have to be fixed solidly into the strata below. In *Railways: The Pioneer Years,* Malcolm Fletcher and John Taylor state:

> There are three types of river-bed which the bridge builder will encounter. In some cases, the river-bed is made of hard rock and can bear heavy loads without any chance of erosion. Frequently, the river-bed is made of gravel, pebbles, sand, compacted clay or tuff (volcanic ash). Again such ground cannot be compressed further and will carry heavy loads. There is, however, a possibility that it may shift. Basically, if the land is peat or mud, it can be eroded and will not support a heavy load. In such cases, the foundations have to be sunk down until they reach a solid base. Where this method proves impractical, compressible soil can still be built on. This is done by increasing the surface size of the base so that the pressure the soil must withstand is reduced.

The most ancient bridge style is the *girder* (or beam) type. In its simplest use by railways it is composed of beams spanning an open space, supported by an *abutment* at each end. This type can easily be adapted for multiple-span use. The *arch* style is also ancient in origin, and was commonly used during the period of the Roman Empire. There are a number of forms of arches, including the segmental, semicircular, elliptical, parabolic, and pointed.

Arched bridges are often called *viaducts.* However, this term also includes *trestles.* These structures typically consist of a series of short spans supported by piers. In the nineteenth century they were built from wood or iron, and later from steel. *Pile* bridges depend, for their support, on rows of wooden piles driven into the ground at short intervals. Support beams, or "stringers," are placed on the piles; the roadbed is then laid on the stringers.

Trussing provides extra strength for the girder type structure by adding support framing. Trusses absorb vertical forces, acting at vulnerable points, and shunt them to supported locations. A number of forms of trusses have evolved over the years. These include the King's Post, Bowstring, Warren, Whipple, Howe, Fink, Bollman, McCallam, and Parabolic Arch. A number of these types can also be *inverted* so that the support framing is constructed beneath the level of the roadbed rather than above. Another type of truss is the *lenticular,* which has both upper and lower framing, with the chords, or outer perimeters, curved in either a circular or parabolic manner. *Lateral bracing* was developed to reduce bridge swaying in high winds.

A *cantilever* bridge employs lower trussing that is projected, bracket-like, in both directions from a supporting pier. It derives its strength from the structure, in which each cantilever balances the weight of its opposite. A span is formed when cantilevers from two piers are connected. This term is also used to describe a method of bridge building. *Trussed arch* bridges are framed below with the truss taking the form of an arch. This arch supports the entire weight of both the structure and its load.

Lattice bridges are simple in design and represent another method of providing a girder structure with additional support. Aside from minimal framing, lattices require only bracing material of uniform size. Typically, in the past, wooden planks (or iron bars) were laid at a 45 degree angle at regular intervals, intersecting other bracing at a right angle. A "treenail" or pin of hard wood was fastened at each crossing point. Over time, however, it was found that wooden lattice structures were vulnerable to warping. Thus, later, these were often reinforced by some type of trussing.

The *suspension* bridge, as we think of it today, was dependent on the development of the metal wire cable, and is thus of nineteenth-century vintage. However, this type has much older roots in simple rope-supported structures. Among all styles of bridges, it is the best suited for long spans. The supporting cables, with ends securely fastened, are passed over towers, and from these the roadbed is suspended below. While this type seemed ideal for railways when first introduced, it later proved inadequate. The subsequent operation of faster and heavier trains resulted in dangerous levels of vibrational stress. Thus engineers turned to the steel truss for solving similar bridging problems.

The styles discussed above have to do with the manner in which bridges are constructed. Another set of categories relates to moveable bridges designed to accommodate the periodic movement of water vessels. These types are the bascule, vertical lift, and swing or turn bridge. The bascule style traces its origin to the medieval drawbridge. In its common *single-leaf* form one end is hinged while the opposite end is raised and lowered as needed. In the rare *double-leaf* version two arms are employed. These are hinged at opposite ends and meet in the middle. *Vertical lift* bridges employ mechanical means whereby a bridge section is raised in its entirety. The arms of *swing* bridges are counterbalanced and pivot on a pier in the middle of the water.

In the construction of iron bridges, British railway builders inherited a preexisting technology. Cast iron was already in general use, and the first all-iron bridge had been built across the Severn River in 1779. This immediate locale came to be called Ironbridge. In 1796 a larger all-cast iron structure was constructed at Sunderland, across the Wear River. Early in the nineteenth century a number of similar bridges were built by the road builder Telford.

The world's first *railway* bridge was made of cast iron and built by George Stephenson. It was located at West Auckland on the Stockton and Darlington Rwy. The first iron bridge *of great size* was built by Robert Stephenson at Newcastle over the River Tyne. Completed in 1850, it had a length of 1,338 feet.

When Robert Stephenson was appointed chief engineer of the Chester & Holyhead Rwy. in 1845, he was confronted with perhaps the most difficult bridge assignment ever given to that time. He was to construct a double-tracked bridge across the Menai Strait separating the Isle of Anglesey from the mainland. The British Admiralty would only allow this to be done in such a way that there could be no possible interference with shipping. Stone piers could only be set at one point in between, on a small rock that protruded from the sea. After difficult calculation and experiment, this problem was resolved by prefabricating the two main iron tubular spans, each 460 feet, and floating them out below. Both 1,500-ton sections were then hoisted up to a height where they were connected to the bridge piers. This difficult feat was ultimately accomplished, and the Britannia Tubular Bridge was inaugurated in October 1850.

The Crumlin Viaduct, inaugurated on June 1, 1857, was at the time the largest of its class in the world. Built by T. W. Kennard on the Newport, Abergavenny & Hereford line, it had a length of 1,700 feet and a height of 200 feet. Thus at the time it was the highest railway bridge in England, and the third highest in the world. The August 29, 1857 issue of *Illustrated London News* gives the following description of this structure: "There are, altogether, ten spans or openings; but they are unequally divided by a tongue of land on which are placed masonry and earthworks, about 50 yards in length, making in reality, as far as the ironwork is concerned, two separate viaducts. The larger viaduct has seven spans and six piers; the smaller three spans and two piers."

The Royal Albert Bridge was built by I. K. Brunel over the Tamar Estuary at Saltash, and opened in May 1859. It was perhaps Brunel's greatest accomplishment in railway construction. As in the case of Stephenson's Britannia Bridge, Brunel operated under similar restrictions from the Admiralty. He solved his dilemma by prefabricating lenticular double tubular arches using riveted plate metal. These were then hoisted into place and connected. In *Our Iron Roads* Frederick S. Williams describes the great effort necessary for the creation of the foundation of the main pier for this structure.

> On the great main pier, in the center of the river, all the strain and pressure come, and nothing short of the solid rock would suffice for its foundation. To reach this, however, was a matter of no ordinary difficulty, as some 70 ft. of water, with 20 ft. of mud and concrete gravel, lay between Mr. Brunel and the stone on which he wished to build. An immense wrought iron cylinder, 37 feet in diameter, 100 ft. high, and weighing 300 tons, was made and sunk exactly in the spot where the masonry was to rise. From this the water was pumped out and air forced in; the men descended, and, working as in a huge diving bell at the bottom of the river, cleared away the mud and gravel till the rock was reached. Steam air pumps were necessary to keep the men supplied, and, as a matter of course, they worked at a pressure of upwards of 38 p.s.i. At first this affected them severely; many were seized with cramps, faintness, and insensibility, and one died. But after a time, forty laborers could remain at work with little inconvenience. All, however, were glad when the solid column of granite built inside the cylinder rose at last above the water's edge.

Four octagonal cast iron columns were set upon this structure. Each had a 10-foot diameter and 100-foot length, and weighed about 150 tons.

In contrast to Britain, America's situation was much more primitive. In the early years, at major river crossings, railroads were often built to the water's edge with ferries providing intermediate transportation. However, shorter gaps required the building of bridges. Because wood was cheap, readily available, and the easiest material to fashion into a workable structure, it was initially preferred over metal or stone. Most wooden bridges were of the trestle type, but covered bridges were also constructed until the Civil War era. Wood structures were always considered temporary, however, as they were vulnerable to both fire and rot. Famous trestles included the Erie's Genesee River bridge at Portage, New York, and the Union Pacific's viaduct over Dale Creek in Wyoming. However, both structures were later replaced by iron bridges. As time passed other materials were increasingly used for spanning longer distances, but trestles remained popular for shorter lengths. The *Railroad Gazette* reported in its May 25, 1888 issue that the total length of American bridges was roughly 3,200 miles. This included 2,127 miles of wooden trestles and 1,086 miles of truss bridges, three-quarters of which were iron or steel.

Some lines, like the Baltimore & Ohio and New York & Erie, put an early effort into creating more permanent structures. The B & O's Carrollton Viaduct over Gwynn Falls consisted of a single stone arch. Completed in 1829, it is still in use today. Another B & O structure was the Thomas Viaduct across the Patapsco River at Relay, Maryland. The first bridge in America to be laid on a curving alignment, it had eight granite arches, and was completed by Henry Latrobe II in 1835. The New York & Erie's Starrucca Viaduct near Susquehanna, Pennsylvania, had 17 bluestone arches, and was completed in 1847.

America's first iron bridge was built by Earl Trumbell over the Erie Canal in 1840. That same year Squire Whipple built the first bowstring truss and published a study of stresses in bridge trusses. The first iron truss *railroad* bridge was built by Richard R. Osborne on the Philadelphia & Reading line, and opened in 1845. This company apparently decided to use iron after having two major wooden structures torched by disgruntled canal men. The use of trusses greatly improved bridge-building technology, making possible much longer spans. Early in the twentieth century a single span could reach 720 feet.

The first suspension bridge for railways was constructed by John A. Roebling across the Niagara River gorge for Canada's Great Western Rwy. It was begun in 1851 and completed in 1856. In these efforts Roebling initially followed in the footsteps of Charles Ellet. Ellet was a famous bridge builder of the 1840s who had experience with suspension bridges. However, he lacked self-discipline and had a flamboyant daredevil side that thrived on public attention. Commissioned circa 1845 to build a Niagara suspension bridge, he wasted his opportunity by, among other things, indulging in various spectacles of self-promotion. One instance of this type was described by Richard J. Cook in *The Beauty of Railroad Bridges:* "after a plank catwalk had been strung, he drove in a small horsedrawn carriage across a swaying span that had no guardrails. Standing up, Ellet drove it like a charioteer. The crowds watched aghast, and women fainted."

After this undertaking collapsed and lay dormant for a while it was put in the hands of Roebling. In contrast to Ellet, Roebling carefully planned each step of construction. Working continuously, he inspired his crew to maintain a similar discipline, even during the severe winter months. Two floors of the bridge were designed, the top for railway traffic and the bottom for use by animal-drawn vehicles. Timber trussing connected the two floors and gave the structure more firmness. Additional stability was provided by extra bracing cables, dropped from the bridge and connected to both sides of the bank.

America's first steel-trussed arch bridge was built across the Mississippi River by James B. Eads. Connecting St. Louis with East St. Louis, Illinois, it opened in 1874. In the 1850s both Roebling and Ellet had made proposals for building a St. Louis bridge; however, both were turned down. Eads, who had never built a bridge, began his own efforts in 1867. Compensating for his lack of bridge-building experience was an unusual and intimate knowledge of this wide, powerful, and dangerous river. Eads had traveled extensively over it, but also had worked beneath it, using diving bells to salvage sunken steamers.

For constructing foundations a newly developed "pneumatic caisson" method was employed. Cylindrical enclosures of iron-reinforced timber, 75 feet in diameter, were constructed with cutting edges at the bottom. These were sunk at the desired locations, and after initial excavations, masonry piers were built up inside. As weight increased the cutting edges were forced deeper into the riverbed. As caissons sank lower into the water, rings were successively added to prevent submersion. The masonry in each caisson had at its center a cylindrical opening allowing men and supplies to reach a number of working chambers below. From these sand pumps were used to remove excavated debris.

As work progressed compressed air was supplied to the crew. However, at 60 feet painful cases of the bends began to appear, and at 76 feet the first incidence of crippling. Shift periods were reduced, first to four hours, and then to two hours. On March 19, 1870, men began dying. No more deaths were reported after a new policy was implemented that required workers to take more time in coming up from the depths. Once foundations had been sunk into the bedrock, stone piers were built above them. From these a cantilever method of construction was used whereby the trussed arches, built from the piers, were extended toward each other. This bridge was formally tested on July 2, 1874. Fourteen locomotives, each averaging 40 tons, were positioned in various ways to test each of the three spans, and the bridge was found to be safe. Its completion resulted in St. Louis becoming one of America's most important rail centers.

The first American *cantilever*-style bridge was built on a rail line between Lexington, Kentucky, and Danville, Virginia. Another of this type was the New York Central's 1,910-foot bridge across the Niagara River. It was located only a few hundred feet upstream from the International Suspension Bridge, and opened on December 20, 1883. The Southern Pacific built a cantilever bridge across Texas's Pecos River in 1892. Supporting only a single track, it had a thin, "spidery" appearance, and had the dual distinctions of being one of the world's highest, while at the same time, one of America's longest.

As British and American railway contractors competed in Latin America, a rivalry emerged in iron bridge construction techniques. The *Railroad Gazette* commented on this competition in its issue of January 10, 1874:

> The English bridge is made in heavy parts, whose transportation costs a great deal. When these parts have reached their destined site, about as much labor has to be expended in riveting them together as in making the parts. The difficulties of transporting skilled workmen, and the inconveniences of riveting on scaffoldings, and in awkward places, are very great.
>
> The American bridge, being all fitted by machinery at the works, is put together by a few unskilled men by driving a few pins through holes accurately fitted to receive them. The economy of weight is chiefly due to the fact that the dead weight, or parts that do no supporting work are reduced to a minimum.
>
> The final cause of the difference between the two systems is this: English bridges are sold by the ton, and it is for the interest of their manufacturers to use the lowest priced iron, and the most of it. American bridges are sold by the span, and it is for the interest of their makers to use less iron, and by accurate workmanship and a better quality of material to accomplish the same or better results. The larger the span the more marked the difference between them.

This was hardly the only competition, however. In Africa and elsewhere a similar rivalry existed between the British and French, and to a lesser extent with other European countries. French-built bridges of this period, both at home and abroad, were often done in the lattice style. The Germans also favored this type, and lattice structures were built at Cologne, Offenburg, and Kehl.

Alexandre Gustave Eiffel, famous for the steel tower in Paris bearing his name, was also known for building steel-trussed arch bridges. His beautiful Maria Pia Bridge, over the Douro River in Oporto, Portugal, opened in 1877. Another of his creations was the Garabit Viaduct on the Marvejols-Neussargues line in central France. Opening circa 1888, it had a central arch supporting a span of 541 feet. The track over this structure was about 400 feet above the valley floor. A different type of structure was the Viaur Viaduct on the Carmaux-Rodez line in France. Having three hinge joints, it was a clever combination of cantilever and steel arch technology.

A remarkable bridge for its time was the 85-span straight girder structure designed by Sir Thomas Bouch and built over the Firth of Tay. Opening on June 1, 1878, it cleared the water by 100 feet and extended for 2 miles, 364 yards. However, during a severe storm on December 28, 1879, much of it collapsed, taking with it an entire train and 78 lives. Afterward it was determined that Bouch's calculations for lateral pressures exerted by high wind were faulty, that the cast iron columns were of poor quality, and that the contractor had done poor work. A subsequent lattice steel structure was designed by W. H. Barlow, and son. Built under the direction of Sir William Arrol, it was completed in 1887.

Arrol began this project for the North British Rwy. in 1882. That same year he received another commission from the line, a bridge across the Firth of Forth. The chosen path for this route presented peculiar problems. To successfully resolve them, the designers, Sir John Fowler and Sir Benjamin Baker, chose to build this bridge on the cantilever principle. The resulting cantilevers were much more massive than their American counterparts, as they used extensive overhead reinforcement. This impressive structure, considered the queen of Britain's railway bridges, opened on March 4, 1890.

PLATE 234

I. K. Brunel's famous Royal Albert Viaduct over the Tamar Estuary, Saltash, England, 1859.

Steel bridge, of truss type with lateral bracing, across Garonne River, Bordeaux, France, P.O. Rwy., 1860: Top: Partial view of span. Bottom: Inside view.

PLATE
235

PLATE 236

John A. Roebling's International Suspension Bridge across Niagara River, connecting Canada and the United States, Great Western Rwy., 1864.

PLATE 237

Double-headed train crossing iron span over Wyoming's Dale Creek chasm, Union Pacific R.R., 1877. This permanent bridge, the largest on the line, was constructed in 1876 to replace a wooden trestle that had deteriorated. The Stone & Browner Co. of Chicago had built that structure in 1869. It was assembled in 30 days from precut material.

PLATE 238

James B. Eads' steel arch bridge over Mississippi River, St. Louis, Missouri, 1874.

Scene on the Firth of Tay Bridge, Scotland, prior to disaster of December 28, 1879, North British Rwy.

PLATE 239

PLATE 240

Bismarck Bridge over Missouri River, Bismarck, Dakota Territory, Northern Pacific R.R., 1883.

New York Central line's cantilever bridge over Niagara River, 1885. Roebling's International Suspension Bridge can be seen in the background.

PLATE 241

Counterweighted lift bridge across Berry's Creek near Rutherford, New Jersey, Erie R.R., circa late 1880s.

PLATE 242

PLATE 243

Double-tracked swing bridge crossing Thames River, New London, Connecticut; New York, Providence & Boston R.R., 1889. At the time this structure was the largest of its type in the world.

PLATE 244

Cantilever bridge across the Firth of Forth, Scotland, designed by Sir John Fowler and Sir Benjamin Baker, North British Rwy., 1890.

PLATE 245

Steel arch bridge over Wupper River, Mungsten, Germany, Prussian State Rwys., 1897. This massive structure, between Remscheid and Solingen, was the highest in Europe at the time of its completion.

CHAPTER

19

Mountain Railways

Center-Rail Adhesion, Cog or Rack Systems, and the Great Tunnels

From the earliest days of railway construction, hills and mountains presented special problems. Depending on the kind of traffic that was anticipated, the available locomotive power, and the construction funds at hand, engineers would determine the steepest gradient to be allowed and then build accordingly. English railways, inspired by the precise engineering standards of the Stephensons and usually blessed with favorable terrain and ample capital resources, typically kept their gradients below 1 in 50, or 2%. It was usually sufficient for these lines to utilize bridges, viaducts, and tunnels to achieve their objective.

Elsewhere in the world, things were not so easy. As low-budget American railroads built westward in the 1840s and 1850s, tunnel building was kept to a minimum, and new techniques were employed. The earlier introduction of four-wheel bogies allowed American engineers to construct their lines with tighter curves. This made possible later innovations like the use of horseshoe curves, loops, and reversing stations, which stretched out lines to gain necessary altitude. Engineers from countries like Austria, Switzerland, and Russia came to America to study these developments, and then returned home to build lines there. These techniques also were increasingly utilized in other developing countries as railways were constructed. In the meantime locomotive power was increased, and this allowed construction of lines with steeper gradients.

In 1863 John B. Fell took out two patents relating to the center rail principle (see chapter 29). Fell, an engineer, was an associate of Thomas Brassey, the famous English contractor. While building railways for Brassey in Italy, Fell proposed the idea of constructing a special line over Mt. Cenis Pass. Railway tunneling through the mountain had begun circa 1858 and was expected to continue for many years. In the interim, Fell's proposed route could provide needed rail service between France and Italy and supply an important communications link between the tunnelers on each side.

This idea was well received by Brassey, and so Fell began making preparations. Between 1863 and 1864, he conducted experiments on the Cromford & High Peak Rwy. in Derbyshire. For his line a 3′7½″ gauge was chosen and laid over a short route that included an incline and tight curves. The raised smooth center rail was 7½ inches higher than the outer rails. A 0-4-0 tank engine, equipped with an auxiliary pair of horizontal driving wheels, was built for the trials by Canada Works of Birkenhead. By means of levers and springs, pressure was applied to these wheels which forced them against the center rail, increasing their adhesion. In the experiments that followed, it was determined that the engine could draw loads of from 24 to 30 tons up the slope when the auxiliary drive was operating. When it was disengaged, the engine could only pull loads up to 7 tons.

By 1865 the French and Italian governments were both interested in the Fell-Brassey proposal. However, Fell was first required to build a 1¼-mile line at Lanslebourg to demonstrate the practicality of his system. The average gradient of this new 3′7½″ gauge line was similar to that of the previous railway, being 1 in 13, with maximum incline of 1 in 12. Trials took place for three months in the spring of 1865, and were attended by representatives from France, Italy, England, Austria, and Russia.

Although Fell's original locomotive was shipped to France for the trials, a second 0-4-0 tank engine of improved design had been constructed by J. Cross of St. Helens and gave a better performance. From a footplate the driver could control the pressure exerted by the horizontal drivers on the center rail, with a maximum of 2½ tons applied from each side. This locomotive was able to draw three wagons, weighing 16 tons, up the incline at a speed of 11 m.p.h. The original locomotive had four cylinders—two outside for powering the vertical drivers and two inside for powering the auxiliary system. However, it had trouble maintaining a sufficient level of steam. The second engine only had two inside cylinders; thus the overall drive system had to be more precisely coordinated.

The trials were successful, and on November 4, 1865, Emperor Napoleon III issued a decree authorizing construction of the line from St. Michel de Maurienne to the border of Italy at the summit of the pass. Construction on this railway began on May 1, 1866. However, floods the following September caused damages that delayed progress. It was built in a zigzag fashion, and when completed to Suse, Italy, in 1868, reached a length of 48½ miles. Travel over the route, including stops at stations, averaged six to seven hours.

While a primary purpose of Fell's railway was to improve communication between the tunnelers, it was also open for business. Situated in a strategic location, it quickly became the route over which the lucrative English-Indian mail traveled, which then went on to Brindisi. The line also carried approximately 100,000 passengers in its short 3½ years of operation. Construction on the Mt. Cenis Tunnel had originally been predicted to continue until circa 1875; however, advances in drilling techniques hastened progress considerably. The tunnel opened for traffic on September 17, 1871, and Fell's line became obsolete overnight. Service on the railway, as stipulated in its charter, was shortly terminated.

The rails over Mt. Cenis Pass did not rust away, but instead were shipped to Brazil for use on the Cantagallo Rwy., which opened in late 1873. This 7¾-mile line, using the same gauge as before, was built into a coffee-growing area, and had inclines of up to 1 in 12. Five four-cylinder tank engines with enlarged boilers were built for this railway—three by Manning, Wardle & Co. of Leeds and two by Ernest Gouin of St. Denis, France. Later, 0-6-0T adhesion locomotives, fitted with center rail brakes, were built by Baldwin for the line.

Fell's system was more vulnerable to the weather than the rack (or cog) rail systems then in development, and as these technologies emerged, they were increasingly favored over his. However, the center-rail system was found to be especially useful in providing braking power for heavy trains on a descent; thus a few more lines of this type were built. Among these were the Rimutaka Incline line in New Zealand and the Snaefell Mountain Rwy. on the Isle of Man.

In 1852, a wealthy Chicago meat packer, Sylvester Marsh, made a visit to the top of Mt. Washington, New Hampshire. This mountain is the tallest among the scenic White Mountains, reaching an altitude of 6,293 feet. While climbing this mountain, Marsh and a companion were overtaken by a severe storm and almost perished. It shortly occurred to Marsh that the mountain might be more attractive and accessible to tourists if a railroad could be built to its summit. However, such a line would have to ascend grades far beyond the capacities of adhesion locomotives. Thus some new system of operation would be required.

Marsh had not been the first to receive such inspiration. Earlier, Herrick Aiken of Franklin, New Hampshire, had conceived the idea of a "cog" railway for the mountain and built a small model. Marsh later studied this model as well as nine patents relevant to the subject, which dated from 1829 to 1839. Perhaps the most helpful was a patent for a cogwheel system for propelling cars up inclines, issued in 1832 to William Bent of Philadelphia. (It should be noted here that these patent-holders probably owed something to Blenkinsop, who built the first cog railway in 1812.)

Marsh may also have drawn motivation from another source. He was perhaps aware of the rack section operated by the Madison & Indianapolis R.R. on the Madison Incline. Extending 7,012 feet at an average grade of almost 6%, this rack line was designed by Andrew Cathcart and operated from circa 1848 to 1868. Its first combination rack and adhesion engine, the *M.G. Bright,* was built by Baldwin, Vail & Hufty and delivered in December 1847. A 0-8-0T type, it had two sets of cylinders, with one pair working the drivers in the usual manner. The second pair, by intricate means, powered a center cogwheel. Solid traction was provided when this engaged a rack rail mounted along the center of the track. Unfortunately, Cathcart's system was primitive and expensive to maintain. Whether at the top or bottom of the incline the center rack had to be engaged by the engine each time a trip was made, and there was no mechanism to facilitate this. Thus rack castings were broken and cogwheels quickly worn down. There were also no safety brakes to stop the train if the cogwheel became disengaged from the rack. On occasion this happened, resulting in a runaway train that usually jumped the track on a turn at the bottom of the slope.

In 1866 the M & I R.R. was consolidated with the Jeffersonville R.R., creating the Jeffersonville, Madison & Indianapolis R.R. By 1868 the master mechanic of the new line, Reuben Wells, had constructed a 0-10-0T adhesion engine that showed more power on the slope than the rack locomotives. Use of the rack system was immediately terminated.

Whatever his sources of inspiration, Marsh eventually produced designs for a "rack and pinion" cog railway and an "atmospheric" brake, for which he soon received patents. Next he petitioned the New Hampshire legislature to grant a charter for his railroad. Although this application was received with some skepticism and merriment, it passed and was signed by the governor on June 25, 1858.

Marsh's plans were delayed by the American Civil War, and rail construction did not begin until May 1866. The base of the line was situated at an altitude of 2,668 feet and building commenced from there, ascending a grade of approximately 25%, or 1,300 feet to the mile. The most difficult portion of the incline rose 1,980 feet to the mile, and included an exceptionally steep area known as "Jacob's Ladder." Due to severe winter weather, building only took place in the summer months.

Marsh was ably assisted at the time by Walter Aiken, son of Herrick Aiken, who in addition to being an inventor and machinist was the most important stockholder in the company. The gauge for the line was 4′8¾″, and running down the center of the track was the cog rail, formed by two parallel angle irons 4 inches apart. Between these angle irons, round steel bars of 1½-inch diameter were connected every 4 inches to form a rack. This rack provided a sufficiently strong support upon which the locomotive cogwheel could operate.

Over most of the route, the rails were attached to a strong trestle underframe, with timber interlaced and bolted and resting on solid rock. The line was completed on July 3, 1869, and extended for a distance of 3½ miles. Trains immediately began operation on it, carrying passengers to the Tip-Top House, a hotel that had been built there in 1853. At the time of its completion, it was considered to be the greatest feat of American railroad engineering.

The original locomotive of the line, *Hero* (later renamed *Old Peppersass*) was built by Campbell, Whittier & Co. of Roxbury, Massachusetts, and was first tested on August 29, 1866; it successfully pushed a flat car loaded with passengers up a 15% grade. Its vertical boiler was freely suspended so that the firebox would always stay level. Pistons from the two steam cylinders drove a transverse crank shaft that was fitted with a drive pinion and a brake drum. The pinion, in turn, propelled a spur gear attached to a cogwheel shaft. The gear reduction that resulted, about 5 to 1, increased power but sacrificed speed. Later a few other vertical boiler engines were built, but, beginning in 1880, a more powerful type with slanted horizontal boilers was introduced.

A unique feature on the *Hero,* which was retained on later locomotives, was the "atmospheric" brake. The engine had a mechanical brake, but it was seldom used. Instead, when descending, steam to the cylinders was shut off, and special air lines connected there were opened. This essentially converted the cylinders into air compressors. By controlling the flow of air, the engineer had a way of selecting the pressure most suited for slowing his descent.

Trains on the line always consisted of a locomotive and a single car. The car was not directly connected to the engine, but was left detached and placed in the uphill position. While ascending, the car was pushed ahead of the locomotive, and while descending, it trailed behind.

Marsh's line was a success, and it helped to draw many tourists to the area. By 1875 railroad lines were built in from both east and west to service this attraction. In 1883 the Boston & Maine acquired the line, and kept it until 1931. Sold at that time to a private party, it continues to remain in service to the present day.

While Marsh was developing his system in America, a Swiss engineer, Niklaus Riggenbach, was working on a similar idea. In 1863, he received a patent for an invention described as "a new track and locomotive system, which will enable trains to cross mountains." It, too, featured a locomotive with a center cog driving wheel, which operated on a rack system. When Riggenbach heard about the line that Marsh was constructing, he took a transatlantic journey to see it, and made drawings of the machinery, track, and other details. Upon returning to Switzerland, Riggenbach began working out plans for his own system; to test his ideas, he built a short experimental line in a quarry area near Berne.

After these efforts proved successful, Riggenbach, along with some associates, sought out a popular mountain site, where construction of an uphill line might be profitable. The location finally chosen was Mt. Rigi, near Lake Lucerne. This mountain was close to the city of Lucerne and was a popular tourist area. The line began at Vitznau, situated on the lakeshore, and ascended to an elevation 4,000 feet higher, a short distance beyond Rigi-Kaltbad, at the boundary of the Canton of Schwyz. It climbed up an average grade of 20%, and included an inclined tunnel 225 feet long, immediately followed by a lattice girder bridge with three spans over the Schnurtobel River.

Gauge size for the line was 4′8½″. Delivery of rails was delayed by the Franco-Prussian conflict, but the line finally opened on May 23, 1871. There was considerable interest in this railway and it received coverage in the press of a number of European countries. The Canton of Schwyz, which had originally blocked construction farther up the mountain, finally came to an agreement with the railway. Thus in 1873 it was extended to the summit, reaching a total length of 4¼ miles. The original locomotives had vertical boilers, and were similar in design to those of the Mt. Washington line. These were later replaced with slanted engines having horizontal boilers. As on Marsh's line, a train consisted of an uncoupled engine and passenger coach.

Riggenbach's rack system was considered to be an improvement over Marsh's in one important respect. E. L. Cornwell, in *History of Railways,* gives this explanation: "Both are in the form of a steel ladder, but whereas Marsh used rungs of circular section, Riggenbach had rungs of tapered section, designed to ensure better locking of the locomotive's pinion-wheel teeth with the rack, and to counteract any tendency of the teeth to climb out of the rack." Riggenbach also provided a safety device for the passenger coaches: a pinion with which to engage the rack in the event of any emergency.

The Rigi line became so popular that a second railway, the Arth-Rigi, was built on the opposite side of the mountain and opened in 1875. A number of other lines were built using the Riggenbach system, both in Switzerland and elsewhere. Among these were the Kahlenberg Rwy. in Vienna, completed in 1874; the Rorschach-Heiden line in Switzerland, which opened in 1875; and, in Brazil, the Prince of Grand Para Rwy., a meter-gauge line inaugurated in 1883.

In its issue of May 7, 1886, *Railroad Gazette* described a dangerous mishap that took place on the Arth-Rigi line:

> A serious accident occurred on Oct. 20, 1885, on the rack railway going from Arth, at the foot of the celebrated Rigi to the top of that mountain. The last train of the season consisted of one locomotive, two loaded freight cars, and a small passenger car, probably all coupled together in order that the cars might pass more easily over the less severe grades, which were covered by thick layers of snow. This coupling together of cars and engine is strictly prohibited on the Rigi Railway and was not justified in the present case, but this rule is often violated. The train in question was a work train under charge of the Roadmaster. The freight cars were loaded with lumber in such a manner that the brake-wheels could not be reached. A fireman stood at the brake of the passenger car, but jumped off before putting on the brake.
>
> The upper part of the road, which was covered deeply with snow, was traversed quite easily by the train, but when it reached the lower end of the Kraebelwand, where no snow was lying, and which is 1¼ miles from the lower station, the men on the train heard a heavy crack, followed immediately by a great increase of the speed of the train. The laborers standing upon the platform of the car jumped off shortly after they had heard this noise, and farther on the Roadmaster did the same. The train ran away, passed with lightning speed and deafening noise a distance of about 2,100 ft., including several curves, and left the track a short distance below the watering station at Kraebel, going over the precipice to a meadow below. The locomotive turned three times, once lengthwise and twice crosswise, and stood then again upright on the wheels. The freight cars probably followed the engine and were thrown around it in a circle of 1,000 ft. The broken coupling was found between the engine and the first freight car. The coupling of the passenger car was also torn off, and the car probably left the track with one wheel soon after the accident; it was thrown entirely off the track further down and stood on its wheels in front of the freight cars; the roof and sides of the car were scattered around. The boiler and the cylinders of the engine remained intact, the rest of the engine being almost entirely destroyed.

This was the first accident in 10 years of operation on the line. No mention was made in the article of any injuries or fatalities.

Riggenbach also experimented with mixed terrain railway systems, which utilized both rack and adhesion methods, and designed a locomotive for this purpose. It had a horizontal boiler and three axles—two for adhesion, and in the middle, one for the cog; which whether on the rack or not, continued to turn. In the rack portions of the line, adhesion provided an additional boost to the power supplied by the cog.

One railway, still in operation today, which uses this method is Switzerland's Brunig Pass line, inaugurated in 1889. Built in the meter gauge, it connects the town and Lake of Lucerne with the Lake of Brienz and the Bernese Oberland, by crossing Brunig Pass. It is 36.4 miles in length, and consists of three sections: a flat portion beginning at Lucerne that follows along the lake, a mountain section over Brunig Pass with a maximum grade of 12%, and another relatively flat zone extending from Meiringen to Brienz. Other railways of this type were built elsewhere. Among the most unusual, perhaps, was the Langres Rwy. in France, which opened in 1887. Its purpose was to connect a railway station on the Est line with the nearby town of Langres, situated 430 feet higher on a plateau.

Roman Abt, an engineer from Lucerne and former employee of Riggenbach, devised a slightly different rack system. He had rack slots machined into an upright steel plate called the "blade." This made possible more precise meshing than in the previous system. His arrangement had either two or three blades laid close together in the center of the track. The teeth of each blade were intentionally set out of alignment with those of the other blade(s). These blades were then engaged by a similar number of cogwheels. This innovation provided a smoother operation and improved speed, power, and safety. On lines with higher gradients and tighter curves, the three-blade system was found preferable to that of the two-blade.

Abt's system was first tested in 1885 on the Blankenburg-Tanne Rwy. in the Harz Mountains of Germany. It was later used on other lines in that country, and in Switzerland, Italy, Bosnia, Venezuela, Africa, India, Japan, and elsewhere. The Manitou & Pike's Peak line in Colorado, which opened on June 1, 1891, used the Abt system. This line became the world's highest rack railway, reaching an altitude of 14,110 feet.

A later modification of Abt's system was devised by the engineer Strub, and tried on Switzerland's Jungfrau Rwy. in 1899. This line, the highest in Europe, reached an altitude of 11,333 feet. Strub's innovation simplified the Abt rack by replacing the blades with a flat-bottomed Vignoles rail having rack slots cut into its head. This rail could then be laid directly on the sleepers, unlike the Abt blades, which required a spacer.

A completely different type of rack system had to be devised for the 2¾-mile Mt. Pilatus line, near Lake Lucerne, which opened on June 4, 1889. Built by two Zurich engineers, Colonel Edouard Locher and M. E. Guyer-Freuler, it began at Alpnach-Staad and ascended to an altitude of 6,965 feet. The situation for this railway was quite unique as the average grade over the route was 40%, with the steepest section reaching 48%. A gauge of 2′7½″ was chosen for the line.

After testing existing rack systems and finding them unsuitable, a method was improvised that borrowed from both rack technology and Fell's center-rail adhesion system. The center rack consisted of a horizontal rail having rack teeth on each side. Engaging these were two horizontal cogwheels, set opposite each other as in Fell's system. A second pair of cogwheels was used for braking during descent.

Other design innovations included an inclined locomotive whose short horizontal boiler was placed sideways, and a passenger carriage with staggered compartments. This system proved to be very effective. The line still exists today and no accident has ever been recorded. Like most of the other rack railways in Switzerland, it was electrified in the twentieth century.

In its March 6, 1896 issue, *The Engineer* reported that beginning with Blenkinsop's line of 1812, a total of 70 rack railways had been built. Of these, 17 were in Switzerland, 14 in Germany, 12 in Austria-Hungary, 4 in France, and 3 in Italy, with the remainder in the United States, England, Spain, Greece, Portugal, South America, Asia, and Australia. The aggregate total of these lines was about 500 miles, of which 188 miles were built in the Abt system. About 300 locomotives were operated over the entirety.

During the nineteenth century tunneling evolved from primitive pick and shovel methods to much more refined techniques. Critical to this progress was the development of the first compressed air drills circa 1860 and the invention of dynamite about a decade later. As the century proceeded, tunnel distances increased dramatically, and ever greater efforts were made in their construction. Tunnels were excavated, not only to penetrate impassable land masses, but also to provide routes beneath rivers, and in some cases, beneath cities.

By the early 1850s the Baltimore & Ohio R.R. completed the 4,100-foot Kingwood Tunnel west of Cumberland, Maryland. About the same time the Blue Ridge Tunnel near Waynesboro, Virginia, was being excavated. When completed, it extended for a distance of 4,273 feet. A much more difficult undertaking was the building of the 4.69-mile Hoosac Tunnel in northwestern Massachusetts. Construction began in 1855, but was interrupted several times and was not completed until 1875. Excavation of this tunnel was begun under the direction of Herman Haupt, who later gained fame in America's Civil War.

Although important tunnels were built in England, America, and elsewhere, the truly great tunnels of the period were built in the Alpine areas of Europe. The natural land barrier formed by this high mountainous region was truly formidable. It blocked what would have been a considerable flow of commerce between economically productive countries with large populations. As wealth increased in these nations and railway technology improved, political pressure grew to tunnel through these obstacles and open up existing economic potentials. At the time, these were undertakings of an unprecedented magnitude, and they required great sacrifice and courage; however, the ultimate payoffs were to be very large indeed.

The first of the great Alpine tunnels was the double-tracked 7½-mile Mt. Cenis Tunnel. It was authorized by King Victor Emmanuel II of Sardinia, on August 31, 1857, to connect Bardonecchia, in Sardinia, with Modane, in Savoy (which was under Sardinian control). Such a tunnel had first been proposed in 1842, but at the time was considered impossible. In 1860 France annexed Savoy, but agreed to allow the tunnel work to continue. For the first three years about 2,000 men with picks and shovels were employed on each side, progressing only about 45 feet a month.

Then Germain Sommeiller, one of the three chief engineers on the project, introduced compressed air drilling equipment, and progress improved to an average of 250 feet per month. This drilling had the added advantage of bringing badly needed fresh air into the tunnel, thus improving ventilation. As the deeper portions of the mountain were penetrated, temperatures grew warmer, eventually reaching 30 degrees Centigrade. The tunnelers finally dug through to each other on Christmas day, 1870. Work continued for the next 8½ months as the tunnel and its railway reached a final state of completion. The line was inaugurated on September 17, 1871, with great ceremony and fanfare. It took the inaugural train 21 minutes to travel the distance of the tunnel.

In September 1869 representatives from Germany, Switzerland, and Italy met at Bern to discuss creation of a rail line through St. Gotthard Pass. It was agreed at the conference to build the line, with all three countries sharing in the cost. At the time, France objected to this idea, fearing that it would divert traffic from the Mt. Cenis route. However, the country's subsequent loss to the Prussians in their 1870–71 conflict silenced this protest.

On November 1, 1871, the Gotthard Rail Co. was formed at Lucerne. Louis Favre, who had worked on the Mt. Cenis Tunnel, was made its chief engineer. By this time, improvements in drilling equipment, made by McKeon and Prof. Colladon, greatly speeded operations. For blasting purposes, black powder was replaced by dynamite, which was about 10 times more powerful. Later, horses and mules, which transported men and supplies and removed debris, were replaced by compressed air locomotives.

The St. Gotthard tunnel was actually a series of 13 tunnels, which extended for a distance of 9 miles, 562 yards. These were described in *Railways: the Pioneer Years,* by Malcolm Fletcher and John Taylor, as follows: "The main feature of the St. Gotthard series of tunnels . . . is the spiral shape of 7 of the 13. The line rises up the sides of different valleys and eventually reaches the main tunnel at the top. As the valleys get higher, they follow each other in quick succession, with the result that there is no room for the train to turn and it has to cross over a bridge or viaduct to the opposite side of the valley. There it enters a tunnel which rises and turns to emerge at a higher point still. From there, the train crosses another bridge or viaduct and re-enters the hill it has just left."

With a constant crew of about 2,500 workers excavating from both directions, these tunnels were completed in a little over eight years. However, in the process, 177 lives were lost, including that of the chief engineer. Layers of rock were encountered that were harder than those at Mt. Cenis, and interior temperatures were warmer, sometimes reaching 35 degrees Centigrade.

The tunnelers finally made contact with each other on February 28, 1880. On December 23, 1881, the tunnel was ceremonially inaugurated. It opened for goods traffic on January 1, 1882, and passenger traffic on June 1, 1882. Additional space along its route had been provided for double tracking, and soon, as traffic flow increased, this improvement was made. The opening of the tunnel completed construction of the St. Gotthard Rwy. This line is still in existence and extends from Lucerne to Chiasso at the Italian border, providing a vital link between Italy and Germany. Over its entire length the line has 101 bridges and 80 tunnels.

On November 19, 1883, work on another important European tunnel, the Arlberg, was completed. Situated in the Tyrolean Alps of Austria, it lay between Innsbruck and the Swiss border, and provided a direct route from Vienna to eastern Switzerland. It was begun in June 1880, and when completed, extended a distance of 6 miles, 650 yards.

As the nineteenth century waned, Switzerland was looking for a more direct rail connection with Italy. Within Switzerland a number of rail lines converged on Simplon Pass, and these were finally incorporated into the Jura-Simplon Rwy. In 1893, the two governments agreed to build a 12½-mile tunnel that would connect Brigue, in Switzerland, with Iselle, in Italy. However, it was only in August 1898 that the Brandt and Brandau Company of Hamburg finally initiated construction.

As drilling progressed deeper into the mountain, temperatures became much hotter than any previously experienced. At 5¼ miles in, a temperature of 53 degrees Centigrade (127 degrees Fahrenheit) was recorded. Thus powerful ventilating fans had to be used and ice cold water was pumped into the drilling areas. Crews could only work in these areas briefly before they had to be replaced. Cave-ins and the penetration of a hot spring further delayed progress. However, work was finally completed and the tunnel was opened on May 17, 1906.

Fell's center-rail adhesion system: Top: Fell's original center-rail adhesion engine, Canada Works, Birkenhead, England, circa 1864. Bottom: Winding course of Fell's railway over Mt. Cenis, circa 1868.

Fell's Mt. Cenis Rwy. connecting France and Italy: Top: View showing early type of engine in use, 1868. Bottom: Train ascending grade after leaving Lanslebourg, 1869.

PLATE
247

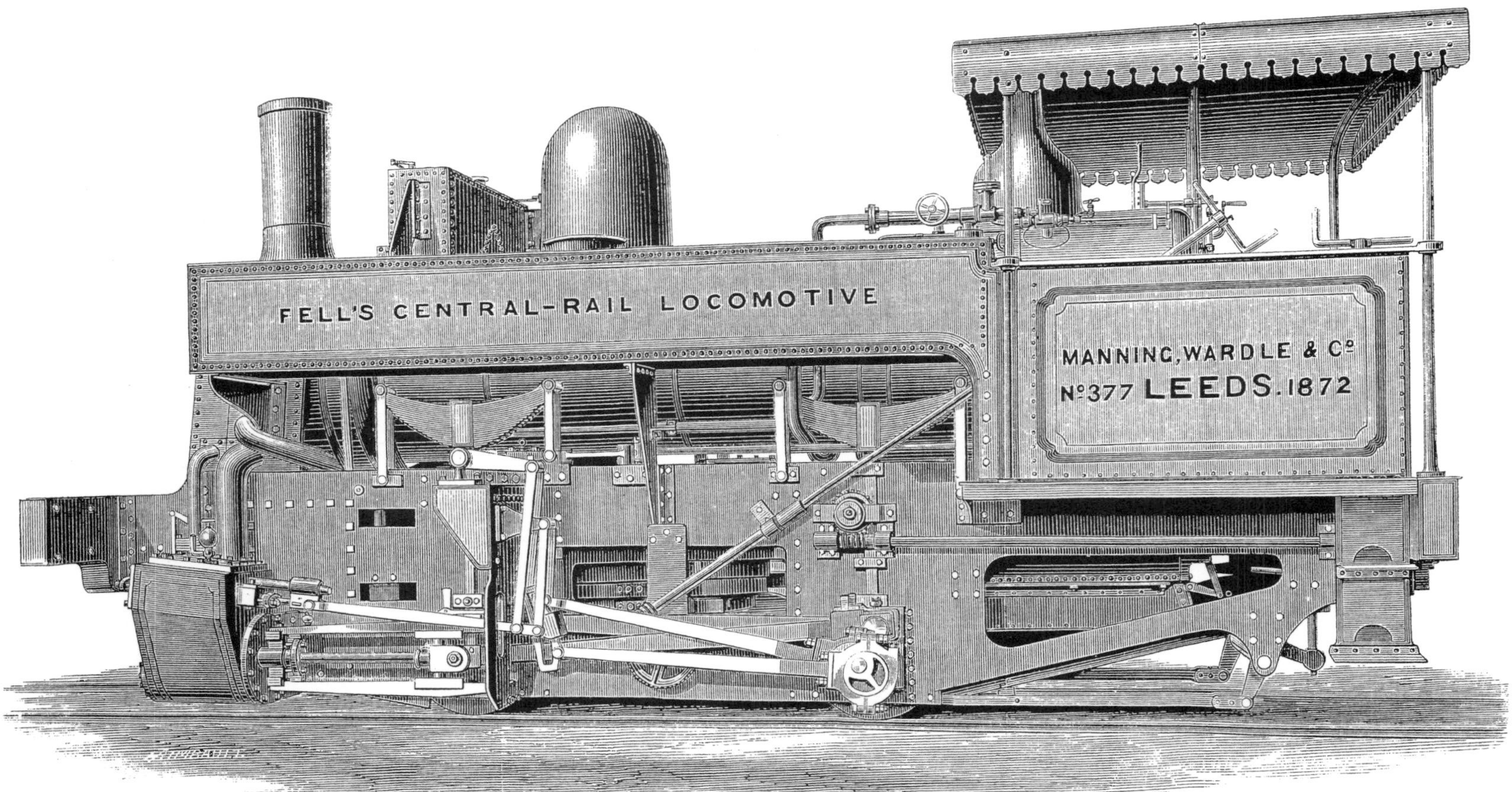

Other Fell center-rail adhesion locomotives: Top: Engine built by Manning, Wardle & Co., Leeds, for the Cantagallo Rwy., Brazil, 1872. Bottom: Engine unidentified as to make and time period, but probably used on the Mt. Cenis Rwy.

Cog or rack railways: Top: Scene on Sylvester Marsh's Mt. Washington R.R., New Hampshire, circa 1868. Bottom: The original style of engine used on Switzerland's Rigi Rwy., 1870.

Scenes on Switzerland's Arth-Rigi Rwy.: TOP: Train on Rothen Flur Bridge, circa 1880. BOTTOM: Train on the Kraebelwand, circa 1885.

TOP LEFT: Combination rack and adhesion 0-4-0T engine, Langres Rwy., France, 1888. TOP RIGHT: 0-4-2T rack locomotive, Abt system, Wengernalp Rwy., Switzerland, 1894. BOTTOM: Baldwin-built 0-4-0T rack engine, Riggenbach system, Prince of Grand Para Rwy., Brazil, 1886.

PLATE
251

Top: Scene near summit on Switzerland's Rigi-Kulm Rwy., circa 1870s. **Bottom:** Locomotive and carriage on Switzerland's very steep Mt. Pilatus Rwy., 1886. Note unique sideways position of boiler.

French engravings showing two railway tunneling techniques, circa early 1860s: Top left, top right, middle left: Grand section method. Middle right, bottom left, bottom right: Divided section method.

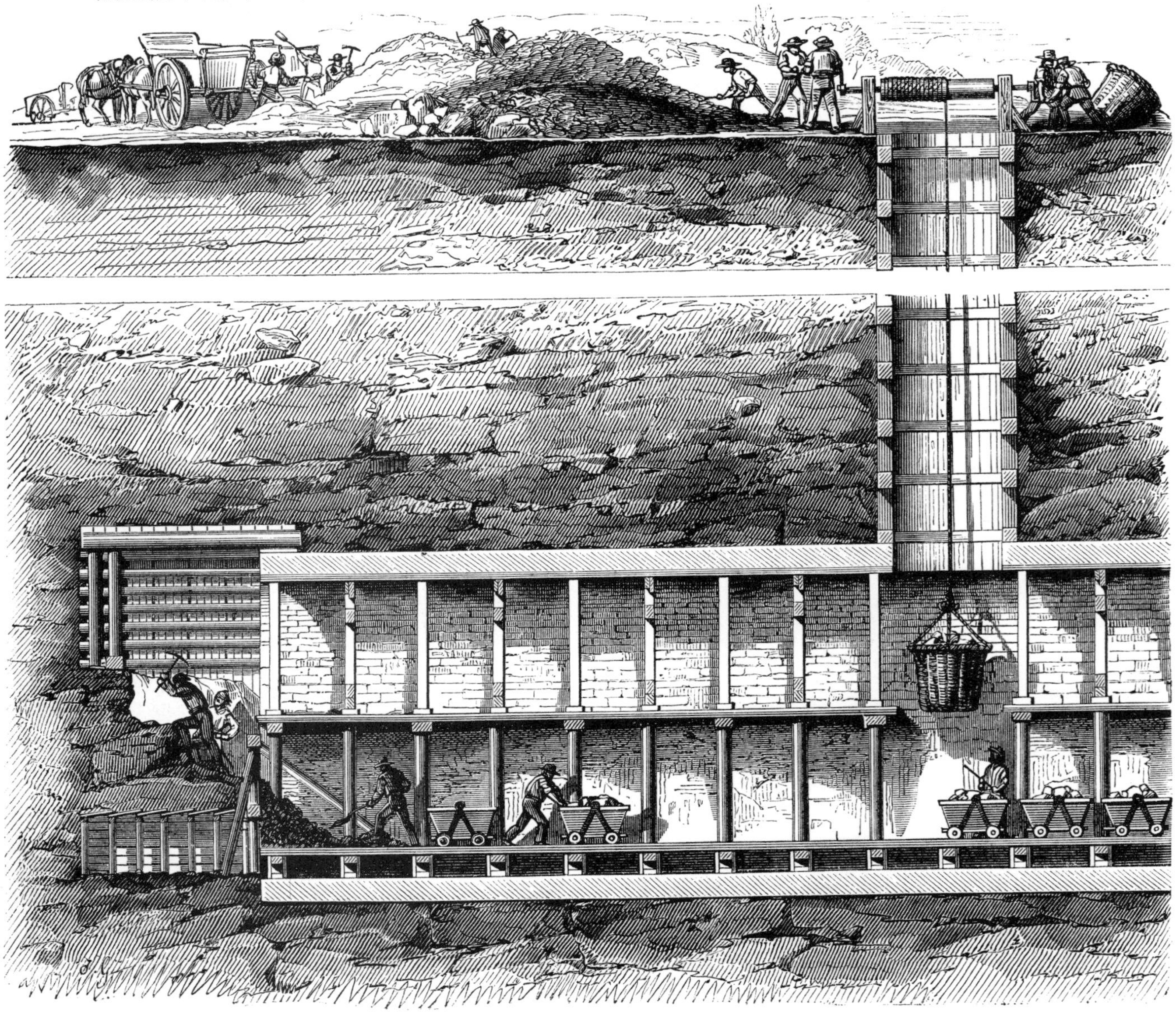

TOP: Cross section of double-tracked railway tunnel showing ventilation shaft and method of drainage, circa 1860s. BOTTOM: Longitudinal section of tunnel showing various facets of excavation, circa early 1860s.

The Mt. Cenis Rwy. Tunnel: TOP: Drilling with compressed air. BOTTOM: Inaugural train, September 17, 1871.

PLATE 255

Excavation on Switzerland's St. Gotthard Tunnel: striking water, 1882.

PLATE
256

Top: Train crossing Maderan Valley Viaduct on St. Gotthard Rwy., 1882. Bottom: Inauguration of Austria's Arlberg Tunnel in the Tyrolean Alps, Vorarlberg Rwy., November 19, 1883.

PLATE 257

CHAPTER

(20)

Military Railways

EARLY in the development of railways, people in positions of political or military power began to see the potential for using this technology in serving their strategic needs of defense, as well as for realizing their dreams of territorial expansion or colonial conquest. Railways could be used for the purposes of putting down civil insurrections, accelerating mobilization in case of war, carrying troops and supplies more speedily to their destinations, gaining strategic advantages over an enemy, and removing wounded personnel. As time passed, a few countries, like Prussia, kept these considerations in mind when planning their railways. This led to the creation of *strategic* lines. While not exclusively intended for military purposes, these were built with a special view to military utility in time of war, either for defensive or offensive operations. In most nations railways were constructed primarily to serve economic needs. However, these lines could be used on occasion when it suited the purposes of the military. During the 1842 Chartist agitation in England, for example, Parliament passed a Railway Regulation Act that required railways to carry troops upon government request.

The Prussia of these years, not to mention the German States, was haunted with the memory of the Napoleonic Wars. By virtue of its central position in Europe, Prussia felt vulnerable from almost every direction. A rail line for defensive purposes between Minden and Cologne had been urged as early as 1833. Later, the economist Friedrich List, who had previously visited the United States, became a passionate advocate of an inner network of German railways. It was his belief that this would make the country invulnerable to any outside attack, as defensive forces could be quickly moved to any threatened position. However, railways could also be used with aggressive intent. This was demonstrated by the Prussians in 1846 when they moved 12,000 troops by rail to take over the Polish Republic of Krakow.

The first actual building of a military railway in wartime occurred during the Crimean conflict of 1854–56. During the severe winter of 1854–55, a British supply breakdown occurred between the port of Balaklava and Sebastapol, where English forces were besieging a Russian fortress. Supplies piled up at the port, while close by, troops suffered from starvation and frostbite. Soon military operations came to a complete standstill. Finally, the English railway contractors, Thomas Brassey and Samuel Peto, volunteered at their own expense to help construct a line to assist this military effort. Arriving in March 1855, they eventually constructed a double-tracked line 29 miles in length. At first, horses were used to pull the wagons, with the additional help of a stationary steam engine on one steep grade. Soon small locomotives arrived to expedite the operation. Supplies were brought forward while the sick and wounded were evacuated. Shortly the British captured their objective, bringing an end to the war.

To further its expansive colonial ambitions, Britain built many railways during the course of the nineteenth century, employing them when needed for military purposes. In 1857 one of these lines was used to move troops for suppressing the Indian Mutiny. This conflict revealed the true vulnerability of colonial garrisons and the importance of railways in maintaining power. Later, many strategic railways were built in India so that troops could be quickly rushed to any point of conflict.

During the Austro-Sardinian War of 1859, France gained an advantage over Austria by sending a force of 76,000 men and 4,450 horses by rail to the aid of its ally, Piedmont. This rapid troop deployment from Paris into Italy only took 10 days.

The French forces won at Magenta and Solferino, but were unable to pursue their defeated opponents because of their lack of additional supplies.

The American Civil War of 1861–65 became the first event that fully demonstrated the permanent new changes that railways brought to the art of war. Perhaps the first thing worth noting in this conflict was the total contrast in rail mileage of the two adversaries. The North, with 20,814 route miles, had a relatively complete network connecting its entire area. The system in the South, totaling 9,819 route miles, was incomplete and fragmented. The North had an even greater advantage in its numbers of locomotives and rolling stock. It also had a more developed system for the manufacture of railroad equipment. While it had a dozen or more locomotive builders, the South had only one. Once the war started, this company halted its engine production and began manufacture of military goods.

The North's four major Atlantic port cities, Boston, New York, Philadelphia, and Baltimore, all had rail connections that penetrated through the mountains into the Midwest. The South had only one equivalent route, the Memphis & Charleston, which went through the Blue Ridge Mountains. Thus, aside from this line, the eastern and western parts of the Confederacy were largely cut off from each other.

The first rail-related military maneuver occurred on April 18, 1861, when a Confederate strike force, led by Capt. Imbolden, moved troops over three rail lines to capture Harper's Ferry, Virginia. Soon, transporting troops and supplies by rail became a common practice on both sides. Such movements supported operations both large and small. Thus, almost immediately, tracks, locomotives, rolling stock, rail personnel, bridges, tunnels, junction points, stations, and other facilities became targets in the conflict. It quickly became a prime military objective for each side to defend its own rail system while trying to inflict as much damage on opposing operations as possible.

The Confederates, under General Stonewall Jackson, showed early brilliance in adapting their tactics to this new reality. On May 23, 1861, Jackson's troops captured 56 locomotives and over 300 cars at Harper's Ferry. The nearest Confederate railroad, however, was some distance away, and only a few locomotives and cars were removed over roads to this line before Jackson was forced to retreat. Thus the remaining engines and rolling stock had to be destroyed. In the first battle of Bull Run, General Joseph E. Johnston's troops were rushed in by rail to bolster Gen. Beauregard's forces, and helped him in gaining a decisive edge. Shortly, the Confederate cavalry leaders Forrest, Stuart, and Mosby became expert in attacking enemy railroads at their most vulnerable points.

During the Civil War, Washington, D.C., was served by a single rail line connecting it to Baltimore. This left the city in a vulnerable position. In contrast, Richmond had four railroads. Thus it could be more easily supplied in the event of attack. Early in the conflict, Washington had been in an even more delicate spot. There was considerable southern support in Baltimore, and it took time for Federal troops to safeguard the city. One Confederate sympathizer arrested at the time was the locomotive builder Ross Winans. He had invented a steam-powered cannon, which authorities confiscated.

While the South had become tactically brilliant in attacking railroads, the North, in turn, acquired a genius for promptly rebuilding them. Herman Haupt, formerly a chief engineer of the Pennsylvania R.R., was recruited in early 1862 to repair a line near Fredericksburg. In rainy weather, and having only untrained volunteers with hand tools, he constructed a bridge 400 feet long and 100 feet high over Potomac Creek in 12 days. This feat brought him to the attention of President Lincoln and soon Haupt was put in charge of what became the Construction Corps. The Construction Corps was given responsibility for repairing, building, maintaining, and operating Union railroads and for destroying Confederate lines. At its greatest strength it employed 17,000 men. Later, Brig. Gen. Haupt was allowed to return to civilian duties and was replaced by Brigadier General Daniel McCallum.

One famous story of the Civil War was that of Andrews' raid of April 12, 1862. On that date James J. Andrews, a Union spy, and a group of 20 Ohio volunteers disguised as civilians, seized a northbound Confederate train on the Western & Atlantic line. Pulled by the locomotive *General,* this train had stopped for breakfast at Big Shanty, Georgia. Andrews' idea in stealing this train had been to return north toward Chattanooga, destroying bridges and tunnels along the way. Unfortunately, his group encountered a number of unscheduled trains that delayed their progress; in the meantime a determined pursuit was begun by the train conductor W. A. Fuller and the engineer Jeff Cain. The two began their effort in a handcar, until they found an old locomotive on a siding. With this they proceeded on to Kingston, where they borrowed a newer engine. This was abandoned at a point where the rails had been broken. Continuing on foot they soon found the locomotive *Texas* and resumed the chase, going at speeds up to 60 m.p.h. As the gap was closing, two cars were released from the *General* to delay pursuit, but these were quickly pushed to a siding. Another car was set on fire and left on a bridge, but this too was pushed aside. Running low on fuel, Andrews and his men abandoned the *General* and set out on foot, but most were soon captured. Shortly, Andrews and seven others were hung.

Another interesting railway incident occurred during the Seven Days Battle in late June 1862. Following a loss at Turkey Hill, Gen. McClellan's troops began a retreat to the southeast. His goal was to gain the protection of Union gunboats positioned along the James River. The July 26, 1862 issue of *Frank Leslie's Illustrated Newspaper* provides more precise details of this story:

> Perceiving the impossibility of saving all the ammunition and stores, Gen. Williams gave orders to Lt. Harry Powers to destroy everything that couldn't be removed. The lieutenant, accordingly, loaded 16 railroad cars with all the shells, torpedoes, ammunition, and other stores, and headed them (on the Richmond & York River line) towards the Chickahominy bridge, where the rebels were scattered all along the line. Everything being ready-steam up-the engineer set off at full speed towards the enemy. The sight was at once grand and novel, the powder, shells, etc. exploding, as the apparently infuriated train of monsters rushed along. As the bridge had been burnt, when the unconscious mass of raving machinery arrived there, the whole train precipitated itself into the river with a fearful crash.

Union trains were sometimes used to transport wounded soldiers back to hospitals. Initially any available cars were used. However, beginning in 1862, passenger coaches were sometimes converted into ambulance cars. These had bunks, lockers, a stove, and seats for medical personnel, and were initially coupled to regular trains. Later, they were sometimes assembled together into special hospital trains. These also had coaches for the surgical staff, and cars serving as kitchen, dispensary, and storehouse. Such trains operated on a number of railroads, including the Orange & Alexandria and Nashville & Chattanooga lines. Following major battles, they sometimes played important roles. After the battle of Gettysburg, 15,000 wounded soldiers were transported by rail to hospitals in Harrisburg, Philadelphia, Washington, and New York.

In 1863, following the defeat at Gettysburg, Confederate Gen. Longstreet moved his troops from the Virginia theater of operations to the west just in time to help win the battle of Chickamauga. This daring maneuver required movement over 10 railroads and covered a distance of about 1,000 miles. Shortly, his Union counterpart, Gen. Hooker, moved his own troops by rail westward to Tennessee.

As time passed, the North gained an increasing advantage over the South in the effectiveness of its railroad system. Perhaps out of concern for states' rights, the Confederacy never developed a clear military strategy for railroads. There was great difficulty in replacing locomotives, rolling stock, and other rail equipment, not to mention personnel. Later in the war less important lines were cannibalized in order to keep other railroads functioning. This, of course, further reduced the South's ability to wage war, and it contributed to the state of economic desolation that was to be the legacy of this conflict.

In 1864 Gen. Sherman left Chattanooga in a campaign to take Atlanta. He skillfully maneuvered past battlefield situations favorable to the Confederates and continually cut rail lines behind them, forcing their retreat. At the same time the Construction Corps kept the rail line to his rear functioning smoothly. After taking Atlanta, Sherman destroyed its rail facilities, and abandoning his own support system, began his famous march to the sea. Along the way two rail lines were destroyed. This not only left Georgia in ruins, but seriously weakened Gen. Lee's position in Virginia, as it greatly reduced the flow of his supplies. Furthermore, it effectively removed the Confederate option of transferring the seat of war from Richmond to a point farther south. Shortly, the remaining rail lines around Richmond and Petersburg were cut off by Gen. Grant's troops, and Lee was forced to surrender.

Besides introducing many new military tactics, the Civil War gave us the first armored trains, hospital trains, and a specialized military rail corps. During this conflict both sides were regularly visited by foreign military observers eager to gain new insights into the art of waging war. Among these was the Prussian military genius, von Moltke, who had previously been a director of the Hamburg-Berlin Rwy. Perhaps as a result of his visit, the Prussian General Staff set up a railway subsection in 1864. This became a full department in 1866. After the war Herman Haupt was invited to London to give lectures on the rail effort, and Brig. Gen. McCallum's official report on U.S. military railroads was published.

In the war of 1866 between Prussia and Austria, railways proved to be a decisive factor in the conflict. Despite a late mobilization, Prussia was able to take advantage of the fact that it had five railways to use in its troop deployment while Austria had only one. Thus Prussia was able to field a force of 250,000 troops along a front of approximately 270 miles, quickly gaining an edge that proved disastrous to the Austrians.

During the Franco-Prussian War of 1870–71 the Prussians took advantage of their rail system to quickly assemble a force of 462,000 men, with horses and artillery, to invade their neighbor. The order to mobilize was telegraphed from Berlin on July 15, and within three weeks, three large German armies had crossed the French border. Prior to this conflict troop movements had been mapped out, and corps commanders were assigned specific instructions for transport of their soldiers by rail. Forty trains a day operated over six different rail routes in carrying troops and supplies to their destinations. However, this rapid deployment did not take place without some difficulty. Rail traffic congestion between Frankfurt and Cologne disrupted the flow of supplies, resulting in brief food shortages at the front.

In theory, the French railway system was adequate for supporting the military. However, no joint military-railway exercises had ever been held. Thus the troop deployment, limited to two routes over the Est Rwy., the Paris-Strasbourg and Paris-Mulhouse lines, was often chaotic and frequently interrupted by conflicts between military and rail personnel. There were long delays in leaving Paris, and at times, due to poor traffic control, soldiers arrived at the wrong destinations. Supplies were also misdirected. In spite of this, troops, artillery, and supplies all managed to get to the front. Once there, however, French troops faced superior artillery and better leadership, and soon suffered a succession of defeats.

Part of the French problem lay in the geography of its border region. The Vosges Mountains separated a portion of Alsace from the rest of France. Thus the rail lines were forced to go around these mountains, with the Paris-Strasbourg going to the north and the Paris-Mulhouse to the south. These routes were separated by a distance of about 100 miles. In July 1870, a party of Prussian lancers penetrated French lines and blew up a viaduct on the Paris-Strasbourg Rwy. This seriously disrupted military communications and prevented the reinforcement of Marshall MacMahon's army at a most critical moment. When MacMahon's troops were defeated at Worth, near Strasbourg, the Prussians blocked his retreat to the northern rail route. Thus his forces were taken out of action for a critical two-week period until they could regroup at Châlons-sur-Marne. In the meantime the Prussians focused on attacking the only other major French army in the region, commanded by Marshall Bazaine.

The advancing Prussians used cavalry to cut French rail connections. They also used captured French railways to their own advantage, finally bringing munitions and other supplies forward to aid in the bombardment of Paris. During this siege, all rail connections around the city were cut, leaving it stranded. Aiding in this task were Prussian *Eisenbahntruppen,* soldiers specially trained in the building, repair, operation, and demolition of railways.

In *Wonders and Curiosities of the Railway,* William Sloane Kennedy describes an earlier incident involving the use of these troops: "When the garrison of the fortress of Metz interrupted railway traffic on the line from Saarbrücken through Pont-à-Mousson to Paris, and by Nancy to Strasbourg, General Von Moltke directed a railroad 25 miles long to be built, uniting the Metz-Saarbrücken and the Metz-Paris lines. Three thousand miners working night and day, amid the roar of cannon, built the road in half a week or less, and the railway communication of the Prussian army was reestablished."

The advancing Prussians were not really prepared for the extent of rail damage they would encounter in the wake of the French retreat. Nor did they anticipate that little rolling stock would be left behind for their use. French forces had strongly defended rail junctions, and in their retreat systematically destroyed tunnels, bridges, and track on a scale much greater than that of the American Civil War. French irregulars, operating behind Prussian lines, also were successful in blowing up a few troop and supply trains. While France lost the conflict, its military was provided with certain useful lessons, and this led to the adoption of railway policies similar to those of the Prussians.

The Prussians felt that their own railway efforts in the war had been far from satisfactory. Thus it was decided to reorganize the existing structure of field railway companies and to raise the overall standard of training. About the same time, France established a corps of railway engineers, setting up its headquarters at the Engineer School in Versailles. Soon also, the Italian government initiated efforts in this direction, attaching two railway companies to each of its engineer regiments.

History of Railways, edited by E. L. Cornwell, suggests that other lessons were learned from this war as well: "As a result of the Franco-Prussian conflict, the British government gave itself the power to take possession of the railways in time of war, and governments throughout Europe began to grasp the basic implication of railway transport—that it made possible the mobilization and deployment of armies unprecedented in size. Mass conscription and a general increase in international tension was the result. Another result was strategic railway building at a pace that might almost be described as frantic. France greatly extended her network along the eastern frontier, neglecting the west; Germany intensified her system to the west, neglecting that of the east; and Russia began to build on a massive scale."

About the same time as the Franco-Prussian War, narrow-gauge railways began to gain popularity. Because there was less expense and effort in constructing these lines, they quickly began to be adapted for strategic purposes. Furthermore, narrow-gauge systems were more adaptable for difficult terrain, with the locomotives, cars, and track being smaller and lighter. Thus they were better suited for military campaigns, as materials could be transported easier to railheads and set up quicker. When France's Decauville portable railway system appeared, it was quickly adopted for military purposes. The Russians under Gen. Annenkoff used Decauville equipment in support of their Balkan campaign against the Turks in 1877–78. It was at this time that Russia's first railway battalion was formed. Later, the general had this material shipped to Mikhailovsk, on the Caspian coast, where he began building the first portion of the Trans-Caspian Rwy.

French armies used Decauville equipment in Vietnam and Madagascar, and the Italian military used it near the Red Sea. Britain adopted this technology after hearing of its use by the advancing Russians in Central Asia. In its order with the Decauville firm, Britain specified that its equipment should be of the same type and gauge as that sold to the Russians. The object of this request was probably to have interchangeable equipment should any Russian lines later be captured. This material was received in the mid-1880s and transported by elephants to several points along India's Bolan Pass. At some point the firm of Orenstein & Koppel began providing similar rail equipment for the German army.

After the Franco-Prussian War, the German army railway corps constructed a 25-mile military line from Berlin to a new artillery range at Zoszen. Besides transporting soldiers to and from the range, the line was used for various training exercises involving both destruction and reconstruction of railways. After the Russo-Turkish War of 1877–78, and the subsequent Congress of Berlin, a chill developed between Russia on one side and Germany and Austria-Hungary on the other. In 1879 the latter two nations formed the "Dual Alliance" and were later joined by Italy and Romania. Russia, for its part, made stronger ties with France. These tensions led to an increase in strategic railway building, especially along the German and Austro-Hungarian frontiers with Russia.

Railroad Gazette, commenting on this development in its October 5, 1883 issue, stated:

> This alliance has borne some practical results in the enlargement of the Austro-German railways, especially near the Russian frontier. Germany has tried very hard to have a double track for every railway likely to carry troops in case of a Russo-German war, and Austria is trying to perfect its railway net so as to make it more efficient, should Russia try to invade Galicia. A railway runs from Cracow via Czernowitz into Romania, and is substantially parallel with the Russian frontier. This road may be thought a rough periphery, the center of which is Budapest, the capital of Hungary. The radii which connects the center with the periphery are not numerous, and at the request of the military authorities the number of them is increased. These radii are to tie Galicia, purely a Slavic country, more closely to the heart of the monarchy from which its people are separated by high mountains, by language, and by the natural preferences which draw them to Poland or Russia.
>
> From Berlin the railways run toward the Russian frontier in the shape of a fan, and are so numerous that a few days would suffice to send hundreds of thousands of combatants to almost any point near the frontier. In fact, taking these military railways and the fortifications together, Russia may consider itself in constant danger, unless its neighbors are above suspicion. Meanwhile Russia cannot but take precautionary measures, and the wonder is that more has not been attempted to place the vast empire in greater security. In Germany and Austria the plans for a war with Russia are complete, and every railway car shows by a legend how many men, how many horses, how much baggage, or how many centals of war stores it is expected to carry.

Fortunately, at the time, the destined conflict lay farther off in the future. For the rest of the century the story of military railways was mainly one of European powers expanding or consolidating their positions in the third world. The British had built a number of lines in Egypt, much as they had in India to further their ambitions there. Perhaps the primary intention was the defense of the Suez Canal, completed in 1869, through which much commerce passed between Britain and India. Later, during an insurrection, British armored trains patrolled this area on a rail line that ran parallel to the canal. Another British goal was the capture of Sudan, which had defeated previous British expeditions led by Gen. Hicks in 1883 and Charles Gordon's force in 1884. In 1885 Lord Wolseley led a fleet of gunboats up the Nile, valiantly crossing over the various cataracts in a desperate attempt to relieve Gordon. Finally reaching Khartoum, the expedition found no sign of him and made a hasty retreat.

Because of other distractions it was not until 1895 that Britain was in a position to launch another invasion. Years earlier, rail lines had been constructed from Alexandria to Cairo, on the Nile, and then up the river as far as Aswan. To support this new military effort, led by Herbert Kitchener, a new railhead was begun at Wadi Halfa. From there a line was constructed 100 miles farther south up the Nile to Akashe. A second line, one of 232 miles, was extended from Wadi Halfa across the desert until it once again touched the Nile at Abu Hamed and continued to Atbara. Both of these lines were of 3′6″ gauge, the same gauge as that used in the Cape. It was there that Kitchener intended to get his locomotives and rolling stock. Directing this effort was a Canadian, Lieutenant Edouard Girouard, who previously had worked on the Canadian Pacific Rwy. The greater portion of this second line was built in 1897. When Atbara was reached the next year, a sizeable gunboat flotilla was launched toward Omdurman, farther up the Nile. There the Khalifah's forces were finally defeated on September 2, 1898.

During the Spanish-American War of 1898, railroads were used to supply the U.S. expeditionary fleet that was being loaded at Port Tampa, Florida. Unfortunately, the existing rail facilities were totally inadequate for the operation taking place. A single track, 10 miles in length, was used to move the main concentration of troops and supplies from Tampa to the embarkation point at Port Tampa. Limited trackage at the piers prevented rapid unloading, and empty cars had to be returned against traffic flow moving forward. Thus considerable delays and confusion resulted.

By this time the U.S. Army no longer had any hospital trains. Anticipating casualties, the surgeon general of the army requested the Pullman Company to assemble a number of sleeping coaches, along with a dining car, for use as a hospital train. This train was put together in Washington and then sent south to Tampa, arriving on June 19. From there it began to carry both sick and wounded passengers to various military hospitals. When this service was terminated on March 4, 1899, the train had traveled 44,110 miles and carried 4,704 passengers.

At the beginning of the Boer War in September 1899, a number of railways connected most of the major cities in the South African region. The Cape Town–Bulawayo line extended northeast from Cape Town to Kimberly in the diamond region, adjacent to the Orange Free State, northward to Mafeking, near western Transvaal, and on to Bulawayo in Rhodesia. The Cape Elizabeth–Pretoria route extended northward from Port Elizabeth in the Cape, through Bloemfontein in the Orange Free State, to Johannesburg and Pretoria in Transvaal. There was a connecting line between these two railways in the central part of the Cape. The Durban-Johannesburg Rwy. began in Natal Province at Durban, extended northwest to Ladysmith, and then went on to Johannesburg. A final route was unique, in that it was out of the immediate reach of British control. This was the Delagoa Bay & Transvaal Rwy., which reached from Pretoria to the port of Lourenco Marques in Portuguese East Africa (Mozambique).

In the initial invasion by the Boers, the targets chosen were all points having rail lines. Among the reasons Kimberly and Mafeking were attacked was to preempt the possibility of the British using the railway to launch a counterstrike against Pretoria. The line south of Bloemfontein was struck for similar reasons. The fourth invasion route followed the Durban-Johannesburg Rwy. into Natal, where the British were forced back as far as Ladysmith. The Boers were successful in the first portion of the war, but failed to take advantage of important opportunities at critical moments.

Shortly a British invasion force under Gen. Buller was shipped from Cape Town to Durban, and then moved by train to Chieveley. Unfortunately for the British, the Boers occupied favorable terrain in this region. Buller's forces fought and lost three successive battles, the last on February 5, 1900, and then retreated. The February–June campaign by Lord Robert's army was much more successful. Kimberly was recaptured, and then Bloemfontein, Johannesburg, and Pretoria successively fell. Following this there was a general belief that the war was almost over.

During this period the British made several faulty attempts to sabotage the Delagoa Bay & Transvaal Rwy. At one point there were three different groups independently scheming to blow up the bridge at Komatipoort. Boer intelligence must have learned of these plans because a strong garrison was soon placed there. Later, on June 16, 1900, a group of irregulars, under the former Prussian officer von Steinaecker, blew up the less important Malelane Bridge. Following this incident, an approaching train failed to see the bridge damage in time, and plunged into the chasm.

Retreating from the British advance, President Kruger of Transvaal established a temporary headquarters in a passenger train near the border with Portuguese East Africa. As the British continued to push forward, Kruger left for Europe and the Boers fled into the interior. Shortly, however, the

Boers began a resurgence, primarily by conducting successful guerrilla raids.

Under the leaders Christiaan De Wet and Louis Botha, operating in the Orange Free State and eastern Transvaal, respectively, British trains were continuously captured or blown up. After one of De Wet's successful raids, he sent a message to the British suggesting that he had acquired enough supplies, but that he would like for them to send some horses. In another of De Wet's strikes, his chief scout, Danie Theron, attacked a train near Leeuwspruit and almost captured Lord Kitchener. In September, Lord Roberts responded by threatening to destroy all farms situated near sabotaged rail lines. Shortly Kitchener began to experiment with another idea that subsequently proved successful. He began to construct small block house fortresses, first with stone, and then, in 1901, using prefabricated metal structures. These were placed alongside railways for defense purposes, and over time were situated at increasingly closer distances to each other, creating fire zones difficult to penetrate.

Other British tactics included the use of armored trains and rail-borne heavy artillery. The earliest trains, equipped only with Maxim machine guns and rifles, were vulnerable to heavy Boer firepower. Beginning in late 1900, quick-firing naval cannons were added to these trains, and they became effective instruments of war. By the end of the conflict, the British were operating about 20 of these trains. In one famous incident of this war, Winston Churchill was captured after his armored train was derailed on November 15, 1899. However, he subsequently was able to escape.

By the end of the century it was clear that the introduction and improvement of railways had totally revolutionized the art of war. In many respects, the military campaigns of Napoleon had not changed much from those of Julius Caesar. Both were limited by the speed of animal-drawn transportation. However, railways, and other technological developments of the century, had totally changed things, making possible the expanded dimension of warfare that found full expression in World War I.

PLATE 258

Scene during construction of Balaclava Rwy., Russia, 1855. The first military railway ever built, it facilitated the English siege of Sebastapol during the Crimean War.

The American Civil War: Top: Federal troops of 28th Pennsylvania Regiment rebuild bridge over Goose Creek near Manassas Gap, Virginia, Manassas Gap R.R., circa April 1862. Bottom: The 16th Regiment of Ohio Volunteer Militia being transported over Tray Run Viaduct en route to Bowlesburg, Virginia, B & O R.R., July 1861.

Top: Union ammunition train explodes amid Confederate patrols before plunging off burned out bridge into Chickahominy River; Richmond & York River R.R., June 30, 1862. **Bottom:** Destruction of bridge at Bull Run, Virginia, by Confederate forces, Orange & Alexandria R.R., circa September 1862.

PLATE
260

PLATE 261

TOP: View in Atlanta as Federal troops begin their evacuation, November 1864. Note boxcars with U.S. Military R.R. markings and engine house to the left. BOTTOM: Scene showing Gen. Sherman's troops destroying track of Macon R.R. between Rough & Ready and Jonesborough, Georgia, circa September 1864.

Armored rail batteries: TOP: One of a number of batteries used by Union forces during the American Civil War. BOTTOM: Battery used in siege of Paris during Franco-Prussian War, 1870. Rail batteries were also used during suppression of the Paris Commune, which shortly followed.

The Franco-Prussian War: Arrival of a train carrying wounded French soldiers, Est Rwy., 1870.

PLATE 263

Top: German war games: Infantry troops receive training in making a night assault on a train, 1875. **Bottom:** Departure of evacuating German artillery unit, Nogent, France, Est Rwy., September 1871.

PLATE

264

PLATE 265

Russian artillery troops loading train at Kitcheneff Station, Moldavia, 1872.

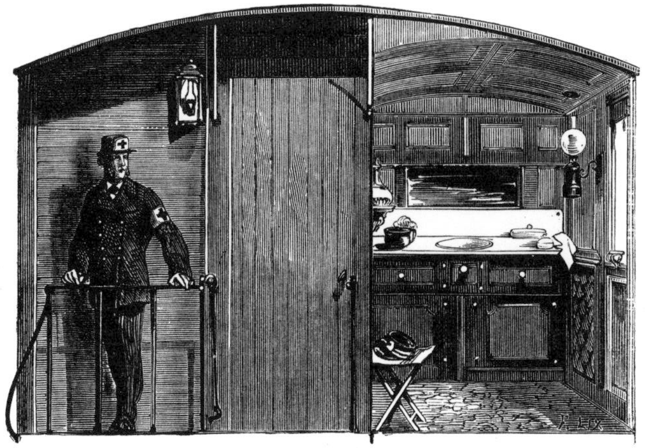

Intérieur de wagon du médecin. — Côté du cabinet de toilette.

Intérieur de wagon du médecin. — Côté du bureau et du lit-couchette.

Intérieur d'un wagon-réfectoire (36 personnes).

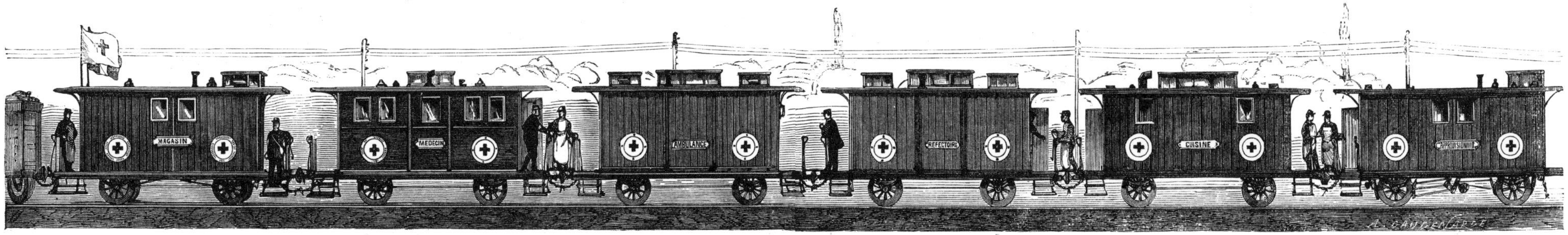

Ordre et composition du train sanitaire.

Locomotive. Wagon-magasin. Wagon-médecin. Wagon-ambulance. Wagon-réfectoire. Wagon-cuisine. Wagon-approvisionnement.

Intérieur d'un wagon-ambulance pour blessés légèrement (44 hommes assis).

Intérieur d'un wagon-ambulance à 10 ou 15 lits.

Intérieur du wagon-cuisine, pouvant alimenter 700 personnes.

French military hospital train exhibited at the Vienna Exposition, 1873.

PLATE 266

British troops transporting portable Decauville railway equipment to railhead site in Bolan Pass on India's northwestern frontier, 1885.

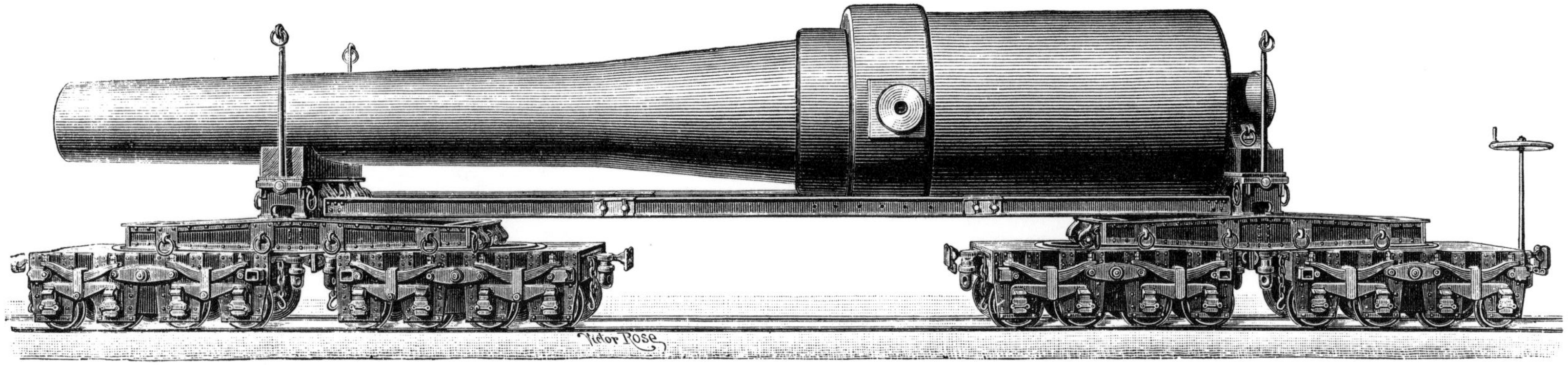

PLATE 268

Top: 48-ton gun mounted on special Decauville artillery wagon, circa 1893. Bottom: Krupp gun on special railway wagon, 1888. This scene shows the massive gun at Essen, Germany, prior to its shipment to the port of Antwerp. From there it was conveyed by ship to Italy.

Construction of narrow-gauge rail line by a brigade of German *Eisenbahntruppen*, circa 1899.

Top: British armored train used during the Boer War in Transvaal, South Africa, circa 1900. **Middle:** British armored train used in Egyptian campaign of 1882. This train carried a Nordenfeldt gun as well as Gatlings. **Bottom:** Spanish troop train in Cuba, 1898.

PLATE 271

The Boer War: British armored train on patrol in the Transvaal, 1899. On November 15 of that year Winston Churchill was briefly captured before escaping after derailment of a similar train on which he was riding.

CHAPTER

21

Hazards and Difficulties of Rail Travel

FROM the beginning rail travel had certain attendant risks. This was especially true in America and other less developed parts of the world. All kinds of things, from the trivial to the tragic, might happen to distract the traveler from the pleasant reveries of the journey. There were always dangers from mechanical breakdown, derailments, boiler explosions, runaways, unsafe bridges, and human error, but nature also provided problems. In an age when few streams or rivers were dammed, floods were frequent and often washed out track or bridges. During winter months, snowstorms could block and strand trains for days. Travel in northern areas, like Canada, Scandinavia, and Russia, could be especially difficult in winter.

As rail lines extended farther into remote areas of the western United States, there were many instances of trains becoming stranded. In the worst cases, crews and passengers alike sometimes had to improvise to stay alive. Following severe blizzards, haunting illustrations of stranded American trains sometimes appeared in *Harper's* or *Leslie's,* and on occasion, were even published in Europe. Various kinds of snowplows were devised in an increasingly successful effort to keep the tracks clear. These types included the wedge, the wing, and the rotary. The Central Pacific R.R. pioneered the use of snow sheds in the Sierra Nevada Mountains. Other lines tried the use of snow fences.

In its January 30, 1875 edition, *Railroad Gazette* gave an interesting account of an arduous journey over the Kansas Pacific route, between Kansas City and Denver. While this journey did not quite fall in the worst-case category, it still provides an intimate glimpse of travel in that period, and also of contemporary railroad practices. This high plains region had been struck by a severe northern blizzard on Thursday, January 7. A number of westward bound trains, having left Kansas City on successive days, were all held up at Salina. On Sunday night a long train departed Salina heading west. It was combined from trains arriving there Thursday through Sunday.

Reaching Russell, 60 miles farther, the train halted as the weather became increasingly severe. The temperature dropped to 22 degrees Fahrenheit below zero, and even those occupying the two sleeping cars had a miserable night. For the next few days, as "the wind blew a gale and the snow was wafted in sheets," the train remained there. Finally leaving on Thursday, it pushed on to Ellis, where eastbound passengers, including a bridal party, were blockaded. Departing at 3:00 p.m. Friday, the train continued to Park's Fort, where it met a snow brigade, consisting of shovelers, a snowplow, and five engines. This group had come eastward from Wallace. As they reported the track to be clear, the train left Park's Fort, but only traveled 2 miles before it ran into a snowbank. It backed up to the fort for the night.

By this time there was no food available on the train or at the fort, but on Saturday morning a relief train arrived from Ellis. It had cooks, cooking fixtures, and an ample supply of provisions. In addition, a special shipment of oysters, in the express car, was made available for general consumption. Later in the day, a snow brigade started westward to clear the track. The train, with a snowplow and three engines at its front, followed behind.

Although frequently encountering snowdrifts, they still managed to travel 80 miles, before stalling 5 miles from Sheridan. It had been growing colder all day, and that night the men refused to work. Instead, they stayed close to the stove and engine furnaces, but resumed their efforts in the morning. By 1:00 p.m. Sunday, the train reached Wallace, encountering two eastbound trains that had left Denver on the 11th and 12th. Continuing on from Wallace, it arrived in Denver at 1:00 a.m. on Monday, the 18th. While this was just one incident, it may have been indicative of a very frigid winter that year. At about the same time, Canada's Toronto, Grey & Bruce Rwy. closed its line to Owen Sound for 44 days.

In the British Isles rail traffic was also vulnerable to winter blizzards. On December 6, 1882, a train on the Welsh narrow-gauge Festiniog line was lost in the snow. Becoming stalled, the train was soon covered by snowdrifts, in some places to a depth of 20 feet. It was only after waiting 36 hours that passengers were finally rescued. In its March 13, 1886 issue, *Illustrated London News* reported on severe rail disruptions occurring the previous week, in the hill districts of both England and Scotland. A number of trains transporting cattle and sheep had become stalled, with many animal deaths resulting. Between Berwick and New Castle a total of 12 trains were blockaded. Among them was a train consisting of 10 passenger carriages and 11 fish wagons. It had left Edinburgh on Monday evening and did not arrive in Newcastle until Thursday. Among its first-class passengers were the Marquis of Tweeddale, Lord Elgin, and Gen. Hutchinson, the government railway inspector.

At times trains encountered other natural forces, including prairie and forest fires, severe winds, hurricanes, floods, sandstorms, and probably lightning. The most spectacular railroad damage in American history occurred on May 30, 1889, in an event known as the Johnstown Flood. This happened in the Allegheny Mountains of Pennsylvania, when the dam at Conemaugh Lake, then the nation's largest reservoir, broke under the pressure of a torrential downpour and sent a wall of water, 40 feet high and half a mile wide, down the Conemaugh River. It inundated the main line of the Pennsylvania R.R., which ran alongside it, sweeping away track, locomotives, and cars, destroying rail yards and a roundhouse at East Conemaugh. The railroad bridge at Johnstown, built of solid masonry, held, but it became a collecting point for all kinds of gruesome debris.

Sandstorms and high winds occasionally posed risks for the traveler. In its issue of January 19, 1883, *Railroad Gazette* reported on a train in the Mojave Desert that had gone through a major sandstorm. The storm had begun in the early morning, and by the time the Southern Overland Express had reached Sumner, in Kern County, it was raging.

> The wind swept across the sandy wastes with such fury that the train swayed and rocked under the violence of the blasts, and seemed ready to plunge from the track. The moon had become overcast in the early part of the night, and the journey was continued in a darkness that rapidly increased until the day began in Stygian gloom. The passengers, who had been aroused from their sleep by the fierce assaults of the wind, and the dashing of the sand against the windows, looked anxiously for the appearance of the sun, but no gleam of light relieved the forbidding darkness of the east. Night maintained her sway, and the blackness of the heavens grew intense with the morning, until the strong headlight of the locomotive almost failed to pierce it. The small portion of the desert exposed by the engine's lights only served to discourage the travelers. The track was lost under the billows of sand that were being tossed across the rails by the angry storm. The desert moved like a sea, and when the waves of sand struck the shivering sides of the train they scattered like spray and filled the air with a dust, which made free breathing impossible.
>
> The travelers' fears of being stopped by a sand drift were soon realized. After leaving Sumner, the train moved cautiously for ten miles through the shifting waste and then stopped with a crash. The alarmed passengers hardly dared to face the driving storm to learn the cause of the unpleasant halt. The few intrepid persons who ventured into the blinding tempest found that their express train had run into a freight train, which had stopped in an impassible sand drift.

The train's slow rate of speed enabled the engineer to stop in time to prevent a serious accident. The collision was only sufficient to knock the locomotive from the track. The passengers remained in the detached train for five more hours until a relief engine arrived. It pulled the train back to Sumner, where the passengers waited out the storm. After an additional delay, caused by a broken piston rod, the train arrived at Oakland Wharf, 30 hours behind schedule.

On the afternoon of April 6, 1886, a passenger train of the Port Huron & Northwestern R.R. was passing over the flats near Frankenmuth, Michigan. At this point a force of wind struck it with such violence that all the cars were blown from the track and fell in a ditch. Only the locomotive was left on the rails. The cars were wrecked and three trainmen and seven passengers were seriously injured.

In October 1890 a Chicago & Northwestern passenger train arrived in Fargo, North Dakota, just in time to encounter a tornado. Because heavy winds were removing the roofs of the freight house and machine shop, the engineer thought it best to avoid stopping at the station. However, the train was compelled to stop at a crossing with the Chicago, Milwaukee & St. Paul line. Having stopped, it started, with great difficulty, to proceed again. Moving slowly, the train's three baggage cars and nine heavy sleepers suddenly blew over, leaving only the engine and tender on the track. The speed of the train was so slow that none of the passengers were seriously injured; however, the experience was very frightening for all concerned.

Earthquakes, landslides, and avalanches could also threaten rail travel. One such incident occurred near Potenza, Italy, on October 21, 1888. A train was returning 400 people from festivities in Naples when it was struck by a tremendous landslide. Ten cars were crushed, and casualties were listed as 19 dead and 55 injured.

On rare occasions trains were known to either catch fire or happen upon one. In the latter case these were usually of two types: naturally caused prairie or forest blazes, or fires along the permanent way, usually ignited from cinders, cigarettes, or other burning debris. Especially vulnerable to sudden combustion were wooden trestles and snow sheds. The latter were erected by the Central Pacific and Canadian Pacific companies to provide protection from avalanches and snow accumulation. However, in dry summer months they were fire hazards. The story of one such blaze appeared in the December 1872 issue of *Industrial Monthly.* Told in rather colorful language, and perhaps only a "tall tale," it was attributed to the *Gold-Hill News* of Nevada.

> Johnny Bartholomew, one of the best engineers that ever mounted a locomotive, had on this particular occasion happened to change off with Charley McCormick, the regular engineer of the "Reno," and therefore was the hero of this occasion. There was no trusty watchman or anybody else to give warning of the impending danger, and the train was around the curve and into the tunnel before the fire, which was raging there, could be seen.
>
> The tunnel is 650 feet in length, and, the fire being at the opposite end, there was little time for thinking, or in which to weigh chances, and certainly not enough for stopping. The deep-red roaring flames and dense black smoke were ahead, and it looked like attempting a mad rush through Tophet; but there was no other way for it. The lives of the passengers and all hands were in the care of Johnny Bartholomew, and nobody knew it better than he. His first impulse was to reverse the engine and whistle down brakes; but his quick conclusion was that that wouldn't do, as he would be liable to fetch up in the very midst of the raging flames; so, with a resolute jerk of the lever, he "threw her wide open." The "Reno" gave an angry snort, and rushed madly forward like a freed racer, with whirlwind speed, right through the fiery mass. Johnny held fast to the lever, closed one eye, held his breath, and tried to think of some prayer, for he knew that the chances of life and death were about equal just then. He imagined he could feel the hot rails bending like lead under the wheels, and both felt and heard the blazing timbers, and other fallen obstructions on the track, being dashed aside by the trusty cow-catcher. The women passengers screamed, and even the stoutest hearted men quivered with sudden fear as they dashed through the fierce, hot light; but in another instant the gauntlet was passed, and they safely slowed down in the cool gray morning daylight beyond.

In the early 1880s a train fire occurred on America's Buffalo, Corry & Erie R.R. This train was of the mixed type, consisting mainly of freight cars but having two passenger coaches in the rear. Immediately behind the engine were six cars carrying 50,000 gallons of oil. One of these ignited just before the train began descending a long grade. When the whistle signal was given the passenger cars were uncoupled from the rest of the train and their brakes set. The oil cars were also uncoupled but no one set the brakes. As a result these followed the engine down the incline, gaining speed the whole way. Twelve miles downhill was a switch for a branch line where the endangered train could be diverted to an ascending grade. Luckily its frantic whistles were heard and the switch made in time. It traveled over a mile farther before finally coming to a halt.

Railway passengers were also vulnerable to criminal activity. While we may immediately think of train bandits, like Jesse James, robbing an express car, these activities were more often directed against individual passengers and go back to the earliest days of rail service. The small compartments of English-style passenger coaches were ideal for criminals, in that they were afforded a certain degree of privacy while victimizing fellow passengers. Over the years, many crimes, from pickpocket thefts to rapes and murders, were committed in such circumstances. In *Wonders and Curiosities of the Railway,* William Sloane Kennedy commented: "In looking over the index of the London *Times,* the writer of these pages was astonished to find that during a period of twenty years, there was not a single year in which many outrages, attempted murders, and attacks by madmen were not reported to have occurred in the closed compartments of English cars." Kennedy went on to report that when known pickpockets traveled on trains, the police placed them away from other passengers in separate locked compartments.

On July 9, 1864, in a first-class compartment of a North London train, Thomas Briggs was robbed and murdered by Franz Muller. His body was thrown from the train with no other passengers taking notice. After this incident came to light, the London & South Western and a few other lines cut circular portholes in the partitions between compartments to discourage future crimes of this sort.

One frightening incident, occurring in a compartment,

was reported in the August 23, 1877 issue of *Railway Age.* Quoting London's *Times,* the article stated:

> Yesterday a terrible scene occurred in one of the carriages of the up Scotch express on the Midland Railway just before arrival of the train at Bedford Station, which it is timed to reach at 6:41 a.m. A pauper lunatic, named Richard Donnelly, about 5′6″ in height and of considerable muscular strength, was being conveyed from Glasgow to his place of settlement, St. Luke's, London, in charge of Mr. Charles N. Lang, workhouse master of the former place. It appears that when nearing Bedford the lunatic suddenly attacked Mr. Lang with a razor as the master was in the act of reading a newspaper. Mr. Lang immediately warded off the attack with his right hand, but the assault was so sudden that in the endeavor to close with the madman he received a gash across one ear, and also across the hand between the thumb and forefinger, from which the blood spurted around. He then closed with his assailant, whom he overpowered, not however, before his coat collar and back portion around his waist had been cut with the razor, the thickness of the material fortunately preserving him from further wounds. The lunatic now finding himself foiled, suddenly drew the blade across his own throat, which he severed in a most frightful manner from ear to ear, falling immediately to the floor in a pool of blood. The train had now entered the station, where the officials took prompt steps to secure medical aid. Mr. Sharpin, surgeon, was speedily in attendance, but the madman had died in the railway carriage, which was all besmeared with blood.

The illustrated newspaper *Le Petit Parisien* was one of a small number of publications in the nineteenth century that thrived on stories of scandal, misconduct, and tragedy. Its relatively high-quality illustrations were sensational, were frequently lurid, and left little to the imagination. In its issue of June 27, 1897, the paper reported on a scandalous murder that had occurred on a passenger train in Bulgaria. After the body of a young Hungarian woman was found on April 21, in Philippopoli, an investigation was begun which soon revealed a fascinating story. The woman was the mistress of the senior military aide of Prince Ferdinand of Bulgaria. This gentleman had wedding plans with a woman of more suitable social status, and was trying to disentangle himself from his lover. When these efforts failed he turned to a friend, the prefect of police in Philippopoli, to arrange for her murder. Shortly after she boarded a train in Sofia, her assassin, a policeman, found her in her cabin, chloroformed her, and threw her from the train. All three participants were indicted in the murder.

America's famous rail robberies were probably inspired by military attacks on trains during the Civil War. By this time many shipments of money, gold, and other valuables, which previously had traveled by stagecoach, were going by rail. Perhaps the earliest incident was one reported in the February 27, 1864 edition of *New York Illustrated News.* A Baltimore & Ohio night express had left Baltimore's Camden Station for Wheeling. At Kearneysville Depot, near Harper's Ferry, it was surrounded by an armed band who entered the cars and robbed the passengers. These intruders were initially thought to be Confederates. However, subsequent investigation revealed them to be "penitentiary birds" from the loyal states, and some were arrested. A more famous incident occurred on May 5, 1865, when a train on the Ohio & Mississippi line was derailed near Cincinnati. The thieves, believed to be former Confederate irregulars, gathered the loot and escaped across the Ohio River in small boats.

Another Ohio & Mississippi train was robbed on October 18, 1866, near Seymour, Indiana. In this heist, the bandits entered the express car from the passenger coach immediately behind it and took $13,000. About a year later, the Reno Brothers Gang hit a train in the same area. Shortly afterward, John Reno was captured and went to prison. Although this group was suspected in the Seymour robbery, no one was ever tried. The remaining Reno brothers, and their gang, preyed for a while on trains of the Ohio & Mississippi and Cincinnati, Hamilton & Dayton lines. Later, three gang members were captured by lawmen, but subsequently were dragged away by vigilantes and hung. In November 1868, three Reno brothers and an accomplice were captured and jailed at New Albany, Indiana. They also became victims of vigilante justice.

The most famous American train bandit was Jesse James. Jesse and his brother Frank were born in Missouri, and were sons of a Baptist minister. During the Civil War they served with William C. Quantrill, an infamous leader of Confederate irregular forces. In 1870, the James brothers formed a gang with other outlaws, and robbed banks, stagecoaches, and, beginning in 1873, trains. These operations were mainly carried out in Kansas and Missouri, although the gang occasionally ventured elsewhere. The first train holdup occurred at Adair, Iowa. Subsequent robberies included hits on a Kansas Pacific train at Muncie Siding, Kansas, in 1874, and on an Iron Mountain train at Otterville, Missouri, in 1876. Between robberies, the gang hid out with relatives and friends in Missouri. At home they were quiet, peaceful, and generous, and Jesse acquired a sort of "Robin Hood" reputation. While successfully staging robberies for over a decade, the gang lost some members, including the Younger brothers in 1876. They were wounded in a bank robbery and sent to prison. Jesse was killed in 1882 by two new accomplices, the Ford brothers, who decided to collect his reward bounty.

In railroad heists, various methods were used to halt the train or otherwise gain access to its valuables. In an incident reported in the August 10, 1883 issue of *Railroad Gazette,* two bridges were set on fire. This took place west of Fort Worth on the Texas & Pacific line at its Brazos River crossing.

> The engineer of the east bound through express from El Paso, shortly after midnight, discovered the fire just in time to stop within a few yards of the structures. The stringers leading to the main abutments were almost burned through, and had the train gone on, it would undoubtedly have been hurled to the bottom of the canyon, nearly 100 ft. below. The engine was detached, and ran across the western bridge to a tank some distance beyond, where it obtained water, and running back, put out the fire with the hose. No sooner was this done than the eastern bridge, over which the engine had not passed, was also seen burning. This fire was soon extinguished, and after two hours hard work both bridges were repaired, and the train passed over safely. Parties who intended to rob the train were deterred from making the attempt by the discovery that a detachment of U.S. troops, and 8 or 10 Texas Rangers were on board, who showed themselves when the train stopped at the first bridge. Several mounted and armed men were seen to emerge from a clump of bushes nearby and ride rapidly away into the mountains.

A Mexican train robbery was described by the same publication in its November 23, 1883 issue:

> A bold train robbery was perpetrated at 7 p.m. last night on the Mexican National Rwy., 20 miles south of Laredo. As the train from Saltillo, bound for Laredo, rounded a curve on a lonely spot, not far from La Jarita, a small station, the engineer discovered that a rail had been removed from the track. He applied the brakes and reversed his engine as soon as possible, but too late to save the train, which was thrown from the track. The engine and the first two cars were wrecked, and others were somewhat damaged. The fireman was killed, falling beneath. The engineer was considerably bruised. Simultaneously, as the train left the track, about 40 masked men issued from adjoining woods and made an attack. They compelled the conductor to give up the money in the express car, $6,000 in silver, and were about to relieve the passengers of their valuables, when it became known that one of the passengers had escaped and gone to La Jarita for assistance. The robbers thereupon took fright, returned to their horses, and retreated into the woods.

On May 31, 1891, a large Greek bandit gang robbed the legendary Orient Express, derailing the westbound train near the Turkish village of Cherkes Keui, about 60 miles from Constantinople. They stole a large amount of money and luggage and also took with them about 20 hostages whom they planned to hold for ransom. The train was wrecked in a remote location, and after the bandits' departure it was some hours before the rest of the world knew of the event, and longer still before a relief train finally arrived at the scene.

After striking an area of damaged track, the engine of the ill-fated train had fallen down an embankment, dragging with it the tender, a baggage van, a third-class local carriage, and a second-class coach. Remaining on the track were three first-class Wagons-Lits Carriages. The bandits first rounded up all the Wagons-Lits employees, 15 in all, and bound them. The first-class passengers were then forced from their compartments to receive instructions from the bandit leader, Anasthatos, a roughly charismatic giant with a black flowing beard. The thieves then rifled through all the passenger possessions, gathered hostages (among whom were some prominent German businessmen), and then rode away. The estimated value of the loot was $160,000.

Ransom messages soon arrived, carried by two released hostages. These demanded the sum of 8,000 gold English sovereigns with the threat that the captives would be shot if the amount was not delivered. This quickly resulted in an international incident as the German government put strong pressure on the Ottomans to pay the ransom. It even went so far as to talk of sending German troops into Ottoman territory to find the bandits. This strong-arm maneuver forced the Ottomans into compliance, and soon negotiations were held with the bandits. After some further delays the ransom money was delivered and the hostages released. They soon arrived in Belgrade to a hero's welcome. The bandits made good their escape and were never found, nor was any loot or ransom money ever recovered.

An efficient lone wolf operator was Oliver Curtis Perry, who successfully robbed the express car of a New York Central train on September 30, 1891. He tried to duplicate this feat the following February, and in the process wounded the express messenger. When the train made a regular stop at Lyons, New York, he tried to make his escape, but was recognized and pursued. Having previous railroad experience, he uncoupled the first car of a waiting coal train, pulled his gun on the crew, and made off with the engine. Armed railway employees uncoupled the express engine, and gave pursuit on a parallel track. The faster passenger engine overtook Perry, and shots were exchanged. However, no one was hit.

At times the engines ran side by side, and Perry sometimes tried to disrupt his pursuers by reversing wheels. However, because he had no time to shovel coal, his steam gave out, and he was soon captured.

Although American railroads became increasingly effective in preventing robberies, these still remained somewhat common to the end of the century. One group of train bandits operating at the time was the "High Fives," or as they were later called, the "Black Jack" Gang. This group operated in Texas, New Mexico, and Arizona, and its members were all young cowboys. The original leader, Cole Estes, was killed in 1896 during a robbery attempt on an Atlantic & Pacific train west of Albuquerque. Later, after another robbery, Sam Ketchum and other accomplices were pursued by a posse. Ketchum was wounded in the shoulder, and after being jailed, died of blood poisoning.

Ketchum's brother was Tom "Black Jack" Ketchum. On August 16, 1899, he appeared alone at Folsom, New Mexico, and gambled away about $1,000. That night he held up a Colorado & Southern train, apparently after somehow hiding on the tender. After the train stopped, Ketchum made some effort to have the engine and tender disconnected from the train, wounding the mail clerk in the process. As two train holdups had recently occurred in the area, the conductor had brought along a double-barrel shotgun. In an exchange of fire, he wounded Ketchum, who staggered off into the darkness. The train departed, and the following day, Ketchum was captured in the immediate area of the holdup. It took a while for him to recover, and also for criminal evidence against him to be assembled. After conviction, he was hung at Clayton, New Mexico, in April 1901. This hanging was notable in Western annals, because the rope was too long for Ketchum's weight and he was decapitated.

Another risk of rail travel was that of becoming a victim of mischief or sabotage directed against railways. In America, such occurrences were more common than is generally known today. An early incident of this type was mentioned in the March 12, 1859 issue of *Frank Leslie's Illustrated Newspaper*. It reported that passenger trains leaving Charleston, South Carolina, received rifle fire from an unknown party on two successive days. In both instances, passengers were narrowly missed.

During the 1870s public hostility increased toward railroads, and malicious acts against them rose in number. In its edition of November 2, 1872, *Railroad Gazette* reported 11 occurrences of this type. These had all happened in the 3 previous months, and resulted in 5 accidents, 3 deaths, and 4 injuries. In one instance, a drunken man on the Newark & New York line was put off the train for failing to pay his fare. He took revenge by throwing a switch at West Bergen, hoping to cause an accident. Fortunately, he was caught in time. A second incident occurred on the Delaware, Lackawanna & Western R.R., when a track walker found the bridge-tender's cabin unlighted. Retracing his steps to alert railroad authorities, he found large stones piled on the track. As a special train was nearly due, he tried removing the stones, but found them too heavy. Returning to the bridge-tender's cabin, he found the tender tied up inside, with a bag over his head. He had been seized and bound by six men, who evidently placed the obstruction.

In its February 8, 1877 edition the *Railway Age* reported on another event of this type:

> A lunatic did his best to wreck a train at Waverly, on the Erie R.R. last Saturday night. About ¾ of a mile east of the town there is a bridge over a creek, and a few rods further on there is a switch. He forced open the switch by breaking the lock, pried the rails out of place, and left large spikes standing upright. The locomotive baggage car, smoking car, and some of the passenger coaches left the track, but remained upon the embankment, which is about 40 ft. high. Investigation showed that other obstructions had been placed upon the track further along. Five persons were arrested on suspicion, and among them was a young man named Samuel J. Bailey, about 19 years old, who confessed his guilt. He had lost his reason through religious excitement. He said that he had been reading about the late P.P. Bliss and the Ashtabula railroad calamity, and he had a great desire to witness a similar disaster. To gratify himself, therefore, he had destroyed the switch, and placed other obstructions upon the track, and then seated himself upon a freight car to enjoy the scene.

People had various other reasons for attempting such acts of sabotage. The May 4, 1883 issue of *Railroad Gazette,* provided some insight into what was behind one incident. William Purselly, his wife, and their son were arrested 15 miles east of Springfield, Missouri, and charged with placing obstructions on the track of the Kansas City, Springfield & Memphis line. It was alleged that this family had made other unsuccessful attempts to ditch trains on the road. They were angry at the railroad, as their farm, which had lain in its path, had been seized during construction under the rule of eminent domain. As required by law, the line had deposited settlement money with the appropriate authorities. However, it was not mentioned in the article if the Purselly family had received this money, or whether this compensation was adequate.

Other dangers involved sticking one's head or limbs too far out a window or moving between cars. In one instance on the Erie line a man was decapitated when he looked out a window to view some passing object. Unfortunately, he failed to see an obstruction immediately ahead. The May 27, 1892 issue of *Railroad Gazette* reported on a passenger who was blown from the platform into a ditch while trying to move from the smoking lounge to the rear coach. This happened near Mason City, Iowa, on the Chicago, Milwaukee & St. Paul R.R. His injuries were fatal. This type of accident must have occurred with some frequency until platform gates, and later, vestibuled cars were universally adopted.

Another type of risk involved crossing tracks to either board a passenger coach or make an exit. In one such incident, reported in the October 8, 1899 edition of *Le Petit Parisien,* a woman and her young child disembarked from a carriage only to step into the path of an oncoming train. She passed her child safely to an onlooker but was unable to save herself. This accident occurred at the Belleville station on Paris's Ceinture Rwy.

A final concern for the railway traveler was the occasional potential for encountering rowdy individuals. The following story, appearing in the January 2, 1885 issue of *Railroad Gazette,* relates a tale about a passenger train that was terrorized by drunken cowboys. On December 23, 1884, an eastbound train on the Southern Pacific line was approaching the bridge crossing the Pecos River in Texas. A number of torpedoes exploded on the track, causing the train to stop.

> Six cowboys boarded the train, and with drawn revolvers compelled the conductor to stop at Langtry, the next small station. At Langtry 20 more cowboys took possession of the train. They were all more or less drunk, and amused themselves by shooting out all the lamps, and terrorizing the passengers. As the train passed small stations cowboys would shove their heads out of a broken window and shoot and yell. A passing freight train was saluted in similar manner. Several hundred shots were fired, and the roofs of the cars riddled. They robbed the train boy of his stock of peanuts and candies, but injured none of the passengers, and did no other damage beyond breaking a few windows. The gang left the train at Uvalde, after riding 80 miles.

Top: A train between Nottingham and Lincoln derails in a flood near Newark, October 22, 1875. Bottom: British train trying to proceed through debris of a storm, 1884.

Stalled train in a flood in Alabama, 1886.

PLATE 273

PLATE 274

Scene of a "washout" on line of New York Central & Hudson River R.R., 1885. While repairs are being made, passengers and baggage are being moved from one train to another.

TOP: Trying to clear a stranded train at night, greater London area, 1854. BOTTOM: Attempt to free a stalled train near Glasgow, 1867.

Clearing a snowbound train near Ogden, Utah, Union Pacific R.R., 1872.

PLATE 277

Trying to dig out a stalled German train, 1883.

Top: Near Valence, France, a PLM express bursts into flames after striking a large container of alcohol that dropped off a previous train, 1884.
Bottom: The great snowstorm in Northumberland: A stalled train near Acklington, North Eastern Rwy., 1886.

Train going through forest fire in cedar swamps of Maine, Eastern R.R., September 17, 1865.

PLATE
279

Top: Unknown sniper firing into train, South Carolina R.R., 1859. Bottom: In a love triangle conspiracy, a Hungarian woman was chloroformed in her compartment and then thrown from train by a Bulgarian policeman. He was acting under orders of her reluctant lover, a top military aide of Prince Ferdinand, 1897.

"The Modern Dick Turpin," 1892. Illustration of a typical American railroad holdup, making reference to an infamous English highwayman of an earlier time who was finally brought to the gallows.

PLATE 281

Fatal accident on Paris's Ceinture line, 1899. At the Gare de Belleville a woman is struck after stepping off a carriage into the path of an approaching train. Her child was saved.

PLATE
282

CHAPTER

22

Rail Accidents and Disasters

As England led the way in the development of railways, it also was the first country to experience train accidents. These occurred for a number of different reasons. Locomotive *boiler explosions* were an early problem, until engineers made refinements in their construction. *Derailments* could be caused by open switches, obstacles in the tracks, loosened or broken rails, or broken wheels or axles. *Head-on* or *rear-end collisions* might result from faulty dispatching, improper signaling, brake failures, poor visibility due to bad weather conditions, or a failure to pay attention. *Side collisions,* which were very rare, could occur where rail lines crossed one another. An additional hazard existed at road crossings, where vehicles, people, or animals might be encountered. *Bridge collapses* and *train runaways* were other sources of potential disaster. Once a major accident occurred, there was the additional danger of fire setting in.

The first recorded accident with fatalities took place July 31, 1815, when a boiler exploded on William Brunton's *Steam Horse* at the Newbottle Mine near Newcastle. It killed three instantly, including a small boy, and injured about 50 others. In 1828, on the Stockton & Darlington line, an engineer was killed when the boiler of Stephenson's *Locomotion* exploded. On September 15, 1830, at the inauguration of the Liverpool & Manchester Rwy., Stephenson's *Rocket,* recent winner of the Rainhill trials, struck William Huskisson M.P., and he died shortly after. On December 3, 1830, a derailment took the lives of three people at Great Corby on the Newcastle & Carlisle line.

America's first fatal *rail accident* occurred June 17, 1831, on the South Carolina R.R. The fireman, an African American, died from injuries shortly after the boiler of the *Best Friend* exploded. The first deadly *passenger train accident* happened November 9, 1833, between Hightstown and Spotswood, New Jersey, on the Camden & Amboy line. Two deaths and 12 injuries resulted when an axle broke on a coach, causing it to overturn. Among the injured was Cornelius Vanderbilt, the future railroad magnate. America's first fatal *bridge accident* occurred on March 20, 1837, when a South Carolina train struck a burned section of trestle, killing the engineer. The country's first *train collision* with fatalities occurred August 11, 1837, near Suffolk, Virginia, on the Portsmouth & Roanoke R.R.

The world's first *major train disaster* took place on the Ouest line at Meudon, France, on May 8, 1842, and claimed 57 lives. The cause of the accident was a broken driver axle on the lead engine of the double-headed Versailles-Paris Express. In the derailment that followed, passenger carriages piled up against the engines and caught fire. Many of the passengers, including several important figures, died because they were locked in their carriages from the outside. This was the normal procedure at the time, but it was subsequently outlawed. In 1846 a train derailed on the Nord Rwy. at Fampoux, and fell into a river.

A survey of rail accidents in France determined that from the introduction of railways through the end of 1853 there had been a total of 1,869. Most of these were minor, and involved no injuries or deaths. However, among the more serious, a total of 642 people were killed and 1,112 injured. A similar study indicated that between 1835 and 1856, 224,345,769 passengers had ridden French trains. It determined that there was only one death per 2,021,133 passengers and one injury per 558,071. Similar figures for horse-drawn coaches indicated one death per 355,453 and one injury per 29,871.

An unusual freak accident occurred in April 1841, on the Birmingham & Gloucester Rwy. A Norris-built engine, the

Boston, blew a tube plug, which killed the company's locomotive superintendent. It was later determined in England that during the 1840s, 56% of railway accidents resulted from broken wheels and axles, 18% from broken rails, 14% from open switches, 5% from obstructions, 3% from cattle on the track, and 1% from boiler explosions. The November 1887 issue of *Railroad and Engineering Journal* reported another type of rare incident: a boiler explosion taking place while the engine was in motion. This occurred on April 7, 1856, when an engine on the Caledonian line exploded near Carlisle, killing both the driver and stoker. "The crown-sheet of the firebox gave way, and the engine was lifted up from the track and turned over six times."

Despite a number of accidents, early American railroads had surprisingly few passenger fatalities. This has been attributed to the light traffic, slow speeds, and infrequency of night activity at the time. It was not until 1853 that a number of growing trends, increasing the risk of accident, finally coalesced in train disasters two weeks apart, which claimed 21 and 46 lives, respectively. After the sudden shock of these events, tragic rail accidents became common.

In its issue of April 6, 1872, *Railroad Gazette* listed a number of train disasters occurring during that period. It should be sufficient here to only mention a few:

> On July 5, 1854 a collision between an excursion train of twelve cars and an accommodation train took place on the Baltimore & Susquehanna R.R. The excursion train was being pushed at the time of collision by its locomotive, and some six cars were crushed. Twenty-nine persons were killed and over fifty injured . . . On Nov. 1, 1855 an excursion train of eleven cars on the Pacific Railroad of Missouri, on the occasion of the opening of the road, was crossing the bridge over the Gasconade River, when the bridge gave way, precipitating the train thirty feet into the water. Twenty-two were killed, and some fifty injured. On July 17, 1856 a collision took place on the North Pennsylvania road between an excursion train of ten cars, containing some five hundred children with their teachers, and a regular train. Five cars were burned, sixty-six persons were killed, and over one hundred injured.

Similar accidents began happening in Canada, although less often. The previous *Railroad Gazette* article gave brief descriptions of two: "On October 27, 1854 a passenger train from Niagara on the Great Western R.R. ran into a long gravel train backing toward it. Forty-seven persons were killed and some sixty injured . . . On Mar. 12, 1857 a locomotive ran off the track of the Great Western R.R. while approaching the bridge over the Des Jardines Canal. The train fell through the bridge and crushed through the ice into 18 feet of water. Sixty persons were killed and a large number wounded. In this case it was thought that a self-coupler would have saved the rear cars of the train, which were, however, both linked and chained to the forward cars and dragged over by them."

By this time, in both the United States and Canada, daytime rail activity had increased considerably, and more trains had begun moving at night. Passenger train speeds had increased to a range of 15 to 30 m.p.h., although freight trains continued to run much slower. Both the increased activity and higher speeds put more stress on poorly built and maintained roadbeds, track, and bridges, as well as locomotives and rolling stock.

This problem was not unique to North America, but existed wherever under-financed railways were constructed. By comparison, European lines usually were built through areas having greater populations and longer traditions of commerce. Such railways, especially those in England, were usually better capitalized, and their infrastructure built more carefully, of better materials, and maintained with an intention of permanence. These lines typically handled a greater volume of traffic, and trains were often operated over them at higher speeds. European railways were not immune to accidents, but they occurred with much less frequency than in America.

After America's Civil War ended in 1865, public indignation grew following a whole new series of tragic disasters. One that drew national attention occurred August 16, 1865, on the Housatonic R.R. near Bridgeport, Connecticut. In this accident, a new engine, out on a test run, rear-ended a passenger train backing toward it. The colliding locomotive telescoped through the rear car of the train ahead, at which point its boiler exploded. Although the casualty figures of this accident, 13 dead and 20 injured, seem relatively modest, both *Harper's Weekly* and *Frank Leslie's Illustrated Newspaper* gave the story prominent coverage and showed illustrations of the wreck. During that same month 17 fatal accidents occurred, taking a total of 88 lives.

Another fatal rear-end collision took place November 21, 1867, on the Atlantic & Great Western line near Lockland, Ohio. After a freight train became disabled and stopped at a bridge, the New York Lightning Express was halted immediately in its rear. Unfortunately, another train was following, and before it could be signaled, it hit the Express. The rear coach, a sleeping car, was telescoped and, splitting almost in two, was ignited by the engine's kerosene headlight. A number of passengers perished in this wreck, among them four young sisters.

Before the end of the year, a more famous tragedy occurred near Angola, New York, on the Lake Shore & Michigan Southern R.R. The rear coach of the New York Express derailed while passing over a bridge, became disconnected from the train, and fell 40 feet to the bottom. The car's hot stoves immediately sparked a blaze that consumed most of the remaining survivors. In all, 42 lives were lost. Although hot stoves instigated this fire, many others were ignited by the kerosene lamps used for lighting.

In England, around the end of August 1861, an excursion train on the North & South Western Junction Rwy. struck a ballast train in the process of being shunted onto another track. This happened near Kentish Town, and followed closely on the heels of another fatal disaster occurring on the London & Brighton Rwy. After striking the middle of the ballast train, the passenger engine and several carriages left the track and rolled down an embankment. A crowd quickly gathered and rendered aid when possible; however, 12 people lost their lives.

Following the report of this accident in its September 7, 1861 issue, the *Illustrated London News* made the following editorial comments:

> It is not our custom to minister to any morbid taste for the delineation of horrors; but in this instance we believe that no means should be left untried to fix upon the minds of the people of this country the now obvious and absolute necessity of legislative and administrative interference in the regulation of railways. Fully admitting the principle of self-government in the conduct of our affairs, and deprecating, as a rule, an organized system of government intermeddling in undertakings which are the result of our commercial genius and the natural issues of our national wealth, yet, when it has become clear that the management of railway companies is such that it is inadequate to the due safety of the lives of those who commit themselves to its care, the question arises whether it is not absolutely necessary to create responsibility somewhere. At present responsibility is as shifting as quicksand, and it is impossible to fix it on any person or body so as to produce any hope of a better state of things. With fatal regularity at this period of the year accidents occur on railways, mostly in the case of excursion trains, accompanied by absolute slaughter of passengers, and we have lists of killed and wounded that read like bulletins of great battles. Coroners' inquests follow, in the course of which practiced legal ingenuity is allowed to turn personal or corporate responsibility into a will-o'-the-wisp which evades the most active pursuit, and in the end perhaps some wretched engine driver—if one survives—or some miserable pointsman (switchman) or signalman, has a verdict of manslaughter recorded against him; the money of the shareholders is appropriated to pay a certain amount of compensation—in the case of excursionists probably computed at a minimum rate—and the matter passes into the limbo of nine-days' wonders, leaving the system exactly where it stood . . . The matter has really come to this, that if private companies cannot provide, and cannot, in the strict sense of the term, be made responsible for the safety of persons for the conveyance of whom they have a practical monopoly, we must give up this branch of the theory of self-government and substitute for it the despotic authority of a Government department.

On June 9, 1865, a South Eastern Rwy. crew removed a 42-foot section of track to make repairs on a viaduct over the River Beult. Unfortunately, in directing this work the foreman referred to the wrong timetable, and a flagman was not positioned an adequate distance away for halting approaching trains. As a result, a passenger train en route from Folkestone to London struck the gap, with the engine followed by six first-class coaches plunging off the bridge. Dangling above was one remaining first-class carriage, still coupled to the rest of the train. Among its passengers were the distinguished author Charles Dickens and two female companions—Ellen Ternan, apparently his mistress, and her mother. After escaping the carriage Dickens escorted his friends from the scene. He then returned alone to render assistance to the injured and dying. In all there were 49 injuries and 10 fatalities. Dickens avoided the subsequent inquiry in order to protect his reputation and marriage. However, he was traumatized by this experience and died on its fifth anniversary.

A frightening accident took place on August 9, 1867, when a passenger train of Ireland's Dublin, Wicklow & Wexford line derailed on the Ramscalp Bridge and plunged off its side. This occurred by the sea at Bray Head. Fortunately, while a few carriages were dragged off by the engine, others remained on the track, and only a few lives were lost. Had the locomotive fallen off the opposite side it probably would have carried the entire train into the sea. On August 20, 1868, the London & North Western's *Irish Mail* struck two goods wagons loaded with paraffin that were rolling backward down an incline. The resulting blaze consumed the first three passenger carriages and claimed 33 lives.

On March 17, 1869, the czarevitch of Russia boarded a special train at Kursk, which then departed for Kiev. While en route, the train passed over a wooden trestle bridge at the Staminj River, and then continued on its way. In the 50-minute interval between the czarevitch's train and the express that followed, a portion of the bridge gave way. As the express approached it, the engine driver noticed the problem ahead, but could not stop in time. The engine, the tender, and a baggage wagon plunged into the river. In jumping, the driver caught hold of a board, leaving him suspended over the water. After his strength gave out, he fell into the river, but was shortly rescued. The stoker and baggage employees perished. The 300 passengers were luckier, only suffering some injuries and the loss of their luggage. In the subsequent investigation of this accident, evidence of sawing was discovered, suggesting sabotage. However, it was determined that light building materials and poor maintenance also contributed to the accident.

In America on the evening of August 26, 1871, the Eastern line's Beverly accommodation train departed from Boston, a half-hour late. Leaving the Revere Station, 6 miles distant it was struck from behind by the Bangor Express, traveling at full speed. The rear coach was telescoped and broken lamps in the forward part of the train ignited, causing a conflagration. Two prominent clergymen were among the 30 fatalities and there were 29 injuries.

Figures for Russian railway accidents in 1874 are interesting. Besides listing 184 employee, 9 passenger, and 104 "other" deaths, it was reported that, while traveling, 14 people died of natural causes and 5 others committed suicide on the tracks. In all, a total of 261 passenger train derailments and 80 collisions occurred. At the time the country had about 10,725 miles of track, upon which 25,000,000 passengers rode annually.

Austrian statistics for the same year showed that the country had 5,950 miles of track on which 1,620 accidents occurred. In these, 107 railway employees and 86 others perished. Among the latter, it was thought that 31 were suicides. It was also reported that accidents damaged or destroyed 105 locomotives, 15 tenders, 24 passenger carriages, and 415 goods wagons.

A semaphore signal, frozen in place, was the primary cause of a twofold accident occurring in early 1876 on the Great Northern line at Abbots Ripton. In this instance a southbound coal train made a late departure from Peterborough toward London during a severe snowstorm. Rapidly overtaking it from behind was the *Flying Scotsman* express. The coal train somehow missed warnings at Holme, and continued to Abbots Ripton, where it was instructed to shunt onto a siding. Stop warnings were set there to halt the *Flying Scotsman*. However, due to the signal malfunction, the express continued on, striking the rear part of the coal train, which had not yet cleared the track.

The engine of the Scotch express jumped the opposite track and fell on its side against the bank. Remaining attached were the tender, guard's van, and about three carriages. These crossed the northbound track and were thus in a very vulnerable position. Efforts were just being made to summon assistance when the northbound Leeds express, failing to notice frantic warnings, plowed full speed into this wreckage.

A brief description of this scene comes from the January 29, 1876 issue of *Illustrated London News:* "The effect of this second collision was most calamitous. The engine literally cut its way through the massive tender of the Scotch train lying across the down rails, glancing off on to the bank with its tender. The leading carriages of the train mounted the wreck of the Scotch express and added further ruin to the already terrible disaster. Here, it is evident, occurred the great sacrifice of life among the passengers of the Scotch express, while no deaths took place among the travelers of the Leeds express." Considering the possibilities, casualty figures for this wreck were surprisingly low: 13 dead and over 20 injured.

A particularly disturbing tragedy occurred December 29, 1876, when a bridge collapsed on the Lake Shore R.R. near Ashtabula, Ohio. Happening during a storm, the accident took the lives of 83 people. The February 3, 1877 issue of *Scientific American* provides the following description:

> The calamity was occasioned by the sudden breaking down of the iron bridge over the creek near Ashtabula station, while the westward bound express train was crossing the structure.
>
> The train consisted of eleven cars, carrying one hundred and seventy-five passengers, and was drawn by the two engines, the "Socrates" and the "Columbia," the former leading. The train had stopped at all stations between Erie and Ashtabula except three, and at the time of the disaster was running slowly. As the first engine was passing over the bridge, the engineer felt the structure slowly settle down. He was then about two car lengths from the western end. In an instant he opened wide the throttle, the drawbar connecting the two engines snapped in two by the sudden jerk, and the "Socrates" shot ahead, while the "Columbia" fell through the bridge, and turned bottom up. The express, baggage, and passenger cars followed, the sleeping-car swinging over to one side, and a moment later catching fire from the stove.
>
> As the engineer of the "Socrates," who alone was in a position to see the disaster in all its terrible details, reports, the entire wreck was a mass of flames in two or three minutes. The engineer of the "Columbia" was thrown head first through the window of his cab, and severely, but not dangerously injured. Fed by the fierce wind, the fire made swift progress, and so lighted up the ravine that the neighboring people who had heard the fall hastened to the scene. The snow finally ceased falling, and a colder wind whistled through the snow-bound ravine. At midnight the fire was smoldering among the ashes and ironwork which it could not burn.

> Ashtabula station was about one-fourth of a mile away. Thither the survivors went as soon as possible, assistance for the injured being rendered by the people of the vicinity and those who went to the scene from the depot. By this means the survivors were lodged in various hotels, when a train with the surgeons of the road on board arrived from Cleveland. Early on the following morning men repaired to the wreck, and began the search for the bodies of those who perished.
>
> But with few exceptions they were either wholly destroyed or burned beyond recognition, except by articles of apparel, jewelry, or the contents of their pockets. Over fifty persons lost their lives, and nearly as many more were badly injured. The depth of the ravine and creek, spanned by the bridge, was 75 feet, and the cars fell that distance, going through the ice to the bottom.
>
> . . . The bridge was a Howe truss, built entirely of iron, at a cost of $75,000, and was eleven years old. It was sixty-nine feet above the water, and had an arch one hundred and fifty feet long in the clear, the whole length of the bridge being one hundred and fifty seven feet. It had been tested with six locomotives, and at the time of the disaster it was considered in perfect condition.

The aftermath of this event was tragic as well. In *The Beauty of Railroad Bridges,* Richard J. Cook reports that the line's chief engineer, Charles Collins, committed suicide a day after the accident. This was apparently in response to the immediate outrage directed against him. This may seem strange, considering that at the time of the bridge's construction, Collins objected to its design and refused to oversee its building. Nonetheless, he was overruled by Amasa Stone Jr., the company's president. Stone, as it turned out, was a brother-in-law of William Howe, the inventor of this bridge style. A subsequent inquiry led to a coroner's verdict against Stone. He was blamed for building a bridge "without the approval of any competent engineer and against the protest of the man who made the drawings . . . assuming the sole and entire responsibility himself." Haunted by this and other reversals, Stone committed suicide in 1883. As a final footnote it might be mentioned that the Lake Shore's owner, the ailing Commodore Vanderbilt, passed away shortly after the bridge crash.

The 1¾-mile bridge, built by the North British Rwy. across Scotland's Firth of Tay, was considered to be one of the great engineering feats of its time. Built in the straight girder style, it had 85 separate spans. During a severe storm on December 28, 1879, almost three years to the day after Ashtabula, this bridge collapsed as a train crossed over it. Plunging into deep water, the entire train with its 78 passengers and crew was lost. This bridge was not replaced until 1887. One interesting sequel to this story is worth noting. After laying on the bottom of the River Tay for three months, the engine was fished up, repaired, and then returned to service.

On June 27, 1881, a temporary wooden bridge on Mexico's Morelos Rwy. collapsed under the weight of a double-headed military train, with the engines and cars tumbling down a precipice. The June 30, 1881 issue of *Railway Age* gives the following description of events: "For a moment after the plunge of the cars from the bridge all was darkness. Suddenly 100 barrels of brandy loaded on two of the vans caught fire and the burning fluid inundated the mass of passengers struggling in the ruins. Only a few, not disabled or hopelessly wedged in the timbers, escaped. Dead and living were wrapped in a sheet of flame and slowly burned before the eyes of the survivors, who were unable to save them." There were differing reports regarding the enormity of the casualty count. However, approximately 17 officers, 197 enlisted men, and a much smaller number of women and children lost their lives.

The *Railroad Gazette* edition of September 28, 1883, gave the following report of a frightening accident that occurred at Steglitz, a village about 3 miles from Berlin, on the Potsdam-Magdeburg-Cologne line.

> The fine Sunday weather had tempted to the place, as to all other suburban resorts, crowds of pleasure-seekers of the poorer class, including some patriotic societies, to celebrate the anniversary of Sedan. At the close of the day the station was besieged by several hundred of the excursionists, eager to get home by the first train from Potsdam.
>
> The station at Steglitz is furnished with insufficient waiting accommodation, and passengers going to Berlin have to cross the rails in order to enter the up trains. Until the up trains arrive, it is usual to detain Berlin-bound passengers on the down platform, by means of a moveable barrier, such as is used at most crossings; and behind this barrier last night, toward ten o'clock, an impatient crowd of 200 or 300 people were waiting for the train that was to take them home to Berlin.
>
> The train arrived, and eager to secure seats, the excursionists either burst, or jumped, or pushed back the barrier poles, and began to stream across the rails like a flock of sheep. This they did in spite of the shouts, remonstrances, and lantern waving of the railway officials—of whom, however, it is said that there was an altogether insufficient number—who saw, to their horror, that the Cologne Express from Berlin was tearing down upon the station at the rate of a kilometre per minute. There was one heartrending and unanimous shriek which silenced even the thunder of the approaching wheels, the heavy engine dashed in among the

wedged and struggling mass, and in another instant the line was strewn with the horribly mutilated and dismembered bodies of some 40 human beings.

For a distance of 100 yards the line was covered with bodies, all shockingly mutilated, and human fragments. Of one family of five persons, only one boy escaped . . . The driver of the express train had to be relieved, being unable to proceed after witnessing the scene of carnage. His locomotive was covered with pieces of flesh, bones, rags, and drenched with blood. The line remained blocked for hours, while detachments of firemen and soldiers gathered the remains of the victims by torchlight.

Belgian statistics from 1886 showed that there had been a great improvement in rail safety since 1880. In 1880 the country reported 124 passenger and 234 goods train accidents. By 1886 these figures had been reduced to 30 and 83, respectively. In 1880 there had been one derailment or collision for every 53,800 train miles, but by 1886 the incidence was only one every 183,500 train miles. Belgium reported only 3 passenger deaths in 1886, but 53 "other" fatalities were recorded.

On July 1, 1886, a head-on collision occurred on the Prussian State Rwy. between Würzburg and Rottendorf. Through a dispatching error, the Stuttgart Lightning Express was routed on a track over which a mail train, running late, was approaching. A report from the August 6, 1886 issue of *Railroad Gazette* provides the following details:

> About two miles from Würzburg the road crosses the Faulenberg by a very sharp curve. On this account it was impossible for the enginemen to see the train approaching them in time; but the engineman of the express, which was ascending a grade, was able to bring it nearly to a stop before the collision, while the mail, which was going down hill, could not be checked.
>
> What occurred then defies description. With a crash like a thunder clap, but much louder, the two trains came together. Some idea of the violence of the collision may be formed when it is known that the two engines made a single snarl of ruins. Naturally the mail train fared worst. The tender with the baggage car shot up into the air, turned over, and fell upon the engine, with the wheels on top. The first two cars were literally reduced to splinters. The front compartment of the fourth car was broken in, and the other cars more or less damaged and thrown from the track . . . The sight which offered itself to the survivors was indescribable and unforgettable. The chaos of ruins, the cries of children for their parents, and of parents for their children, the horribly mutilated corpses, the frightful wounds of the injured.

This collision resulted in 13 fatalities and 37 injuries.

In 1887 the United Kingdom reported only one major passenger train accident. This occurred at Hexthorpe on September 16, and took 24 lives. By this time a number of changes, designed to improve safety, were beginning to have their effect. This included widespread adoption of the interlocking block system, the construction of stronger rolling stock, permanent way improvements, and better braking systems. However, employee deaths remained somewhat high. This included 99 struck on the permanent way; 93 hit while walking, crossing, or standing on the line during duty; and 8 through accidents involving moving trains. In addition, 336 others lost their lives when struck by trains, among these, 70 suicides.

While 1887 was relatively safe for the United Kingdom, it was not for the United States. That year a number of major catastrophes occurred. Included among these was the worst American disaster since Ashtabula. This happened August 10 on the Toledo, Peoria, & Western line near Chatsworth, Illinois. There, a double-headed 15-car excursion train, en route to Niagara Falls, crashed through a burning trestle. The train, which was carrying 950 people, promptly caught fire. Apparently the engineer of the lead engine saw the fiery bridge about 500 feet away. However, he was in a freight engine, unequipped with either air brakes or driver brakes. Thus the engineer in the second engine had insufficient warning for applying the air brakes of the passenger cars. It was perhaps fortunate that the drop into the dry creek bed was only about 15 feet. The death toll was 82.

On January 4, 1888, two fatal collisions occurred on separate continents. At Meppel, Holland, a collision of express trains resulted in the deaths of 26 and injuries to many more. On the Canadian Pacific line near White River, Ontario, two freight trains met in pitch darkness on a trestle 90 feet high. Five railway employees were killed as both trains fell. On October 10, 1888, America's worst rear-end collision to that time occurred at Mud Run, Pennsylvania, when a crowded excursion train was hit by a double-header. On that day there was a celebration of mainly Roman Catholics at Hazleton, Pennsylvania, which drew thousands of people from a large surrounding territory. For this event the Lehigh Valley line ran eight excursion trains from Wilkesbarre, via Penn Haven Junction. On the return trips at night, the cars were crowded, and some of the passengers under the influence of alcohol. The trains were instructed to maintain an interval 10 minutes apart. When the sixth train arrived at Mud Run it encountered a stop signal. It pulled out about 170 feet from the station and then stopped. A brakeman returned as far back as the station with red and white signal lights. The train waited about 8 minutes and then was struck from behind.

The seventh train had two engines, the first being a freight locomotive. It only had a steam brake and no brake connection to the second engine. That locomotive, as was usual by then, had an automatic air brake connection with the cars behind. This train was moving about 25 m.p.h. when the collision happened. The lead engine telescoped through most of the rear car, and this car, in turn, smashed half its length into the one ahead. It was thought that there were about 200 people in these two cars. Few people escaped injury. The total number of dead was listed at 63. In testimony afterward, it was reported that the engineer of the colliding locomotive had worked 13 hours the previous day, was given 4 hours off, and had worked for 14 hours more at the time of the wreck. The fireman and other crew members had been on duty somewhat longer. There seemed to be much conflicting testimony as to whether or not the brakeman had signaled a warning.

Although rail disasters, like Mud Run, continued to plague America, the rate of incidence, by 1890, had begun to decline. The cumulative effect of a number of safety changes was starting to show. One of the earliest improvements was the replacement of rails, wheels, and axles, made from iron, with steel. This change began in the 1860s. When automatic braking systems appeared in the 1870s, railroads were quick to adopt them for passenger trains, but resisted them for freight use. After the Ashtabula tragedy, railroads became more careful in bridge building. Greater efforts were also made to improve tracks and roadbed. Safety also improved when telegraphic dispatching became a general practice and automatic couplers gained wider adoption.

During the 1880s metal began to be used in the construction of passenger cars. Sometime in this period, also, coal stoves were replaced with steam radiators, and kerosene lamps were supplanted, first with gas, and later with electricity. Continual improvements in signaling eventually resulted in the automatic block system, but this innovation did not gain wide use in America until after the turn of the century. Employees, unfortunately, continued to work long hours, and this practice did not change until 1907.

In 1889, on China's Tientsin Rwy., an engine driver proceeded out of a siding before a train traveling in the opposite direction had cleared. As a result 20 people lost their lives. In its July 5, 1889 issue, *The Engineer* recorded this story in the tragedy's aftermath:

> Meanwhile, along the countryside, in the neighborhood of the accident, legends are circulating respecting it. According to one, the people of the village near to which the disaster took place in the dead of the night heard the piteous wailings of the spirits of the victims, which had been forced from their earthly habitations with such awful suddenness. The cries rose and fell, sounding sometimes near and sometimes far away, as if the spirits were hovering through the air looking for a place of refuge. The people lay awake listening to this wailing, which was sometimes long and piercing, sometimes broken off abruptly, and their hair stood on end with fear. Again, people looking up towards the bright moon and stars saw a ghostly train sweeping on at full speed, then suddenly stop, and then vanish altogether. These and other legends circulating round the accident show what a profound effect the latter has had on the minds of the people.

On June 12, 1889, a tragic accident occurred on the Great Northern of Ireland Rwy. near Armagh. An excursion train consisting of 13 carriages, with brake vans at front and rear, failed in an attempt to reach the summit of a grade. It was decided to divide the train in two, so that each part could be pulled up separately. Unfortunately, the rear section was inadequately secured and it began rolling backward. Quickly gaining speed, it traveled 1½ miles before colliding with a regular passenger train approaching at a slow speed. The descending train's three rear vehicles were totally destroyed, with the debris falling down the side of an embankment. The locomotive of the passenger train was knocked from the track, and its trailing carriages began rolling backward. However, these were soon stopped when hand brakes were applied from a brake van. The death toll of this accident was 80, including 22 children age 15 or under. About one-third of the 260 seriously injured were also children.

In Gwalior, India, on May 3, 1890, a freak accident resulted in the loss of two lives and six passenger carriages. As a storm blew into the city a string of nine empty coaches was set in motion. Its journey was halted 2 miles away when the carriages slammed into the front of a ballast train. In late May 1890, a train on a railway connecting Oakland with San Francisco struck the end of a raised drawbridge. The engine, the tender, and a first-class coach crashed through and were catapulted into the water. The engineer apparently missed a warning signal while rounding a curve, and finally seeing the bridge could not stop in time. There were 13 fatalities as well as many injured.

On June 14, 1891, the Moenchenstein Bridge, between Belfort and Basle, on the Est line collapsed in an incident claiming 110 lives. Built in 1872, this bridge had been designed for local traffic having a maximum speed of 18 m.p.h. Later, however, Paris- and Calais-bound expresses were rerouted over this line, often traveling at speeds over 40 m.p.h. The ill-fated train, en route to Calais, was mainly carrying passengers bound for a choral festival at Moenchenstein. It consisted of 2 engines, 2 luggage vans, and 9 passenger carriages, each with approximately 70 passengers.

In July 1891 a tragic accident occurred near Paris on the Vincennes line. A passenger train from Joinville-le-Pont arrived at St. Mande, where a fair had been going on. Many Parisians, ready to return home, quickly boarded this train. As it was about to depart it was rear-ended by an extra train, also sent from Joinville-le-Pont. Two rear carriages and a brake van were crushed by the engine, which then overturned and ignited. Spreading quickly to the brake van and rear carriage, the fire took other lives before finally being doused. The engine driver at fault was arrested for ignoring the posted signals. In all, there were 50 fatalities. About this same time other accidents occurred in France, including one in November 1891, at Grenoble on the PLM line. In the popular imagination these letters came to signify "pour la morgue," suggesting the potential fate for all those daring to ride this railway.

On January 11, 1893, a train on Russia's Slatoust-Samara Rwy. was traveling at full speed, carrying 200 military recruits en route to their new duty station. A fire started in the forward portion of a long carriage and quickly spread its length. The train was stopped as quickly as possible and the coach, enveloped in flame, uncoupled from it. Those surviving made a rush to exit through the doors and windows. However, many became wedged in a passage to one door and were unable to escape. It was subsequently determined that 41 perished from burns and 8 from smoke inhalation.

A unique accident occurred on October 22, 1895, at the Ouest line's Gare Montparnasse in Paris. In this incident, the Granville train had a brake failure and entered the station at an unusual rate of speed. It went through the buffers, mounted the platform, and smashed through the station's front wall, the engine falling to the sidewalk below. There, a woman selling newspapers was crushed. She was the only fatality. On December 30, 1895, an evening mail train carrying passengers departed Johannesburg. Bound for Durban on the Natal line, it derailed on a sharp curve between Dannhausen and Glencoe, completely overturning. Thirty-seven passengers were killed and 50 more injured.

On the night of September 22, 1897, a southbound locomotive and five carriages plunged into a flooded river after a bridge collapse. This accident occurred on the Bangalore-Mysore section of the South Maharatta Rwy. near Maddur. There were at least 150 fatalities, making it at the time India's worst train disaster. America's most well-known train wreck was probably that taking the life of engineer, Casey Jones. Made famous by a popular song, it occurred April 30, 1900, on the Illinois Central R.R. at Vaughan, Mississippi.

Top: Near Bridgeport, Connecticut, a passenger train is struck from the rear by a locomotive, which telescopes into the back car, killing 13, Housatonic R.R., August 14, 1865. **Bottom:** After a flood weakens foundations, a bridge between Ahmoodpore and Ramporehand collapses under weight of a goods train, East Indian Rwy., October 24, 1862.

PLATE

283

TOP: Bridge collapse on Russia's Kursk-Kiev Rwy., March 17, 1869. Sabotage was suspected in this incident. The unlucky train followed 50 minutes behind a special train carrying Russia's czarevitch. BOTTOM: Passenger train plunges off bridge after derailment on Ramscalp Bridge at Bray Head, Ireland; Dublin, Wicklow & Wexford Rwy., August 9, 1867.

PLATE
284

Top: Scene near Willow Tree Station after accident on Long Island R.R., April 23, 1869. **Bottom**: Near Angola, New York, the rear car of New York Express leaves the track, plunges into a ravine, and catches fire, killing 42, Lake Shore & Michigan Southern R.R., December 1867.

The Abbots Ripton disaster, Great Northern Rwy., January 1876: Top: Raising a locomotive with break-down cranes. Bottom: Clearing debris from track the evening of the collisions.

PLATE 287

Scene after Atlantic Express and subsequent rescue engines all derail in blizzard near Fairport, New York; New York Central & Hudson River R.R., January 3, 1879. The line was blocked for more than a week.

PLATE 288

A passenger train en route from Paris to Boulogne strikes a derailed goods train near Poste du Landy, Nord Rwy., May 1884.

Top: Scene after second engine of a double-headed train has a boiler explosion, throwing it on top of lead locomotive, Strommen, Norway, December 22, 1888. **Bottom:** Scene in Charenton, France, after halted passenger train is struck from rear by a second passenger train, PLM Rwy., September 6, 1881.

Terrible train disaster at La Hulpe, Belgium, 1889.

PLATE

290

At Paris's Gare Montparnasse, brakes fail on a train and it plunges through front of terminal, killing one, Ouest Rwy., October 22, 1895.

PLATE 291

CHAPTER

((23))

Horse Car Tramways

HISTORICALLY, a tram was a shaft of a cart or carriage. It later became a common name for a coal wagon. Thus in time, a road laid with narrow tracks of wood, stone, or iron for trams or wagons became a tramway. The word *tramway* was later associated with horse railways and other systems employing mules or oxen. Aside from the mode of traction, these differed primarily from railways in that they either were temporary, or had no special right of way and coexisted with other traffic on streets or roads. The former category mainly included short lines built for industrial, agricultural, or construction purposes. The latter classification chiefly existed for the transportation of passengers, and is the focus of this chapter. By the end of the nineteenth century, the word's definition had been expanded to include any street railway system, whether its power was provided by horses, steam, cable traction, overhead or ground conducted electricity, storage batteries, compressed air, or oil- or gas-burning engines.

The horse-drawn street railway had its origins in America. In the early 1820s horse-drawn omnibuses made their first appearance there, operating on the rough roads and streets. These large stagecoaches functioned as the first inner-city common carriers, and ran over certain routes picking up and dropping off passengers. More economical than the hack cabs already in use, they gained great popularity. Omnibuses were quickly redesigned to carry more passengers, and they soon took on a style of their own.

With the introduction of railroads in America, it may have been apparent to some that this new technology offered possibilities for urban use. However, this development did not occur all at once, but began from a situation peculiar to the historical circumstances of that time: that most big cities preferred to keep steam railroads out of their central districts. As a result terminals were usually located at the periphery of a city, or at best, at the edge of its downtown area. Passengers and businesses, however, were unhappy with this arrangement and clamored for more direct service. This led to compromises where central access was allowed for full-size passenger and freight cars drawn by horses. This was done on track sunk into the streets so as to avoid interference with other traffic. It was with such an understanding that John Mason received a charter on August 25, 1831, allowing construction of the New York & Harlem R.R.

This line was initially built along Bowery and Fourth Aves., between Prince and Fourteenth Sts. Inaugurated on November 14, 1832, it was soon extended south to City Hall, where a terminal was established. Construction also began northward up Fourth Ave., reaching Thirty-second St. in June 1833, and after a difficult cut through Murray Hill, Eighty-fifth St., in May 1834. In late 1837 the line finally arrived at 125th St. in Harlem. However, progress had been difficult, forcing the company to build a 596-foot tunnel between Ninety-second St. and Ninety-fourth St. Earlier, for his original rolling stock, Mason had commissioned John Stephenson, a young coach builder, to provide him with two passenger cars. The first coach, named after Mason, had three compartments, each with the capacity for 10 passengers. Like most cars of its day, it was essentially a stagecoach with flanged wheels. Standard passenger cars were later used for many years afterward.

Mason's New York & Harlem R.R. soon thrived, extending northward beyond Harlem and acquiring its first locomotive in 1837. Steam engines were initially allowed to run as far south as Fourteenth St.; however, laws were passed in 1844 and 1858 raising this barrier northward to Thirty-second St. and Forty-second St., respectively. These laws helped to increase the city's dependency on horse-drawn transport. Information on the early stations of this line can be found in chapter 14. In 1863 the New York & Harlem Co. was sold to Commodore Vanderbilt. Stephenson's small coach-building firm was also destined for greater things. In later decades, it became the world's largest builder of tramway cars, exporting them to many different countries.

Another early horse car line was opened on Magazine St. in New Orleans in 1834. Elsewhere there was initial resistance to this idea, and it was not until the 1850s that the mood of the country changed. However, by this time the paradigm had shifted. A horse railway was no longer imagined as an adjunct to an intercity steam operation, but as an independent system ideal for transporting inner-city passengers. At the time horse car designs changed as well. Influenced by omnibus styles, they became shorter and lighter.

In 1852 construction began for lines on New York's Second, Third, Sixth, and Eighth Aves. Tramways were opened in Boston and Cambridge in 1856, Philadelphia in 1858, and Baltimore, Pittsburgh, Cincinnati, and Chicago in 1859. Shortly they were operating in many other American cities and towns. The first rails used for tramways had grooves, or iron gutters, for guiding the wheels. Unfortunately, these protruded somewhat above street level, and caused problems for carriages and other vehicles. This problem was solved with the invention of the low step-rail, which was first introduced in Philadelphia in 1855.

In 1872 Boston had four horse car companies: the Metropolitan, Middlesex, South Boston, and the Union line (in Cambridge). That same year in Chicago, tramway companies were discussing the new concept of transfers, whereby a passenger might pay a single fare at any point on one division and ride to any point on any other division.

Railroad Gazette's November 30, 1872 issue gave a brief but interesting glimpse of street railway activity in Chicago:

> The West Chicago Railway Company commenced running cars on East Van Buren St. on Monday, Nov. 25, crossing the Van Buren Street Bridge. The Chicago City Railway Company has a large force of men laying track on South Clark St. from Clark Street Bridge to 12th St. At the bridge the road connects with the North Division cars, and when the South Division section is completed, cars will run from 12th St. to Lincoln Park. The track is double, and is not an object of pleasant contemplation to many old fogy owners of real estate in the South Division. A new company has been formed to construct a street railroad on Ashland Ave., from Archer Ave. southward to 55th St., which is the line of the boulevard to connect the south with the west parks. The southwestern portion of the city has been growing rapidly and promises soon to be a very prominent manufacturing point. The design is to connect also with the Union Stock Yards, which are fast growing in importance.

In its October 19, 1872 issue, the same publication gave a detailed description of a street railway being laid out on 23rd St. in New York:

> The pavement being removed, an excavation of the necessary width and about 18″ deep is made, and this is filled for a depth of some 12″ with a coarse gravel, such as is used as a bed for ordinary stone pavements. In this gravel are bedded the cross-ties, which are 7″ wide by 5″ thick, and about 7′ long. In this case the ties are placed about 5′ from center to center; but we believe that it is not usual to place them so close together, 6′ being a more usual interval. On these ties are laid the longitudinal stringers, which are 5″ wide, 7″ deep, and run about 25′ long. They are secured to the cross-ties by means of cast iron brackets, which are spiked to both the stringer and cross-tie by ⅜″ spikes about 3½″ long. Two of these brackets are used at each cross-tie, except the one at which the joint between the stringers comes, where four of the brackets are used. The substructure is now ready to receive the rails, which are spiked to the stringer by ½″ spikes, about 5″ long, driven about 21″ apart. Under the joint between each two rails a flat, wrought-iron plate, about 16″ long, the same width as the rail and ⅜″ thick, is placed, to prevent the ends of the rails from being bent or driven down into the wood . . . The weight of the rails is about 50 lbs. per yard, and the rails average about 26′ in length. The rails being in place, nothing remains to be done but to fill in to the required depth with gravel, rammed down hard, and to put down the pavement.

Because of their operation on rails, streetcars offered less drag or resistance to the horses than omnibuses and thus had the advantage in speed. This advantage could be traded off for greater size, carrying capacity, and comfort. Horse-drawn tramcars generally averaged a speed of 6 m.p.h. At the end of their run they had to be turned around. This was typically done using a small turntable. Later a swiveling mechanism was devised where cars might be turned around on their chassis. This development, however, did not gain wide acceptance. In the 1860s streetcar design improved. Weight was cut down and the driver changed his position from the high perch to a front platform. In 1865, the smaller "bobtail" car was introduced in New Orleans. It required only one horse and was intended for servicing areas of low passenger volume. Other specialized cars made their appearance about the same time: open cars for use in good weather, and double-decker cars for use on lines of high passenger volume. Better lighting and ventilation were provided in 1873 with the adoption of the "monitor" roof, a split-level design borrowed from contemporary railway passenger cars.

Fares were originally collected by the driver, but later conductors were employed for this purpose. Their work was long and arduous. A typical shift lasted 12 to 16 hours a day, six days a week. Although horses were occasionally portrayed in the media as being overworked, they generally had an easier time of it. Horse teams usually were driven a few hours each day, going about 15 miles before finishing their shift. Single horses that drew the smaller "bobtail" cars were often retired after 10 miles. Street railway companies generally had 5 to 10 horses for each of their cars. It worked to their advantage to employ all the veterinarians, hostlers, and ferriers necessary to keep their operation functioning smoothly. Typically, horses were permanently retired when they had a few good years left and could be resold. Those lines that used mules had a more difficult problem in this respect. Tramway companies also derived extra income from the sale of manure.

In 1853, the French engineer M. Loubat, who had built the 6th Ave. Rwy. in New York, returned to Paris and began construction of the first tramway there. It extended from Place de la Concorde to Passy, in the Avenue de la Reine, and had a gauge of 5′½″. Circa 1854 lines from Place de la Concorde to Sèvres and Boulogne were built by the Compagnie Generale des Omnibus. In 1855 a route was built between Sèvres and Versaille. The Tramways Nord Co. constructed a line between the Arc de l'Etoile and the Trone, and later, between Porte Maillot and the Pont de Neuilly. A Tramways Sud system also developed in Paris, and included a line from St. Germaine-des-Pres to Montrouge. Tramways built after Loubat's line all used the standard 4′8½″ gauge. Le Havre, Lille, Nantes, and other French cities also built horse car lines.

In England, street railways were first promoted by the flamboyant American George F. Train in 1857. In 1858 Parliament rejected a proposal by Train and James Samuel for laying tramways in some of the metropolitan thoroughfares and in a few provincial towns. In May 1860 Train was authorized by the Commissioners of Birkenhead to lay out a line there using the standard gauge. In 1861 he was allowed to build other lines on London's Bayswater Road between Marble Arch and Nottinghill Gate, in Westminster, and on Kennington Road. In 1863 Train built the Staffordshire Potteries Street Rwy., a tramway in the Potteries district between Burslem and Hanley. Unfortunately, on the macadamized road surfaces, Train's step-rails did not always remain at the same level as that of the streets. This led to broken carriage wheels and other accidents that sparked public outrage. The lines in London were ripped out, and elsewhere, flat grooved rails were substituted.

In November 1865, a line on Castle St. in Liverpool was inaugurated that used a "crescent" rail. Although more successful than the step-rail, it also had problems, and four years later was replaced by a type with a flat level surface, having a narrow groove to receive and guide the wheel flanges. In 1868 Parliament authorized the construction of several tramways in the city of Liverpool: the South line, from the Exchange to Dingle; the North line, from Old Haymarket to Spillow Lane and Whitechapel St.; and the Aigburth Road line. These tramways, opening between 1869 and 1871, totaled about 7 miles in length. In 1869 a number of other street lines were authorized elsewhere in the country.

In 1870, the London Street Tramways Co. was authorized to lay rails in the north of London, from Lower Holloway to the south end of Hampstead Rd., and from Kentish Town to King's Cross. By the end of 1872 London's functioning system was 42 miles in length. That year, in England, 16 new tramways received authorization for construction. By June 30, 1876, England and Wales had built 132.22 miles of street lines, Scotland 41.3 miles, and Ireland 25.09 miles. Tramway growth in Britain continued. By 1889 total distance for the 14 companies operating in the County of London was 106 route miles.

The first tram line in Geneva opened in 1863. The Vienna Tramway Co. began its operation in 1865. The earliest line in Brussels, inaugurated in 1869 between Shaerbeek and the Bois de la Cambre, was one of 4⅜ miles. By the end of 1874 Brussels had 4 tramway companies with a combined total of 23⅛ miles of track. These were Belgian Street Rwy., Compagnie Bresilienne, Compagnie des Voies Ferrees Belges, and Compagnie Becquet. All lines were built to standard gauge. At the same time Antwerp had 6.16 miles, Liège 4.78 miles, and Ghent 4.66 miles. On May 18, 1872, the first section of the Leipzig Tramways opened. It was built in standard gauge and consisted of the Promenade Line around Leipzig with branches to Rendwitz and Connewitz. Shortly a line was opened to Lindenau, at which time the system's total mileage was 8⅔ miles.

In August 1874 the first section of a tramway in Moscow was opened, using the standard Russian gauge of 5′. By late 1875 the total distance of this system was 60 miles. Its chief engineer was Col. Syrtenko. During the same year, the St. Petersburg Tramway Company conducted trials for a number of cars built by English and Belgian makers, as well as New York's John Stephenson firm. Among European builders at the time, the Starbuck company of Birkenhead and the Francq firm of France were considered the best.

Bad weather made the operation of tram lines difficult. In America, the earliest method devised for dealing with the heating problem was to cover the floor with hay. Later, small coal stoves and other heating devices were tried out, but nothing proved very effective. When snow became packed, cars were easily derailed. Snow plows and rock salt were effective in reducing this problem, but created a rough path for the many horse-drawn sleighs. Thus for years some cities prohibited these practices. Tramway lines turned to the use of sleighs themselves on occasion, but this policy was not widely adopted. Cold weather also brought on health problems for both drivers and horses. In late 1872, a severe epidemic of equine influenza wiped out thousands of tramway horses in America. As a result service was curtailed or halted on numerous lines until replacement horses could be acquired.

Because street railways offered public transportation, the rider, on occasion, had to contend with rude, drunken, or obnoxious individuals. Thieves and even more dangerous criminals could also be encountered. In its September 21, 1867 edition, *Frank Leslie's Illustrated Newspaper* described the typical operation of pickpockets on New York's city lines.

> Take for example the instance of the gentleman whose adventure with them we have illustrated. He is a quiet man of business, whose office is in the lower part of the city, so that to reach his house he uses a convenient line of cars. Having arrived at the cross street in which he lives, he motions the conductor to stop the car, and hastens to get off. The conductor most probably takes no notice of his request, and finding himself being carried beyond the place he wishes to get off, he is in a greater hurry to get out. Hastening to the platform he finds it occupied by a group of four or five persons, who show no disposition to give place, and who are so arranged that it is impossible to pass them without pushing his way through.
>
> Proceeding to do this, while perhaps asking them politely to make way, he presses against a man who stands with his back toward him, and who apparently angered at being touched, turns partly round, and asking savagely with an oath what he means, at the same time thrusts an elbow violently into his side, while at the same time another man, on the other side of him, pushes him in the opposite direction. Astonished and bewildered for a moment at such coarse violence, his first impulse is to get out of such rude company, and he steps quickly to the street, while … The moment of bewilderme… tering around him have robbe… of value. While one on each side … takes his breast-pin, and a confederate… hand round and captures his watch; his … be gone, if he carried it in any exposed p…

Many tramways were built elsewhere in the … the city of Buenos Aires had about 70 miles of ho… either already operating or under construction. Th… pal companies there were the Billinghurst, City, La… Mendez, National, and Southern Tramways. The earli… lines in Buenos Aires had employed "trumpeters," men on horseback who rode ahead of the trams warning away any other traffic. Much had changed since that time.

While attention here has been given only to passenger use of horse car lines, these same tramways in America were sometimes used for hauling mail or various kinds of freight. At least a few other street railways existed solely for such purposes, but little is known about them today. Passenger tramway companies also sometimes offered specialized services. Streetcars and other specialized cars could be rented for special occasions of various kinds. These included excursions and funerals. As time went on different categories of passenger service emerged; some cars offered transportation exclusively to women; others provided extra-fare parlor car comforts.

The American horse railway reached its zenith in 1888. As the century waned, alternative forms of traction increasingly became preferable over that of the lowly horse. Among these, the electric trolley finally gained prominence in offering substantial improvements in both speed and cost-effectiveness, and it provided the coup de grace in the demise of the horse cars. This trend was already apparent in a report from the July 10, 1891 issue of *The Engineer*. It noted that in the United States and Canada there were 957 street railways. Of these 589 were still worked by horses. However, 49 were operated by cable, 73 by steam, and 246 by electricity. The last American horse car line shut down in 1917.

Early American horse car scenes: Top: Cars on Brooklyn's Fulton St. near the ferry terminal, circa 1855. Bottom: View of cars on Tremont St., Metropolitan R.R., Boston, 1856.

“Pickpockets at work on the city railway cars,” New York, 1867. Trying to exit, the man in the light hat finds his way blocked. When elbowed by the man in the top hat, he is sufficiently distracted to be set upon by other accomplices who strip him of his diamond stickpin, pocket watch, and wallet.

PLATE 293

PLATE 294

"Started at last," a scene in New York, 1872.

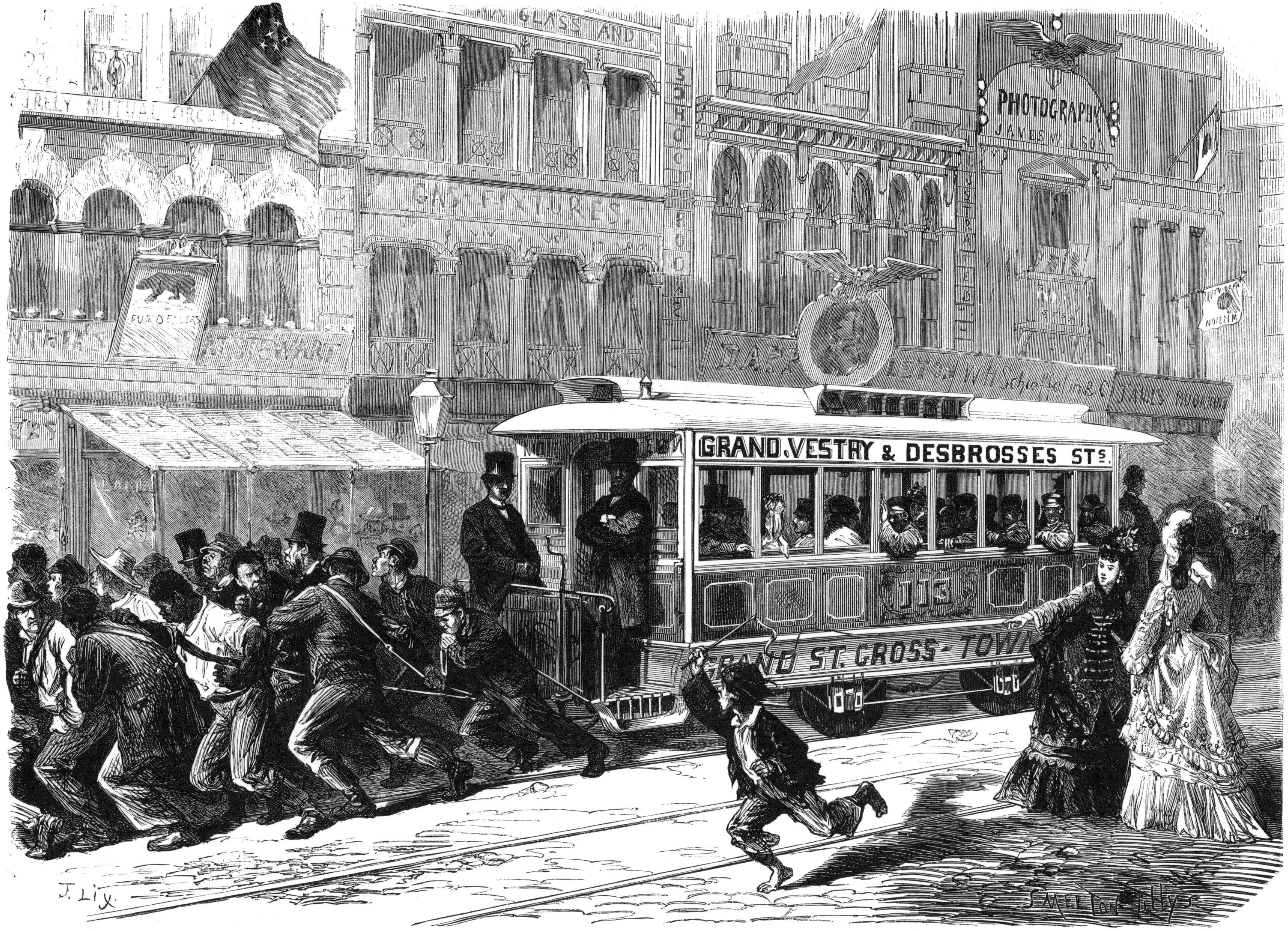

PLATE 295

The great horse epidemic of 1872 devastated the street railway business in the United States. This scene in New York shows the unusual efforts some companies made to continue providing service.

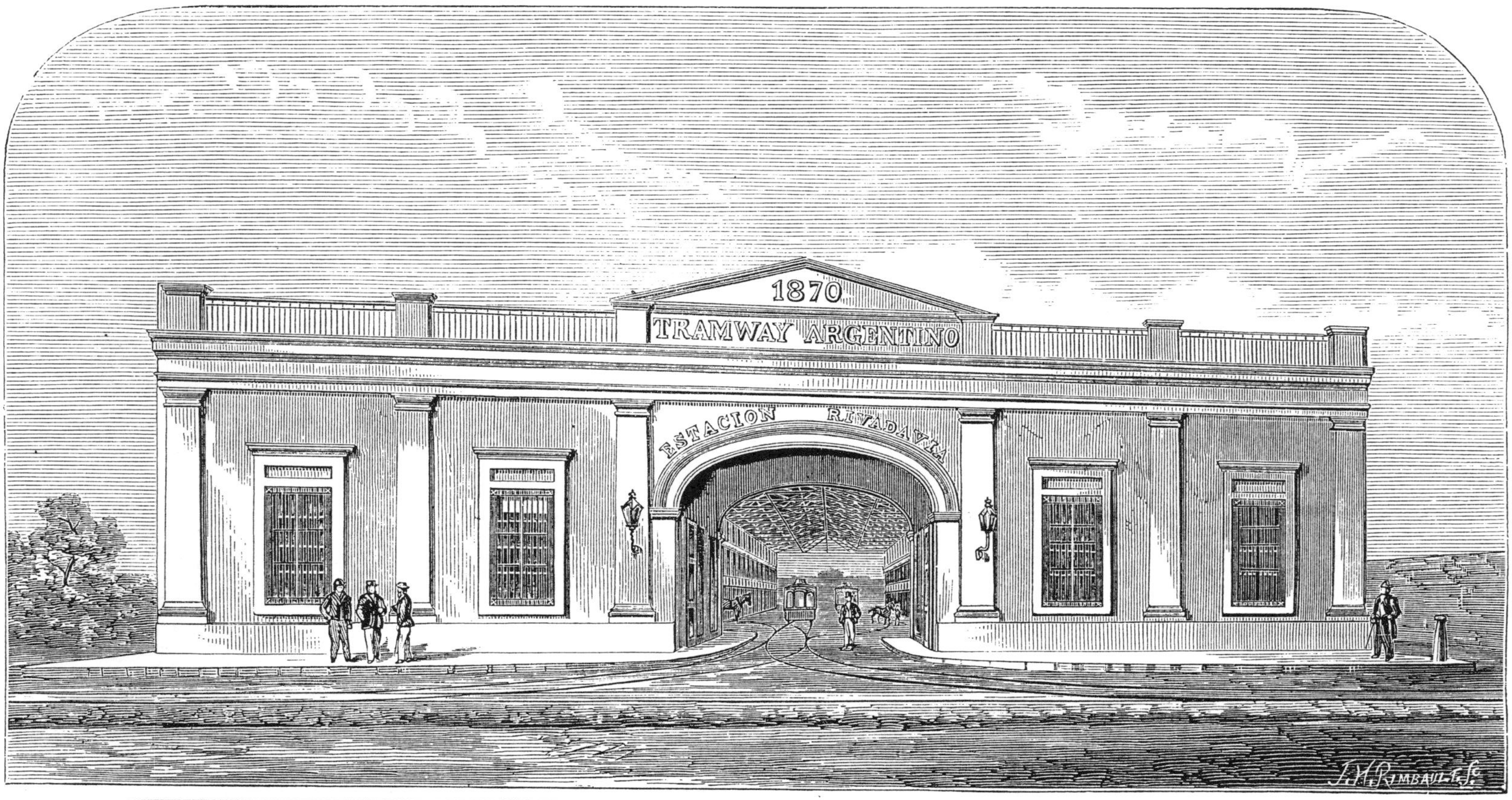

Facilities of the Buenos Ayres Tramway Co., 1872: Top: Facade of depot, Rivadavia Station. Bottom: Stables in rear.

PLATE 296

European horse car lines, circa 1870s: Top: Scene in le Havre, France. Bottom: Double-decker car on Berlin-Charlottenburg Rwy.

PLATE 298

While crossing Paris's Ceinture line at Rue de Montreuil, a horse car is struck by a train, 1878.

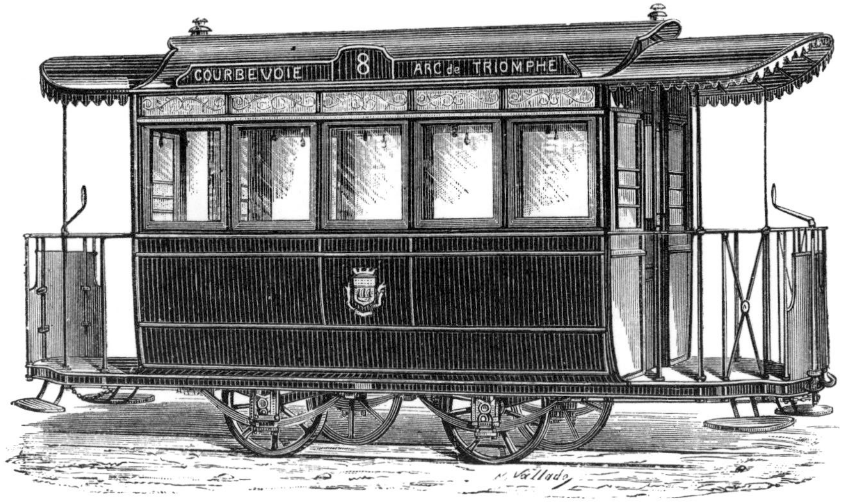

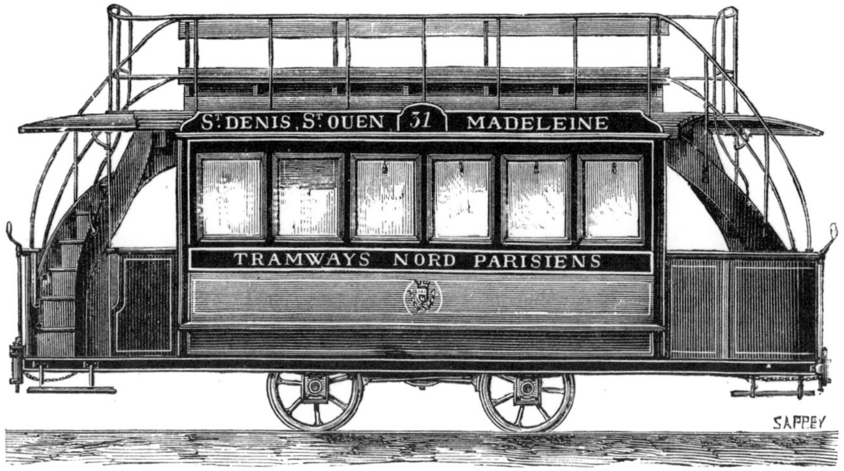

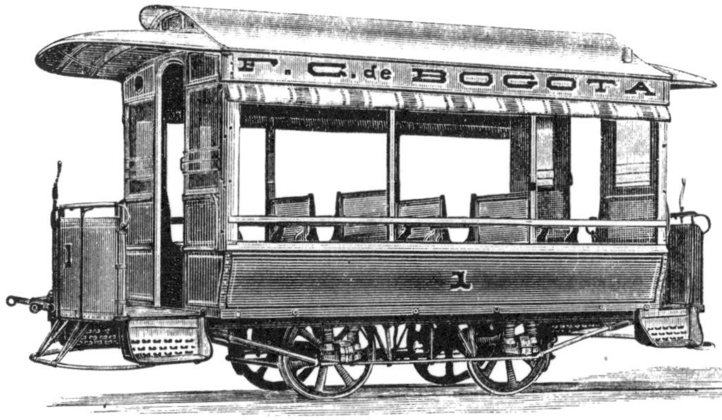

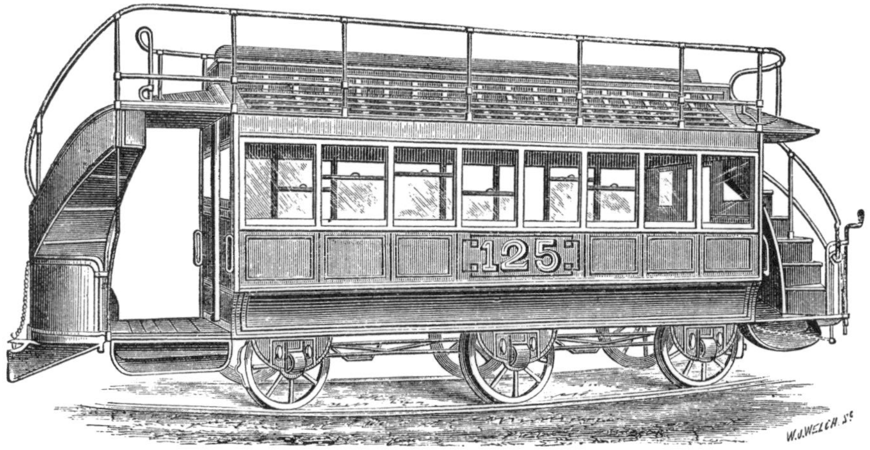

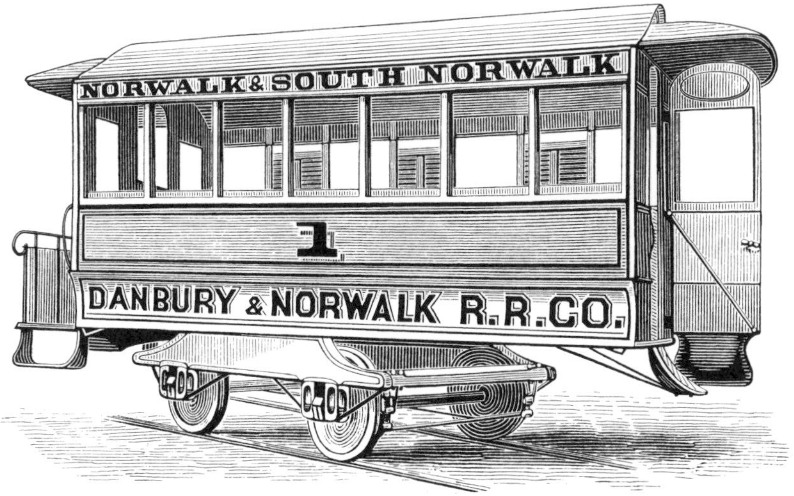

PLATE 299

TOP LEFT: Paris street car, MM. Delettrez Co., 1878. TOP RIGHT: Double-decker car, MM. Delettrez Co., 1878. MIDDLE LEFT: Open car built by America's Brill Co. for service in Bogata, Colombia, circa 1880s. MIDDLE RIGHT: Three-axle double-decker car built for use in Dublin, Ireland, circa 1875. BOTTOM LEFT: Horse car in Vienna, circa 1874. BOTTOM RIGHT: American car that could be swiveled in opposite directions, circa 1855.

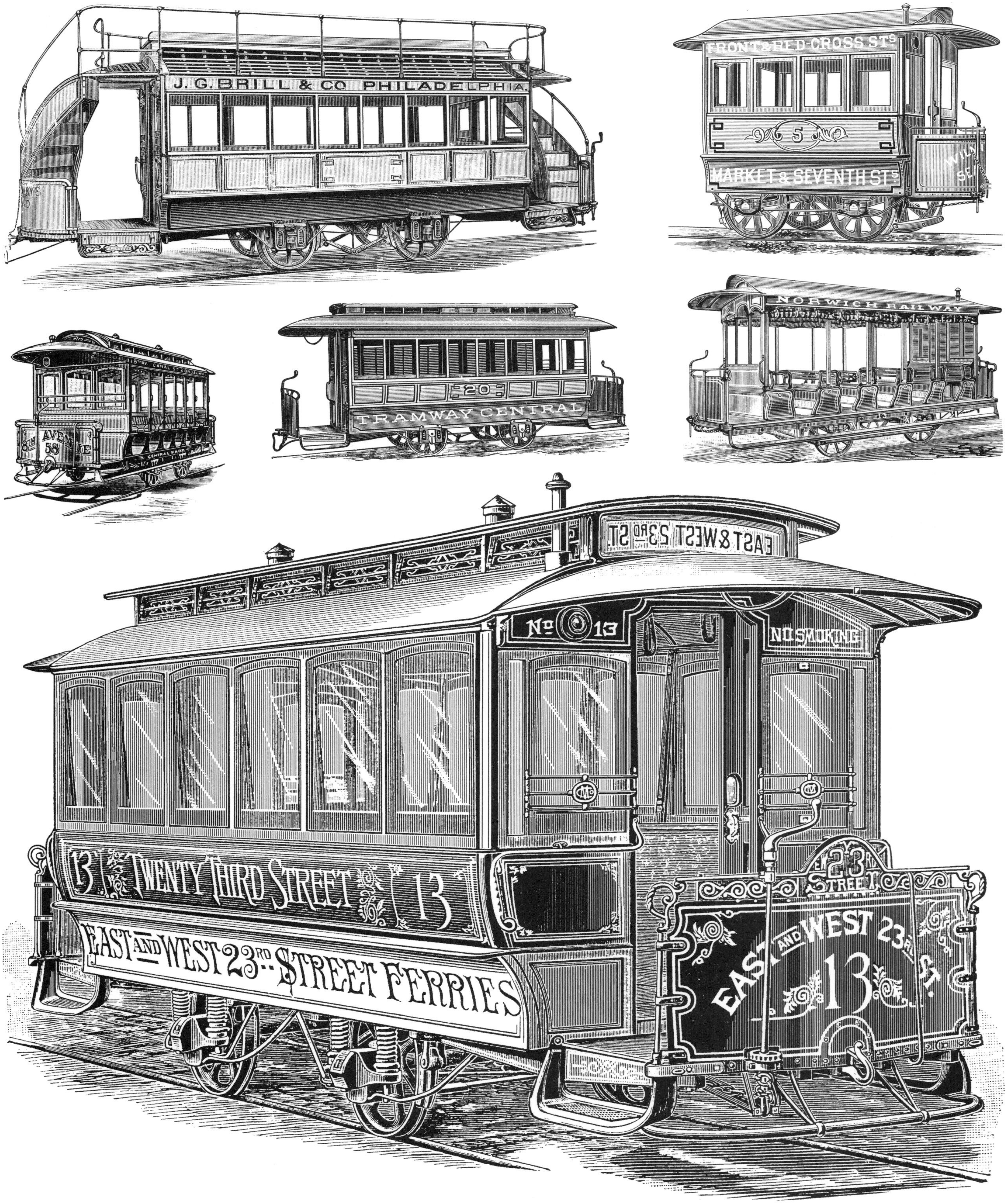

Top left: Double-decker car, J. G. Brill Co., Philadelphia, circa 1870s–80s. **Top right:** Bobtail car, Brill Co., circa 1870s. **Middle left:** Open car of Brooklyn's Lewis & Fowler Co., circa 1880s. **Middle center:** Closed car, Brill Co., circa 1880s. **Middle right:** Open car, Brill Co., circa 1870s–80s. **Bottom:** Closed car, John Stephenson Co., New York, 23rd Street Railway Co., Manhattan, circa 1890.

PLATE 300

CHAPTER

((24))

Atmospheric, Pneumatic, and Compressed Air Traction Systems

THE atmospheric railway system was a new type of traction technology tried by a number of railways during the mid-1840s. The original inspiration for this idea came from the English engineer Medhurst, who took out a patent for a primitive vacuum system in 1810. In a pamphlet entitled *A New System of Inland Conveyance,* published in 1827, he proposed an airtight tunnel traction system in which suitably designed carriages running on iron rails could be moved, piston-like, by means of air pressure from behind. Medhurst also suggested an alternative system in which a long airtight tube of smaller diameter, with an open slot running along its top, could be laid between two rails. Moving through this tube would be a piston, connected by a curved arm extending through the slot to a rail wagon. Thus the traveling piston would function as a propelling appendage of the wagon. Both systems depended on a stationary steam engine to create the necessary vacuum suction.

In 1835, an expatriate American in London, Henry Pinkus, received a patent for a vacuum tube propulsion system. Basically employing Medhurst's alternative idea, his longitudinal slot was to be made airtight by employment of a "valvular cord of yielding substance." The tube, originally conceived to be 3 to 4 feet in diameter, was soon reduced to a more practical 10-inch diameter. This was to be laid along the center of the tracks, and held in place by stone blocks laid end to end. Pinkus displayed a working model at Cavendish Square. This soon resulted in the formation of the National Pneumatic Railway Association. After receiving a second patent for improvements in 1836, Pinkus built a full-size trial line near the Kensington Canal, but his efforts were not entirely successful.

The engineers Samuel Clegg, Jacob Samuda, and Joseph Samuda received a patent for a similar system in 1839. They replaced Pinkus's longitudinal valve, made from cord, with leather, hinging one side and fastening down the other. Both leather strips were coated with a wax and resin compound that acted as a sealant. By 1840 they constructed a trial line at Wormwood Scrubbs, and in test runs, up a 1 in 120 gradient, a 5-ton rail carriage attained a speed of 30 m.p.h. Shortly they were invited by the Dublin & Kingstown Rwy. to try this system on an extension between Kingstown and Dalkey. This line was short, extending for only 3,050 yards, but it had a 1 in 57 gradient, making it a difficult ascent for the locomotives of the time. After conversion to the atmospheric system, trial runs began on August 19, 1843. The line became relatively successful, remaining in operation until April 12, 1854, at which time it was incorporated into a through route and returned to steam traction.

Just at the time when "railway mania" was beginning in England, many notable individuals took up the cause for atmospheric traction. A select committee was appointed in March 1845 to investigate the practicality of this system. Respected engineers, such as I. K. Brunel and Charles Vignoles, testified in favor of its merits, while others, like Robert Stephenson and Joseph Locke, gave dissenting views. The committee concluded by taking a favorable position on the matter. The main advantages of the system appeared to be the following: economy of operation, better performance on grades, less track damage (resulting from removal of heavy locomotives), smoothness of ride, impossibility of collision, and elimination of smoke annoyance.

In 1845 the London & Croydon Rwy. converted a line to the Clegg-Samuda system between Forest Hill and West Croydon. In tests shortly afterward, a speed of 70 m.p.h. was attained by a train consisting of six carriages, and 30 m.p.h. by a train of 16 carriages. Public service was inaugurated on

October 27, with all remaining steam operations discontinued by January 19, 1846. Soon the line was extended from Forest Hill to New Cross, the inauguration taking place on January 14, 1847; 50 h.p. pumping engines were used to create the necessary vacuum. These were housed in Gothic-style structures built at Forest Hill, Norwood, and later at Croydon and New Cross. The chief engineer of the company, Cubitt, vigorously advocated atmospheric traction for a number of other lines, including one from London to Portsmouth. Elsewhere, other individuals actively promoted atmospheric railway schemes.

While the Croydon line's experiment was initially successful, a major problem arose during the hot summer of 1846, when its valve-sealing compound deteriorated. This allowed an intrusion of air into the vacuum tube and greatly disrupted function of the system. Although an improved compound was quickly developed, a greater problem arose when the Croydon line was amalgamated into the London, Brighton & South Coast Rwy. The management of the new company, immediately hostile to this expensive experiment, shut it down on May 4, 1847, and most of the apparatus was sold at public auction.

I. K. Brunel made an even more ambitious attempt to implement the Clegg-Samuda system when he tried it on a 41½-mile portion of the South Devon Rwy. This line, between Exeter and Plymouth, required considerable tunneling and the building of numerous embankments and sea walls. Brunel began plans for the line in 1844, employing a system differing in a few details from that of the Croydon Rwy. The pumping stations, situated about every 3 miles, were as impressive as their Croydon counterparts, but done in the Italianate style. After considerable construction delays, atmospheric service began between Exeter and Teignmouth on September 13, 1847.

However, the system had problems, among them the breakage of essential parts and destruction of cupped leathers on the piston when passing through inlet and outlet valves. These were added to the piston in an attempt to make it more airtight. A greater problem, however, was rapid deterioration of the longitudinal leather valve. The greasy protective compound quickly broke down through exposure to weather extremities of hot dry temperatures, rain, and winter freezing. The exposed leather soon cracked from dryness, swelled from water saturation, became rigid in freezing temperatures, and, on occasion, was eaten by hungry rats. This resulted in poor functioning of the system, as air (and sometimes water) leakage became an increasingly severe burden. Fuel expenses grew as pumping engines worked harder to compensate for the air loss. It was soon calculated that the cost of traction was about nine times what it had been on the Kingstown & Dalkey line, and about 2½ to 3 times the expense of operating steam locomotives.

While some improvements were made, in the end they were too little and too late. By July 1848 Brunel determined that the entire valve would have to be replaced between Exeter and Newton. It was also suggested that the iron portion next to the valve be painted or galvanized, as it had been found that direct contact between wet leather and iron had hastened decay. When the directors of the railway fully considered the matter, it was decided to abandon the experiment until Samuda and his associates could fully demonstrate the system's effectiveness. Thus atmospheric operations were terminated and steam service inaugurated on September 10, 1848.

Although the South Devon experiment failed, it had shown real potential. Passengers were happy with the lack of smoke, smoothness of ride, and speed. On one occasion a 28-ton train on the line had reached a top speed of 68 m.p.h. and had averaged 64 m.p.h. for a distance of 5 miles. Clegg and the Samuda brothers were more successful with their St. Germain Rwy. in France. This short line, on a slope between St. Germain and Vesinet Woods, opened on August 14, 1847, and lasted until 1859. Elsewhere in France atmospheric railways were tried by Hallette in Arras, 1844; Chameroy in Paris, 1844; and Hediard in St. Ouen, 1848. A short time later on Switzerland's Geneva-Plainpalais line, Turettini experimented with a similar tube system. However, rather than exhaust air ahead of the traveling piston, he propelled it by pumping in compressed air behind it using a stationary steam engine.

A different compressed air–atmospheric system was demonstrated by Christopher Nichols at York Road, Waterloo, in 1845. In the place of a traveling piston, it instead depended on two solid drums that were attached to a rail carriage and positioned like horizontal guide wheels. These rotated against a hollowed-out center rail, and were pushed from behind by the inflation of long elastic tubes attached on each side of the rail. Each set of tubes, made from leather, canvas, and Gutta Percha, extended for a distance of 50 yards and then was separated by an interval before the next set. It was assumed that the carriage would gain sufficient momentum from its initial start to reach each succeeding set of tubes.

Buried beneath the center rail for its entire distance was a large cast iron pipe, which functioned as a compressed air reservoir. At frequent points, small vertical pipes extended upward from the reservoir into the third rail. Each had, at its top, a stop-cock, which when triggered by a "skate" on the underside of the carriage, quickly opened the air valve and inflated the tube. In Nichols' demonstration this inflation occurred so rapidly that his carriage quickly gained good speed, leaving those who observed the performance suitably impressed. The advantages he claimed for his system were perfect speed control, quick stopping, and a smoother ride. Nichols intended for cars to be propelled individually, rather than together in a train.

T. W. Rammell returned to Medhurst's original idea when, as engineer for the Pneumatic Despatch Company, he constructed an experimental 2′6″ pneumatic railway. Built in 1861, it extended ⅔ of a mile through Battersea Park. Shortly, however, it was removed and then laid between Euston Station and the North Western District Post Office at Eversholt St. Opening in February 1863, it was used to transport mail.

A more ambitious project was the opening of Rammell's 600-yard pneumatic railway at London's Crystal Palace grounds in 1864. For this purpose he constructed a brick tunnel, allowing just enough room in it for the passage of a standard size passenger carriage. At one end of the coach a disc-shaped shield was attached around it. Having an outer edge of bristles, it fit snugly against the tunnel's interior wall. Thus it served to both seal off air leaks and function as a wind collector. On its "down" trip the car was released on a slight incline and coasted past two airtight doors. Once past this point, it received air from the rear, forced in by a fan of 22-foot diameter, which propelled it to its destination. On the "up" trip, the same fan was reversed to create a vacuum ahead of the car, giving it a sufficient speed to ascend the 1 in 15 gradient and return to its original starting point. This short line quickly became a tourist attraction, with rides taking 50 seconds and costing six pence.

In late 1865 the Pneumatic Despatch Company opened the first portion of a 4′6″ diameter line that was to extend from Euston to the General Post Office. Although the line was not intended for passenger service, the company allowed the Duke of Buckingham and others of his entourage to ride through the tunnel. Hopes of the company were very high at the time; however, subsequent problems forced abandonment of the project.

In 1869 Alfred E. Beach began construction of a short pneumatic line in New York City. Only extending for about a block, it ran beneath Broadway from Warren St. to just beyond Murray St. A tunneling shield, the first used in America, was employed for the project. Beach's passenger coach was cylindrically shaped and perfectly fitted for the tunnel it operated in. Like its London counterpart, Beach's line became an instant tourist attraction, opening on February 26, 1870. However, he had larger ambitions in mind, seeing his short line as a prototype for a mass transit system, one perhaps adaptable for elevated use. Dr. Rufus H. Gilbert, a later builder of elevated lines in the city, lent support to this idea for a while. However, Beach's dreams failed to materialize and his line was shut down in 1873.

The first, independently operating, compressed air rail vehicle was designed and built by the great French engineer Andraud and his associate, Tessie du Motay, at the Chaillot Iron Works in 1840. Successfully tested that year on July 9, it was essentially a small 2-2-0 eight-seat coach. With air tanks fitted underneath, it employed a heating device to increase air expansion in its two cylinders. These functioned essentially the same as in a steam locomotive. Later, on September 25, 1844, Andraud tested a 5-ton 2-2-2 compressed air locomotive. Having a single 106 cubic foot tank holding an air pressure of 300 p.s.i., it averaged a speed of 17 to 20 m.p.h. over a distance just exceeding 2 miles.

The *Illustrated London News* of February 28, 1846, reported on a patent, granted to Mr. Parsey, for a 2-2-2 compressed air engine. Parsey's design had an air tank at each end, positioned over the front and rear carrying wheels. The controlling apparatus and two cylinders, for powering a crank axle, were situated above the drivers. Parsey hoped to construct air tanks capable of holding a pressure of 1,000 to 2,000 p.s.i., which could drive his engines up to speeds of 100 m.p.h. In the railway he envisioned, pumping stations would be situated every 30 miles for refilling. There was considerable interest in this idea at the time, and Mr. Parsey was allowed to brief Prince Albert on his proposal.

In 1872 the engineer Mekarski began experiments on a compressed air rail vehicle at Le Creussot, France. Besides incorporating the standard elements used up to that time, he added a reservoir container, filled with hot water. Air released from the storage tanks was passed through this reservoir before reaching the driving cylinders. Mekarski's purpose in heating the air was to compensate for any heat loss due to expansion as air was released from the tank. This also maintained a higher temperature in the cylinders than would otherwise be possible. This water reservoir could be reheated quickly by connection to a stationary boiler.

In the early excavation of Switzerland's St. Gotthard Tunnel in 1872, removal of the resultant debris was a major problem. Steam locomotives could not be used due to the poor ventilation. Horse- and mule-drawn cars operated slowly and at a high cost. Finally, in an experiment, two ordinary locomotives were placed at each side of the tunnel. Their boilers were drained of water and filled instead with 4 atmospheres of compressed air. This air played the part usually done by steam. It passed into the slide-valves, entered the cylinders alternately on each face of the pistons, which it set in motion, and then escaped as exhaust into the atmosphere.

As it was soon apparent that this supply of compressed air was inadequate, the next modification was to supply this by creating a "tender" to follow each locomotive. In each case, an air tank was mounted on trucks and connected by air hose to the engine. The locomotives operated in this way for about two years, each being able to draw a train of 12 wagons 600 meters, and then make a return trip before exhausting the air supply. A later innovation allowed the compressed air to flow at a fixed pressure, regardless of the remaining pressure in the tank.

Finally, circa 1875, genuine 0-4-0 compressed air tank locomotives were constructed. The tunnel's chief engineer, M. Ribourt, devised further modifications, which included a mechanism that cut off air to the cylinder before full expansion occurred. This helped to conserve the supply of air. These locomotives performed well and fully demonstrated the practicality of this technology for certain situations. Thus, from this time on, similar engines began to be used in mines and tried out for other purposes.

An early locomotive of this type was tested in the United States at a coal mine near Pittsburgh. The March 30, 1883 edition of *Railroad Gazette,* quoting the *Coal Trade Journal* as its source, gave the following report:

> What is undoubtedly the first practical attempt to use compressed air as an underground motor in a coal mine in this country is meeting with success at the Old Eagle pits of W.H. Brown Sons, 27 miles up the Monongahela above Pittsburgh. The new motor was built at the Baldwin Locomotive Works, and is a most singular looking affair. The available height above the pit rails being only 5′10″, the air locomotive had to conform thereto. The air receivers (tanks) are 27 ft. long, and 38 in. in diameter, and made of sheet steel. These are filled with air compressed to 400 p.s.i., forming the actuating power of the machine. These air receivers rest on four wheels, driven by a pair of locomotive cylinders, gearing, etc. just as in a railroad engine, the air taking the place of steam. The originator of this idea, Capt. Harry Brown, expressed himself as more than satisfied with this locomotive. It does the work of a score of mules, requires the attention of only one man—who also operates the air compressing machinery, and can haul 55 loaded cars (60 tons) up a gradient of 100 ft. to the mile.

Baldwin, soon followed by the H. K. Porter Co., would make compressed air mining engines through the end of the century. Mekarski perfected his own mining design in France, and similar locomotives were manufactured elsewhere in Europe. Later, however, most mining lines were electrified.

Compressed air engines were tried out on the New York elevated system in 1879, 1882, and 1896. The 1882 test involved a locomotive, designed by Mekarski, which had a front cab and a 2-4-0 wheel configuration. Its three tanks, containing air at a pressure of 600 p.s.i., were placed to the rear, while the engineer's cab was positioned in front. Air was passed through water heated to a boiling temperature, and reached the cylinders as a mixture of air and steam. Mekarski's engine employed an additional feature, borrowed from Riggenbach, which allowed the driving cylinders to function as brakes on downhill descents. However, tests determined that the reservoir pressures were insufficient for the job.

The 1896 tests were apparently more successful. For air cylinders a Scotch engineer, Robert Hardie, employed Mannesmann rolled steel tubes that could carry a much greater pressure. Thirty-seven of these, 9 inches in diameter and 15 feet, 6 inches long, were laid horizontally in the forward portion of the 0-4-2 locomotive and covered with a cylindrical metal jacket. Viewed from the front these air cylinders were arranged in the shape of a hexagon. In the rear was the cab that included a vertical reheater holding 50 gallons of water. The reservoir pressure after charging was 2,500 p.s.i., and this could be depleted to 450 p.s.i. before a recharge was necessary. These engines were designed for a trainload of 130 tons. Tests were run over the 4½-mile Sixth Ave. line between Rector St. and Fifty-eighth St., a route having 14 intermediate stops. The July 23, 1897 issue of *The Engineer* reported that after 10 months of operation these locomotives had run over 29,000 miles and carried over 180,000 passengers.

The April 19, 1890 issue of *Scientific American* reported on a system, patented by Victor H. Tomlinson, which could convey compressed air or steam from a power station directly to an operating engine. It employed a pipe or tube, mounted alongside a track like a third rail, from which, via a sliding receiver, the motive force could be fed into the cylinders of the locomotive. While Tomlinson's receiving apparatus was no doubt sophisticated, this whole idea was perhaps too closely akin to the original slot concept of atmospheric railways. Thus, one would suspect, it received little support.

At a time when alternative forms of traction were sought to replace horses on tramway lines, compressed air offered an attractive possibility. In 1875, a compressed air tramcar, built by Mr. Scott-Moncrieff, was tested on the Vale of Clyde Tramway in Scotland. At each end of the car, three air tanks, each having a pressure of 350 p.s.i., were attached to a frame beneath the floor. The tram had two cylinders, and a cutoff device was employed to conserve air pressure. When empty, the car weighed 6¾ tons, and when fully loaded with 40 passengers, 10½ tons. Tests determined that operation ceased once the air pressure dropped to 110 p.s.i., and that air tanks had to be refilled every 3 miles.

In early 1876, Mekarski tested a tram of his own design. He installed his apparatus in a standard horse car of the Compagnie des Tramways. Being 11½ feet long, it was designed to carry 20 people inside and 14 outside, on the rear platform. Tested on the Courbevoie line of the Tramway Nord, in Paris, its operation was found to be free from smoke, steam, noise, and fumes. The compressed air, of 25 atmospheres, was stored in eight connected metal tanks placed under the carriage.

Borrowing an idea from his earlier locomotive experiments, Mekarski placed an upright water reservoir, superheated to a temperature of 340 degrees Fahrenheit, on the front platform of his car. As before, this was used to reheat the air on the way to the cylinders. From there it was wire drawn down to a pressure of 5 atmospheres before reaching the cylinders. In 1879, after further improvements were made, a tram of Mekarski's design began regular service on the Doulon-Chantenay line in Nante.

In 1881 a dummy tramway engine, powered by compressed air, was tried in England on the Stratford and Epping Forest branch of the North Metropolitan Tramways. This engine was designed by Col. Beaumont R.E., and contained an air tank built to withstand a pressure of 1,000 p.s.i. Air was pumped out of a building in Stratford and then piped 250 feet to a station on the street. At this point a flexible hose was connected to the engine to fill the tank, an operation that took about 15 minutes. The engine then drew a tramcar 2¼ miles up a grade to Leytonstone, and upon returning, was reconnected for a refill.

Similar experiments in tramway traction were made in the United States. These included trials of compressed air streetcars in New York; on the Second Ave. Rwy. in 1878, and on the 125th St. line in 1896. A short time later, compressed air trams were operated on the Twenty-eighth St. and Twenty-ninth St. lines of Manhattan's Metropolitan Street Railway Company. For this purpose a large compressed air plant was erected at Twelfth Ave. and Twenty-fourth St. These cars could operate about 15 miles before their charge of air was exhausted.

On October 1, 1890, the city of Berne, Switzerland, inaugurated a 1.86-mile line having an 11.6% grade over its entire length. The city's choice favoring compressed air traction proved successful as tramcars averaged speeds of 7.5 to 9.3 m.p.h. While electrified systems were becoming increasingly popular at the time, innovations in compressed air technology allowed it to remain competitive, at least in the short run. One new change was the positioning of stations along routes where trams could get easy refills.

A further refinement of this idea, the Popp-Conti system, was first tried in Paris during the mid-1890s. An underground main, running beneath the tracks, received air from a pumping station. The main, in turn, supplied a number of refill stations, housed in metal boxes, sunk into the roadway. These would automatically recharge tramcars stopping above them. Initiating this process was a trip lever, situated just inside a rail, which was activated by passage of the car's front wheel. This opened a valve, raising a plunger, which pushed aside the housing's hinged-flap covers before reaching full extension. At the end of this plunger was a nozzle. It engaged a connection allowing replenishment of the car's reservoirs. Upon completion, this mechanism dropped back into its housing, and the hinged flaps folded over, restoring the surface of the street.

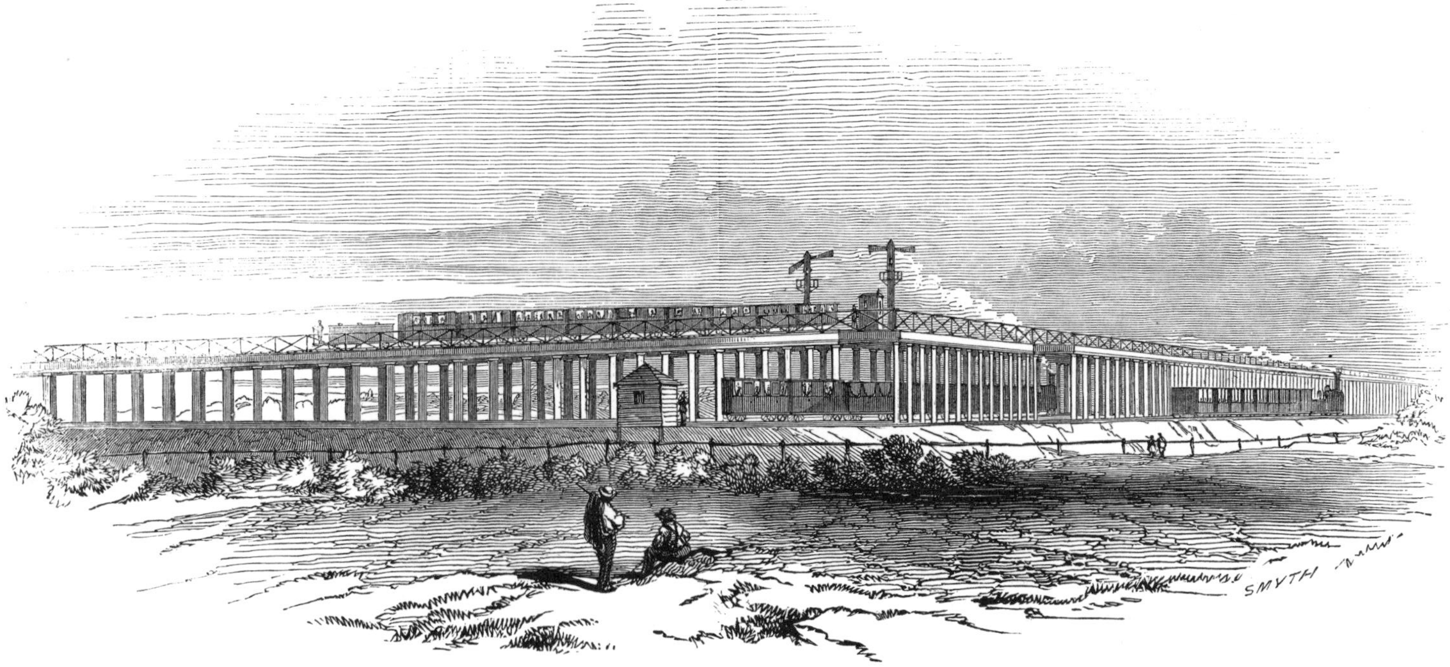

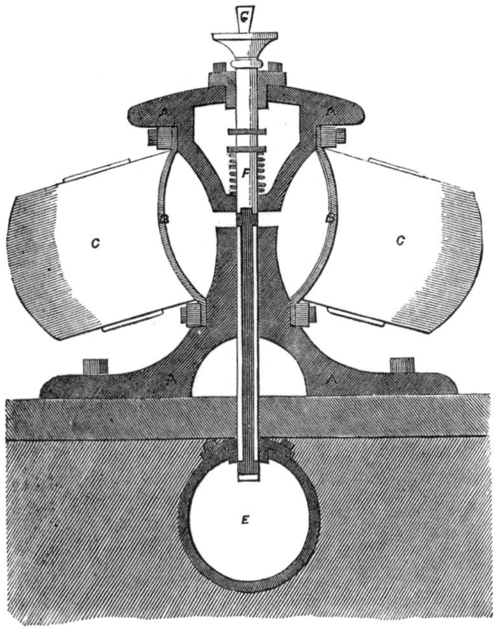

PLATE 301

Atmospheric railways: Top left: Scene on the Kingstown & Dalkey Rwy., Ireland, circa 1843. Top right: Gothic-style pumping station, Croydon, England, 1845. Bottom left: Viaduct, between Norwood and Croydon, where Croydon Atmospheric Rwy. crosses the Dover and Brighton lines, 1845. Bottom right: Cross section of Nichol's system center rail showing inflation valve open, 1845.

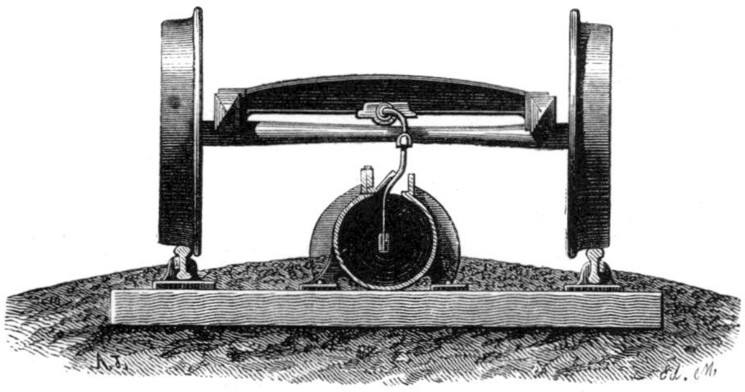

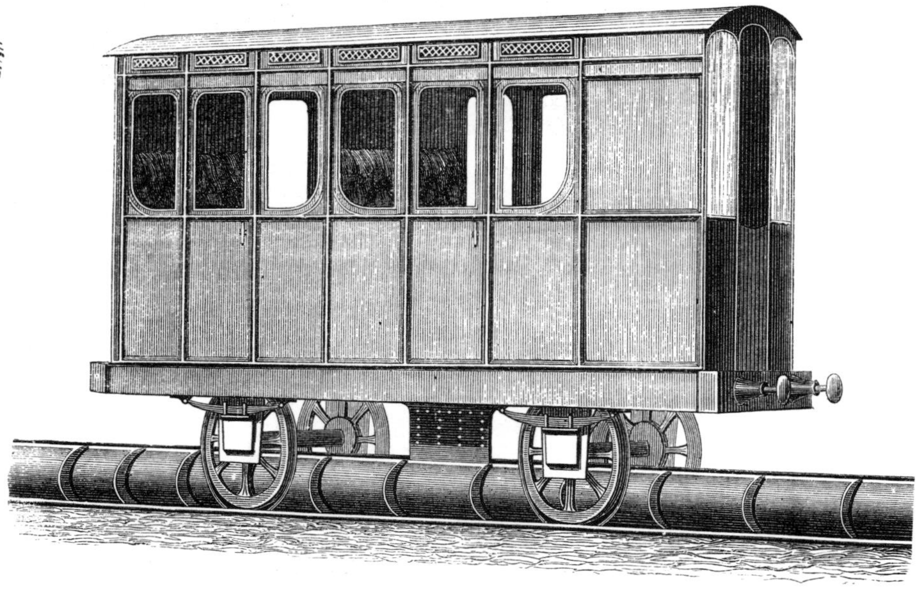

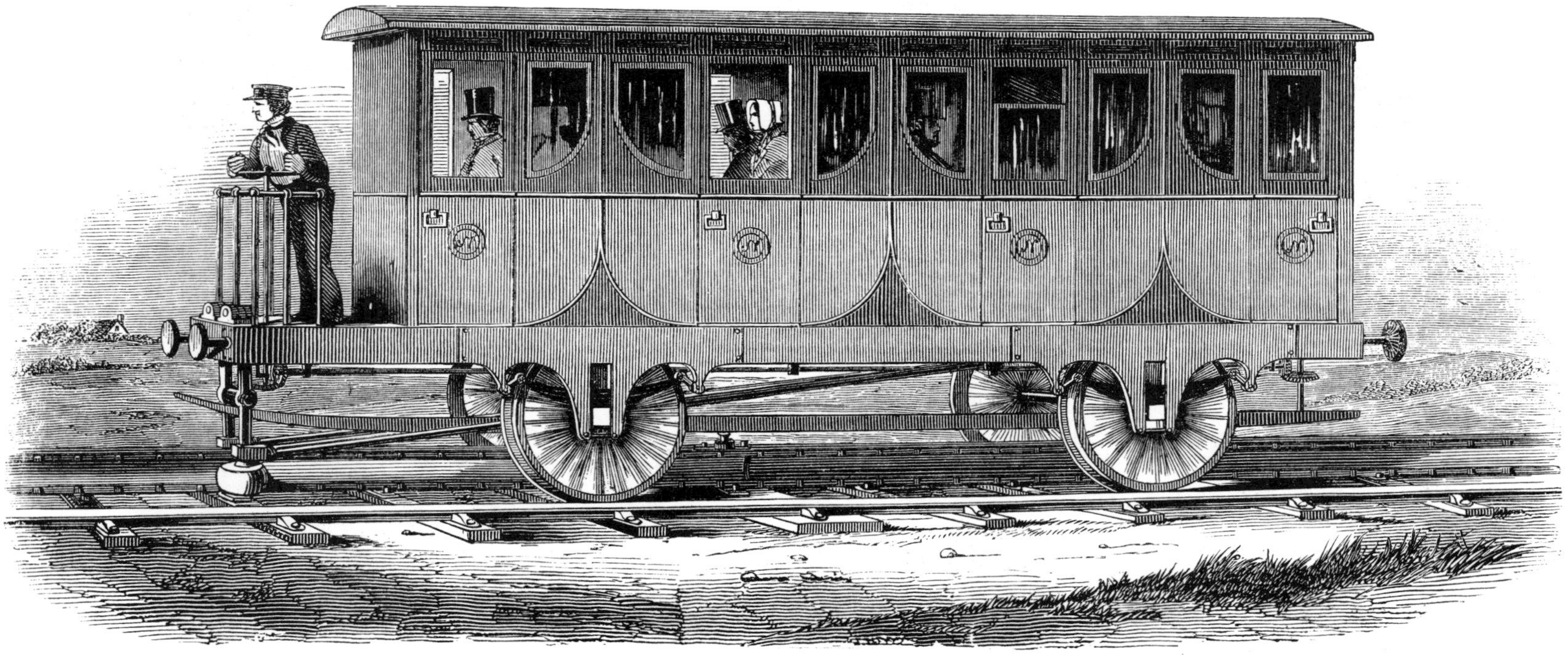

Atmospheric railways: TOP LEFT: Lead carriage of train, St. Germain Rwy., 1847. TOP RIGHT: Lead carriage, Pilbrow's system, 1845. UPPER MIDDLE RIGHT: Cross section of track and tube, Clegg-Samuda system. LOWER MIDDLE LEFT: Clegg-Samuda carriage showing operation within propulsion tube. LOWER MIDDLE RIGHT: Carriage of St. Germain Rwy., circa 1850s. BOTTOM: Carriage of Nichol's system, 1845.

Pneumatic railways: Top: Passengers disembark from pneumatic car at London's Crystal Palace, 1864. Bottom: Inaugural dispatch of mail bags in pneumatic cars operating between postal facility at Eversholt St. and Euston Station, 1863.

PLATE
303

Compressed air traction: TOP: Parsey's patent air engine, 1846. BOTTOM: Locomotive used in construction of St. Gotthard Tunnel, Switzerland, circa 1880.

Compressed air locomotives: TOP: Mekarski-designed mining engine, circa 1880. MIDDLE: Porter mining engine, 1892. BOTTOM: Engine built by Swiss Locomotive Works, Winterthur, for use in excavation of the Simplon Tunnel, circa 1900.

PLATE 305

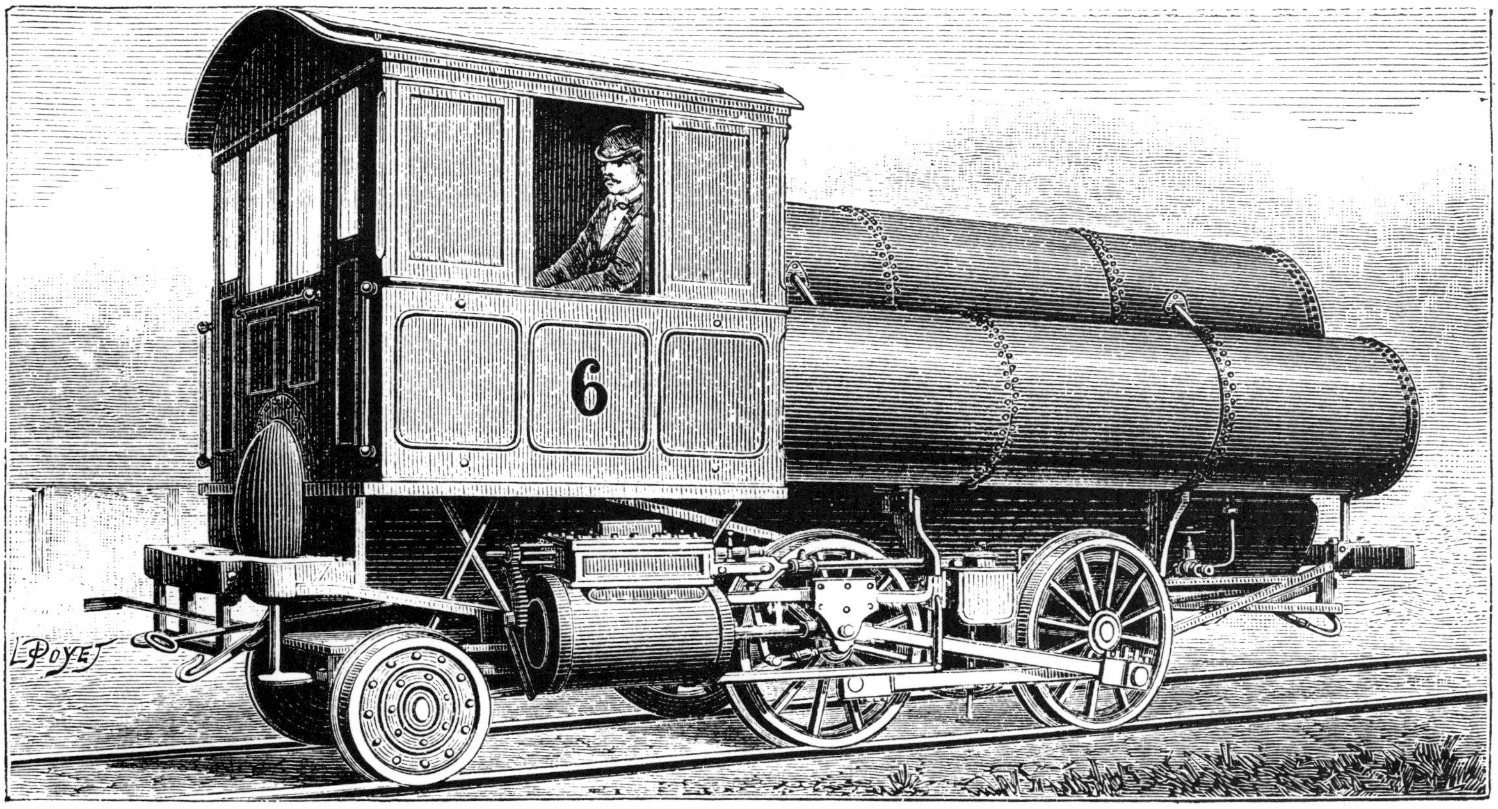

Compressed air locomotives tested on New York's elevated railway system: Top: Mekarski engine, circa 1882. Bottom: Hardie locomotive, 1897.

Top, middle left, middle right: Tomlinson's system for supplying compressed air to locomotive, 1890. **Bottom:** Compressed air locomotive for underground haulage built by Grange Iron Co., Durham, England, 1887.

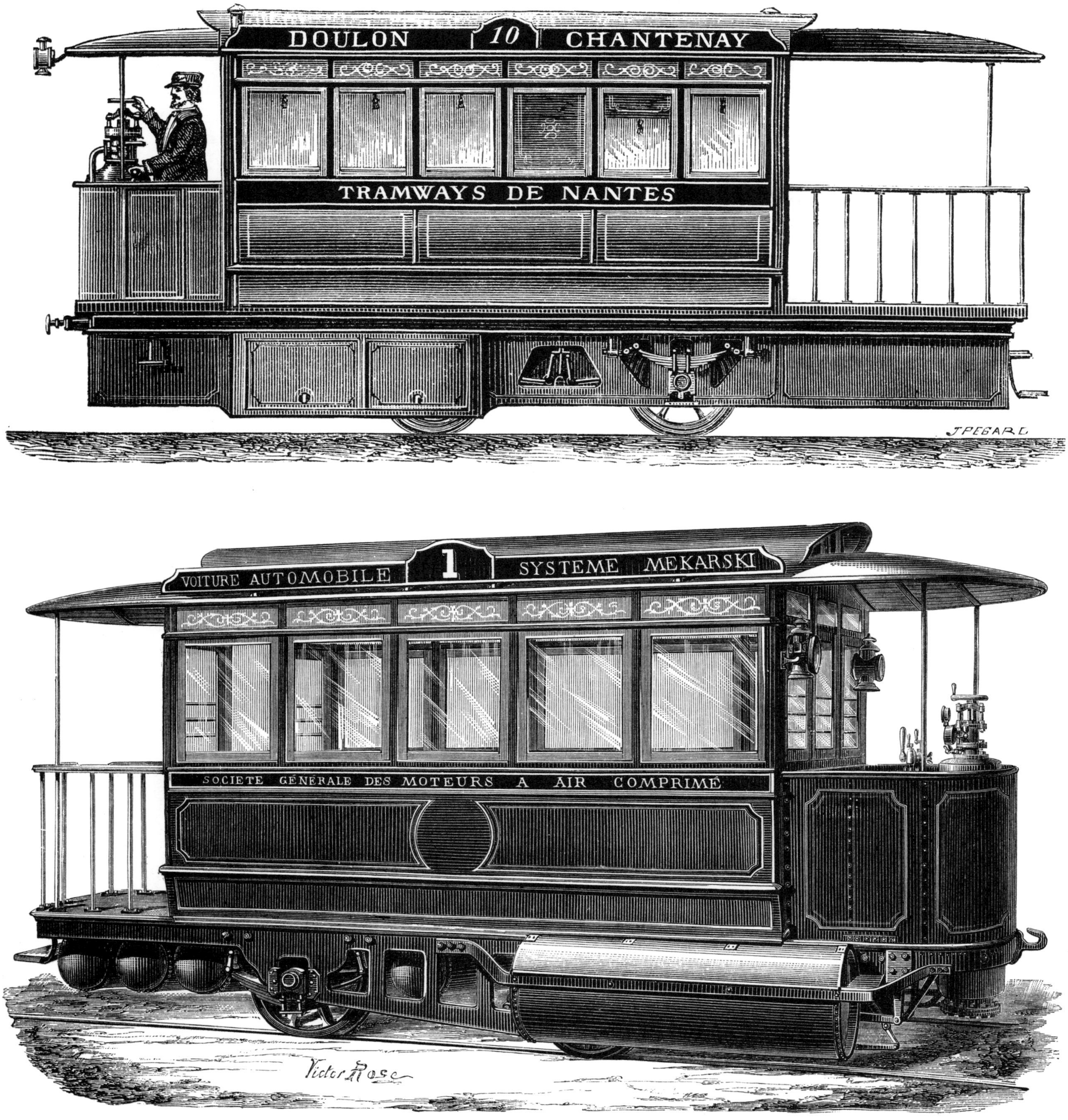

Mekarski compressed air streetcars: Top: Tram car in Nantes, France, 1879. Bottom: Experimental car operated on Paris's Tramways Nord Courbevoie line, 1875.

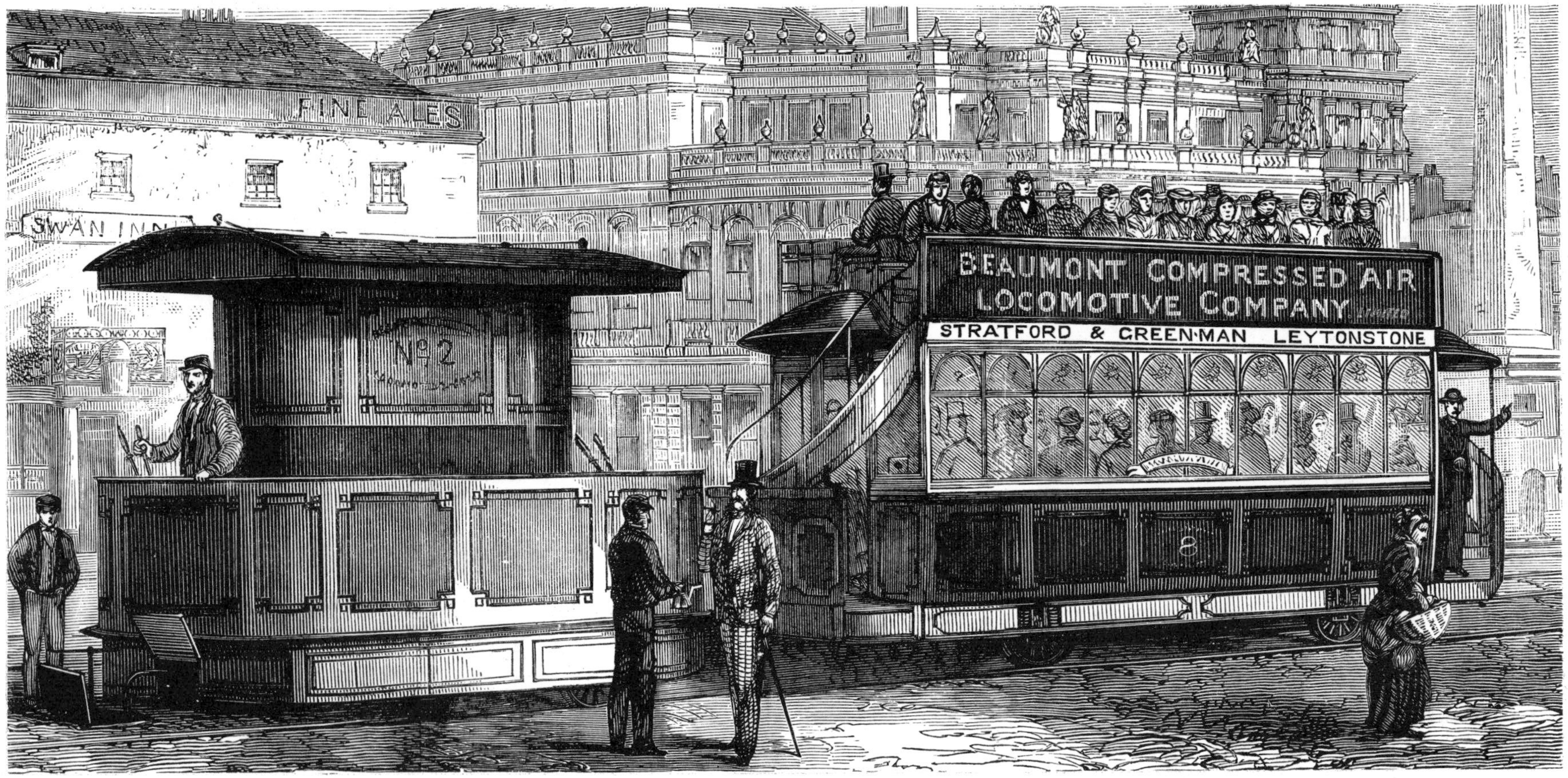

Compressed air locomotives for street railway use: TOP: Mekarski engine pulling double-decker street car, Paris, 1878. BOTTOM: Col. Beaumont's locomotive tested in London, 1881.

CHAPTER

((25))

Steam Traction

Trams, Tramway Locomotives, and Railcars

AFTER the advent of horse-drawn streetcar lines, it was not long before individuals began to dream of improving these systems, either by designing small locomotives for pulling the trams or by incorporating a small boiler and firebox into a car to provide its own means of self-propulsion. Another potential lay in employing a similar method for powering a passenger car for service on short branch lines or commuter routes. In 1847 Charles H. Gregory, engineer of the Bristol & Exeter line, saw a tiny locomotive that had been built for the purposes of track inspection. Convinced that such an engine might power a light steam carriage, he placed an order with the builder of the original engine, Adams & Co., of Fairfield Works. His broad-gauge coach was designed in such a way that the vertical boiler engine was bolted ahead at the front, allowing for easier maintenance. Turns were made easier by allowing lateral movement of the axles in the journals. Once completed, Gregory's *Fairfield* underwent trials on the West London Rwy. The carriage was designed for carrying up to 22 first-class passengers and 48 of the second class. In tests without passengers it exceeded a speed of 35 m.p.h. going up a 3-mile grade of 1 in 100.

On December 2, 1851, the partners Joseph H. Moore and William P. Parrott, both of Boston, received a patent on a steam car they had built. This vehicle had an upright boiler and two four-wheel trucks. Cylinders were mounted on the outside frame of the lead truck and connected by flexible steam pipe to the boiler. Although this car had mechanical problems, it could reach a speed of 40 m.p.h., and was apparently operated for a while on the Boston & Worcester R.R.

After Moore became superintendent of the Pittsburgh, Fort Wayne & Chicago line, he decided to try his idea once again. In 1860 he took over a car manufacturing business in Massillon, Ohio, with the intention of building an improved version of his original machine. His first new car, 77 feet long, was designed to carry 94 passengers and 3 tons of baggage. Moore took this vehicle for a demonstration run during the fall of 1860, carrying the Ft. Wayne road's president from Pittsburgh to Chicago. After two other cars were built, all three were requisitioned by the U.S. government for Civil War service in the South.

About the same time similar experiments were being carried out elsewhere. In 1860 G. F. Train, the American builder of early English tramways, patented a steam car having a Bissell truck at one end and a pair of drivers at the other. In June 1860 the Baldwin company tested the *Novelty,* a six-wheel steam car, which it subsequently delivered to the Ft. Wayne line. In January 1860 Trenton car builders Joseph Grice and Robert H. Long received a patent for a six-wheel steam car. After failing to interest any street railways in this venture, they began to solicit main line companies. One early rail car was sold to the Camden & Amboy line in 1861. Designed for short-distance commuter traffic, it had a rear truck and single front drive axle. This was powered by a two-cylinder engine having a vertical boiler, with intervening spur gearing. The company sold similar cars to other railroads, and introduced an eight-wheel version in 1863. Another early builder of steam cars was the Darker Co. of Philadelphia.

After the Civil War steam trams were tried on the Second Ave. line in New York. Heavier rail was laid to accommodate the additional weight. Service was improved for a while because the trams were faster than their horse-drawn counterparts. However, they all began breaking down about the same time, and the replacement costs were so prohibitive that the experiment was abandoned. Circa 1869 the South Side R.R.

of Long Island tried similar experiments in Brooklyn. These steam trams proved unpopular, however, and in 1870 service was discontinued.

In 1869, the English designer Robert Fairlie displayed at the Hatcham Ironworks a steam rail carriage intended for short-distance operation. Its length was 43 feet and it was designed to carry 16 first-class passengers and 50 of the second class. In front, the locomotive section had its four small drivers so close together that it functioned as a bogie. The passenger area of the carriage rested on a bogie, and was connected to the forward part by a center pin. This carriage was seen as a prototype for a larger one that was intended to carry 100 passengers and reach a speed of 40 m.p.h.

Despite the earlier setbacks in America a small number of steam trams and railcars continued to be built. In 1872, the Baxter Steam Street Car, constructed by the Remington Co. at Ilion, New York, underwent extended trials on New York's Bleecker Street Rwy. It was successful in carrying up to 70 passengers over all the grades and curves on the line. The *New York Times* downplayed the positive trial results, however, and took a position opposing steam trams in the city. On July 23, 1872, Henry E. Knapp of New York was granted a patent for a steam-powered car. Knapp's design was unique as it placed the furnace, boiler, and all other connecting parts of the engine beneath the vehicle, leaving the platform unobstructed and allowing easy entry and exit. Coal was to be supplied from the platform into a chute, and the car also had an exhaust filtering system for removing carbonic acid. In 1873 Philadelphia's Baldwin Co. began production of steam-powered streetcars.

At some point, probably in the 1860s, it was discovered in America that small locomotives could be disguised to resemble streetcars or boxcars, so as to make them less threatening to horses. These small engines came to be called "dummies," and they were soon employed for regular railroad use as well as for working on street lines. Following the horse epidemic of 1872, which severely crippled service on many lines, tramway operators took an increased interest in "dummy" locomotives. Manufacturers of these engines included Baldwin, and later, Porter.

In England in 1871, Leonard J. Todd of Leith designed a small locomotive specifically intended to address the unique problems of tramways. His requirements were that it be specially adapted for operation through common roads, be quick to stop and start, and have reduced noise, smoke, and steam emissions. His primary innovation, a large reservoir of water with a small fire grate, anticipated the subsequent development of "fireless" technology. Water was an ideal heat accumulator, thus if a large store of it was available at high temperature, a locomotive could operate with less attention paid to the boiler, and produce less smoke along its route. Todd eventually constructed a locomotive for the Tramway Via de Santander, capable of drawing two streetcars having a combined total of 76 seats.

Todd's experiments drew attention to the problems associated with steam traction. Aside from the fact that heavier and more expensive track was required, pollution from smoke, soot, and noise was a constant public irritation. There was also a general fear of boiler explosions. Despite these drawbacks, however, people continued to experiment with this system. In 1872 John Grantham had a steam tram of his own design constructed. It was tried in November 1873 on a portion of the London Tramways, but did not perform well. The 27-foot car was subsequently rebuilt. Changes included replacement of the two small boilers, previously situated in the center, with a large one moved to one side, thus allowing passage between the front and rear portions of the car. The car then began operation on the Wantage Tramway. In 1876 Grantham designed a steam car for use on the Vienna Tramways. It was constructed by the Starbuck Co. of Birkenhead and had the boiler near the front.

In 1875 the Societe Metallurgique in Belgium constructed a three-cylinder tram locomotive, 7 feet in length, which in external appearance resembled an omnibus. In August 1875 a tramway locomotive built by A. Kohl, a Danish engineer, was tested on the American Omnibus Co. line in Copenhagen. It successfully drew two streetcars filled with passengers. Another Danish tram locomotive, built by Smith and Mygind, began operation that December.

In 1875, the English firm of Merryweather & Sons began regular manufacture of a tram locomotive of their own design. By 1878, 46 of these were operating in Paris on the 4½-mile line between the Bastille and Gare Montparnasse and on the line from the Bastille to St. Mande. The Merryweather firm also provided locomotives for tramways that opened at Kassel, Germany, in August 1877; Rouen and Barcelona in November 1877; and Wellington, New Zealand in February 1878. By 1877 two other British companies, Kitson of Leeds and Hughes of Dublin, were manufacturing tramway engines. In trials, a locomotive of the latter company successfully drew three loaded cars.

In 1876, at the time of America's centennial celebration, steam trams began operation on Philadelphia's Market St. line. The following year they received trials on New York's Third Ave. Rwy., between 130th St. and its depot at Sixty-fifth St. Also about this time, following successful trials, five steam cars took over all work on the Carondelet line in St. Louis.

Earlier in America, in 1872, an experimental "fireless" or hot water locomotive, invented by Dr. Emile Lamm, began service on the New Orleans & Carrollton R.R. Its general principle of operation was an interesting one. At standard atmospheric pressure water will boil at 100 degrees Centigrade or 212 degrees Fahrenheit. If atmospheric pressure rises, however, then water has to reach a higher temperature to boil. If it is in a pressure-resistant container, and there is no available space for it to expand and become steam, then it can only continue to increase in temperature and pressure. When an escape outlet is opened, steam is spontaneously formed from the water, and can be released to a cylinder to power a piston. Lamm's locomotive contained a large "reservoir" for holding 60 cubic feet of water. Prior to a trip, it would be filled with cold water and then connected to a steam pipe from a stationary boiler. The cold water was quickly heated to a high pressure, the connection was then uncoupled, and the locomotive was ready for its journey.

By 1875, in England, Todd had designed a "fireless" double-decker tramcar, similar in concept to Lamm's. His two reservoirs and all the required machinery were placed beneath the floor of the vehicle. In 1875, Bede & Co. of Belgium built a hot water steam car for the Societe Generale de Tramways. It operated daily at speeds up to 10 m.p.h. on level ground. Its supply of hot water was replaced every 2 hours, in a process taking 15 minutes. In 1875, a French engineer, Leon Francq began experimenting with the "fireless" or "vapeur sans foyer" system. By 1878 a locomotive of his design, built by Cail, had been put into service on the line between Rueil and Marly. Francq's engine was only able to operate a distance of about 9 miles before its steam was exhausted. Its advantages were claimed to be the following: (1) Steam was produced at low cost in a fixed boiler using coal straight from the mine (with standard firebox systems the more expensive coke often had to be burned to reduce smoke). (2) Dead weight was reduced as no fuel needed to be carried. (3) Higher pressure could be applied for ascending inclines. (4) Manpower needs were less because a single person could both operate the engine and keep a lookout.

In July 1883 a traction system of this type was opened at Batavia (now Jakarta), Indonesia. By 1885 the Batavia Steam Railway Co. had two sections. One consisted of a 4.97-mile line between Batavia and Kramat. The second was a 2.80-mile route from Kramat to Muster Cornelis. While the first section was double-tracked with Demerbe grooved rails, a single track of Vignoles rails served the second. Two stationary boilers were situated at Batavia and three at Kramat. Serving the entire system, these were built by Lokomotiv Fabrik Hohenzollern, Dusseldorf, as were the company's 21 Lamm-Francq engines.

This technology continued to evolve. In its June 19, 1890 issue, *The Engineer* reported on a new generation of these tram locomotives operating on a line from the Arc de Triomphe to St. Germain. Each engine pulled two passenger carriages and a luggage van. Later on fireless engines were found to be ideal for serving certain specialized needs of industry.

While tramway technology is primarily associated with urban transportation, it was adapted in many countries for providing suburban and interurban service. This trend was observed by *Railway Age* in its November 15, 1877 edition. Quoting the English publication *Railway News,* it commented: "The tramway, constructed upon the common roads or upon the sides of turnpike roads, as is the case at Swansea, land now lying waste and unprofitable, may be made at a cost infinitely less than the ordinary railway, and these may be made to serve as most useful tributaries to the great trunk and other railways of the country. The successful application of steam to tramways constitutes a new era in the history of locomotion in this country, and it is impossible to predict the limits to which it may be applied, or the benefit which it will ultimately confer upon our railways." Holland was one country that built many such lines, most of which were narrow-gauge. Similarly, steam tramways became very popular in Italy. In 1891 the country had 124 lines extending for a total distance of 1,577 miles. Of this mileage, all but 200 were laid out on public roads. Only three of these lines, however, were in narrow gauge. The longest of these tramways was the 43½-mile Mantua-Brescia route. The speed of these small trains averaged about 12½ m.p.h.

In 1877 a steam carriage designed by A. Brunner of Winterthur, Switzerland, began operation on the Lausanne and Eschellens Branch Rwy. This was an 8.86-mile line of meter gauge, and was laid on the common road—a "Chemin de Fer Régional." For inspiration, Brunner had gone back to Fairlie's steam carriage of 1869, although making many modifications. It had two four-wheel bogies, was 42 feet, 3½ inches in length, and provided three sections for passengers—the saloon seating 24, the pavilion at the rear seating 7 (the smoking section) and, after ascending the staircase, the imperial or upper story, having seats for 30. In tests the car carried up to 120 passengers. It had a normal operating speed of 9 m.p.h., but reached speeds of up to 19 m.p.h. on level ground. Also in 1877, a steam carriage designed by W. R. Rowan of Copenhagen was built. Rowan was inspired by the double bogie design of typical American passenger cars. His carriage was 32 feet long, most of it built from teakwood, and was designed to seat 60 passengers. It later operated on a line of the Copenhagen Tramway Co.

In its December 23, 1882 issue, *Scientific American* briefly reviewed a new steam car, designed by C. Thomas, which was operating in Berlin.

> The European practice is in many respects quite different. For the suburban traffic of Berlin, for example, the cars are much heavier than American cars for similar work, and usually are double deckers . . . yet it is evident that they are successful and apparently popular . . . (Thomas' car) is said to be especially adapted for short suburban roads and for connecting pleasure resorts. The car can be detached from the locomotive, but generally they remain together. The engine, which is capable of hauling two or more cars, has two 7″ cylinders with 12″ stroke, and can develop 100 horsepower. It has a speed of 38 m.p.h., and may be quickly stopped and started. It burns from 5 to 6½ pounds of coal per mile and can carry fuel for a run of 125 miles and water enough for 50 miles. The ground floor of the car has seats for 20 first and second class passengers; the upper story has 40 third class seats. There is also standing room for 10 or 12 passengers, space for baggage, a closet for mail bags, two drop boxes for letters, a closet for the conductor and another for tools and implements. The car is heated by steam, and is provided with both hand and steam brakes.

In the tramway trials, held at the Exposition Universelle at Antwerp in 1886, the Rowan steam carriage was awarded a gold medal for its class. For this competition Rowan adopted the principle of reducing the dead weight by the use of an engine whose power greatly exceeded its adhesion, and utilized the weight of the passengers to make up the difference. The locomotive portion of this car functioned as the front bogie. It was inserted into the front of the carriage in such a way that it could be removed for inspection, cleaning, or repairs. Perhaps to cut down on weight, Rowan provided only two wheels at the rear. By the time of this competition steam rail cars were operating in France, England, Belgium, Italy, Switzerland, Germany, Holland, Denmark, Sweden, and elsewhere. The December 14, 1888 issue of *Railroad Gazette* mentions three steam tramway lines in Berlin: the 3¾-mile Kurfürstendamm line, which used Rowan equipment; the Teltow Rwy.; and a line between Rollendorfplatz and Steglitz.

In America steam traction experiments continued to the end of the century. However, despite these efforts, steam cars never gained the popularity that they did in Europe. In *The American Railroad Passenger Car,* John H. White Jr. cites these reasons for the failure of this technology in the United States:

> The vertical boiler was a major problem because of its limited steaming capacity, at least in sizes practical to enclose within the confines of a car body. Also, deposits settled on the lower tube sheet of such boilers, causing them to overheat and thus loosening the tubes. To maintain steam, small boilers were worked very hard, resulting in rapid deterioration. Too much time was required to water, coal, and fire up a cold boiler. And there was the inherent defect of the small steam plant, where efficiency seems to be directly related to size. The most productive steam units are ocean vessels and power plants, which do not have width, height, or weight limitations. Steam cars are relatively small units that proved high in maintenance expense and low in thermal efficiency.

By 1890 electrical trolley systems had reached a level of refinement that made American steam trams obsolete. However, when trolley companies began to compete with branch lines for short-distance passenger traffic, there was a renewed interest in steam railcars. In a single year, during the 1890s, the New England R.R. lost 350,000 passengers to trolley competition. After a careful study of the problem, it was determined that the local Boston trains, consisting of a locomotive, two cars, and a full crew, averaged only 47 passengers per trip.

As this work could be done cheaper by a steam-powered rail car, the company decided to consult the Schenectady Locomotive Works. After some experimentation a boiler and engine truck were constructed and installed in an old dining coach. Named the *Composite,* this car began service in October 1897. After failing on one line because of the gradient, it was operated successfully on the Cape Cod route, sometimes pulling a number of other cars. However, about this time the New Haven R.R. took over the New England company and no further cars were built.

The Erie R.R. also had problems with trolley competition, and hearing of the New England experiment acquired a similar Schenectady steam unit. After installation in an old immigrant coach, the car began service, running 14 trips a day over a 4-mile branch line. While costs were reduced, this car suffered from frequent breakdowns, and gained a reputation for having a rough ride. At the same time the Baldwin company made renewed efforts in this area, in this case experimenting with compounding. Three cars were built and sold to railroads; however, none proved successful.

In France, Serpollet's invention of a new type boiler in the late 1880s made possible a fresh generation of lighter, faster, and more economical steam vehicles. For this he employed capillary water tubes that allowed instant steam generation. In 1897 the Purrey boiler provided further refinements, allowing additional weight reduction by cutting down on the water supply. By 1900 Purrey system tramcars were in use on a number of lines in Paris, in some cases replacing Blot-type cars powered by electrical storage batteries.

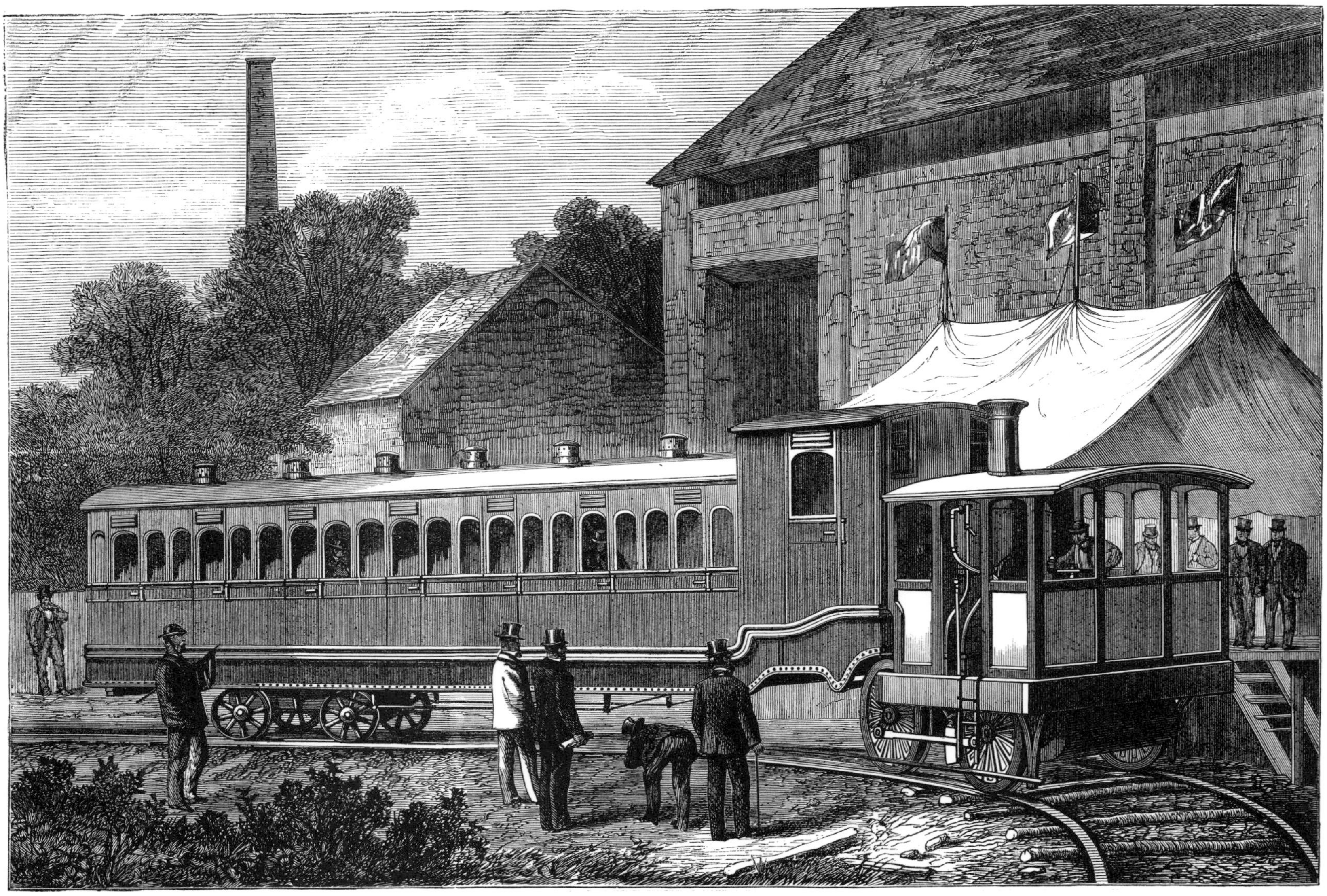

English steam railway carriages: Top: Charles H. Gregory's "Fairfield," 1848. Bottom: Robert Fairlie's steam carriage, 1869.

American steam-powered streetcars: Top: The Baxter car tested on New York's Bleecker St. line, 1872. Bottom: Steam tram design patented by A. Irwin of Pittsburgh, September 2, 1862.

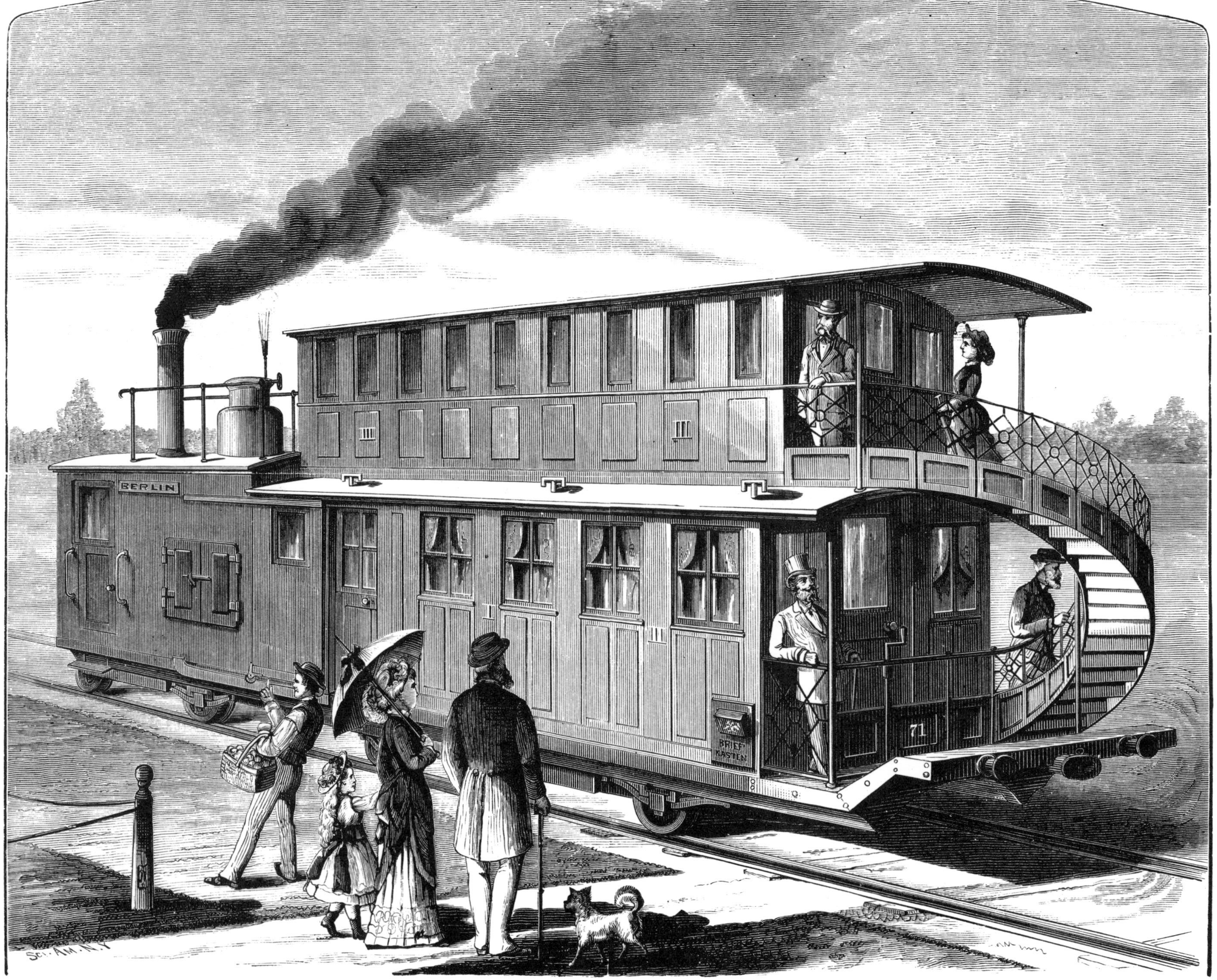

PLATE 312

Steam-powered streetcar designed by C. Thomas, Mainz, Germany, 1882. This 100 h.p. tram could reach a top speed of 38 m.p.h., burning 5 to 6½ pounds of coal per mile.

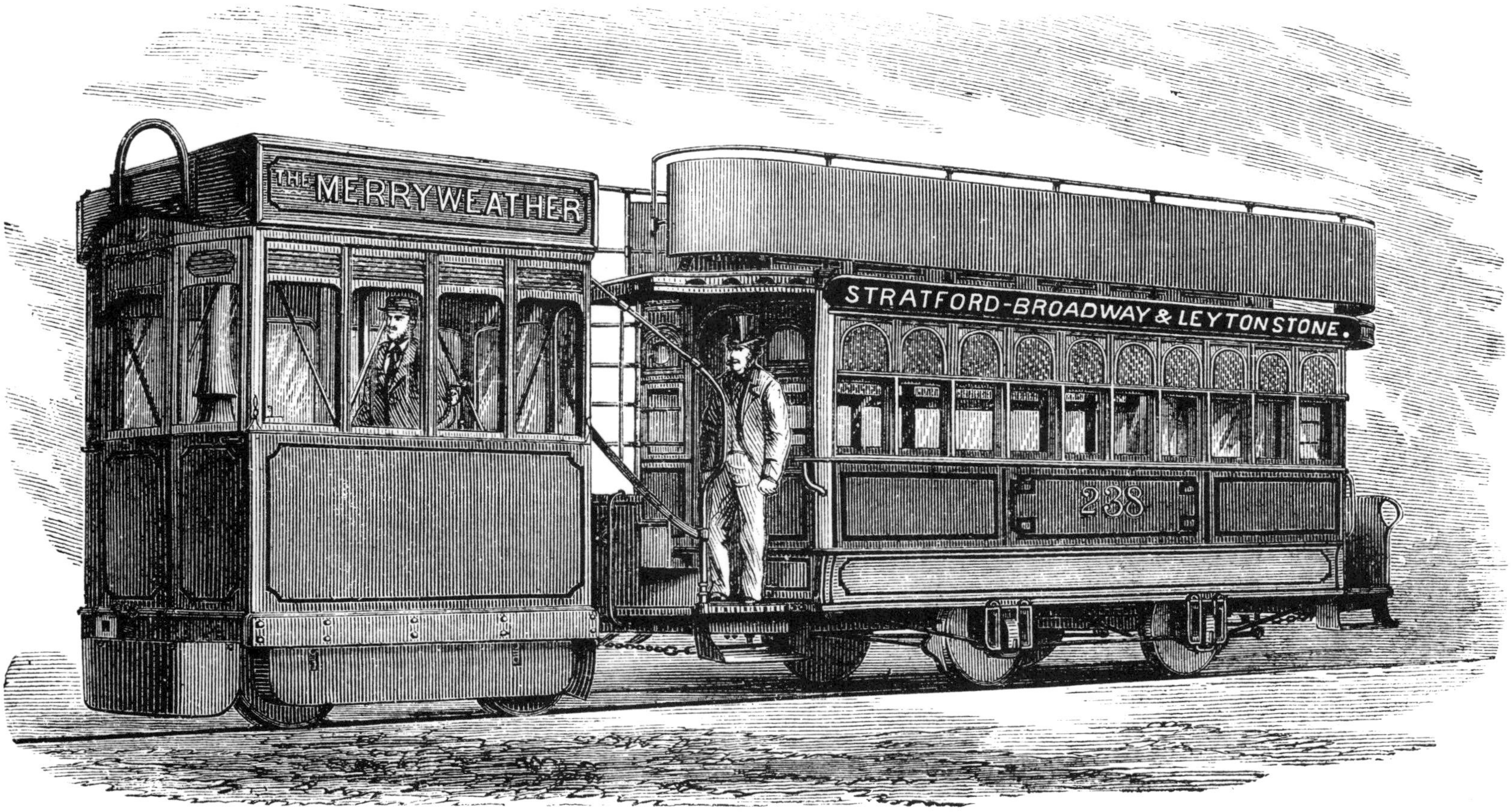

Steam-powered dummy locomotives for street railway use: Top: Engine built by England's Merryweather Co., 1875. Bottom: Merryweather locomotive in operation on Paris's Bastille–Gare Montparnasse line, 1878.

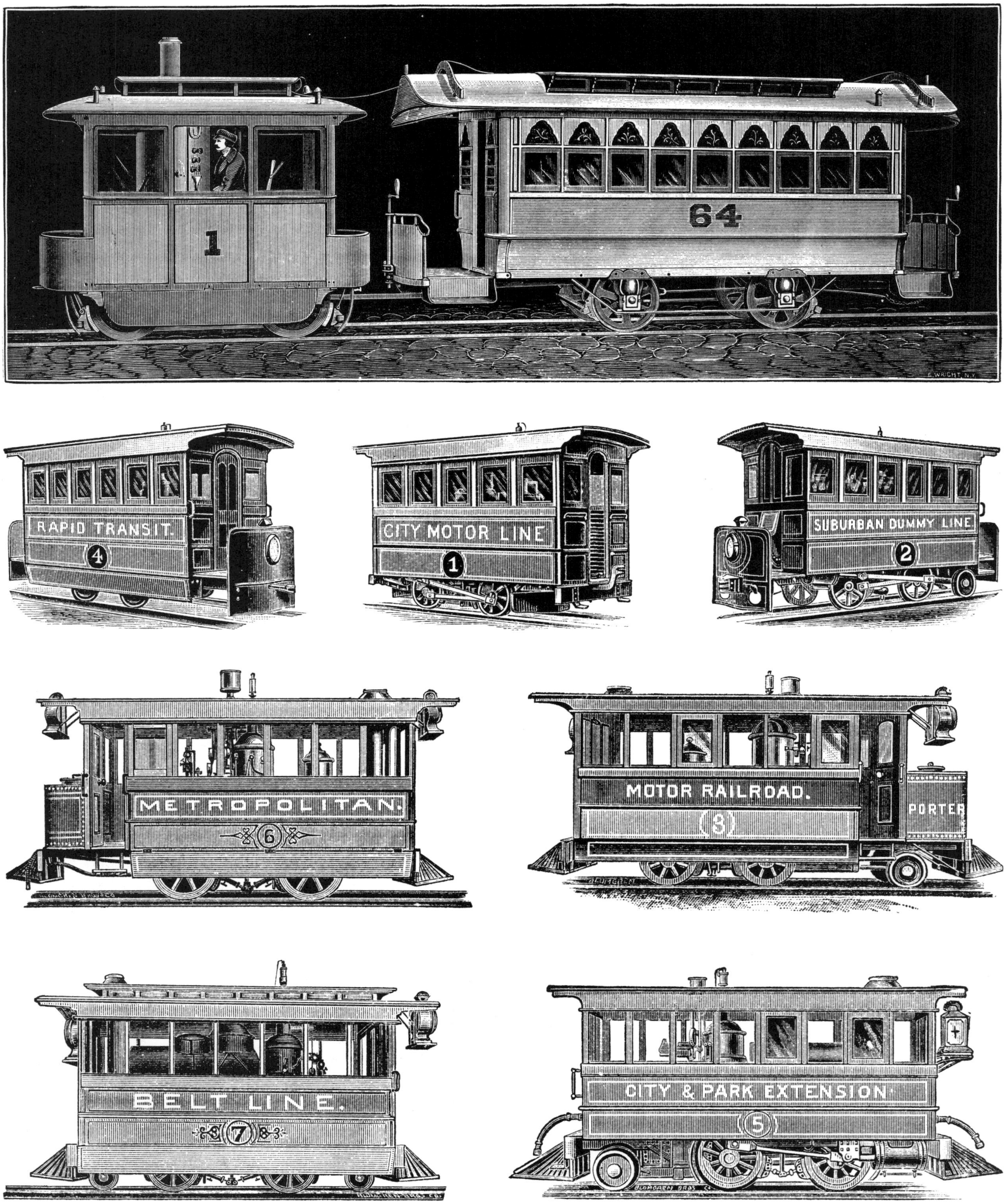

American "dummy" locomotives: TOP: Baldwin engine pulling streetcar, 1876. Below are various engines, all built by H. K. Porter & Co., Pittsburgh, as shown in the company's 1889 catalog.

PLATE 314

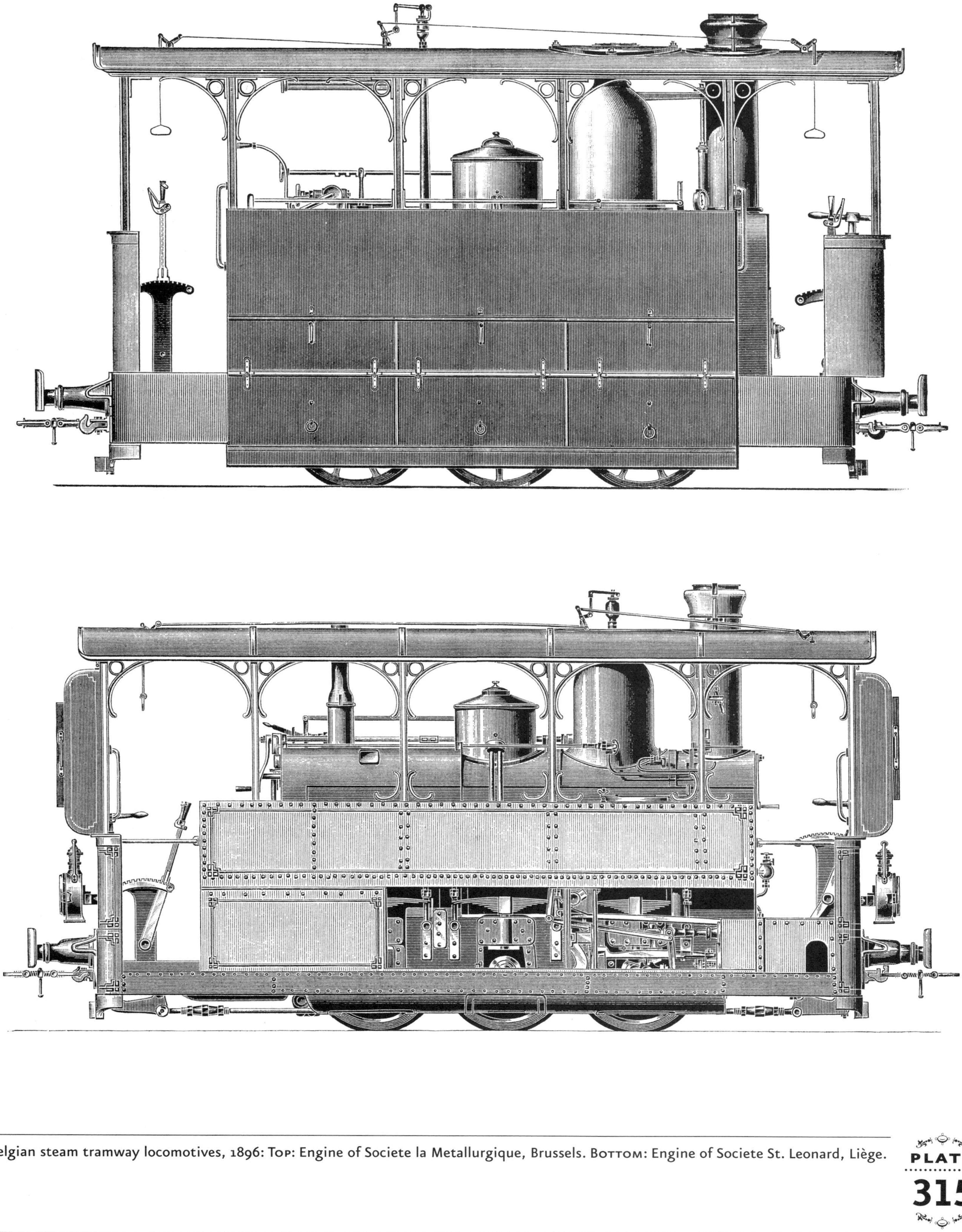

Belgian steam tramway locomotives, 1896: Top: Engine of Societe la Metallurgique, Brussels. Bottom: Engine of Societe St. Leonard, Liège.

PLATE 315

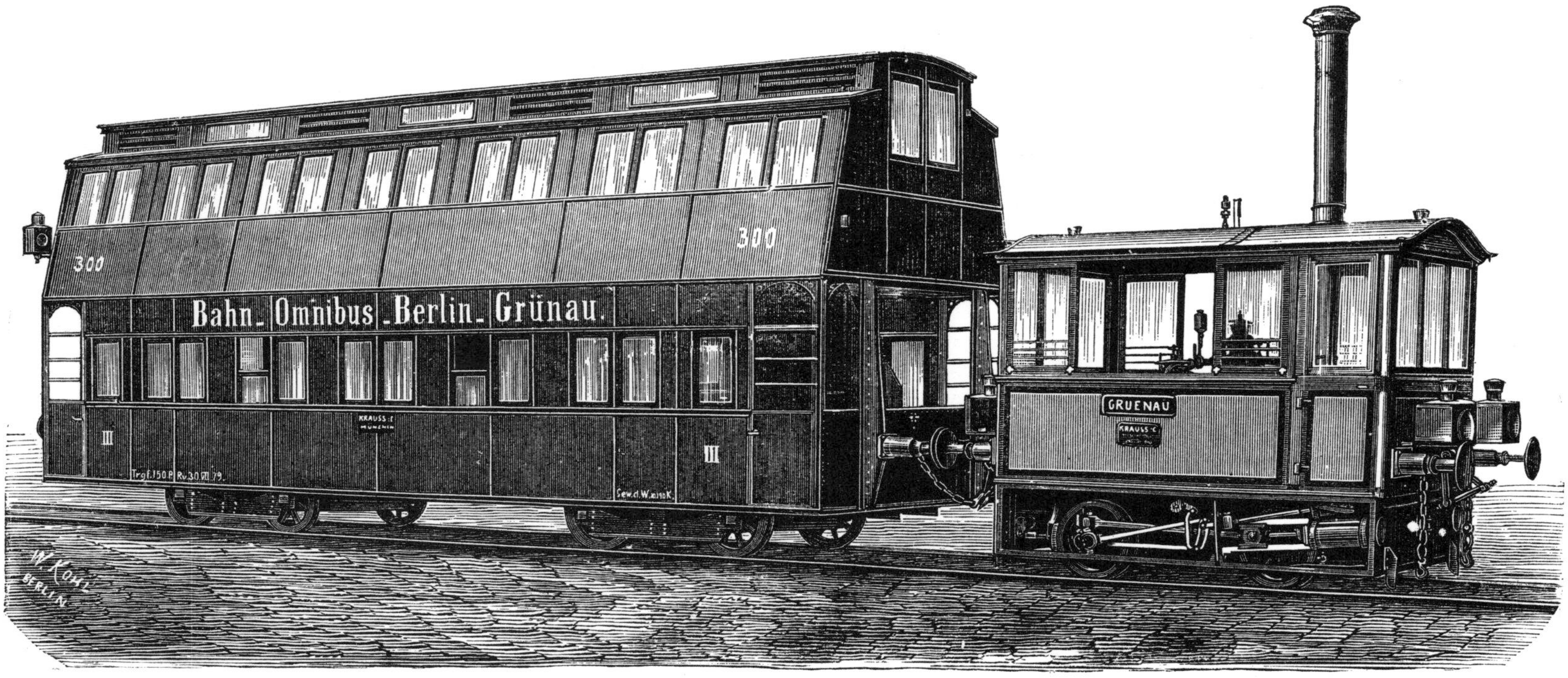

Top: Tram engine and double-decked carriage built by Munich's Krauss & Co. for operation in Berlin, circa 1890s. **Bottom:** German tram engine, circa 1890s.

Fireless technology: Top: Lamm's fireless tram locomotive, New Orleans & Carrollton R.R., 1872. Bottom: Lamm-Francq fireless engine, Paris, 1878.

Fireless technology: Top: Todd's fireless tramcar, England, 1875. Bottom: Fireless Lamm-Francq tramway engine built for service in Batavia (now Jakarta) by AG für Lokomotivbau Hohenzollern, near Dusseldorf, 1882.

CHAPTER

26

Cable Traction

Inclined Planes, Street Railways, and Funicular Systems

THE practice of using rope, and later, wire cable for hauling heavy loads goes back a number of centuries. This technology, often involving pulleys and counterweights, had been previously used for mining purposes, for carrying materials across rivers, and for transportation up and down inclines. Its first application for tramway use occurred about the end of the eighteenth century. Through use of rope and pulley a funicular system was improvised, so that the descent of loaded wagons provided the necessary force to pull up unloaded cars on a parallel track. More sophisticated funicular developments are discussed later in this chapter.

In 1808 the Englishman S. Cooke devised a system for hauling mineral wagons up a slope by using a stationary steam engine for winding purposes. His engine station was built at the top of a hill near Inspeth Mine at Birthy Fell in Durham County. From there a rope was connected to cars ready to ascend the grade. This practice soon became common in the Newcastle area and was later exported elsewhere.

In *Railways: The Pioneer Years,* Malcolm Fletcher and John Taylor make this observation: "These engines positioned at the top of the hill worked in two ways. Either they turned two drums which simultaneously drew in up both slopes ropes which, passing through guide pulleys, hauled up the trucks (wagons), or they operated a length of rope which turned continuously in a full circle, the trucks being attached to it."

Sometimes engine houses were set at intervals along a route, with trains being drawn from station to station. In other instances level sections of a line were worked with either animals or locomotives, while stationary engines were used on slopes. This technology was utilized on the Stockton & Darlington, Blackwall, Brusselton, and Greenfield railways. It was employed in Liverpool Tunnel on the Liverpool & Manchester line. The London & Birmingham used stationary engines for moving the first trains between Euston and Camden Town. As locomotives increased in power, the use of this technology declined. Later it was mainly utilized only for the most extreme grades.

Pennsylvania's Lehigh Coal and Navigation Co. employed this system in 1845. Previously mules had been used for drawing cars uphill while loaded cars were coasted downhill. However, activity had increased so much that the uphill traffic seriously impeded the flow of that going downhill. A plan was devised whereby empty cars could be returned for reloading by an alternative circular route. After inclined planes were graded on the eastern slopes of both Mt. Pisgah and Mt. Jefferson, tracks were laid and stationary engine houses were installed at the top of each. Empty cars were assembled for return at the base of Mt. Pisgah and then drawn to its top. From there they were coasted downhill and then drawn upward to the top of Mt. Jefferson. From that point they descended by gravity to loading points near Summit Hill.

At Lyon, France, probably in the 1860s, two inclined plane railways were built to provide passenger and goods service to nearby villages. One line ascended to Fourvieres and St. Just, and a second line to La Croix-Rousse; 150 h.p. stationary engines provided the necessary traction.

The idea of propelling a surface car by means of connecting it with a gripping device to an endlessly moving underground cable was first proposed by W. Brandling in 1845. In 1858, E. S. Gardner, of Philadelphia, suggested the use of a slotted underground conduit through which the propelling cable might move. The following year, Foster and Brown advocated the possibility of adapting such a system for an elevated railway. In 1866 Charles T. Harvey came up with the first practical idea for a grip. A forked device could be lowered from the car that might catch any one of a series of discs attached to the cable at regular intervals. To stop the car, the grip only had to be raised and brakes applied.

In 1867 Harvey began construction of New York's first elevated rail line. Originally called the West Side & Yonkers Patent Rwy., it extended from lower Manhattan up the east side of Greenwich St., and later up Ninth Ave. to Thirtieth St. A stationary steam engine powered the cable, which could draw cars at speeds up to 15 m.p.h. However, this system proved inadequate in the long run, and in 1870 it was replaced with steam traction.

Tradition has it that in 1869 a manufacturer of wire cable in San Francisco, Andrew S. Hallidie, witnessed a mishap that was all too common in the city. While struggling to pull a streetcar up a hill, a team of four horses stalled. When one slipped, the entire team was dragged backward downhill for some distance. All the horses were badly injured, and had to be destroyed. Hallidie, who was familiar with cable techniques employed in mines, became resolved to try out similar methods to provide a better solution to this problem.

While this story may be true, the first franchise granted by the city for such a line was in 1870 to Benjamin H. Brooks and three associates. Although this group worked out many details of the proposed system, they were unable to gain financial backing. Thus they soon sold their franchise to Hallidie.

Their system employed the endlessly moving cable, which was to be operated in some sort of housing beneath the street. Power from a stationary engine would be supplied to the cable by the use of pulleys. Rather than connect a streetcar directly to the cable, a special car was devised, smaller in size and closer to the ground, which was called the "grip car." Modeled perhaps on the "dummy" engines of the period, it was intended to draw the streetcar. The grip car had a device, called the "grip," which was an improvement on a cable grip invented by former Confederate Gen. Beauregard in 1869. It extended downward through a thin slot in the street, where it could engage or disengage the moving cable. When the driver, known as the "gripman," wanted to move forward, he would engage the cable. For stopping he would disengage from the cable and apply the brakes.

Forming the Clay Street Hill Railroad Company, Hallidie soon began excavations for his tramway. In the course of this activity many gas, water, and sewer lines were uncovered and these had to be relocated. The average grade on the route was 580 feet to the mile, with the steepest portion being 850 feet to the mile, or 16 in 100. Experimental operations on the line began on August 1, 1873, and it soon became successful.

A report on this tramway, in the April 17, 1875 issue of *Scientific American,* indicated that the route was double-tracked, with the endless cable drawing traffic downhill on one side and uphill on the other. Upon arrival at either end, the streetcar was disconnected from the grip car, and coupled to a team of horses, where it continued its journey on level ground. The grip car was then turned around on a turntable for its return trip and coupled to another tram. The line extended for a distance of 1,100 yards and had a gauge of 3½ feet. The journey in either direction took approximately 11 minutes, and averaged 4 m.p.h.

About the same time construction began on other cable car lines in San Francisco. The Sutter Street Rwy. opened in 1877. In April 1878, the California St. line was inaugurated with much fanfare and highly decorated cars. The Geary Street Park & Ocean Rwy. opened in 1879. The steepest gradient on the California St. route was 18 in 100, while maximum grades on the Sutter and Geary lines were 8.7 in 100 and 9.8 in 100, respectively.

The Market Street Cable R.R., which opened on August 22, 1883, was a much larger system. From the Embarcadero, where it connected with a number of ferries, it went down Market St. through the heart of the city, and then split off on three branch routes leading out to the suburbs: the McAllister St. and Haight St. lines, both of which terminated at Golden Gate Park, and the Valencia St. route, which crossed over the Southern Pacific tracks and ended at Mission Road. There was substantial money invested in this enterprise. Its officers included Leland Stanford, president, and Charles F. Crocker, vice president, both prominent officials of the Central Pacific R.R.

Railroad Gazette, in its November 23, 1883 issue, quoting the *Mining and Scientific Press* as its source, published a lengthy article on this line, and reported the following: "The building of this road has had the effect of enlivening, so to speak, a large tract of this city. For some years, south of Market St. and out by Valencia, the city has been rather slow, the northern and northwestern portions improving much more rapidly, and having improvements of better character. The length of time it took to get out toward the Mission or thereabouts, or even on Market St. to Tenth and Twelfth Sts., was detrimental to those portions of the city. The cable cars now go from the ferry to the end of Valencia Street in 34 minutes and run frequently."

While previous cable car companies followed Hallidie's example, and used grip cars for pulling trams, the Market St.

line improvised combination cars, adding forward extensions to old horse cars. Single cars were created, each having an open "grip" section in the front and a closed section in the rear. This type, patented by the company's chief engineer Henry Root, was later called the San Francisco or California pattern. The cars were run at faster speeds than before, averaging, with stops, 7½ m.p.h. An engine house, for providing motive power to the cables, was constructed on the south side of Market St. at its intersection with Valencia. Car houses were built at the end of Valencia St. and on Haight St. between Shrader and Stanyan. Upon completing its run at the wharf, each car was turned around on a turntable, and then, depending on its outbound destination, moved to the appropriate sidetrack. There it received new passengers and shortly resumed operation.

Circa 1884 an alternative cable system was devised by Orel D. Orvis and Nelson B. Adams. It employed subsurface rails and two cables, one for low-speed operation (4 m.p.h.) and one for high-speed (8 m.p.h.). Replacing the middle slot were two slots positioned above the underground rails. Operating on these rails, and also underground, were double-flanged wheels. These were connected to the streetcar above by metal supports sufficiently thin to pass through the slots. Parallel cables, supported by pulleys, ran on each side. These pulleys were situated just inside each rail but at a slightly higher elevation. Like the rails, they were bolted to special housing units spaced at regular intervals. These housing units also supported the slot frames. The United States Cable Road & Subway Co. of New York promoted this system. However, it is unknown if any lines using this technology were actually built.

While expensive to construct, cable railways provided a smoother, faster, and cleaner service than horse car lines. Furthermore, once built, they were cheaper to operate. Thus, while originally designed for inclines, they also proved more economical in areas of high traffic density. Soon many of America's largest cities found it expedient to construct their own systems. Cable traction was introduced in Chicago in 1881, in Philadelphia in 1883, and on the Brooklyn Bridge the same year. By the end of 1885 other lines had begun operation in Los Angeles, Kansas City, St. Louis, and Cincinnati.

Of these new lines, the Kansas City Cable R.R. encountered the steepest gradient, climbing an 18.3 in 100 incline along Ninth St., from the Union Depot to the main business district. It was also unique at the time in having a duplicate cable set up for immediate operation. Thus repairs could be made without any interruption of service. Although the New York & Brooklyn Bridge R.R. operated only the short distance across the bridge, it was probably the busiest of all cable routes. Its operation at each end was facilitated by the use of saddle tank switch engines. Circa 1897 these were replaced by electric motor switchers built by the Pullman Co. New York later added other cable lines: the Tenth St. and 125th St. railways in 1886, and lines on Broadway and Third Ave. in 1894.

The Broadway Cable Rwy. was an extensive undertaking. Beginning at the Battery and South Ferry, it extended northward up Broadway to Forty-second St., and then up Seventh Ave. to Fifty-ninth St., a distance of almost 6 miles. In its construction a vast network of underground pipes (water, sewage, gas, steam, pneumatic, telegraph, and telephone) were uncovered and rerouted. Once completed, the system had four cable divisions and two power stations. One of these, on the block from Fiftieth to Fifty-first St., between Sixth and Seventh Aves., replaced a stable that had previously housed 2,000 horses used for pulling streetcars. The second power station was situated at Broadway and Houston St.

From the beginning, the public was concerned about the strength and durability of cables, whether or not they might break. It became the custom to inspect the cables each night, following the close of service. Although cables were usually only 1¼ to 2 inches thick, they were extremely well made and generally lasted about 75,000 miles, or a year of normal operation. Cable distances varied on different systems, generally being 3 to 4 miles long. However, in Denver a 7-mile cable was operated.

On rare occasions, when one or two strands on a cable might break, a grip could become entangled, thus making the car impossible to stop. One incident of this type was reported in the July 13, 1888 edition of *Railroad Gazette:*

> A grip car, beyond control and running at the rate of eight miles per hour through the business district of the city, caused a lively scene in Chicago one night last week. The result was three badly wrecked cars, many frightened women and children, and some severe contusions, but no fatalities. As a Clark Street car came out of the tunnel on La Salle going south, the driver applied the brakes, but without effect, as a strand of cable had wound itself around the grip, and the grip would not release. Just ahead of this was a Wells St. train filled with passengers. At Monroe Street the Wells Street train was struck and thrown in a heap into the gutter. Two blocks further on, the moving train struck the curve at Dearborn Street. Here the cars became disengaged from the strand, and ended their mad progress by jumping the track.

Cable car systems also had problems with their slots. These typically contracted when it was cold and expanded when it was hot. Narrow buggy wheels occasionally slipped through widened slots, and the slots could pose additional dangers for the unwary public. Cables had a tendency to expand during the course of their lifetimes. "Dead spots" existed where cars made turns, and often, passengers had to assist in pushing a car forward to once again engage the cable. Cables were also vulnerable to juvenile pranksters who devised methods of attaching things for creating raucous noises or causing other havoc.

Although cable car technology was soon exported elsewhere, it never received the widespread enthusiasm that it enjoyed in America. The first foreign-built line was the Roslyn Tramway, which opened in Dunedin, New Zealand, in 1881. This was the first of four lines that would be built there. London's Highgate Hill Cable Tramway opened in 1884, and Brixton Cable Tramway in 1888. Cable operations began at Melbourne in 1885, and at Sydney the following year. In 1888 cable routes were opened in both Birmingham and Edinburgh. Lisbon inaugurated two lines in 1890. The following year cable service began in Paris on a line between Place de la Republique and Belleville. Cable operations commenced at Matlock, Derby, in 1893, on the Isle of Man in 1896, and at Glasgow in 1897.

The engine house of the Paris-Belleville line contained two Corliss steam engines providing together 100 h.p. The steel cable, 30 millimeters in diameter, was composed of six strands and had a total weight of about 7 tons. The cars operated at a speed of about 7½ m.p.h. They made their journey from the center of Paris up a steep grade to Belleville Church in 10 to 12 minutes.

Cable car railway growth continued in America until circa 1890, at which time approximately 5,000 cars operated over 500 miles of track, carrying 400 million passengers annually. By this time the St. Louis system had over 55 miles of track and Chicago's network had become the largest in the world. Washington, Philadelphia, and New York had built large systems as well. By 1890, however, electrical tramway technology had improved to the point that it surpassed cable operations in both speed and economy.

Soon, many cities felt it in their best interest to convert cable systems over to trolley use. Los Angeles began conversion in 1890, but only completed it in 1902. San Diego terminated its cable operations in 1892 and Denver in 1893. Omaha, Philadelphia, and Providence converted their systems in 1895. Pittsburgh's cable service ceased in 1897, and St. Paul's in 1898.

The Baltimore City Passenger Rwy. and Washington's Columbia line were both late to inaugurate their cable operations, only doing so in 1893 and 1895, respectively. Both lines were scrapped in 1899. San Francisco's Market St. system halted cable service in 1906, the same year that Chicago converted its last cable routes. The New York and Brooklyn Bridge R.R. ceased business in 1908. Cable service continued in Kansas City until 1913, Tacoma until 1938, and Seattle until 1940.

Outside of America, cable tramways were not so quickly discarded for new technologies. Although Sydney ceased cable operations in 1900, Melbourne would not do so until 1940. Cable service continued in London and Birmingham until 1911, Lisbon until 1912, Edinburgh until 1923, Paris until 1924, the Isle of Man until 1929, Glasgow until 1935, and Dunedin until 1957. Today, only two lines in the world remain, both in San Francisco.

Funicular railways differ from the previously mentioned systems in that two cars are connected to a cable in such a manner that when one descends the other will automatically ascend. This system has historically been employed on short routes having high gradients. One line of this type was the South Cliff Rwy. in Scarborough, England. Inaugurated in 1874, it operated over standard-gauge rails and, on a 1 in 1.75 gradient, ascended 284 feet in altitude. It employed a water ballast system in which water was added to the car at the summit. This provided sufficient gravity for the car to descend, drawing up its opposite counterpart. A number of other railways of this type were later built in Britain. They were commonly called cliff lifts.

Another funicular operation was the Metropolitan Rwy. of Constantinople (Istanbul). Inaugurated on January 17, 1875, it had a lower terminus in the shipping and commercial area of Galata and an upper terminus in the residential quarter of Pera. It extended for a total distance of 672 yards, of which 634 were in a tunnel. Having a 1 in 10 gradient, it gained 200 feet in altitude.

The Mt. Vesuvius Rwy., which opened in 1880, operated over parallel monorail tracks, but ran its cars in the usual manner. Built on a 1 in 2 gradient, it required the service of two 45 h.p. steam engines that powered the cable from a station at the lower terminus.

In 1884 Switzerland had four funicular railways. These extended for distances of 0.1, 0.2, 0.37, and 1.12 miles, respectively. The longest of these, the Lausanne-Ouchy line, was inaugurated in 1877, and employed a water ballast system. While double-flanged wheels were typically used on funicular railways, this line employed the single-flange variety. This was because, in addition to carrying passengers, it ran coal wagons down to lake steamers at Ouchy. It also usually opted to run three-car trains in opposite directions rather than single cars. In 1884 the line ran an average of 95 train trips daily, carrying 1,328 passengers.

A number of funicular systems were also constructed in the United States. One such line was the Cincinnati Inclined Plane Rwy., built up the side of Mt. Auburn. Opening on May 12, 1872, it became noted for a disaster that took place on October 15, 1889. In an article for the *Cincinnati Historical Bulletin,* John H. White Jr. provides this account of the tragedy:

> Eight passengers bound home for lunch boarded the incline car for the ascent to Mount Auburn. The trip proceeded normally until the car approached the summit, but instead of slowing down it continued at full speed with a fearful crash into the top station. The engines continued to turn, holding the car at the top of the plane, but with such force that both the hoisting and safety cables were pulled loose. The car, now free from all restraints, lingered for an instant and then began its lightning-like plunge to the bottom. The impact sounded like an exploding boiler—the air was filled with flying bits of glass, wood, and dust. The car body wrenched loose from the truck frame and hurled itself into a grocery store across the street; the roof of the car sailed down Main Street a full one hundred feet. Three passengers were killed outright; two more died within a few hours; and a sixth was dead several days later.

Pittsburgh's Duquesne Inclined Plane R.R. was another of the funicular type. Opened in May 1877, it ascended Mt. Washington, at the mouth of the Monongahela River. By September 1, 1880, this line had transported 500,000 passengers. For its 19-hour operation each day, only five employees were required: two engineers, one conductor, one fireman, and one trackman.

The Superga Rwy., near Turin, Italy, was unique in that it combined elements of both cable and rack technology, and successfully surmounted the problems of curves. The line began at the base of Mt. Superga near the village of Sassi, and ascended to the basilica at its summit. Having a length of 3,130 meters, it was begun in March 1883, and inaugurated on April 27, 1884. The engineer, Thomas Agudio, devised a special locomotive, which received power from an endless cable. It transmitted this force by means of wheels and gears to two horizontal cogwheels that engaged a center rack. The cable, powered by two stationary steam engines, was guided by intermediate pulleys, adjusting to the rectilinear direction or curve of the track. Much in the same manner as rack railways, the engine always operated from the lower position, pushing the carriage uphill and holding it back in descent.

While cable railways provided a useful way of ascending steep inclines, aside from Agudio's system, the technology lacked the flexibility of rack railways in adapting to curves and changing grades. Furthermore, construction of these lines was complicated by the difficulty of transporting up the necessary weight of cable. Thus, while use of this technology continued, its possible applications remained limited.

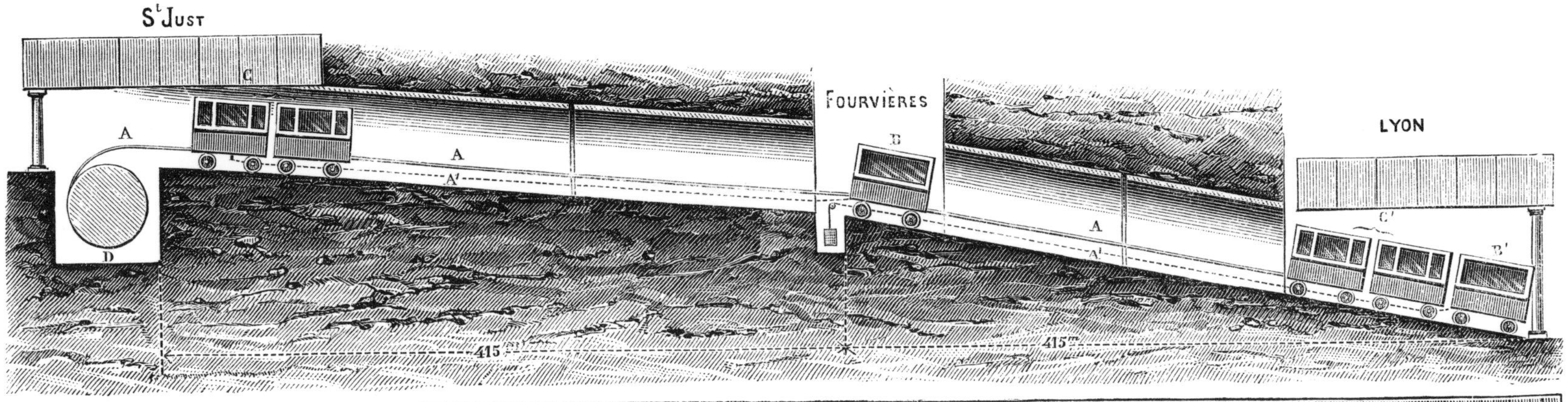

PLATE 319

Inclined plane railways, circa 1870s (?): Top: Cutaway illustration showing operation of line ascending from Lyon to Fourvieres and St. Just, France. Bottom: Train awaiting departure on Lyon–La Croix–Rousse line.

Cable car street railways: TOP: Hallidie's original line in San Francisco, Clay St. Hill R.R., 1875. BOTTOM: Cable system of Orvis and Adams, 1884. In this system there was no middle slot. Instead, there were two slots where the rails would normally be, and for providing power, the gripman had the choice of either a high- or low-speed cable. The car's double-flanged wheels operated underground.

PLATE 320

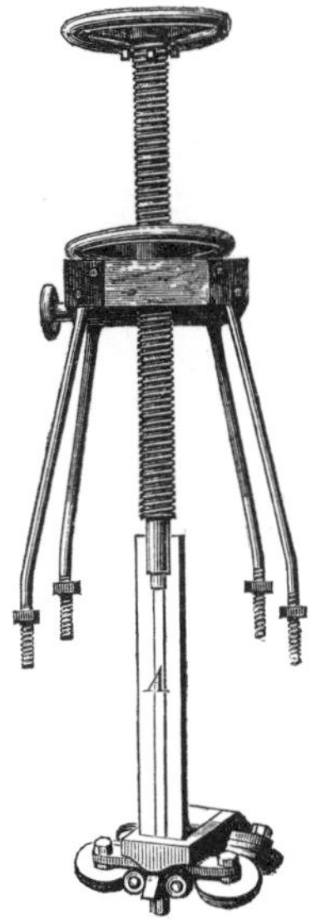

PLATE 321

Top left: Open street car of type used on cable railways, Lamokin Car Works, Chester, Pennsylvania, 1894. **Top right**: Grip device used on Clay St. Hill R.R. **Middle right**: Same device showing lower portion of grip on cable. **Bottom**: Combination-style car and cutaway showing underground cable layout, Market St. Cable Rwy., San Francisco, 1883.

Rolling stock for American cable railways, circa 1890: Top: Grip car pulling streetcar, Metropolitan Street Rwy., Kansas City, Missouri. Middle: Open combination style car, "San Francisco Pattern," Brownell Car Co., St. Louis. Bottom: Open car built by Calumet Car Co., Chicago, for operation in Los Angeles.

PLATE 323

Cable car of Paris's Tramway Funiculaire de Belleville with cutaway showing operation of grip and cable, 1890.

Scene in Pittsburgh showing cars of the Duquesne Inclined Plane R.R., operating on the funicular principle, 1880. This inclined line, one of 17 originally in the area, is still running as a tourist attraction. Originally steam-powered, it was converted to electricity in 1932. Shown passing by the lower terminus is a train of the Panhandle R.R.

TOP: Italy's Mt. Vesuvius Funicular Rwy., circa 1880. BOTTOM: Cable-powered locomotive on Italy's Superga Rwy. near Turin, 1884.

CHAPTER

27

Electric Traction

Trolleys, Railcars, and Locomotives

WHILE steam was totally dominant in the early years of railways, there were a few who began to look to other technologies as possible sources of motive power. In 1834 Thomas Davenport, a blacksmith from Brandon, Vermont, became fascinated by magnets, and he soon constructed what was perhaps the world's first electric motor. Shortly, he devised a second motor that propelled itself around a small track, in the manner of a toy train. In 1835 he exhibited it in Boston and Springfield. In 1839, a Scotsman, Robert Davidson, created a full-size 6-ton electric locomotive. It was powered by a primitive galvanic battery system whereby some 78 pairs of zinc and iron plates were lowered into sulfuric acid to provide power and then raised out again to cut it off. In September 1842 it was tested on the Edinburgh-Glasgow line, and achieved a speed of 4 m.p.h. drawing a 6-ton load. Unfortunately, Davidson's locomotive was destroyed by drunken rail employees and never rebuilt.

In 1847, a New Hampshire professor, Moses Farmer, developed a small electric rail car that carried two people. Galvanic batteries provided power, only this time they were stationary, being placed at one end of the track and connected to each rail. Current was carried through one rail, conducted by the wheels to the motor, and then discharged to the opposite rail. In 1850, Professor Charles G. Page gained funding from the U.S. Congress to build an electric-powered locomotive. For this engine Page improvised a ring-shaped iron bar, apparently the world's first armature. With seven passengers onboard, his awkward creation made its first run April 29, 1851, between Washington and Bladensburg, Maryland, over a 4½-mile stretch of the Baltimore & Ohio line. At best, the ride was jerky and uneven. Cells collapsed and the professor was constantly engaged in rewiring. At times passengers were asked to walk alongside the locomotive as it crept along, but later, at an unexpected point and with passengers onboard, it accelerated to a speed of 19 m.p.h. The trip from Washington to Bladensburg took 39 minutes, but the return journey was a disaster, taking 2 hours. Much improvisation was required on Page's part to complete the trip. Afterward Congress showed no further interest in this project.

In 1850 Thomas Hall, a maker of scientific instruments and employee of Prof. Page, devised a small electric locomotive that took its current from a stationary battery through the rails and wheels. This locomotive was exhibited at Boston in 1851. That same year in France, the Count of Moncel and Geraldy built an experimental electric locomotive, but its trials were unsatisfactory. In December 1864 at Versailles, Louis Bellet and Charles de Rouvre tested a small electric-powered locomotive designed to carry mail. A third rail system was devised for providing current to the motor, and over this ran a tiny two-wheel trolley. The Austrian Miletzer and the Frenchman Adolphe Gaiffe also built electric locomotives about the same time. Unfortunately, although certain useful things were learned in these experiments, they all proved impractical. The galvanic battery was the only known source for a strong electric current at the time. However, the heavy weight of these batteries, their low power and fragility, and the relative costs of zinc versus coal made them unsatisfactory for use on railways. Thus a breakthrough was needed, and providing this was the Belgian Zenobie Theophile Gramme.

In 1870, Gramme invented the first successful dynamo generator, a device that was essentially an iron cylinder enclosed in a coil of copper wire. When driven by the motion of a steam engine or by waterpower, this machine generated current that could be conveyed by wires to any distance. In 1872 Gramme discovered that his dynamo could also function as a motor, receiving generated current and reconverting it into motion. Thus a basic principle of electric traction was established—that essentially two dynamo machines were required, one connected to an exterior power source to generate electricity and the second located in a rail vehicle and connected to its driving mechanism. This dynamo motor would receive the current through the track by overhead wire, or some other means, and power the vehicle.

In 1879 Werner von Siemens, and his partner Johann George Halske, demonstrated the first effective electric railway at the Berlin Trades Exhibition. A primitive 3 h.p. engine, drawing current from a third middle rail, was run around a small circular track. The locomotive pulled its three cars, each with a carrying capacity of 6 passengers, up to a speed of 7½ m.p.h. During the month-long exhibition, 80,000 passengers rode on this train. Siemens had been influenced by the experiments of Felix and Chrétien earlier in 1879, where they attempted to devise a system for using electricity to plow fields.

On May 12, 1881, Siemens opened the world's first public electric street railway, Berlin's 1½-mile, meter-gauge Lichterfelde line. It had a simple two-rail system, receiving current from one rail and releasing it to the other. On it operated a converted horse car that had been fitted with an electric dynamo motor. The line was somewhat successful, but there were complaints of people receiving shocks when crossing it. Furthermore, short-circuits often occurred during rain or snow, and there were power interruptions when the car was in motion.

Later in 1881, Siemens and Halske opened a second line, the Charlottenburg-Spandau Rwy. It had twin elevated wires for conduction. The wires were suspended from poles built alongside the track, and over them ran a four-wheel pulley, with the current passing via a connecting wire down to the car below. The electrical engine of the car was mounted beneath the floor, and the driver operated a single lever controlling the power flow. The same year Siemens displayed another twin overhead system at the Paris International Electric Exhibition. This was described in *Railways: The Pioneer Years,* by Malcolm Fletcher and John Taylor as follows: "Two tubes split along their bottom length took the current to and from the car. Flexible cables were used to attach these tubes to the engine, and they pulled a sliding shuttle along the inside of the tubes through the slots." Siemens and Halske installed this system in 1883 near Vienna on the Modling-Hinterbruhl line. In 1884 they installed a 500-volt system on the Frankfurt-Offenbach line in Germany, and in 1886 a similar system on the Vevey-Montreux line in France.

Harper's Weekly, in its December 24, 1881 issue, made the following comments concerning the advantages of electric rail systems: "The power producing machine is in the depot, and thus relieves the car from all the load of fuel and water which burdens the steam locomotive. There is no annoying smoke, no deafening noise to frighten horses, and no sparks to get into the eyes of passengers. Like all other great discoveries, or rather like all new applications of great discoveries, many improvements will have to be made. That the system will be gradually perfected we have no doubt. The speed, which at Paris was about 6 m.p.h., will be increased, and means of arresting the progress of the cars be rendered more certain. For elevated roads the electric motor is an absolute necessity, which would dispense with all the noise and nuisance with which we are now afflicted."

The invention of the first storage battery by Plante circa 1880 also influenced electric rail traction. On May 25, 1881, the Frenchman Raffard tested a storage battery-powered streetcar on the Vincennes Tramway. It used a Gramme dynamo for a motor. A short time later, storage battery-powered cars were tried out in the London underground. In March 1883 a battery-powered tram was tested on the Acton-Kew line of the West Metropolitan Tramways Co. Later in the decade battery-powered trams began operation in Antwerp and Brussels. During the 1890s storage battery-powered streetcars were used in a number of places, including New York's Fourth Ave. line, and on the Neuilly-Levallois and Madeliene-Courbevoie lines in Paris. On the Paris-Pantin & Aubervilliers line, completed in 1898, batteries were used in combination with an overhead wire system.

In America, the inventive genius Thomas Edison took up the cause for electrical railways when he built his first small dynamo-powered locomotive in 1880. To facilitate these experiments at his research center in Menlo Park, New Jersey, a small narrow-gauge line of about 2 miles was built. Drawing a small passenger car, this locomotive reached speeds of up to 40 m.p.h. Although several experimental locomotives were built, the system itself had a number of problems. Primary among these was the inherent danger from electrical shocks by the two-track conduction system. To reduce the power of the shocks, the voltage had to be cut down. This, in turn, reduced the power of the locomotive and its effectiveness. Unfortunately, at the time, Edison's attention was on the development of the incandescent lamp. Thus his railway project languished and was later dropped.

Other American inventors also attempted to solve the problems of electrical traction. In 1882, Leo Daft, an English-born inventor, discovered that the electrical current passing between the locomotive and track increased its adhesive powers. The following year Daft completed a 2-ton locomotive, the *Ampere,* which had a 25 h.p. motor and employed a third rail conduction system. During 1884–85 small Daft-built narrow-gauge engines pulled passengers at Coney Island, the World's Fair in New Orleans, Mechanic's Fair in Boston, and Point of Pines, Massachusetts. One visitor at Coney Island was Thomas Robbins, an official of a Baltimore street railway. He persuaded Daft to try out his system in Baltimore. For this purpose, three small locomotives were built, and then each connected to a horse car. After a successful trial run, Daft's experimental line was opened on August 10, 1885. Service became much faster, doubling the previous speed of the horse cars. Unfortunately, however, Daft's third rail system, much like the two-rail systems of Siemens and Edison, proved vulnerable in wet weather. Complaints were received of shocks and the occasional electrocutions of small animals. As a result electrical traction was terminated on the line after about a year.

During the Chicago Exposition of 1883, an experimental locomotive, *The Judge,* built on the Field-Edison principles, was operated on a demonstration line. Work on this project was almost entirely carried out by electricians from the United States Electric Co. Running June 9–23, it pulled a single passenger car and carried a total of 26,805 visitors.

In 1884, John C. Henry, an electrical novice, attempted to build a "horseless" streetcar in Kansas City. Previously he had been wiped out in farming by a grasshopper plague, and had since become a telegraph operator. He purchased a dynamo in Chicago and then set it up in a shed, powering it with the steam engine from a thrashing machine. For his experiments he equipped an old mule car with another dynamo and arranged to use a small stretch of track. His conduction system consisted of two bare copper wires that were suspended overhead above the track. Two little "trollers," or wheels, ran above these wires, drawing current down through connecting wires to the streetcar below. In early tests, his car accelerated so abruptly that twice it derailed. Henry was unable to solve his problem electrically, but instead dealt with it mechanically by rebuilding the car with a clutch and 5-speed gearbox. Soon three other cars were equipped this same way. Operating on a small line, they gave uneven and sporadic service, and after several months the project was terminated.

Two partners, Bentley and Knight, also tried to develop an electric tram system in Cleveland in 1884. For conduction they chose to use a slotted ground conduit between the rails through which passed the "plow," a thin blade-like protrusion extending from the car. Unfortunately the conduit was made of wooden boards, which continually loosened, thus making the passage of the "plow," at times, something less than smooth. The manner in which the motor was suspended beneath the car and linked to the wheels was slipshod and precarious. Coiled springs served as drive belts, but continually broke. Derailments were frequent and in these events the plow usually damaged the conduit. Other problems arose as well. Finally, the onset of winter brought an end to this endeavor. The pair did continue their efforts, however, and installed other lines later on, using overhead wires.

In the early 1880s, the Belgian American Charles Van Depoele experimented with an overhead "troller" system that used a single overhead wire with a single "troller" connected to a pole. Unlike other systems, however, this "troller" (or trolley, as it came to be called) ran beneath the wire, and was held against it by the upward pressure of a spring at the bottom of the pole. Van Depoele experimented with other techniques for drawing current as well, and began displaying his inventions at trade fairs. For the Toronto Exhibition of 1885, he set up an electric-powered train that carried visitors the one-mile distance from the railroad station to the site of the event. This train consisted of an electrified car drawing three open cars. It was very popular and in one five-day period carried 50,000 people. In this electrified car, he had returned to his single overhead trolley system.

Later, in his early tram line ventures, beginning at South Bend, in 1885, Van Depoele chose to use a simpler method: the overhead running trolley connected to a trailing wire. Soon he built other lines in Scranton, Pennsylvania; Montgomery, Alabama; Appleton, Wisconsin; and elsewhere. It took a few years, however, before he realized that the upward pushing trolley was the superior method of conduction. For his streetcars, Van Depoele preferred to have the motor located on the front platform where it would be under the watchful eye of the motorman. Unfortunately, these overhanging platforms were not strongly supported, and after a period of use they would begin to give way. Van Depoele's trolley lines also had other flaws, but they sufficiently demonstrated, perhaps for the first time in America, the true practicality of electrifying street railways. In 1888 the Thomson-Houston Electric Co. bought Van Depoele's patents.

In the meantime, Leo Daft had begun experimenting with a larger 125 h.p. electric locomotive, the *Benjamin Franklin.* It was tested in 1885 on New York's Ninth Ave. elevated line pulling a number of passenger cars. After these trials proved disappointing, the locomotive was rebuilt in 1888, and then performed much better, drawing eight-car trains up to speeds of 25 m.p.h. After this, Daft began experimenting with streetcars, using an overhead two-wire trolley system. Like Van Depoele, he used poles connected to trolleys, but the trolleys ran above the wires, depending on the force of gravity to keep them on the track. Often they would fall off, interrupting service each time. Daft preferred to have the electric motors of his trolleys underneath the vehicles, and in this choice he was more successful than Van Depoele. Daft's system was soon tried out on lines in Orange and Asbury Park, New Jersey; Pittsburgh; Los Angeles; and a few more locations.

In the meantime, certain developments were taking place in Europe, although progress generally was slower there than in America. The January 19, 1883 issue of *Railroad Gazette* mentioned two new electric tramways in operation; the 1¼-mile Landvoort & Kostverloren line in Holland and a 6¼-mile line between Bush and Bushhaven in northern Ireland. On August 4, 1883, Magnus Volk inaugurated England's first electrified tram line, the Brighton Electric Rwy. This quarter-mile line had a 2′ gauge and a simple two-rail conduction system. The car ran an average of 7 m.p.h., carrying 12 to 16 passengers per trip. Shortly it was averaging a daily passenger total of 350, and plans were made to extend the line 2 more miles, covering the entire distance of the esplanade wall.

In 1884, the Blackpool Tramway was opened. For conduction it tried an underground central conduit system devised by Holroyd Smith. A conductor drew current by passing along this conduit, with the spent power being released to the tracks. The line, however, was not successful, due primarily to prohibitive maintenance costs.

In Germany several mining lines were electrified, using overhead power sources; these were the Zanckerode (in Saxony) and Neu Strassfurt lines in 1883, and Hohenzollern near Beuthen, Upper Silesia, in 1884. The locomotive on the 0.4-mile Zanckerode line was 32 inches wide and weighed 3,300 pounds. It typically pulled 10 cars, carrying 8 tons of coal, at about 7 m.p.h. In the early 1890s at Rock Springs, Wyoming, the Union Pacific Coal Co. began using electrified engines. Having a 60 h.p. capacity, these were generally run at a speed of 8 m.p.h. On one occasion a locomotive pulled 30 loaded cars (each weighing about 3,000 pounds) behind it and pushed 16 others ahead. This was done without difficulty over the entire length of track.

In the 1880s a true American genius emerged who finally provided the breakthroughs necessary for firmly establishing the legitimacy of electric traction. This person was Frank Sprague. Trained at the Naval Academy in Annapolis in mathematics and engineering, he worked for a time for Thomas Edison. Not being overly impressed with Edison's electric locomotive, he decided to focus some attention on the matter. In 1884 he founded his own company and began the first manufacture of electric motors designed as such. Previously dynamo motors had been in universal use. Soon he began experimenting with installing his motors in railroad bogie trucks. He found that he could mount two in a truck in a way that they could withstand shocks while still powering the wheels. He also discovered the principle of regenerative braking; that while braking, the motors could become temporary generators, having a drag effect upon the wheels, while at the same time putting electricity back into the system. Also during this period Sprague began experimenting with electrified rail cars.

In May 1887 he was given the daunting task of building a complete 12-mile electrical tram system by the city of Richmond, Virginia. In his contract he was given 90 days to lay out the track, construct a central power station, provide 40 trolley cars, and do whatever else was necessary to have this railway functioning properly. Unfortunately, perhaps partly from overwork, he soon became ill with typhoid fever, and was absent for weeks. In the meantime, the track was laid out poorly by an independent contractor and the trolley cars proved to be under powered. The city gave Sprague more time, and it took over a year before the system was functioning smoothly. However, by this time, the Richmond Union Passenger Rwy. had gained national attention, inspiring other cities to adopt Sprague's system. By the end of the decade in America at least 200 lines were already in service or under construction, most licensed under Sprague's patents. At the time, Sprague lines had also begun operation in Germany and Italy.

In 1890, a new extension of Berlin's Lichterfelde line was opened employing the American-style single-wire overhead conduction system. A new sliding bow type of conductor replaced the trolley and improved car performance when passing through junctions. That same year, another American-style trolley line opened in Leeds. Russia's first trolley, using an overhead conducting system, opened in Kiev in 1892.

Unfortunately, the new electrified tramways had their share of accidents. On one occasion, in the late summer of 1890, a car operating between Florence and Fiesole, Italy, ran off the track. There were five fatalities and 20 injured. An even more frightening accident occurred in Cleveland, Ohio, in late 1895. On a cold night of mist and drizzling rain, a trolley fell through an open drawbridge on the Central Viaduct, shattering on pilings 120 feet below. Fifteen people were killed including the conductor. On March 8, 1896, a collision of trolley cars took place on the St. Louis–Kirkwood line. Besides claiming 4 lives, the accident injured 62 others, 12 of them seriously.

On November 4, 1890, the world's first electric underground railway began operation. This was the 3¼-mile City & South London line which connected London's commercial district, via a tunnel under the Thames, with Stockwell. For conduction, the line used a third-rail, 500-volt, direct current system, and on it ran small 100 h.p. locomotives. These were supplied by the Mather & Platt Co., and were capable of pulling the three-car trains at a speed of 25 m.p.h. The cars were short, of awkward design, and had only tiny windows. They quickly became known as "padded cells."

In Liverpool in 1889, construction began on the world's first electric elevated railway. Opening on February 4, 1893, it replaced a horse car line and operated over the 6- to 7-mile length of the Liverpool Docks. It also had extensions at both the north and south ends, connecting with the Lancashire & Yorkshire Rwy. on one end and with the Cheshire line on the other. A U-iron conduction system was devised for the line, with spent current being returned to the rails. Each train consisted of two coupled electrified rail cars. In tests held on January 26, 1893, a speed of 50 m.p.h. was reached.

In France, Claret, who previously had built the Clermont-Ferrand tramway using Siemens' split-tube, twin overhead conductor system, combined with M. Vuillemier to devise their own ground conduction method. This was employed in tramways opened in 1892 connecting Marseilles and St. Louis, in 1893 connecting Bordeaux and Vigan, in 1894 connecting Lyon and Le Havre, and at Paris in 1896 connecting Place de la Republique and Romainville. Often, in both Europe and America, local governments did not like overhead wiring, thus forcing engineers in many cases to try to work out alternative systems using either ground conduction or storage battery methods. Ground conduction was found to be superior, however, as storage batteries, in the long run, proved inefficient due to their heavy weight and lack of power, and the frequency with which they had to be recharged.

In its October 7, 1892 issue, *Railroad Gazette* reported that Armour & Co. had constructed an elevated electric railroad connecting different parts of their establishment at the stockyards in Chicago. Having a 3′ gauge and track 23 feet above the ground, this was probably the first extended elevated electric road to be built in America. At the time of the article, 6,000 feet of the line had been constructed, and there were plans to extend the system for another 2 miles.

In 1889 Frank Sprague conducted experiments with an electric locomotive on the Manhattan Elevated R.R. Unfortunately, these efforts were not entirely successful. After losing control of his company to Edison General Electric, Sprague began working on his own projects, and in 1892, helped design a 1,000 h.p. electric locomotive, which at the time was the world's largest. In 1893 General Electric built a 30-ton prototype engine intended for use on elevated railways. In 1895 the company built an even larger locomotive, which ferried Baltimore & Ohio passenger trains beneath the City of Baltimore to and from Camden Station. It was successful in terms of both power and performance, and also in its ability to cut down on air pollution.

In 1895 Sprague began a project designed to coordinate the movement of a train of electrified railcars. This was called "multiple-unit control," and it allowed a driver to control all the motors on the train from a single location. Sprague was successful in his efforts, and by the end of the century his railcar trains had begun operation in a number of places, including the elevated lines in Chicago. These trains began operation on Brooklyn's elevated lines in 1901, and on Manhattan's in 1902.

In France in 1893, an engineer, M. Heilmann, tested an engine on the Ouest line in what was the first attempt at a self-generating electric locomotive. For this he created a steam engine that powered a dynamo, which then generated current to drive electric motors fitted on the axles. When trials of this locomotive were unsatisfactory, Heilmann began work on a second engine, testing it on the Ouest line in 1897.

The major problem with both locomotives was that they had excessive weight in relation to the amount of power they could generate. Both had a number of advantages over steam engines, including an ease in negotiating curves and, due to the absence of connecting rods, a reduction of noise and rocking movements. In comparison with contemporary electric engines, Heilmann's second locomotive was superior in requiring no conduction system; however, operating costs were higher due to excessive weight. Thus his efforts did not result in success, but perhaps they were a necessary experimental step, contributing in some way to the ultimate discovery of the diesel-electric locomotive early in the twentieth century.

In 1899, the engineer Auvert tested an aerodynamic locomotive of his own design on the Paris-Laroche section of the PLM Rwy. A lowered forward section allowed easy access to the storage batteries providing the source of power. This locomotive reached a speed of 46 m.p.h. while drawing a 100-ton load. Because steam engines of similar weight performed better, this project was dropped.

In America in the early 1890s a new kind of hybrid railway began to emerge. This was what later came to be known as the interurban. Interurbans fit somewhere in between street lines and railroads. They began as trolley cars that operated over longer distances to convey passengers from one town to the next. Perhaps the earliest interurban service was one inaugurated between Portland and Oregon City in February 1893. Other early lines included two in Ohio, one connecting Sandusky to Milan and Norwalk and another linking Akron, Bedford, and Cleveland. It took a while for these railways to become successful, but beginning around the turn of the century they gained greater popularity. Soon they began evolving into extensive networks, especially in the Northeast and Midwest. While interurban service began with the trolley car, it was not long before larger and heavier rail cars were used.

American trolley companies also began to provide other kinds of service. A mail car began operating in St. Louis in 1892. Carrying two postal employees, it made three trips daily on a 5-mile route that extended into the suburbs. The car was fitted in a manner similar to a railway post office, with sorting cases and racks. Other trolley companies entered the freight business. *The Engineer,* in its January 31, 1896 issue, reported that the city of Newark, New Jersey, had passed an ordinance banning street railways from engaging in freight or express hauling. "The electric railroad companies . . . have made extensive plans looking to the transportation of freight, and it is said that the Consolidated Traction Co. already has a number of freight and express cars. An officer of one of the street lines has published a letter warning the city that in prohibiting the transportation of freight on street car tracks it is injuring its own interests, as several manufacturers now looking for sites for factories will come to Newark, if they can have track connections, and will stay away if they cannot."

By 1896 trolley lines were either operating or under construction in Toronto, Montreal, Hamilton, Ottawa, Winnipeg, Brantford, Niagara Falls, and other Canadian cities. At the time the system in Montreal was quite large, aggregating to 74 miles of track laid on 41½ miles of street. In the summer months 163 trolley cars were available for use. In the winter months 145 closed trolleys and 14 snow sweepers were operated. Remaining from the horse-drawn era were 53 open streetcars and 60 closed streetcars. These were sometimes attached to the trolleys.

Up until 1892, direct current was the only type of electrical power used for traction. That year Siemens and Halske began experimenting at Charlottenburg with a three-phase alternating current system. They employed this method on a tramway in Lugano in 1895, and in 1898, to two Swiss rack railways—the Gornergrat line, and Europe's highest, the Jungfrau line. Employing this same system, the Swiss Burgdorff-Thun Rwy. was inaugurated on July 21, 1899, and Italy's Valteline Rwy. in 1900.

In 1899 a number of French tramways were opened that combined both overhead and ground conduction systems. Paris's Bois de Boulogne line used Vedovelli's system. Several lines inaugurated in the city of Tours used a similar system devised by Diatto. These latter were built by the Compagnie Industrielle de Traction.

In 1895 the New York, New Haven & Hartford R.R. conducted experiments with electric traction on its Nantasket Beach line, only operated during the summer months. An overhead wire system was used on the initial experiment. Later, a third-rail conductor was successfully tried at East Weymouth. This led to the establishment of electrified service from Berlin through New Britain to Hartford. Regular operations began on May 11, 1897.

Following the turn of the century, electrification took place on many American suburban and elevated lines, on portions of the New Haven and Pennsylvania systems, and in a number of long rail tunnels. Electricity also powered the New York subway system, the first portion of which opened in 1904. However, electrical traction never expanded much further in this country, and it was left to the later development of the diesel electric engine to finally overcome the dominance of steam. Trolleys and interurbans had their greatest moment in the 1920s, but today exist in only a few cities like New Orleans, Boston, Philadelphia, and San Francisco.

Europe, in contrast, has had to contend with larger populations, greater need for public transport, and higher levels of air pollution. Thus it has been more supportive of electrified rail systems, both intercity and urban. Electrical railways have also proliferated in Japan, other parts of Asia, and elsewhere in the world where similar conditions exist.

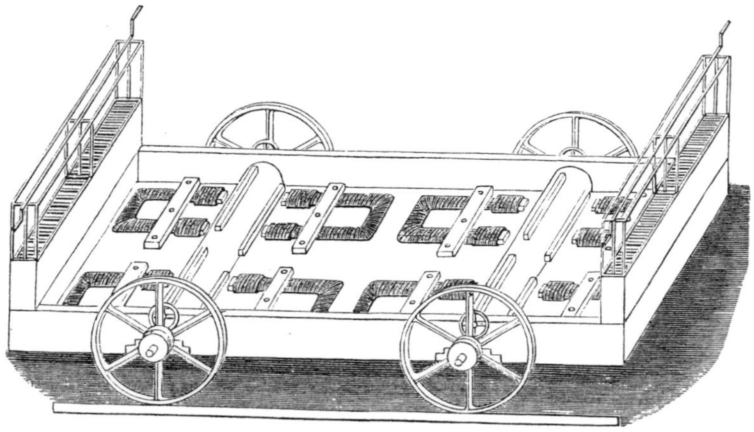

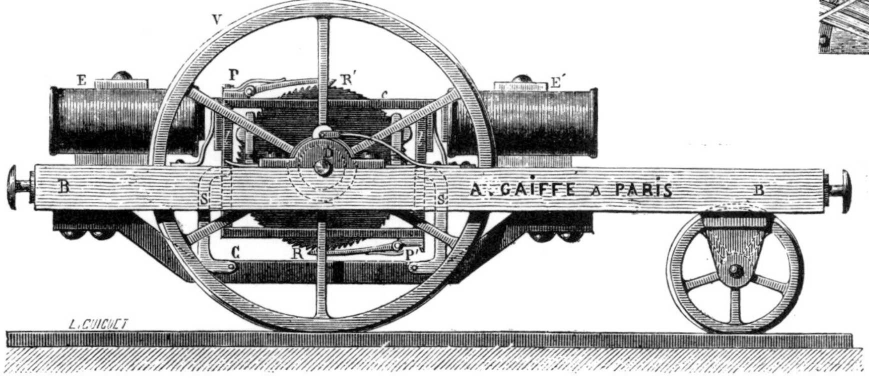

Early electric locomotives: Top left: Robert Davidson's engine, Scotland, 1842. Top right: Leo Daft's locomotive *Benjamin Franklin* tested on New York's Ninth Ave. elevated line, 1885. Middle left: Adolphe Gaiffe's engine, France, 1867. Middle right: Thomas Hall's locomotive, 1850. Bottom: Engine displayed at Chicago exhibition of 1883, Electrical Railway Co. of the United States.

Early European electric locomotives: Top: Werner von Siemen's engine pulls visitors at the Berlin Exposition of 1879. Bottom: Clovis Dupuy's small industrial engine in use at a bleaching ground in le Breuil-en-Ange, France, 1882. The rear tender carried Faure accumulators.

PLATE
327

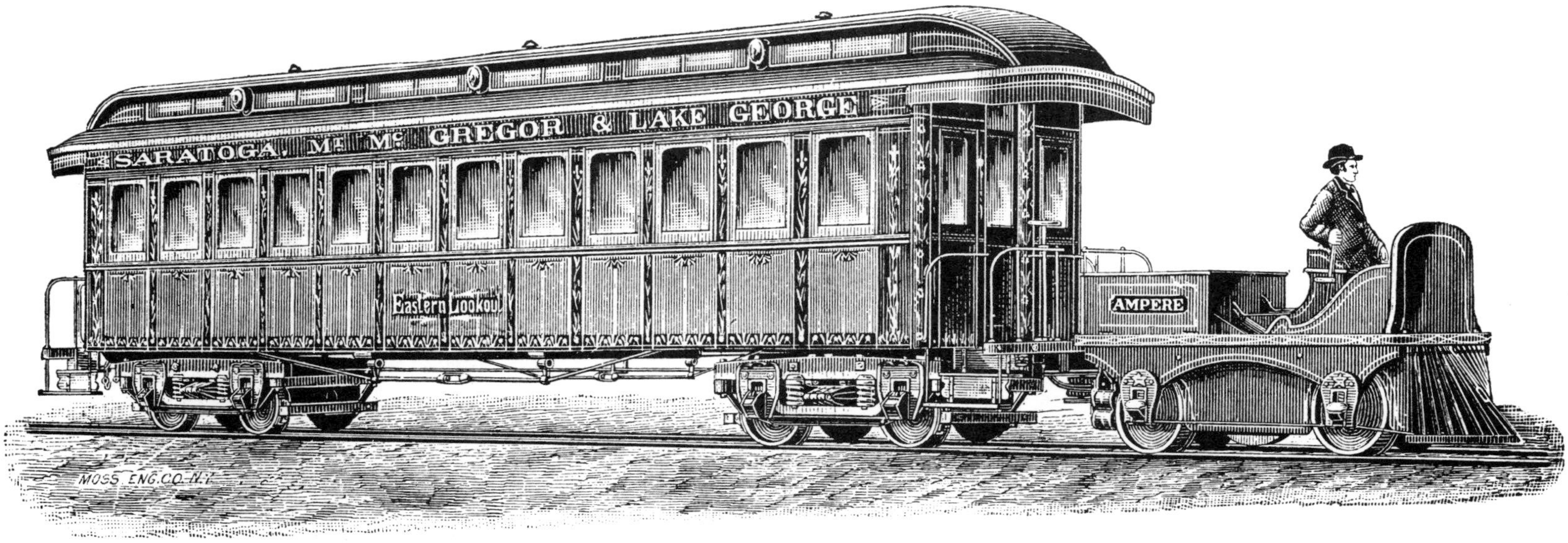

Early American electric locomotives: TOP: Leo Daft's *Ampere*, 1883. BOTTOM: Thomas A. Edison's second engine, Menlo Park, New Jersey, circa 1882.

Early Siemens & Halske electrified tramways: Top: Charlottenburg-Spandau line, Germany, 1881. This route employed a twin overhead wire conduction system. Bottom: Scene on Berlin's Lichterfelde line, circa 1881. For this tramway ground conduction was used.

PLATE
329

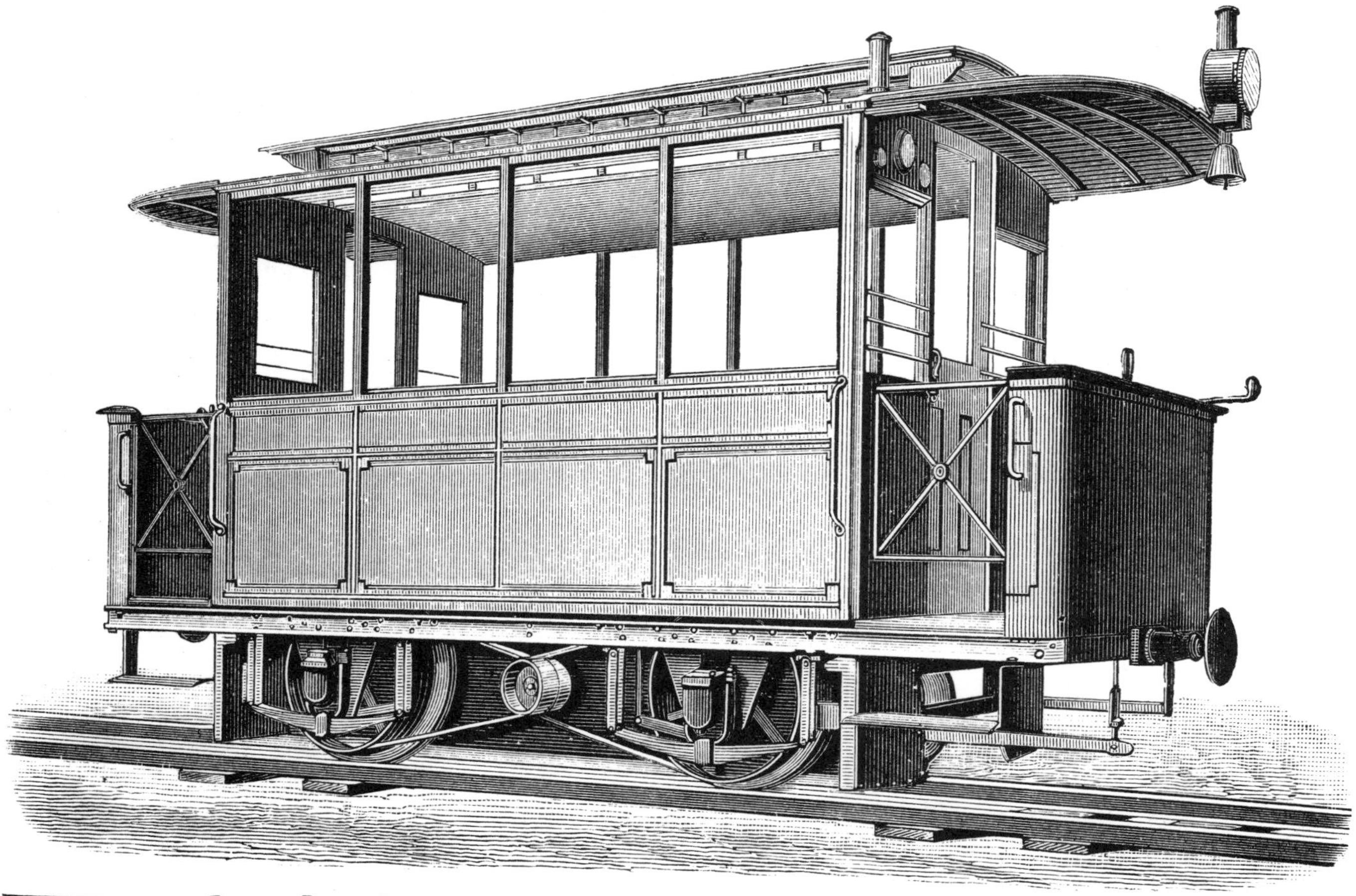

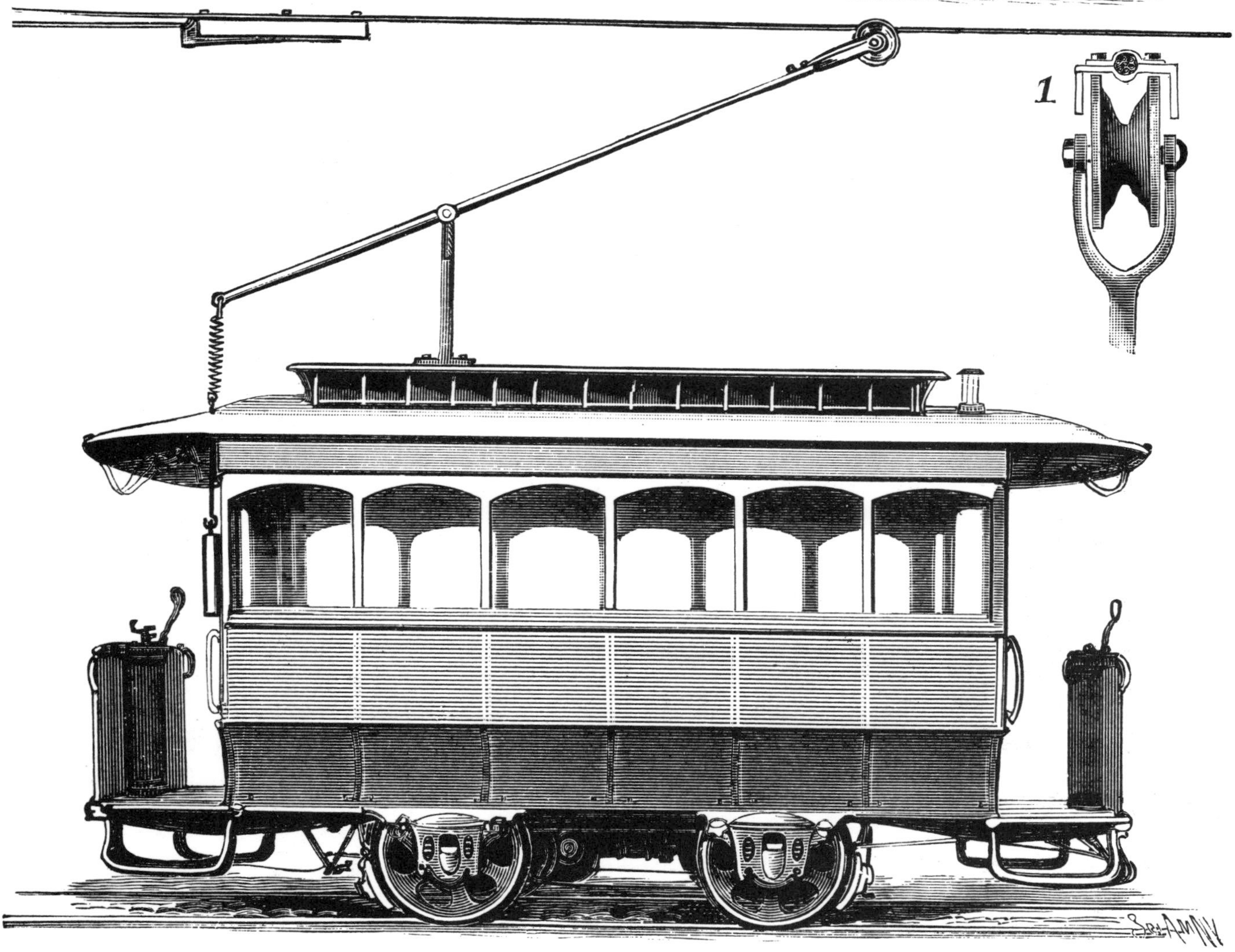

Early electric streetcars: Top: Siemen's ground conducted car, Berlin's Lichterfelde line, 1881. Bottom: Charles Van Depoele's car employing overhead wire conduction system with an under-running trolley, 1883.

PLATE

330

Electric locomotives: Top: Engine used in underground City & South London Rwy., 1890. Bottom: Early engine of America's General Electric Co., 1893.

PLATE 332

The Liverpool Overhead Rwy., 1893: Top: Electrified rail carriage for first- and second-class passengers. Bottom: View of Pier Head Station.

PLATE 333

A vision of the future: Artist's concept of the proposed Chicago & St. Louis Electric Rwy. and Boulevard, 1893.

Heilmann's first steam-electric locomotive, Ouest Rwy., 1893: Top: Side view of engine. Bottom: Pulling a passenger train.

PLATE 334

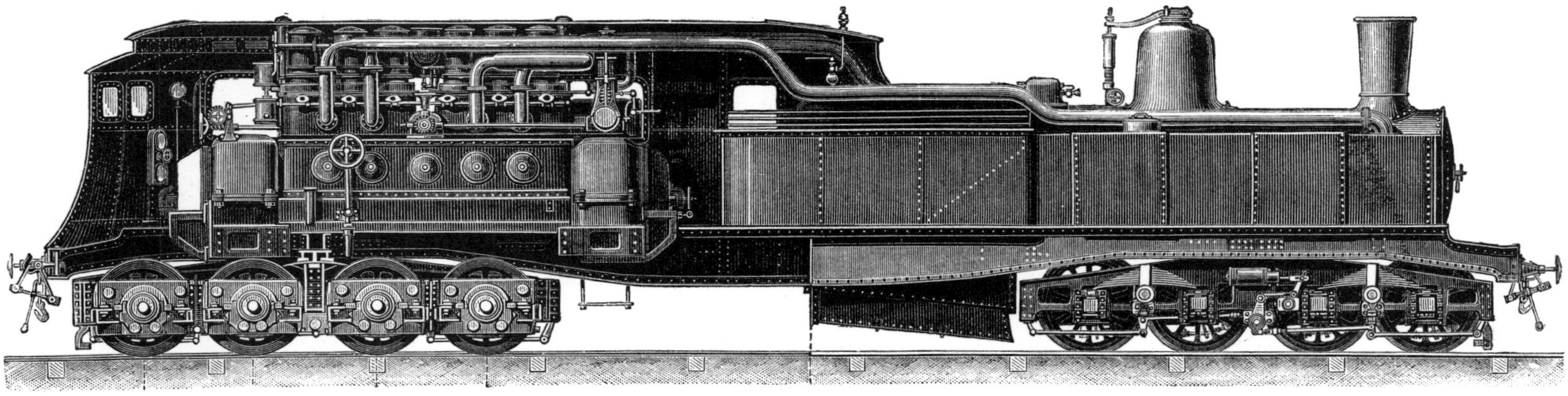

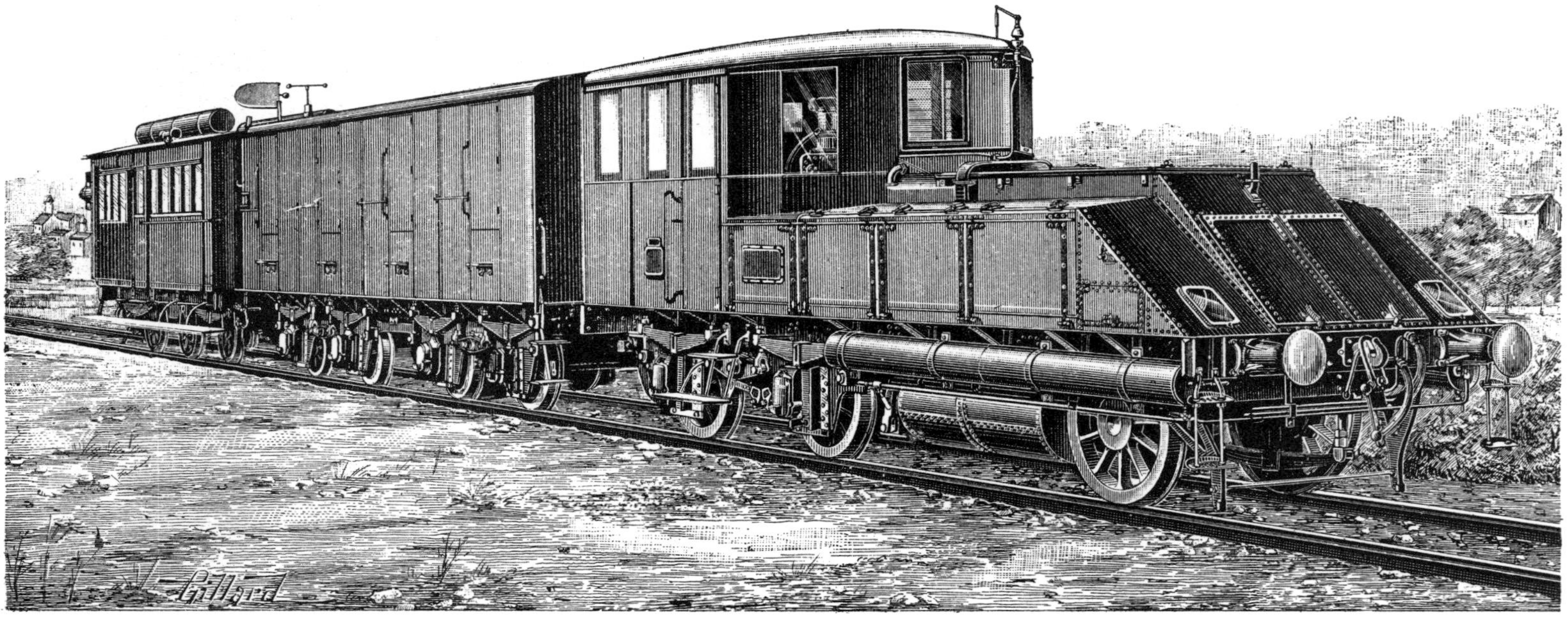

Top: Heilmann's second steam-electric engine, Ouest Rwy., 1897. **Middle:** Side view of same engine with cutaway section. **Bottom:** Auvert's aerodynamic battery-powered locomotive, PLM Rwy., 1899.

PLATE 335

CHAPTER

((28))

Mass Transit Railway Systems

As the industrial revolution gathered momentum in the first half of the nineteenth century, many people were drawn from the countryside into major cities. As a result, a number of these growing urban centers were soon faced with unique and unprecedented transportation problems. London, of course, was at the forefront of this development. In 1831 the population of greater London totaled 1,903,572. People walked a lot in the city, but its cobblestone streets were also crowded with slow-moving hackneys, carriages, buggies, gigs, omnibuses, and various freight and commercial vehicles. When London evolved into the world's first major metropolitan rail center, its traffic problems reached a uniquely new dimension. This resulted ultimately in a swarm of visitors, luggage, and goods traffic being unleashed daily on the city's already packed streets.

Although there was strong opposition in London to railways, the first line reached the city in March 1836, and others soon followed. As the street traffic problem grew ever greater, the city solicitor Charles Pearson suggested the possibility of a subterranean city railway, having a number of different stops, to relieve the situation. This idea, perhaps not unexpectedly, was received with almost unanimous derision and ridicule. It quickly became a great joke in the city.

There was a precedent of sorts for this idea, however. It was a recently finished tunnel that had been constructed beneath the River Thames by Marc Isambard Brunel, father of the famous engineer, I. K. Brunel. This was completed on March 25, 1843, after severe obstacles had been overcome and two previous attempts by others had failed. In devising the unique tunneling shield that was employed for this effort, Brunel had studied the shipworm, or *Teredo navalis*, a mollusk known for penetrating into the hulls of wooden ships. Tunneling proceeded out of the shield. In its wake the newly opened portion was lined with brick. It was a tedious process, however, and the project took 17 years to complete. In 1828 the river had broken through the roof of the tunnel and killed seven workers. Two other incidents of flooding resulted in the loss of funding for seven years. Although designed for vehicular traffic, the tunnel was never employed for this use. In 1866, it was purchased by the East London Railway Co. and utilized for building an underground line between the Wapping and Rotherhithe stations.

As railway traffic into London grew, congestion and noise on the streets reached increasingly intolerable levels. In the early 1850s, it was estimated that an average of 790,000 people entered the city each day over both road and rail. Thus Pearson's idea gained increasing support as a means for resolving this intractable problem. However, when an authoritative outline was announced, proposing a circular subterranean system connecting all the termini of the various railways, strong opposition was immediate.

In addition to those who were anxious for their safety and property along the proposed route, there were others who opposed the idea for religious reasons. Dr. Cumming, a preacher of the time, raised the possibility that this project might have the unintended consequence of opening a road down into hell. He feared that the liberated denizens of that realm would soon be terrorizing London's citizenry. He suggested, in its place, an overhead elevated system, and this idea gained many adherents. Another fear was that tunnels would collapse as loaded passenger trains passed through them.

Perhaps the most realistic concern raised was whether the accumulation of smoke and fumes in the tunnel might asphyxiate the passengers or train crew. John Fowler, who later was to become the Metropolitan Railway's chief engineer, suggested the possibility of building "fireless" locomotives for this purpose. Fireless technology was only theoretical at the time, however, and Fowler's efforts in this direction, which continued into the early 1860s, ultimately ended in failure. In 1853 Parliament granted Pearson and his supporters permission to build an underground line between Farringdon St. and Bishop's Rd. in Paddington. Rather than tunneling deeper in the manner of Brunel, this tunneling was to be done in the "cut and cover" method, which consisted of temporarily cutting through the surface of streets, digging out the tunnel, and then covering over the excavation and restoring the pavement.

The North Metropolitan Railway Co. was incorporated in 1854, but due to opposition, it was not until 1860 that actual construction on the line began. All sorts of new problems had to be dealt with: ensuring the stability of buildings along the route, razing other buildings directly on the path, rerouting mazes of underground pipes, paying compensations, and trying to pacify unhappy businesses, landlords, tenants, and an increasingly impatient public. Once construction commenced, however, the public attitude began to change.

In its issue of April 7, 1860, the *Illustrated London News* reported:

> This remarkable undertaking—which had been so long in abeyance that the public had well-nigh despaired of its ultimate accomplishment—has now been commenced in earnest, and the contractors are proceeding vigorously with the works at various points . . . The need of railways communication between the City and the great series of Railway on the north of the Thames, both for passengers and goods, had been long grievously felt, but the difficulties in the way of carrying a railway into the City appeared to be almost insuperable. To have a railway, after the American fashion, passing through a densely populous district, and crossing on the level our overcrowded streets and thoroughfares, was utterly out of the question; and scarcely less so to carry an unsightly viaduct through the heart of the metropolis. The only alternative was that adopted by the Metropolitan Company—namely, that of an underground communication, by which the most densely crowded districts could be traversed without the slightest annoyance or obstruction to the existing traffic.

Between the Paddington and Farringdon St. termini, stations were planned along the line's route at Edgeware Rd., Baker St., near Trinity Church, Regent's Park, Hampstead Rd., Euston Square, and King's Cross. At Paddington connections would be made with the Great Western Rwy., at Euston with the North Western line, and at King's Cross with the Great Northern. It was intended that light trains should be run at short intervals, perhaps stopping at alternate stations. To strictly maintain safety, the system was also planned so that the arrival and departure of each train at every station was reported by telegraph. This procedure would help ensure an interval of at least one station between trains.

Construction of this line was difficult, especially when various water, gas, and sewer mains were encountered. In June 1862 the Fleet Ditch Sewer burst and filled the tunnel as far back as King's Cross with 10 feet of water. Despite this, however, the project was completed without the loss of a single life. On January 9, 1863, a special day for viewing the line was held for dignitaries including such figures as Mr. Gladstone, the Duke of Sutherland, and the chancellor of the exchequer. A banquet was held at Farringdon and High Holborn Station, and then the dignitaries boarded a train of open wagons to explore the length of the line. The next day the railway was opened to the public, and over 30,000 passengers boarded trains. In the following three weeks the railway carried an average of 29,196 passengers per day. After a year of service 9,455,175 people had traveled on the line.

Its construction had been in broad gauge, and the Great Western Rwy. initially operated it. The locomotives at the time were of the 2-4-0T variety. Soon, rails for standard gauge use were added. In *The Romance of London's Underground,* W. J. Passingham cites the purpose of this dual-gauge system: "The object of the two sets of rails was that a line would in time be opened to operate on other railway systems. In 1875, for example, Metropolitan trains ran into Liverpool Street Station (Great Eastern) before the building of Bishopsgate Station was completed. It was intended at that time to run Metropolitan trains on to Walthamstow over the Great Eastern track."

Rather than being one long tunnel, this railway was actually a sequence of tunnels, with open cuts in between. However, this did not seem to help much in relieving the air pollution that many had anticipated. Experiments with various types of coal and coke eventually led to the selection of a special coke from the Brancepeth Collieries near Durham. To eliminate as much of its sulphurous and other noxious agents as possible, this coke was burned for 120 hours. Generally speaking coke was preferable to coal because it was relatively lighter, cleaner to handle, virtually smokeless, and required less draught for ignition.

Later Bwiffa coal, a medium-hard smokeless type from South Wales, was used. Ventilation shafts were also opened in the tunnels to help relieve pollution. When three passengers became ill in August 1867 and died several hours later, the blame was partially laid on the "suffocating atmosphere." However, the line quickly invited in scientists who gave reassurance that the air was safe to breathe. They suggested that a part of the smoke problem appeared to result from the application of wooden brakes when the trains made stops. However, they did not think that these fumes were harmful.

Some efforts were also made in locomotive design to help eliminate this issue. It was reported in the February 8, 1868 edition of *Illustrated London News* that "with a view to the greatest attainable purity of the air, the engines of this line have some peculiarities of construction, being made to discharge their exhausted steam, not into the air of the tunnel, but into large tanks of water on each side of the engine where the steam is re-condensed. They are made, also, with a very large heating surface, and the damper is kept on the fire, without any draught, while passing through each tunnel, to lessen the consumption of fuel." Despite these innovations, however, the railway remained a smelly and dirty environment, with flying soot adding to the problem.

The October 26, 1872 issue of *Railroad Gazette,* quoting the *London Daily Telegraph,* reported on other defects of the system:

> A correspondent calls attention—not before it was necessary—to the perpetual risk of frightful accident which is run on our underground railways. He speaks especially of the guards, who perform a hundred acts of perilous acrobatism in the course of the day's work, shutting the open doors and leaping into the train while it is flying out of the station. He describes an instance where death was avoided almost by miracle, and we suppose that, upon the present breathless neck-or-nothing system of "conducting," death in its most awful form is the reformer, who will spare us the constant terror of seeing an underground guard crushed into pulp. But we would add a word for the hapless passengers. The incessant hurry at these subterranean stations, the nerve-shaking slamming of every carriage door, the hideous yells of the engines, the difficulty of distinguishing one train from another, or of getting a coherent answer to a question from the fevered and inarticulate officials—all these peculiarities, with the oppressive atmosphere and spine-convulsing way of putting on the brakes, render underground lines as terrible as they are useful.

Perhaps to take the minds of passengers off these imperfections, extra attention was devoted to providing comfortable carriages to ride. Compared with other British railways, the Metropolitan line had the most luxurious carriages of the day, with nice upholstery, carpets, mirrors, and oil lamps to provide light. Three classes of carriages were provided. A typical 6-car train had one first-class, 2 second-class, and 3 third-class carriages. The cars were all 40 feet long, 8 feet wide, and 11 feet high. The first-class carriages had 6 compartments, each seating 8, while those of the second- and third-class had 8 compartments, each seating 10. The seats on the second-class were covered with leather while those of the third-class were painted boards. A no smoking policy prevailed until 1874. While relatively comfortable, these carriages had no heat, and passengers had to plan accordingly. The trains traveled on the line at an average speed of 12 m.p.h.

In 1864 the Metropolitan Rwy. opened its Hammersmith line. From Hammersmith trains ran to Shepherd's Bush, Notting Hill and Westbourne Park, to Bishop's Rd., and then into the city. Moorgate St. was reached by the line in 1865, and in 1868 a western Metropolitan extension was opened between Edgeware Rd. and South Kensington. In the meantime other improvements were made on the existing system. In the Supplement of its February 8, 1868 issue, *Illustrated London News* reported on improvements between King's Cross and the Farringdon Road Station: "This section of the railway, which is 5,192 ft. in length, or 88 ft. less than a mile, is to be used jointly by the Metropolitan, the Great Western, Midland, Great Northern, and London, Chatham & Dover Companies trains, being laid with rails of a mixed gauge, to admit both the broad and standard." It was anticipated that the Clerkenwell Tunnel, as this section was now called, would "become the great central artery for the internal traffic of the whole metropolis, as well as the common link of connection between the great national railways, diverging to the north, south, east, and west of London, and traversing the entire kingdom."

In 1868 the St. John's Wood Rwy. was formed. It shortly opened an underground line from Baker St. to Swiss Cottage. This railway was eventually assimilated by the Metropolitan system and extended to Harrow. In the early 1880s it was extended farther to Uxbridge. During that same decade the Metropolitan made extensions to Pinner, Rickmansworth, and Chesham, and took over the Aylesbury & Buckingham Rwy. Total annual passenger traffic for the system grew from 44,392,000 in 1872 to 88,513,000 by 1894.

Following the early success of the Metropolitan, another group of financiers petitioned Parliament for permission to build additional railways in the city. This group incorporated itself on July 29, 1864, as the Metropolitan District Rwy. This was the first of a group of railways, separate from the Metropolitan system, which were amalgamated in the early twentieth century to become the "UndergrounD." The District line's original purpose was to provide rapid transit across London from east to west. The first portion of the railway was opened on October 1, 1868, between Kensington High St. and Gloucester Rd. The line was extended to Westminster by the end of the year.

After yielding to strong opposition against building beneath the Strand and Fleet St., the District line was instead constructed along the Thames embankment. Within a short time a connection was made between Gloucester Rd. and West Brompton. On May 30, 1870, a Westminster-Blackfriars section was inaugurated. This portion was soon extended to Mansion House and Earl's Court. During this period the system was operated by the Metropolitan Rwy. On July 1, 1871, an inauguration ceremony was held when the District Co. assumed operational control of its system. By then it had nearly 6 route miles of line and 13 stations. A journey over its route from Mansion House to West Brompton took 29 minutes.

The locomotives chosen for use by the District line were the same as those operated on the Metropolitan. These were 4-4-0T engines built by the Beyer-Peacock Co. By 1885 the Metropolitan line had purchased 66 of these engines, and the District, 54. This same type would remain in service until electrification occurred. However, in 1895, the Metropolitan line also introduced 2-4-0T engines built by Sharp, Stewart & Co.

The railway was opened to Hammersmith in September 1874. By 1875 the two major systems were about equal in size, each being about 8 miles long. Joined together at Kensington, they formed an irregular, almost complete ellipse, with each company possessing half. The St. John's Wood Rwy., still independent at the time, was 1¾ miles long. By June 1, 1877, the District line reached Richmond, and over the next few years it made numerous other extensions. On October 6, 1884, the long-awaited completion of the Inner Circle was made when the District and Metropolitan systems met at Aldgate. Following this the District line was extended farther east to Whitechapel, where it joined with the East London Rwy., which operated through Brunel's Thames tunnel to Rotherhithe. In 1872 District trains had carried 20,892,975 passengers. By 1887 this annual figure had increased to 38,965,308.

On October 30, 1883, explosions occurred on both systems, and these were attributed to Irish terrorists. In its November 2, 1883 issue, *Railroad Gazette* reported:

> About 8 o'clock . . . a terrible explosion occurred near the Praed Street underground station on the Metropolitan Rwy. On the first alarm a strong cordon of police was summoned to preserve order, to keep the way clear, and to convey wounded people to the hospital. Passengers who were on the train at the scene of the explosion say that there was suddenly a loud report like that of a cannon, then a sudden darkness; the gas lights from the front to the rear of the train being put out. The glass was broken, and splinters of wood flew about the cars, cutting and wounding many passengers. Above the din were heard the shrieks of the injured and panic stricken people. The train, which was crowded, contained principally country visitors returning from the Fisheries Exhibition. After leaving the station the train traveled steadily ahead, though the concussion smashed nearly all the lamps in the Praed Street station, and caused much damage to the permanent way. When the train reached the next station, Edgeware Road, the first effort was made to rescue the wounded. Some were dreadfully burned, but more were prostrated by fright. All the doctors, surgeries and dispensaries in the neighborhood were called into requisition. The first impression was then that an explosion of gas had caused the disaster, but later inquiries tend to show malice.
>
> The General opinion of the railway officials and of experts is that the disaster was caused by some explosive material probably placed in the rearmost carriage. A searching inquiry is proceeding. It is believed that about 40 persons were injured, some of them dangerously.
>
> Almost simultaneously with the Praed Street affair a violent explosion occurred on the District Rwy. between Charing Cross and Westminster stations. The windows of the signaling stations in the tunnel were shattered, and at Charing Cross the glass roof of the station partly collapsed. The report was like that of artillery. The effects were like those of the Praed Street explosion. Carriage lamps, windows, etc. were smashed. All traffic was suspended for a while. Though both explosions are matters of the greatest mystery, they are generally thought to have been of Fenian origin.

By 1899 both the Metropolitan and District railways had decided on electrifying their entire systems. The following year Charles Tyson Yerkes, a financier from Philadelphia, arrived in London to help the District system in its transition. Yerkes had taken an active role in developing Philadelphia's electrified street railway system, and had gained earlier transportation experience in Chicago. He approached the Metropolitan with the idea of creating a unified system, but was turned down. However, he was later successful in consolidating the District and other railways, such as the North London and East London lines, into the system that became the UndergrounD. In the meantime both major railways rapidly proceeded with electrifying their systems. The last steam trains ran beneath London on November 5, 1905.

Two other subterranean lines worthy of brief mention were the City & South London and Central London railways. These were both of a type called tube railways, which were tunneled in the manner originally pioneered by Brunel. Tunneling techniques, however, had since been improved by Peter Barlow and James H. Greathead, and so the job had become easier. Greathead oversaw construction on the 3¼-mile City & South London line, which was begun in 1886 and completed in 1890. It had both an "up" and "down" tunnel, the two running side by side through most of the line. This railway, originally intended for cable traction, instead became the world's first electrified underground system. Its northern terminus was on King William St., just north of London Bridge. From there it ran south-southwest to Stockwell, with several stations in between. The 5¾-mile Central London Rwy. ran essentially in a straight line from Shepherd's Bush to the Bank. It was opened by the Prince of Wales on July 30, 1900.

Although not of ancient origin like the great cities of Europe, New York was of commercial importance and enjoyed an enviable location at the mouth of the Hudson River, which helped it to grow very rapidly during the nineteenth century. The expanding population quickly filled out Manhattan's narrow physical confines and the crowded situation only grew worse as time passed. While horse-drawn railway service had begun on the New York & Harlem R.R. in 1832, no other lines were built until the early 1850s. However, even this development did little to relieve congested traffic.

About this time publications like *Scientific American* began to publish rapid transit ideas, submitted by various inventors and engineers, for relieving the city's transportation problem. Proposals from the next two decades included steam trains at street level, elevated horse car lines, elevated steam railroads, cable-powered traction, vacuum or pneumatic systems, and underground or "subway" railroads. Another interesting idea submitted was for building arcades, where columns and arches were employed to create two levels, one for rail transportation and the other for ordinary street traffic.

Following America's Civil War, in 1867, Charles T. Harvey began building, along the east side of Greenwich St., what would become the world's first elevated metropolitan railroad. The first section of the line was from the Battery to Dey St. Harvey's mode of traction was unique for the time: an endless cable, powered by a stationary steam engine. It was capable of drawing his cars at speeds up to 15 m.p.h. Construction continued up Greenwich St. and then Ninth Ave., and by the spring of 1870 the line had reached Thirtieth St. Expensive trials were conducted at this point to determine the practicality of the system, and the cable traction was found to be unsuitable. Harvey's investors panicked and their company, the West Side & Yonkers Patent Rwy., was sold to pay off creditors. A new company, the New York Elevated R.R., was quickly organized in its place. A light 0-4-0 dummy engine was tried and found to be better suited for the job. At the time the company's only rolling stock consisted of three horse cars, modified by attaching bogies. On April 20, 1871, the line recommenced its service and quickly gained favor with the public.

In 1873, 723,253 passengers traveled on the railway. By 1874, following an extension to Thirty-fourth St., the line was about 4 miles long. Between its termini there were seven stations, located at Dey St., Franklin St., Canal St., Houston St., Twelfth St., Twenty-first St., and Thirtieth St. At the time 82 trains were run on the system each day, the earliest departing at 6:20 a.m. and the last at 7:55 p.m. Distinctively designed passenger cars had also been added.

The *Railroad Gazette,* in its January 24, 1874 issue, said of the line: "Anyone who will take a ride on it will learn that it is the pleasantist way of travelling he will find in New York. It is high enough above the streets to give a good view, and be out of the dust and dirt. The line has no obstructions of trucks and wagons, which are so annoying down town to the street roads. The Elevated Railroad Company also refuses to take more passengers than it can seat. Its cars are kept clean and comfortable, although, as in nearly all attempts at ventilation, they have ample provision for exhausting the foul air but none for admitting that which is pure."

During the early 1870s there were expectations that Commodore Vanderbilt would build an underground railroad. It was to run between City Hall and Fifty-sixth St. at Fourth Ave. In its September 14, 1872 edition, *Railroad Gazette,* quoting the *New York Times* as its source, gave the following report:

> The depot at the City Hall will be opposite the north end of the Post Office, facing Beekman St., and will be 112 feet long and 50 ft. broad. The track at this point will be sunk about 19 ft. below the street level. There are to be 8 stations along the road. The route of the new road is to be as follows: From the City Hall Park it curves into Center St., near the Register's Office; then in Center as far as Reade; then it runs through the center of the blocks south of Park St., crossing Duane, Pearl, Baxter, Worth, Mulberry, Mott, Doyer, and Peil, the latter situated in the close vicinity of Chatham Square, where it reaches the Bowery. Following the Bowery until Cooper Institute is reached, the road diverges into 4th Ave., passing to the east side of the Grand Central and along to 56th St., where it will connect temporarily with the Harlem R.R.

Unfortunately, the line was never built. The cost of excavation through the city's hard granite subsurface proved to be too expensive at the time.

In 1867 Dr. Rufus H. Gilbert, a former physician, began promoting mass transit in New York. Gilbert had quit the medical profession following the death of his wife, and later, while visiting Paris and London, he became interested in public health issues. In London he observed how the underground lines allowed poor workers to live farther from their places of employment. This resulted in a migration from the crowded slums to nearby suburbs, which improved health for the workers and their families. Thus Gilbert became convinced that mass transit could help Manhattan's poor as well, providing better homes and a healthier environment.

Initially Gilbert favored the idea of an elevated pneumatic system. An experimental pneumatic line had been built beneath Broadway by A. E. Beach in 1870. However, it was only a block long and it became little more than a tourist attraction. This idea was soon discarded in favor of steam traction. Gilbert eventually formed a company, and began building a line extending from the Battery to Central Park, primarily along Sixth Ave. Shortly, however, construction was halted due to opposition from property owners and horse car lines, and it only resumed in October 1877. Gilbert's line, which became the Metropolitan Elevated R.R., followed Harvey's earlier example and opted for dummy-type locomotives. About the same time, however, the New York Elevated began converting its dummy engines to standard 0-4-0T and 0-4-4T Forney-style engines.

Once the Sixth Ave. Metropolitan line was complete, an extension was built from Fifty-second St. westward to Ninth Ave. By then the New York Elevated line up Ninth Ave. had reached Fifty-ninth St. An arrangement was made whereby the Metropolitan R.R. shared tracks from Fifty-second St. to the New York Elevated terminus. From there the Metropolitan built north to 110th St., east to Eighth Ave., and then north again to Harlem.

During the same period the two competitors also began construction of east side routes. Both lines started in lower Manhattan and reached up to Harlem. The New York Elevated route mainly went up Third Ave., while the Manhattan's followed First Ave. and then Second Ave. In May 1879 both railroads were leased to the Manhattan Co. and began operation under a single directorship. Following this, refinements were made on the existing system and new extensions were opened. Over some of these routes trains were run at frequent intervals. At one time, on the Sixth Ave. line, 440 trains were run daily, each way.

On March 25, 1879, two trains had a head-on collision near Forty-second St. on the New York Elevated's east side line. In February 1880 a late hour train carrying off-work rail employees derailed after hitting an unexpected switch. A number of people were hurt and the locomotive was destroyed when it struck the street below. This accident occurred on the Metropolitan Elevated line at 145th St. and Eighth Ave. In January 1891 on the Third Ave. line, a passenger train collided with a switch engine. The switcher, hurled from the track, smashed when it hit the street.

Across the East River in Brooklyn, three elevated lines were built during the 1880s. The first of these, the Brooklyn Elevated R.R., was opened on May 13, 1885. The other two lines were the Union Elevated R.R. and the King's County Elevated R.R. As electrical traction dramatically improved during the 1880s, it became obvious that the elevated lines would eventually be electrified. In 1885 and 1888 tests were made with an engine built by Leo Daft on the Ninth Ave. line. Electrified railcars developed by Frank Sprague also received trials. By the turn of the century, further refinements allowed a single motorman to control a whole train of electrified cars. Brooklyn's lines were electrified in 1901 and Manhattan's the following year. About the same time, New York got the underground railroad it had long dreamed of. In early 1900 ground was broken for the Interborough Rapid Transit (IRT) line. The first portion of this "subway" was opened on October 27, 1904.

Traffic congestion had been a problem in Paris for hundreds of years. However, during the mid-nineteenth century, the city's population exploded, and naturally, its transportation situation worsened. At the time, the city mainly consisted of a maze of narrow, meandering streets, many existing from medieval times. Extensive changes to this system were initiated during the reign of Napoleon III. Under the direction of Baron von Haussmann, wide boulevards were constructed in east-west and north-south directions. These cut through the ancient mélange of streets, opened up the city, and provided a more efficient network for horse-drawn transportation. This also improved access to railway stations located at the city's periphery.

While these and other changes made by von Haussmann were helpful, they were hardly enough. Inner-city railways had been suggested for Paris as early as 1845. That year M. de Kerizouet proposed a cable line for carrying goods between the Gare de Lyon and the central marketplace of Les Halles. Brame and Flachat revived this idea in 1853, but planned it as an underground route. Telle, in 1855, and Brame, in 1856, published proposals for elevated passenger lines. In 1857 Le Hir and Mondot de Lagorce introduced their prospectus for an underground public railway.

During the latter part of the 1860s construction began on the Ceinture Rwy., which was elevated over portions of its route. When finally completed years later, it circled the city and formed connections with all the railways entering Paris.

Although failing to relieve much congestion, it did help to give Paris a slightly more progressive look. Traction was provided by large-wheeled 0-6-0T engines, of a style used by the Ouest line.

In 1871 the city council approved the idea of an urban railway provided that it: (l) connect the different parts of the Department of the Seine in a circular line; (2) link the districts to each other and to the center; and (3) connect the main line stations. This plan went before the French parliament in 1883, but was defeated. The national government was less concerned with the needs of Parisians than with designing a system that would help ensure its own survival. Among state concerns was the paramount need to have ultimate control, and a desire that the system be designed so that troop movements could be made quickly. Paris had long been a hotbed of radical unrest, and fresh in memory was the 1871 repression of its Commune uprising. Thus, basically, the government had a deep mistrust of the city. It favored a plan whereby the major rail lines, whose stations lay at the outskirts, would extend services into the city.

Following this deadlock, there was much bickering between the city and the state, with each trying to advance its own agenda. Within Paris itself, a similar debate raged between advocates of underground and elevated lines. Louis Heuze, an architect, proposed an elevated system that included arcades lined with shops and boutiques. He suggested that, unlike Londoners, Parisians were "people of the sun," and would not easily tolerate the oppressive tyranny of underground railways. Underground advocates, however, feared that elevated lines would destroy the harmony and beauty of the city. Both sides formed associations and enlisted prominent people to aid their cause.

In 1881 Jean Chrétien, a prominent engineer, suggested a double-tracked elevated line between the Bastille and Madeliene. He later designed a railway that employed single-column supports. In 1885 Jules Garnier proposed an elevated system having two levels, with one for outbound traffic and the other for inbound. In 1886, Panafieu and Fabre recommended a monorail system.

Finally the French government gave in to Parisian demands. On November 22, 1895, approval was given for a metropolitan railway system. It was decided that the city would construct all necessary infrastructure, which would include tunnels, stations, and viaducts. A concession would be granted to a private company, which would operate the system and provide rolling stock, power stations, and other superstructural needs.

After public bids were submitted, the Compagnie Generale de Traction became the concessionaire. It shortly formed the Compagnie de Chemin de Fer Metropolitain to operate the system. On March 30, 1898, a law was passed declaring it in the public interest to build five lines for a combined distance of 40 miles. Excavation began in that year on the Bois de Boulogne–Bois de Vincennes line. Stops would include Pte. Dauphine, Pl. Victor Hugo, Pl. de l'Etoile, Pl. de la Concorde, Palais Royale, Louvre, Châtelet, Hôtel de Ville, Pl. de la Bastille, Gare de Lyon, and Pte. De Vincennes. It was to be powered with a 600-volt direct current system that employed a center conductor rail. The first part of this line opened on July 10, 1900.

The different railways entering Berlin were united in 1875 by the Ringbahn, a line that encircled the city. While originally intended to connect these various routes, this railway was soon found useful for carrying suburban traffic. To further improve this service, as well as relieve inner-city traffic congestion, the government decided to build an urban line running directly through the city, which connected at each end with the Ringbahn. This elevated line, the Stadtbahn, was 6¾ miles long, and ran from east to west between the Schlesische and Charlottenburg stations. Inaugurated on February 7, 1882, it had four tracks running its entire length. Two of these tracks serviced local traffic, and the other two carried long-distance and suburban trains. The Stadtbahn also had a military purpose. Troops could be moved rapidly through the city without transfer, and if need be, massed quickly at any point.

In the line's first eight months of full operation, it carried 6,173,700 city passengers. In connection with the railways at its terminus points, it also carried 307,670 suburban passengers and 126,870 long-distance passengers. The April 13, 1883 issue of *Railroad Gazette* reported that trains initially ran at intervals of 10 minutes, but soon this time was reduced to 6 minutes. As a result the total number of daily trains increased from 280 to 392. In addition, 162 suburban and through trains of connecting railways also operated over the line. Thus the total number of trains traveling daily on this route was 554, or 277 in each direction.

In its May 1887 issue, *Railroad and Engineering Journal* provides a brief description of the Stadtbahn's stations: "The stations are two stories high, the lower story, of brick, on a level with the street, containing the ticket office, waiting rooms, etc.; the upper story is of iron, and is an open shed or train-house, covering the platforms. At the larger stations, where through trains stop, there are baggage rooms, and hydraulic elevators are provided for raising the baggage to the platforms. At each station there is a block signal."

Originally 2-4-0 and 0-4-2 tank engines were used for drawing local trains. Coke was burned to reduce air pollution. Later T5 class 2-4-2 tank engines were added. The supports on this system were made of stone, as opposed to the metal fabrications in New York; thus trains were less noisy. This line was electrified in 1902. Also the same year, Berlin inaugurated the first section of an underground railway.

While important developments in mass transit occurred in the four cities previously mentioned, other efforts were made elsewhere. In May 1892 Chicago inaugurated the first portion of its Chicago & Southside Rapid Transit line. Extending from Congress St. to Thirty-ninth St., it was locally known as the "Alley" line. This was because a portion of its route ran down a long alley. Forney-style 0-4-4T compound engines, built by the Baldwin Co., were used until later electrification. The cars were built by Jackson & Sharp Co. of Wilmington, Delaware, and were similar in appearance to Manhattan's elevated cars. They had cane seats and were finished inside with mahogany.

In May 1896 Siemens & Halske opened the Franz Josef Electric Underground Rwy. in Budapest. This was the first electrified subway on the European Continent. The first American subway was opened in Boston in 1897. Trolley cars were operated on this line, which ran beneath Tremont St. The Vienna Stadtbahn, which used steam traction, opened in June 1898. These pioneering efforts led to the creation of much larger transit systems after the turn of the century.

London's underground Metropolitan Rwy.: Top: Proposed design for the Baker St. Station, 1860. Bottom: Inaugural train, January 9, 1863.

Top: Junction of Metropolitan, Midland, and Great Northern railways at King's Cross, 1868. Bottom: Junction of Metropolitan and London, Chatham & Dover lines near Smithfield, 1866.

PLATE 337

Entrance to Clerkenwell Tunnel, 1868. This 5,192-foot tunnel was used jointly by the Metropolitan, Great Western, Great Northern, Midland, and London, Chatham & Dover lines.

Top: Cutaway of Thames embankment showing route of Metropolitan District Rwy., 1867. The tunnel beneath was planned for a pneumatic railway. Bottom: Wapping Station, East London Rwy., 1870. Train is emerging from Thames Tunnel completed by Marc Isambard Brunel in 1843.

PLATE 339

Scene showing "cut and cover" excavation of route for Metropolitan District Rwy., Thames Embankment at Westminster, 1869.

PLATE 340

Top: Early morning workmen's train, circa 1872, from illustration by Gustave Dore. **Bottom:** Metropolitan Rwy. extension to Aldgate: views showing exterior and interior of terminus, 1876.

PLATE

341

Top: 2-4-0T locomotive built by Sharp, Stewart & Co., Glasgow, for London's Metropolitan Rwy., 1895. **Bottom:** Explosion of terrorist bomb seen from Charing Cross Station, Metropolitan District Rwy., October 30, 1883. This blast occurred in the tunnel between the Charing Cross and Westminster stations.

Illustrations showing ideas for proposed mass transit railway systems in New York, 1872: Top: An "arcade" line for Broadway. Bottom: An elevated railway employing columns, fixed on both sides of a street, supporting a double-tracked line running overhead.

PLATE 343

The New York Elevated Railroad: Top: New type of passenger car, 1872. Bottom: View showing the new extension at Ninth Ave. and Thirty-fourth St., 1874.

PLATE 344

TOP: Extension of Metropolitan Elevated R.R., looking north from 110th St. at Eighth Ave., November 1879. BOTTOM: A scene on the New York Elevated line's Third Ave. route, circa 1879.

Franklin Square Station, New York Elevated R.R., 1882. This scene anticipates the completion of the Brooklyn Bridge, which opened the following year.

Forney-style tank locomotives used on New York's elevated railway system: TOP: 0-4-4T Baldwin engine, 1878. BOTTOM: 0-4-4T Rogers engine, built March 1885.

PLATE
347

PLATE 348

Proposed underground railway in Paris (la Villette–Halles Centrales), 1861. This illustration shows Gare de l'Est in background with underground route proceeding south-southeast beneath Boulevard de Strasbourg.

Completed portions of Paris's Ceinture line on the southern periphery of the city, circa 1867: Top left: Gare de Gentilly. Top right: Gare de la Maison Blanche. Middle: Viaduct crossing Rue Michel-Ange. Bottom: Viaduct at Pont-du-Jour.

PLATE 349

Two views of Berlin's elevated Stadtbahn, circa 1882: Top: Viaduct over street showing portion of Alexanderplatz Station. Bottom: Inside train shed at Friedrichstrasse Station.

PLATE
350

CHAPTER

((29))

Monorails and Other Anomalous Modes of Traction

While the main course of rail development proceeded within a narrow range of generally accepted parameters, there were always a few individuals who wanted to break from such constraints and follow their own vision. These efforts, for the most part, accomplished little, but they added to the pool of engineering knowledge and sometimes established precedents that later resulted in greater success. Some of the following ideas were actually built, but others were only constructed as models or proceeded no further than engineering drafts.

Probably the world's first monorail was one built at the London Docks by a Mr. Palmer in 1824. Each of his pannier or binlike monorail cars functioned on two wheels rigidly in line. In 1828 Mr. T. Chapman of Lambeth took out a patent for a monorail that utilized a beam suspension system.

A somewhat different monorail, if it could be called that, was one devised in 1840 by Frenchman Achille de Jouffroy. His three-part articulated locomotive-tender had a single large driving wheel, located in the forward section. This ran along a grooved center rail, pulling the train behind it. Large center pins, allowing easy swivels on turns, connected the three parts of this locomotive and the wagons behind. The engine's forward part also included the cylinders, while the boiler rested on the midsection and fuel was stored in the rear. Aside from the locomotive's front part, the rest of the train was supported in the conventional fashion, although flangeless wheels were used. These operated on flat support beams positioned to each side of the center rail. Jouffroy's invention aroused considerable interest, and by 1844 he had built an experimental engine. However, there was little weight on its single driver, and this resulted in poor traction.

In France in 1846 M. Andraud suggested a monorail that would have employed an atmospheric traction system similar to that demonstrated by Christopher Nichols the previous year in London. His monorail was vertical and girder-like, with support pillars of varying lengths to accommodate minor variations in elevation. Split carriages, separated longitudinally below the roof level, hung saddle-like over the rail. On the lead carriage vertical rollers pressed against the central rail beam. These passed over long rubber tubes attached along the length of this beam on both sides. Air supplied by a stationary pump inflated the tubes, put pressure on the rollers, and forced the forward momentum of the train. A reduced scale model of this system was built at the Champs d'Elysses in Paris and tested out in 1856.

Much different in concept were various center-rail ideas, which received attention for a while. On September 30, 1830, Charles Vignoles and Captain John Ericsson, co-designers of the engine *Novelty,* a contestant in the Rainhill trials, received a patent for a center-rail system designed to improve locomotive traction on inclines. This plan utilized a raised third rail placed in the middle of the track. The locomotive operating on it would have, in addition to its normal wheels, a pair of horizontal wheels situated below, which would grip the center rail from each side and turn on it. Later, others like Seguier, Dumery, Giraud, and Fedit in France, Kraus in Germany, and Teller in America undertook similar experiments.

A center-rail system of this kind seemed to offer several possibilities. It could function as a guiding system to reduce dangers of derailment, and it could provide a means whereby added tractive power might be applied for operating on grades, and it could improve braking on grades. An example of the former was suggested in a letter to the Paris Academy of Arts & Sciences, from Baron Seguier in 1843, proposing the use of center rails for preventing high-speed train derailments. A variation of this idea was for the third rail to provide the only means of guidance. This was proposed by Kollman in England in 1845. The locomotive and carriages on his system each had two pairs of horizontal guiding wheels placed underneath that adhered to the center rail. The outer vertical wheels had no flanges and ran on flat support rails. Later experiments of this kind led toward monorail systems.

Also appearing circa 1844 was Prosser's guide-wheel system. Designed for wooden beam rails of the standard layout, it employed grooved wheels, fixed at an angle, which engaged the inside edges of the rails. These guide wheels were fixed to the front and rear portions of the engine as well as all of the carriages. In between were vertical wheels, apparently flangeless, which carried the weight of the train. Prosser set up a small line at Wimbledon Common to test this system.

Beginning circa 1844 M. Arnoux used a similar idea for an articulated system he was devising for the Paris-Sceaux Rwy. This line had iron rails, and the coaches rested on two axles in the usual manner, with flanged wheels contacting the track. For coupling, he used a rigid rod. In *Railways: The Pioneer Years,* Malcolm Fletcher and John Taylor describe this system further:

> The front set of wheels of the leading coach had four offset guide wheels which made contact with the rails and turned the axle according to the direction of the track. This axle transferred a corresponding direction to the rear axle by means of a crossed chain passing through two pulleys which were of equal diameter and fixed to each axle. The direction to be taken was then transmitted to the second wagon by a crossed chain which passed over two pulleys, one of which had twice as large a diameter as the other. The system was then repeated along the full length of the train. Because it transmitted simultaneously a different angle of direction to the two axles of the coach, the second axle changed its angle just before it entered the curve, and when the direction had been transmitted to axles of the following coach, the angle of these axles was adjusted in time to guide the coach through the curve.

The 2-2-2 engines originally used by the line had guide wheels placed around both the front and rear carrying wheels. A later 2-4-2T Sceaux engine had a similar arrangement. The line was opened on June 23, 1846, and its Paris terminus was at Pl. d'Enfert. While this system was somewhat complicated, it was also effective, allowing trains to negotiate very tight curves. A branch was extended 9.94 miles from Bourg-la-Reine to Orsay in 1854. Three years later the Sceaux line was absorbed into the Paris-Orléans system, which in 1867 extended it farther to Limours. Later work on the line reduced the severity of its curves, allowing a conversion to standard rolling stock in 1893.

Circa 1850 George E. Sellers designed a center-rail system for the Panama Rwy., which had recently begun construction. At the time, it was feared that adhesion engines would be incapable of climbing some of the grades on the line. The 4-4-0 locomotive Sellers conceived for this system had, in addition to its normal driving apparatus, an auxiliary set of horizontal driving wheels situated below, intended to operate opposed to each other on the center rail. These were powered by an extra pair of cylinders. Three engines of this design were ordered, but after the first was completed in 1851, the center-rail plan was scrapped. These locomotives were shipped in 1852 minus the auxiliary power units. Center-rail adhesion finally became a practical technology after further experiments by John B. Fell, and was employed in building a railway over Mt. Cenis. This story is related in chapter 19.

In 1857 M. Grassi received a patent based on application of the Archimedean screw, as a means to propel a locomotive up a steep ascent. Under his plan a beveled wheel was to be fixed to the driving axle so as to connect, by means of one intermediate motion, with the crown wheel at the end of the screw shaft. The screw would turn on a specially constructed center rail. The motion of the engine's drive wheels and screw would be synchronized, with 12 turns of the screw made for each turn of the drivers. It was projected that this engine could achieve a speed of at least 12 m.p.h. pulling a gross load of not less than 100 tons up an incline. To some extent Grassi's idea anticipated the later cog/rack systems of Marsh and Riggenbach. These stories are covered in chapter 19.

A variation on Marsh's center-rack system was patented in 1869 by David Harrison of Fayette, Mississippi. Harrison proposed a system having two auxiliary toothed rails, each placed just inside the standard rails. Toothed wheels, fitted inside the ordinary drive wheels, would engage these toothed rails, providing better tractive capabilities. Harrison's system also employed an extra set of rails mounted beyond the standard ones. These were elevated somewhat off the ground and grooved underneath. Beneath them ran small wheels extended from the engine. These provided greater stability and prevented slippage on extreme grades.

In 1868 M. Larmenjat built a simple and successful monorail line that ran from Montfermeil to Raincy. Larmenjat's locomotive was basically a road traction engine with modifications including a small double-flanged guide wheel at its front. This engaged a center rail, but a driver steered it as well, perhaps for added security. The large drive wheels of the engine had wide tread and ran on support beams, with both these and the center rail mounted on sleepers. Carrying most of the engine's weight, they provided adequate traction. An engraving from *L'Illustration* in 1868 shows a three-wheel locomotive followed by what appears to be a similar arrangement for the passenger carriages. An engraving from the May 20, 1870 issue of *Engineering* shows a slightly different arrangement, with the engine having in addition, a rear guide wheel. The example for the passenger carriage shows a four-wheel vehicle, having front and rear center guide wheels plus a support axle in between. The latter illustrations probably represent improved versions of the original equipment.

In 1873, the Lisbon Steam Tramways Co. began construction on a unique 51-mile system. It consisted of two separate lines: one of 33 miles from Lisbon to Torres Vedras, and a 28-mile section from Lisbon to Cascaes and Pero Pinheiro. Unfortunately, permission was denied for establishing a connecting route, and the enterprise soon failed. It operated from September 4, 1873 to April 8, 1875.

In concept this system borrowed from Larmenjat, using a similar center guide rail with flat support beams on each side. A special tank locomotive, built by the Sharp, Stewart & Co., and prototype carriages were tested at Buckhurst Hill. The constructed portions of this system were laid on the common roads with no special right of way, and train speeds were not intended to exceed 15 m.p.h. It was planned to provide both passenger and goods service.

On France's St. Pierre & Goudal Tramway, probably built in the early 1870s, a raised center rail was used having asphalt strips laid out to each side. These strips supported the wide-tread vertical wheels of the engine and coaches. The engine also had three pairs of horizontal wheels engaging the center rail. The front set functioned as guide wheels, with the latter pairs providing traction in the same manner as the Fell system.

At the Exposition de Lyon in 1872, a ⅔-mile line designed by M. Duchamp was built between the Pont Morand and the Tete d'Or Park. Traction was supplied by cable, and the cars were similar in concept to those of Andraud, being like saddles, with separated passenger compartments, but having a common roof that ran above the rail. Also looking back to Andraud during the mid-1870s, were military authorities in Transvaal, South Africa. They decided to experiment with a wooden "post and rail" elevated monorail system. A small locomotive and saddle-like cars were designed for straddling this rail. The advantages of this system were lower cost, no grading required, flexibility in rough terrain, and easy maintenance.

Circa 1880, M. Lartigue began his own experiments with the monorails, particularly for industrial uses. In 1884, to better acquaint potential customers with his system, he set up a small demonstration line at an agricultural exhibition in Paris, fitting the cars with seats to carry passengers. Lartigue subsequently set up a full-scale working line at the Esparto farms in Algeria.

In 1887 he built a 9-mile passenger line between Listowel and Bullybunnion in County Kerry, Ireland. In this system side rails ran on each side parallel to the main rail. These provided additional support, thus avoiding any problems from load imbalance. Lartigue's locomotive and coaches hung, saddle-like, over the main rail, but were suspended at a lower point than his monorail predecessors. Thus the locomotives and rolling stock were joined from about midway up rather than at the roof line. Each side of the locomotive was identical to the other, having its own firebox, boiler, and smokestack. Passengers on each side were seated on benches extending the length of the coach. A few years later Lartigue built the 10½-mile Feurs-Panissieres line in France. Unfortunately, on its inauguration run in 1894, the main rail collapsed under the weight of the locomotive, causing a sudden jolting halt to the movement of the train. Although no one was badly hurt, local outrage led to a subsequent inquiry and the closing of the line.

At the American Centennial Exposition of 1876 in Philadelphia, a small monorail was displayed. It featured a steam locomotive riding mostly above, but at least partially astraddle, a single elevated track. Two visitors from the oil boomtown of Bradford, Pennsylvania, Colonel A. I. Wilson and E. W. Coddington, resolved to build such a railroad, and they did so between Bradford and Gilmore, a distance of 6 miles. Officially called the Bradford and Foster Brook Rwy., but locally known as the "Peg-leg Railroad," this line was completed in 1878. Initially a rotary-engine locomotive was used, but when this proved unsatisfactory, a somewhat more conventional-looking engine was employed. Unfortunately, the "Peg-leg" line ran parallel to a narrow-gauge railroad, both having stops in the same communities, and was thus only barely able to survive. In a locomotive trial on January 27, 1879, a boiler explosion killed two company officials and three crewmen, and injured others, including a locomotive sales representative. This provided the death blow for the company, which never again operated a train.

David B. James' locomotive, reported in the December 20, 1877 issue of *Railway Age,* was designed to run on a single beam of wood, having 12″×14″ dimensions, upon which was mounted a 2-inch thick flat iron rail. His engine was supported by two wide flangeless wheels, with the larger one, in front, connected to two steam cylinders and serving as the driver. Horizontal wheels, two on each side and operating with spring pressure against the central beam, kept the locomotive on track. For crossing roads James added auxiliary flanged vertical carrying wheels, three on each side. In approaching an intersection the engine would drop slightly, at which point the auxiliary wheels engaged standard rails, with the momentum of the locomotive carrying it across the divide.

The February 17, 1881 edition of *Railway Age* included an article on another novel railway, one designed by G. S. Curtis of Chicago. This system was mainly dependent on "shoe" trucks, which replaced the carrying wheels on the front and back of the engine, as well as all of the bogies of the cars. These shoe trucks functioned as skids that slid over rollers attached to the tracks at set distances apart. Flanges on the inside of the shoe trucks prevented derailment. A single pair of driving wheels propelled Curtis's engine. These drivers had special notches or openings on the rim, set at regular intervals, allowing the wheels to go over the rollers without putting any weight upon them. At the same time, the rollers provided traction for the gearlike drivers passing over them. They were not set opposite each other, but in a zigzag position.

In 1873 Joe V. Meigs, of Lowell, Massachusetts proposed a unique monorail system. Meigs's system employed many interesting design features, among them one of essentially transforming the normal horizontal two-track system to a single vertical one where a number of separate points of contact were made. In 1886, following passage of a bill in the Massachusetts state legislature encouraging cities to reduce the number of grade crossings in populous areas, the City of Cambridge turned to Meigs and began construction of a line based upon his ideas. At least part of this line was constructed and operated experimentally, beginning from the company's shops at Bridge St., East Cambridge. Meigs's locomotive and cars were aerodynamic in design, being somewhat cylindrical in shape. The engine had horizontal drive wheels on each side, which with the aid of hydraulic pressure for better adhesion, rolled against the upper track beam and pulled the train. For additional support and better balance, inclined wheels were provided as well. These ran on small rails beneath the upper track beam. Meigs's train also employed automatic couplers and functioned extremely well on curves.

Another experimental railroad system that might be classified as monorail was invented by Eben M. Boynton in the late 1880s. The first locomotive designed by Boynton was built by the Portland Locomotive Works in Maine and shipped to Gravesend, Long Island, in 1889. There, on a stretch of track abandoned by the Sea Beach & Coney Island R.R., it began extensive trial runs. The Boynton Bicycle R.R., as it came to be called, essentially employed one normal rail laid above ties on the ground, but also had an overhead support system of guiding beams. The locomotive and cars, tall and of narrow width, were designed so that two trains could pass safely in opposite directions, each occupying one rail of a standard-gauge railroad track. The locomotive cab had two stories, with the engineer in the top portion and the fireman occupying the lower part. An improved locomotive was used experimentally in the summer of 1890, carrying passengers between Gravesend and Coney Island. Boynton subsequently began construction of a line intended to cross Long Island between Bellport and the Sound, but probably little of it was ever completed.

At an electrical exhibition in St. Louis in early 1891 a small working model was displayed of the system promoted by the National Unicycle Elevated Railway Co. A car seating two people carried passengers for a short distance on an elevated track. This track had one main rail and two guide rails fixed to a girder-like structure supported by poles. The cars designed for this system were to run above the track, with each having two trucks. Each truck consisted of a special housing in which were mounted one main wheel and four guide wheels. These were all double-flanged to ensure proper adherence to the respective rails. Friction rollers, resting on springs, were placed between the truck and the car bottom. This facilitated the smooth negotiation of sharp curves.

In Germany, during the 1890s, M. Langen conducted experiments with an electrified overhead suspended monorail system. To test his ideas he set up a trial line at Deutz, near Cologne. His design was described, as follows, in the May 25, 1895 issue of *Scientific American:* "The derailment of the wheels is prevented by the friction plates. The wheels are secured to the motor case in the usual way and are supplied with springs to take up shocks. The car is fastened to the motor through the medium of a center pin, which permits a certain amount of lateral play. The car is suspended by springs which render riding easy. Each truck has four wheels. Each car is provided with two motors. Each truck is also provided with brakes which seize the rail at both the top and bottom." Langen thought that speeds of up to 186 m.p.h. might be achieved with this system.

His efforts led to the completion, in 1899, of the 8-mile Barmen-Elberfeld line. One of its unique features was that for 6 miles it ran above the River Wupper. Each car was capable of carrying 50 passengers. The girders supporting the tracks were constructed on the cantilever plan, and were secured to different kinds of supports, depending on whether the line was over land or above the river. This futuristic railway employed an automatic block signal system, allowing cars to run as frequently as 2 minutes apart.

TOP: Kollman's proposed system employed horizontal guide wheels operating on a third center rail, circa 1845. MIDDLE: Prosser's test track at Wimbledon Common, 1844. This system used double-flanged angled guide wheels running along the inside edge of wooden beam rails. BOTTOM: Side view of Prosser's train.

Articulated trains: Top: Front and side views of Jouffroy's locomotive-tender, 1844. Upper middle: Top and side views of Jouffroy's articulated carriage. Lower middle: Locomotive and tender, Arnoux system, Sceaux Rwy., France, 1845. Bottom left: First-class carriage, Sceaux Rwy. Bottom right: Second-class carriage, Sceaux Rwy.

PLATE

352

Top: Grassi's screw locomotive for ascending steep gradients on railways, 1857. **Bottom:** David Harrison's patent double rack system, 1869.

Monorail tramways: Top: Larmenjat's Raincy-Montfermeil line, France, 1868. Middle: Larmenjat's exhibition line at the Trocadero, Paris, 1872. Bottom: Trial at Buckhurst Hill, England of rolling stock built for the Lisbon Steam Tramways Co. of Portugal, 1873.

PLATE 354

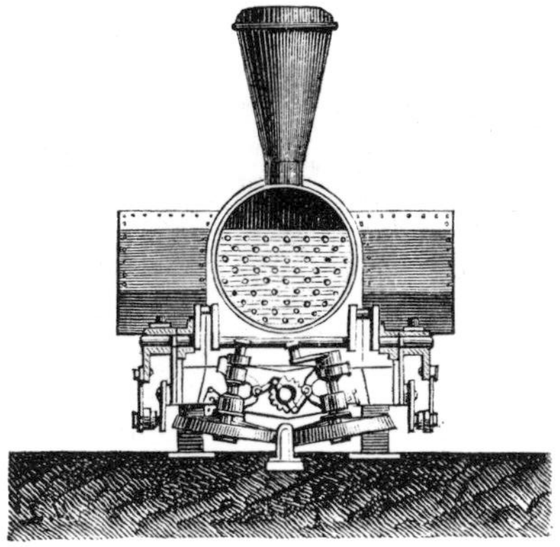

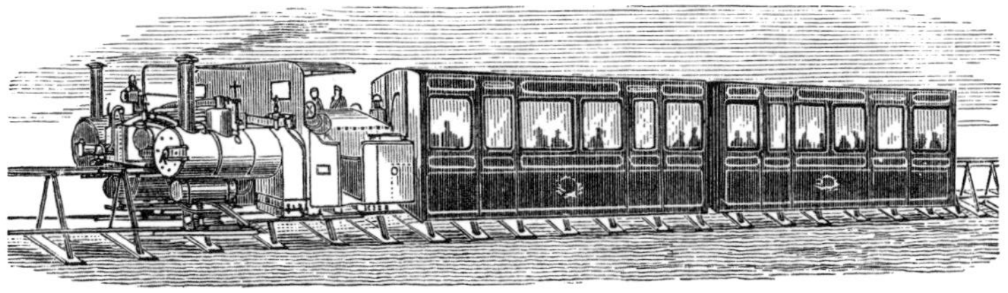

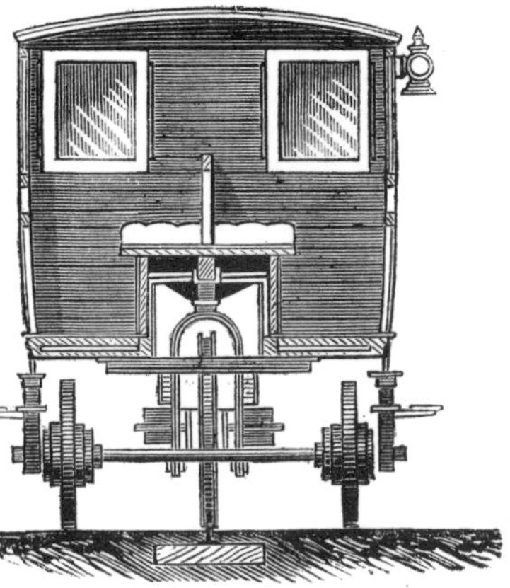

PLATE 355

TOP LEFT: Larmenjat carriage, 1870. TOP CENTER: Frontal section of locomotive and track of France's center-rail adhesion St. Pierre & Goudal Tramway, circa 1871. TOP RIGHT: Side view of same locomotive. UPPER MIDDLE LEFT: Train of Lartigue system, circa 1887. LOWER MIDDLE LEFT: Section showing Larmenjat track, support beams, and sleeper. BOTTOM LEFT: End view of Larmenjat carriage. BOTTOM RIGHT: Larmenjat locomotive, 1870.

TOP: Duchamp's monorail at the Exposition de Lyon, 1872. BOTTOM: Monorail at the American Centennial Exposition, Philadelphia, 1876.

PLATE
356

"Pioneer" elevated monorail, Transvaal, South Africa, 1878: Top: Train crossing a stream. Middle: Carriages and wagons. Bottom left: Engine with two tenders. Bottom right: Method of anchoring support beams.

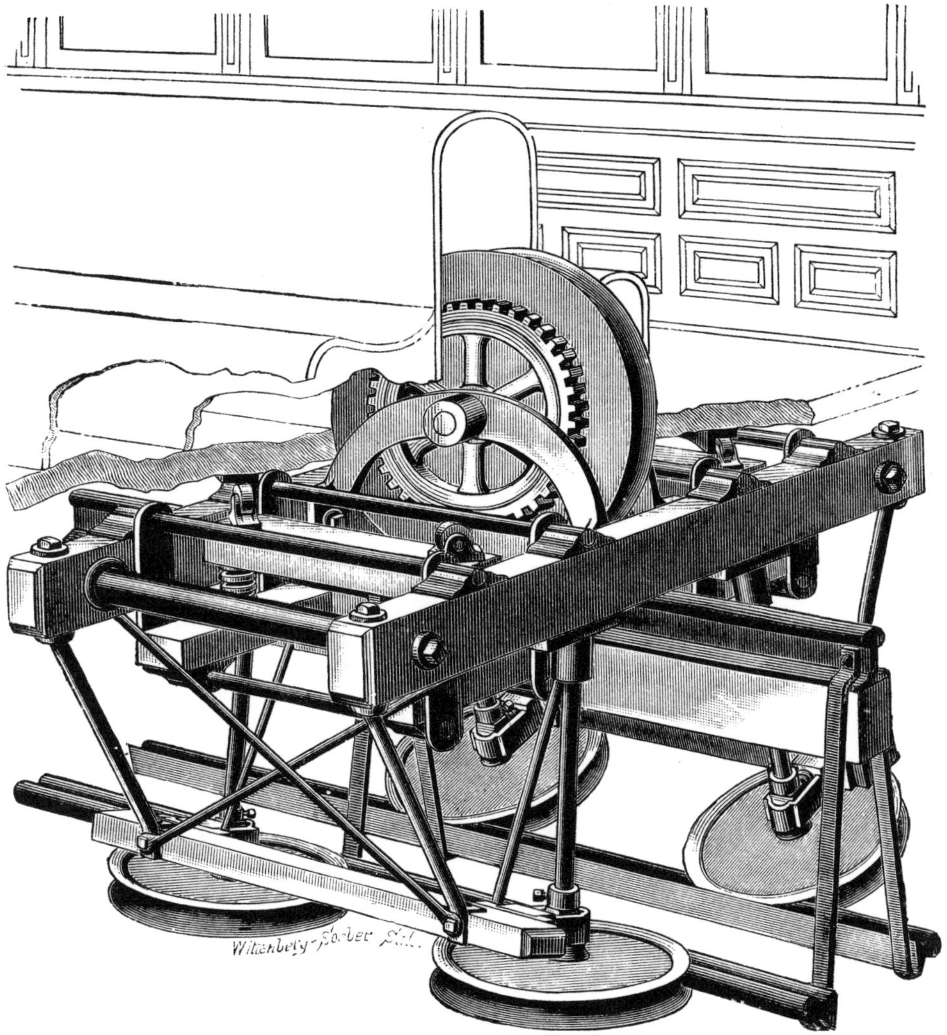

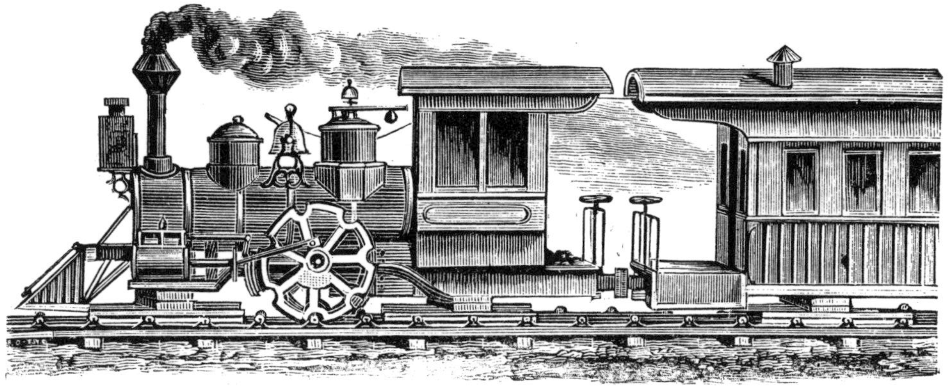

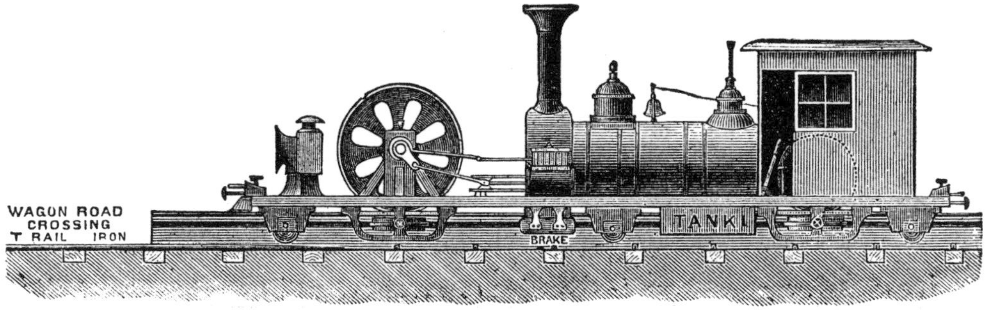

PLATE 358

Top left: View of Unicycle railway system, 1891. **Top right:** Unicycle railway truck, 1891. **Bottom left:** Railway devised by G. S. Curtis using rollers and skids, 1881. **Bottom right:** James' monorail locomotive, 1877.

PLATE 359

Joe V. Meigs' elevated railway system, East Cambridge, Massachusetts, 1886.

Germany's suspended Barmen-Elberfeld Monorail, 1900: Top: Traveling above the Wupper River. Bottom: Station at Doppersberg.

BIBLIOGRAPHY

In researching and preparing this book, I heavily employed the primary materials that I cite. I also used general railway histories and a variety of specialized secondary sources, which I list below. These secondary sources consist mainly of volumes focused on the particulars of rail history—locomotive development, British railways, military railways, train stations, and so on.

Among the most helpful was Malcolm Fletcher and John Taylor's *Railways: The Pioneer Years.* The authors, drawing strongly from French sources, offer a distinctly international point of view. Sarah Searight's *Steaming East* deals with the influence of steam technology on the expansion of Britain's nineteenth-century colonial empire, along with the rivalry in Asia between the British and Russians. While not dealing exclusively with railways, Searight sheds considerable light on rail development in areas other historians largely have overlooked. For example, alone among the works listed here, her volume mentions the French Decauville Company, a builder of locomotives and portable railways that was very active in the nineteenth century. The Russians used Decauville equipment in their military campaigns, and the British ordered identical material for use in India's northwest frontier.

A few nonrailroad titles do not appear below but also proved of value for their specialized information. On Ottoman rail history, I consulted Halil Inalcik with Donald Quataert, *An Economic and Social History of the Ottoman Empire: 1300–1914* (Cambridge: Cambridge University Press, 1994). Graham Robb's *Victor Hugo: A Biography* (New York: W.W. Norton, 1998), recounted the writer's dramatic first encounter with railways in Belgium.

Alexander, E. P. *Iron Horses: American Locomotives 1829–1900.* New York: Bonanza Books, 1941.

Allen, G. Freeman. *Railways: Past, Present and Future.* New York: William Morrow, 1982.

Atkins, Philip. *The Golden Age of Steam Locomotive Building.* London: Atlantic Transport, 1999.

Beebe, Lucius, and Charles Clegg. *Hear the Train Blow.* New York: Dutton, 1952.

Binney, Marcus, and David Pearce, eds. *Railway Architecture.* New York: Van Nostrand Reinhart, 1980.

Bishop, Denis, and Keith Davis. *Railways and War Before 1918.* New York: Macmillan, 1972.

Bobrick, Benson. *Labyrinths of Iron.* New York: Quill, 1986.

British Railway Maps of Yesteryear. Surrey: Ian Allan Ltd., 1991.

Brown, John K. *The Baldwin Locomotive Works, 1831–1915: A Study in American Industrial Practice.* Baltimore: Johns Hopkins University Press, 2001.

Bye, Ranulph. *The Vanishing Depot.* Wynnewood, Pa.: Livingston, 1973.

Cook, Richard J. *The Beauty of Railroad Bridges.* San Marino, Calif.: Golden West Books, 1987.

Cookridge, E. H. *Orient Express: The Life and Times of the World's Most Famous Train.* New York: Random House, 1978.

Cornwell, E. L. *History of Railways.* Secaucus, N.J.: Chartwell Books, 1976.

Day, J. R. *More Unusual Railways.* New York: Macmillan, 1960.

Dilts, James D. *The Great Road.* Stanford, Calif.: Stanford University Press, 1993.

Ellis, C. Hamilton. *British Railway History: 1830–1876.* London: George Allen & Unwin, Ltd., 1954.

———. *The Pictorial Encyclopedia of Railways.* London: Hamlyn House, 1968.

———. *Railways: A Pictorial History of the First 150 Years.* New York: Peebles, 1974.

———. *The Lore of the Train.* New York: Barnes & Noble, 1995.

Faith, Nicholas. *The World the Railways Made.* New York: Carroll & Graf, 1991.

Fink, Kevin. *The Beginnings of Railways in Russia.* Available at www.fink.com/papers/russia.html.

Fletcher, Malcolm, and John Taylor. *Railways: The Pioneer Years.* Secaucus, N.J.: Chartwell Books, 1990.

Forsyth, James. *A History of the Peoples of Siberia.* Cambridge: Cambridge University Press, 1994.

Foster, Hugh F., Jr. *United States Army Domestic Railway Hospital Train.* Unpublished manuscript.

Fry, David. *Staplehurst-1865.* Available at www.danger-ahead.railfan.net/accidents/stplhrst.html.

Greggio, Luciano. *Locomotives a Vapeur.* Paris: Hachette Realities, 1979.

Haine, Edgar A. *Seven Railroads.* New York: A. S. Barnes, 1979.

Harter, Jim. *American Railroads of the Nineteenth Century.* Lubbock: Texas Tech University Press, 1998.

Haywood, Richard Mowbry. *The Beginnings of Railway Development in Russia in the Reign of Nicholas I, 1835–1842.* Durham, N.C.: Duke University Press, 1969.

Henderson, Rick, et al. *Shay Locomotives.* Available at www.shaylocomotives.com/index.html.

Hoyt, Edwin P. *Commodore Vanderbilt.* Chicago: Reilly & Lee, 1962.

Jones, Huw M. *The Delagoa Bay Railway and the Origin of Steinaecker's Horse.* The South African Military History Society. Available at www.rapidttp.com/milhist/vol103jo.html.

Kia-Ngau, Chang. *China's Struggle for Railroad Construction.* New York: De Capo, 1975.

Lee, Emanoel. *To the Bitter End: A Photographic History of the Boer War 1899–1902.* New York: Viking Penguin, 1985.

Loco Profile 11: The Norris Locomotives. Windsor: Profile Publications Ltd., 1971.

McNeely, John H. *The Railways of Mexico.* El Paso: Texas Western College Press, 1964.

Meeks, Carroll L. V. *The Railroad Station: An Architectural History.* Secaucus, N.J.: Castle Books, 1978.

Middleton, P. Harvey. *Railways of Thirty Nations.* New York: Prentice-Hall, 1937.

Middleton, William D. *Grand Central: The World's Greatest Railway Terminal.* San Marino, Calif.: Golden West Books, 1977.

Morgan, Bryan, ed. *The Railway-Lover's Companion.* London: Eyre & Spottiswoode, 1963.

Nock, O. S. *Great British Trains.* London: Pelham Books, 1985.

———. *The Railway Enthusiast's Encyclopedia.* London: Arrow Books, 1970.

———. *Railways of Western Europe.* London: Adam & Charles Black, 1977.

Nock, O. S., ed. *Encyclopedia of Railroads.* New York: Galahad Books, 1977.

Otis, F. N. *Illustrated History of the Panama Railroad.* Pasadena, Calif.: Socio-Technical Books, 1971.

Passingham, W. J. *Romance of London's Underground.* London: Sampson Low, Marston, n.d.

Redder, Gustav. *The World of Steam Locomotives.* New York: G. P. Putnam's Sons, 1974.

Reed, Robert C. *Train Wrecks: A Pictorial History of Accidents on the Main Line.* New York: Bonanza Books, 1968.

Richards, Jeffrey, and John M. MacKenzie. *The Railway Station: A Social History.* Oxford: Oxford University Press, 1986.

Schlichting, Kurt C. *Grand Central Terminal.* Baltimore: Johns Hopkins University Press, 2001.

Searight, Sarah. *Steaming East.* London: The Bodley Head, 1991.

Simmons, Jack. *The Railways of Britain.* London: Guild, 1986.

Singleton, C. C., and David Burke. *Railways of Australia.* Sydney: Angus & Robertson Ltd., 1963.

Snell, J. B. *Early Railways.* London: Octopus Books, 1972.

Solomon, Brian. *Railroad Stations.* New York: Metro Books, 1978.

Stevens, G. R. *History of the Canadian National Railways.* New York: Macmillan, 1973.

Stover, John F. *The Railroads of the South.* Chapel Hill: University of North Carolina Press, 1955.

Sulzer, Elmer G. *Locomotives for the Madison Hill.* The Railway and Locomotive Historical Society, Bulletin 123, Westford, Mass., October 1970.

Teague, Ellen C. *Mount Washington Railway Company.* Princeton: The Newcomen Society in North America, 1970.

Thompson, Joe. *Where and When Did Cable Cars Operate.* Available at www.geocities.com/CapeCanaveral/Launchpad/3518/html/ccwhere.html.

———. *Who Was Important in the History of the Cable Car?* Available at www.geocities.com/CapeCanaveral/Launchpad/3518/html.ccwho.html.

Tufnell, Robert. *The New Illustrated Encyclopedia of Railways.* Edison, N.J.: Chartwell Books, 2000.

Vasser, Ed. *The Climax Locomotive.* Available at www.trainweb.org/climaxlocos/history.html.

White, John H., Jr. *American Locomotives: An Engineering History, 1830–1880.* Rev. ed. Baltimore: Johns Hopkins University Press, 1998.

———. *The American Railroad Freight Car.* Baltimore: Johns Hopkins University Press, 1993.

———. *The American Railroad Passenger Car.* Baltimore: Johns Hopkins University Press, 1978.

———. *The Cincinnati Inclined Plane Railway Company: The Mount Auburn Incline and the Lookout House.* The Cincinnati Historical Bulletin, vol. 27, no. 1, Spring 1969.

———. *Grice and Long: Steam-Car Builders.* Prospects: An Annual of American Cultural Studies, 1976.

———. *Once the Greatest of Builders: The Norris Locomotive Works.* The Railway & Locomotive Historical Society, Bulletin 150, Westford, Mass., Spring 1984.

Wilson, B. G., and J. R. Day. *Unusual Railways.* London: Frederick Muller Ltd., 1957.

INDEX

References to illustrations appear in italics following any page citations and are by plate number.

Picture Source List

Plate numbers appear in boldface.

American Artisan and Illustrated Journal of Popular Science, Brown & Allen, Publishers, New York. **311** top, 7 Dec 1872.

American Railroad Journal and Advocate of Internal Improvements, New York. **26** top right descending (4th from top), 15 Jun 1840; **173** top, 1855.

Ballou's Pictorial Drawing Room Companion, Boston. **30** top, 10 Jan 1857; **215** top, 21 Feb 1857; **292** top, date unknown; **292** bottom, 13 Dec 1856.

Benjamin, Park, ed., *Appleton's Cyclopedia of Applied Mechanics*, D. Appleton & Co., New York, 1891. **238**.

Blätter für den Häuslichen Kreis, Verlag von Hermann Schönlein, Stuttgart, 1870. **284** top.

Brockhaus Konversations Lexikon, F.A. Brockhaus, Leipzig. **81** top left, middle center; **113** top right.

Brown, Robert, *The Countries of the World*, Cassell, Petter, Galpin & Co. London. **180** right.

Brown, William H., *The History of the First Locomotives in America*, D. Appleton & Co., New York, 1874. **23** middle, bottom; **24**; **25**.

Das Buch für Alle: Illustrierte Familien Zeitung, Stuttgart. **97** bottom, 1892; **109**, 1878; **291**, 1895; **357** top, middle left, bottom left, bottom right, 1878.

Catalog Illustré du Decauville, no. 83, Petit-Bourg, France, Mar 1893. **84** top; **88** top, bottom; **174** bottom right; **178** top right; **207** top, middle, lower middle left, lower middle right; **209** top left, top right, upper middle left, upper middle right, lower middle left, lower middle right, bottom; **211** all pictures; **268** top.

Clark, D. Kinnear, C.E., *Tramways: Their Construction and Working*, Crosby Lockwood & Co., London, 1878. **178** middle right; **299** middle right; **313** top.

Clark, Susie C., *The Round Trip from the Hub to the Golden Gate*, Boston, 1890. **38** top.

Corthell, Elmer L., *The Interoceanic Problem and its Scientific Solution*, An address before the American Association for the Advancement of Science, Ann Arbor, Mich., 26 Aug 1885 (Pamphlet). **143**.

Dary, Georges, *A Travers L'Électricite*, Third Edition, Librairie Nony et Cie, Paris, 1903. **335** bottom.

Deharme, E., *Les Merveilles de la Locomotion*, Librairie Hachette et Cie, Paris, 1874. **1** middle right; **53** top right, upper middle right; **65** middle right; **101** lower middle right; **105** top; **299** bottom left; **302** upper middle right, middle left; **352** top, upper middle left, upper middle right; **355** top center, top right.

The Electrician. **326** top left, 9 Sep 1882; **327** bottom, 27 May 1882.

Enciclopedia Universale. **207** upper middle right.

The Engineer, London. **4** bottom left, 7 Mar 1890; **6** top left, 20 May 1892; **7**, 1 Nov 1889; **8** top, bottom, 27 May 1892; **28** bottom, 24 Jan 1879; **48**, 2 May 1890; **54** top left, top right, 1 Feb 1895, bottom, 5 Feb 1892; **54** middle left, 1 Feb 1895; **58** top right, 1 Feb 1895, bottom, 31 Oct 1890; **79** bottom, 11 Oct 1889; **83** top, 18 Apr 1890; **85** top, 14 Aug 1891, middle, 6 Sep 1889, bottom, 13 Mar 1891; **91** top left, 1896 adv., bottom, 22 Jul 1892; **96** top, 4 Sep 1896, bottom, 11 Sep 1896; **130** bottom, 11 Sep 1891; **133** top left, top right, bottom left, 1 Feb 1895; **152** top, bottom, 4 Sep 1896; **156** middle right, bottom left, 1 Feb 1895; **157** top, 1 Feb 1895, bottom, 2 Jan 1874; **164** top, 27 Nov 1885; **209** middle right, 22 May 1891; **210** middle, 7 Feb 1890; **225** bottom, 22 Feb 1895; **228** top, 10 May 1895; **315** top, bottom, 1 May 1896; **332** top, 10 Feb 1893.

Every Saturday: An Illustrated Journal of Choice Reading, Fields, Osgood & Co., Boston, 15 Oct 1870. **263**.

L'Exposition Universelle de 1867 Illustré, Paris, 1867. **16** bottom.

Ferdinand Hirt's Geographische Bildertafeln, Ferdinand Hirt, Breslau, 1886. **252** top.

Fergusson, James, *History of the Modern Styles of Architecture*, John Murray, London, 1862. **185** bottom.

Figuier, Louis, *Les Grandes Inventions Modernes*, Librairie Hachette et cie, Paris, 1876. **46** top; **218** lower middle left; **222** middle right; **225** middle left, middle center.

Figuier, Louis, *Les Merveilles de la Science*, Furne, Juvet, Paris. **2** bottom left; **6** top center (group); **101** top right; **106** top, bottom; **171** top; **218** lower middle right, bottom right; **246** top, bottom; **254** top; **302** lower middle right; **326** middle left.

Figuier, Louis, *Les Nouvelles Conquêtes de la Science: Grands Tunnels et Railways Métropolitains,* Librairie Illustree, Paris. **257** bottom; **345** top, bottom; **350** top.

Figuier, Louis, *Les Nouvelles Conquêtes de la Science: Les Voies Ferrees Dans les Deux Mondes*, Librairie Illustree, Paris. **2** top left, top right; **3** top right; **158** top; **162** middle right, bottom left, bottom right; **169** bottom; **173** middle; **218** top left; **249** top; **262** bottom; **297** top, bottom; **309** bottom; **313** bottom; **319** top, bottom; **325** top; **329** top, bottom.

Forney, Matthias N., *The Car Builder's Dictionary*, Railroad Gazette, 1879. **206** all pictures but top right and bottom; **227** top left; **233** upper middle center, lower middle center, lower middle right.

Forney, Matthias N., *Catechism of the Locomotive*, Railroad Gazette, New York, 1874. **60** bottom.

Forney, Matthias N., *Catechism of the Locomotive*, Railroad Gazette, New York, 1883, adv. **61** top.

Forney, Matthias N., *Catechism of the Locomotive*, Railroad Gazette, New York, 1891. **142** top.

Frank Leslie's Illustrated Newspaper, New York. **32** top, 28 Aug 1869, bottom, 27 Apr 1867; **33** top, 15 Feb 1864; **36** top, 11 May 1878; **37**, 27 Apr 1878; **38** bottom, 23 Aug 1879; **41**, 4 Aug 1877; **42** bottom, 30 Nov 1878; **43**, 29 May 1886; **175** bottom, 4 May 1867; **179** bottom, 15 Jan 1870; **202** top, 29 Dec 1877; **205**, 10 Nov 1877; **215** bottom, 18 May 1867; **217**, 20 Jun 1874; **237**; **259** top, 10 May 1862, bottom, 3 Aug 1861; **260** top, 26 Jul 1862, bottom, 27 Sep 1862; **276**, 17 Feb 1872; **279**, 17 Sep 1865; **280** top, 12 Mar 1859; **283** top, 2 Sep 1865; **287**, 25 Jan 1879; **293**, 21 Sep 1867; **303** top, 8 Oct 1864; **326** top right, 5 Sep 1885; **339** top, 27 Jul 1867; **343** bottom, 14 Dec 1872.

Die Gartenlaube: Illustriertes Familienblatt, Leipzig, 1880. **196** top, bottom.

Gleason's Pictorial Drawing Room Companion, Boston. (Changed to *Ballou's* on 6 Jun 1855.) **30** middle, 4 Mar 1854, bottom, 8 Jan 1853.

Le Globe Illustré, 1887. **102**.

The Graphic, London. **159** bottom, 17 Jan 1880; **179** top, 14 Apr 1877; **342** bottom, 3 Nov 1883 Supplement; **354** bottom, 18 Jan 1873.

Hallberger, ed., *Illustrirte Welt: Deutsches Familienbuch aus Natur und Leben, Wissenschaft und Kunst,* Stuttgart and Leipzig. **127**, 1889; **269**.

Harper's Weekly, Harper Brothers, Publishers, New York. **19**, 6 Jul 1872 Supplement; **33** bottom, 15 Aug 1868; **35** bottom, 29 May 1869; **39**, 25 Sep 1886; **42** top, 21 Sep 1889; **44**, 25 Apr 1891; **114**, 2 Dec 1876; **146** top, 27 Jan 1866, bottom, 7 Apr 1877; **239**, 17 Jan 1880; **261** top, date unknown, bottom, 1 Oct 1864; **273**, 17 Apr 1886; **274**, 29 Aug 1885; **281**, 16 Jan 1892; **285** bottom, 11 Jan 1868; **294**, 24 Feb 1872; **336** bottom, 13 Jan 1866; **341** top, 9 Nov 1872 Supplement.

Harrison, Joseph, Jr., *The Locomotive and Philadelphia's Share in its Early Improvements*, George Gebbie, Philadelphia, 1872. **122** top, bottom.

Hornung, Clarence P., *Two Thousand Early Advertising Cuts*, Dover Publications, New York, 1956. **29** middle, bottom.

The Illustrated London News, London. **4** bottom right, 22 Feb 1845; **9** top, 2 Dec 1843, bottom left, 26 Oct 1844, middle center, bottom right, 22 Feb 1845; **10** top, 17 Nov 1849, bottom, 2 Aug 1845; **11** top, 28 Oct 1848, middle, bottom, 30 Sep 1848; **12** top, 29 Jul 1854, bottom, 1 Aug 1863; **13** top, bottom, 15 Oct 1864; **14**, 27 Apr 1865; **15** top, bottom, 30 May 1868; **17**, 12 Mar 1870; **18**, 28 Jan 1871; **20**, 7 Sep 1872; **21**, 5 Jul 1879; **22**, 10 Jan 1891; **47** top, 16 Aug 1856; **113** bottom, 1 Dec 1860; **124** top, 10 Jul 1886; **131** top, 7 Oct 1854; **138**, 1 Sep 1860; **139** bottom, 11 Jan 1862; **140**, 15 Dec 1888; **153** top, 19 Sep 1863, bottom, 23 May 1863; **154** top, 23 Apr 1870, bottom, 11 Oct 1873; **163**, 9 Aug 1862; **165** bottom, 29 May 1858; **171** bottom, 8 Nov 1862 Supplement; **172** top, bottom, 7 Sep 1861; **185** top, 23 Oct 1852; **186**, 3 Jun 1854; **187**, 18 Apr 1863; **188**, 11 Jun 1864; **189**, 8 Sep 1866; **202** bottom; **204** bottom, 1 Mar 1873 Supplement; **212** top left, middle left, middle right, bottom left, bottom right, 28 Dec 1844, top right, 21 Feb 1846; **213** top left, top right, bottom, 21 Feb 1846; **216** top, bottom, 8 Sep 1866; **223**, 10 Jan 1891; **234**, 4 Jun 1859; **247** top, 1 Feb 1868, bottom, 23 Jan 1869; **258**, 24 Mar 1855; **272** top, 6 Nov 1875, bottom, 2 Feb 1884; **275** top, 14 Jan 1854; **278** bottom, 13 Mar 1886; **283** bottom, 28 Feb 1863; **284** bottom, 24 Aug 1867; **286** top, bottom, 29 Jan 1876; **301** top left, 6 Jan 1844, top right, bottom left, 11 Oct 1845, bottom right, 6 Dec 1845; **302** top right, 20 Sep 1845, bottom, 6 Dec 1845; **303** bottom, 28 Feb 1863; **304** top, 28 Feb 1846; **310** top, 25 Nov 1848, bottom, 14 Aug 1869; **332** bottom, 11 Feb 1893; **336** top, 7 Apr 1860 Supplement; **337** top, 8 Feb 1868 Supplement, bottom, 27 Jan 1866; **338**, 8 Feb 1868 Supplement; **339** bottom, 8 Jan 1870; **340**, 6 Nov 1869; **341** bottom, 2 Dec 1876; **351** middle, bottom, 1844; **353** top, 27 Jun 1857.

The Illustrated News of the World, New York, 4 Jul 1891. **128**.

L'Illustration Européenne: Journal Internationale de la Famille, Brussels, 8 Sep 1895.

L'Illustration: Journal Universal, Paris. **34**, 16 Dec 1871; **51**, 29 Jul 1876; **103** top, upper middle, lower middle, bottom, 1846; **107** top, 11 Apr 1857, bottom, 1865; **110**, 9 Jun 1888; **112**, 24 Jun 1899; **125**, 8 Sep 1888; **126**, 8 Sep 1888; **132**, 1862; **134** top, 1866, bottom, 1863; **135**, 1870; **137**, 6 Dec 1890; **148**, 1872; **161**, 12 Feb 1887; **165** top, 1862; **166** top, 1862, bottom, 1872; **167**, 22 Apr 1893; **169** top; **170** top, bottom, 20 Aug 1898; **178** bottom left, 9 Jun 1888; **181** top, bottom, 7 Jun 1884; **182** top left, top center, top right, bottom, 7 Jun 1884; **184** bottom, 24 Jun 1899; **200**, 1870; **203** bottom, 21 Oct 1871; **219**, 4 Jul 1891; **225** top, 15 Apr 1893; **255** top, date unknown, bottom, 30 Sep 1871; **256**, 3 Jun 1882; **257** top, 3 Jun 1882; **264** top, 17 Jul 1875 (courtesy Mary Evans Picture Library, London), bottom, 30 Sep 1871; **271**, 16 Dec 1899; **295**, Nov 1872; **302** top left, 1847; **308** bottom, 20 Nov 1875; **323**, 22 Nov 1890; **334** bottom, 21 Jan 1893; **349** top left, top right, middle, 1867; **352** lower middle, bottom left, bottom right; **354** top, 1868; **356** top, 1872.

L'Illustration Nationale, Brussels. **101** upper middle left, upper middle right, lower middle left, lower middle center, 9 May 1880; **195** bottom, 9 May 1880.

Illustrirte Zeitung, Leipzig. **3** bottom, 13 Jun 1874; **115**, 9 Feb 1878; **183**, 8 Sep 1888; **194** top, 15 Aug 1885.

La Ilustración Española y Americana, Madrid, 8 Jan 1898. **270** bottom.

Le Journal Illustré, Paris. **1** top left, 15 May 1887; **101** top left, bottom left, bottom right, 15 May 1887; **278** top, 3 Feb 1884; **288**, 25 May 1884; **298**, 3 Feb 1878.

Kennedy, William Sloane, *Wonders and Curiosities of the Railway*, S. C. Griggs & Co., Chicago, 1884. **5** bottom; **23** top right; **26** top left; **270** middle, 328 top, bottom.

Kirkman, Marshall M., *The Science of Railways*, The World Railway Publishing Co., New York, 1903. **1** upper middle left, middle left; **26** top right; **28** top left, top right; **104** top left; **207** upper middle left, upper middle center; **208** middle left, middle center, middle right.

Knox, Thomas W., *The Boy Travelers in Mexico*, Harper & Brothers, New York, 1889. **142** bottom.

Knox, Thomas W., *The Boy Travelers in the Far East*, Harper & Brothers, New York, 1883. **158** bottom.

Light Locomotives, 6th ed., H. K. Porter & Co., Pittsburgh, 1889. **68** middle center; **72** top left, middle left; **74** top left, top right, middle left, middle right; **314** upper middle left, upper middle center, upper middle right, lower middle left, lower middle right, bottom left, bottom right.

Locomotives and Locomotive Building, Wm. S. Gottsberger, New York, 1886 (Rogers Co. catalog). **67** top; **141** middle; **164** bottom.

Le Magasin Pittoresque, Paris. **3** middle left, middle right, 1861; **105** bottom; **124** bottom; **214** top left, top right, bottom left, bottom right, 1861; **220**, 1862; **221**, Aug 1861; **222** top left, 1865, top right, middle left, bottom left; **253** top left, top right, middle left, middle right, bottom left, bottom right; **254** bottom.

Manual of the Corporation of the City of New York, New York, 1870. **191**.

Maw & Dredge, eds., *Engineering: An Illustrated Weekly Journal*, London. **1** bottom right, 9 Mar 1894; **2** middle center, 18 May 1894, middle right, 13 Aug 1894, bottom right, 9 Mar 1894; **4** top right, middle left, middle right, 31 Aug 1894; **16** top, 29 Dec 1871; **23** top left, 30 Jun 1876; **26** top left descending (2nd from top); **47** bottom, 12 Oct 1894; **49** top, 21 Jun 1867; **52** bottom, 17 Feb 1871; **55** top, bottom, 27 Jan 1871; **57** top, bottom; **62** top, 22 Nov 1872, bottom, 4 Jul 1873; **63** top, 17 Oct 1873, bottom, 28 Nov 1873; **68** bottom, 22 Aug 1873; **71** top, 27 Mar 1885, bottom, 1 May 1885; **78** top, 21 Jun 1889, middle, 21 Sep 1883; **81** bottom, 22 Feb 1889; **82** top, 6 Feb 1891; **83** bottom, 3 Oct 1890; **84** bottom, 4 May 1894; **86** bottom, 6 Apr 1894; **89** top, bottom, 16 Nov 1894; **90** top, 11 Jan 1895, bottom, 23 Nov 1894; **91** top right, 15 Jun 1894; **92** top, upper middle, lower middle, bottom, 27 Jul 1894; **93** top, 6 Dec 1895, bottom, 6 Sep 1895; **94** top, 26 Feb 1897, bottom, 14 Dec 1894; **95** top, bottom, 22 Nov 1895, middle, 11 Oct 1895; **97** top, 24 Jun 1898 adv.; **139** top, 8 Nov 1872; **141** top, 1 Dec 1893; **144** top, 8 Mar 1895, middle, 22 Mar 1895, bottom, 8 Sep 1893; **151** top, 12 Jul 1895, bottom, 13 Apr 1894; 156 top, middle left, 29 Mar 1895; **168** top, middle, bottom, 27 Oct 1893; **174** top, 6 Dec 1872, middle right, 16 Dec 1870; **178** top left, 3 Aug 1894; **199** top, bottom, 19 Apr 1889; **201**, 28 Nov 1873; **208** top, 6 Apr 1894; **210** top, 26 Jul 1895, bottom, 7 Sep 1894; **229** top, 14 Nov 1890, bottom, 6 Jan 1893; **231** top, 13 Oct 1893; **248** top, 5 Jul 1872; **296** top, bottom, Dec 20, 1872; **321** top left, May 11, 1894; **326** bottom, Aug 24, 1883; **331** top, 7 Nov 1890; **334** top, 9 Jun 1893; **342** top, 13 Sep 1895; **355** top left, lower middle left, bottom left, bottom right, 20 May 1870.

Meyers Kleines Konversations Lexikon, Bibliographisches Institut, Vienna. **81** top right; **100** bottom; **305** bottom; **316** bottom.

Meyers Konversations Lexikon, Verlag des Bibliographischen Instituts, Leipzig and Vienna. **270** top.

Le Monde Illustré: Journal Hebdomadaire, Paris. **123** bottom, 13 Jun 1863; **160**; **266**, 28 Jun 1873; **348**, 1861; **354** middle, 17 Aug 1872.

La Nature: Revue des Sciences, G. Masson, Paris. **5** top, middle, 5 Sep 1887; **77** bottom, 16 Nov 1889; **80** top left, 5 Dec 1891; **86** middle right, 5 May 1900; **99** top, 5 Nov 1900, middle, 8 Sep 1900, bottom, 13 Oct 1900; **104** top right, bottom left, bottom right; **130** top left, top right, 10 Nov 1900; **162** middle left, 10 Nov 1900; **208** bottom; **251** top left, 11 Feb 1888, top right, 28 Apr 1894; **306** top, 17 Jun 1882; **317** bottom, 9 Nov 1878; **327** top, 24 Jan 1880; **335** top, 11 Dec 1897; **360** top, bottom, 20 Oct 1900.

The New Eldorado: A Summer Journey to Alaska, 5th ed., Houghton, Mifflin & Co., New York, 1890. **36** bottom.

Ny Illustreret Tijende, Christiana (Oslo). **136**, 20 Nov 1881; **197** bottom, 28 Aug 1881.

Parville, Henri de, *Causeries Scientifiques*, J. Rothschild, Editeur, Paris, 1879. **299** top left, top right.

Patch, Olive, *Christmas Fun and Frolic*, Cassell Publishing Co., New York. **v**.

Perdonnet, Aug., Traité *Élémentaire des Chemins de Fer*, 3rd ed., Garnier Frères, Libraires-Éditeurs, Paris, 1865. **6** bottom; **27** top, bottom; **49** middle center, bottom; **53** top left, upper middle left, lower middle left.

Le Petit Parisien, Paris, 1879–99. **111**, 23 Oct 1898; **280** bottom, 27 Jun 1897; **282**, 8 Oct 1899.

Das Pfennig Magazin, F. A. Brockhaus & Bossange Vater, Leipzig, 1838. **26** top left descending (3rd from top).

The Pictorial Times, London. **351** top.

The Popular Encyclopedia: or Conversations Lexicon, Blackie & Son, Glasgow & Edinburgh, 1879. **47** middle left, middle center; **49** middle left; **53** lower middle right; **55** middle right; **56** middle left, bottom; **58** top left; **123** top; **133** bottom right; **156** lower middle center; **248** bottom.

The Railroad and Engineering Journal, New York, 1887. **26** top right descending (2nd, 5th, and 6th from top), Jan 1887; **28** top center, Jan 1887; **67** bottom, Jan 1887; **73** bottom, Jan 1887; **75** bottom, Mar 1887; **150** top, Jul 1887; **197** top, Mar 1887; **347** bottom, Jan 1887.

The Railroad Gazette: A Journal of Transportation, Engineering, and Railroad News, New York. **26** top right descending (3rd from top), 29 Apr 1892; **27** middle left, 24 Aug 1872, middle right, 6 Jul 1872; **59** top, 5 Aug 1871, bottom, 12 Nov 1870; **60** middle, 1873 adv.; **61** middle, bottom, 1874 adv.; **66** top, 1874 adv.; **72** top right, 1874 adv.; **75** top, 1889 adv.; **76** bottom, 19 Feb 1886; **79** top, 2 Jul 1886; **86** top, 16 Nov 1888; **156** bottom right, 24 Aug 1886; **162** top, 28 Dec 1888; **176** top, 23 Sep 1871, bottom, 24 Apr 1875; **178**, 1874 adv.; **184** top, 23 Nov 1888; **192** bottom, 2 Oct 1875; **193**, 18 Dec 1875; **207** bottom, 17 Aug 1888; **218** top right, 1886; **222** bottom right, 1885; **226** bottom, 24 Aug 1886; **227** upper middle, 1886, lower middle, bottom, 1885; **229** lower middle left, 1889 adv.; **230**, 27 Jul 1883; **231** bottom, 6 Apr 1888; **233** top right, middle left, bottom right, 1883; **240**, 30 Nov 1883; **305** middle, 21 Oct 1892; **321** bottom, 23 Nov 1883; **343** top, 17 Feb 1872; **344** top, 17 Aug 1872, bottom, 24 Jan 1874.

The Railway Age, Chicago. **86** middle left, 1883 adv.; **192** top, 1 Feb 1887; **206** bottom, 3 Mar 1881 adv.; **225** top right, 10 Mar 1881 adv.; **226** top, 1881 adv.; **299** bottom right, 1887; **358** bottom left, 17 Feb 1881, bottom right, 20 Dec 1877.

The Railways of America: Their Construction, Development, Management and Appliances, John Murray, London, 1890. **180** left; **224** left, right.

Recent Locomotives, Railroad Gazette, New York, 1883. **72** bottom; **314** top; **347** top.

Recent Locomotives, Railroad Gazette, New York, 1886. **70** bottom; **74** top, bottom; **82** bottom; **150** bottom; **232** top; **251** bottom.

Reports of the United States Commissioners to the Paris Universal Exposition, 1878, vol. IV., U.S. Government Printing Office, Washington, 1880. **65** top; **68** top left; **309** top.

La Republic Illustré. 198.

Richardson, Albert D. *Beyond the Mississippi*, American Publishing Co., Hartford, 1867. **35** top.

La Science Illustrée, Journal Hebdomadaire; A la Librairie Illustree, Paris. **1** top right, 1 Dec 1888; **149**, 16 Aug 1890.

La Science Pratique, Paris. **77** top; **250** top; **305** top; **308** top.

Scientific American: An Illustrated Journal of Art, Science, and Mechanics, Munn & Co., New York. **45**, 17 Oct 1896; **65** bottom, 1875; **66** bottom, 10 Nov 1877 Supplement.; **69** bottom, 1 Dec 1883; **80** bottom, 17 Jul 1886; **87**, 3 Sep 1898; **98** top, 3 Dec 1898, bottom, 19 Feb 1898; **108**, 11 Jun 1870; **145**, 4 Mar 1893; **147**, 22 Aug 1885; **178** bottom right, 28 Aug 1886; **190**, 11 Dec 1869; **204** top, 15 Apr 1882; **226** middle, 30 Oct 1880; **228** bottom, 13 Jan 1877 Supplement; **232** bottom, 27 Sep 1870; **241**, 28 Mar 1885 Supplement; **242**; **243**, 8 Jun 1889; **245**, 10 Jul 1897; **250** bottom, 21 Mar 1885 Supplement; **252** bottom, 23 Nov 1889; **267**, 15 Aug 1885; **268** bottom, 1 Oct 1888; **289** top, date unknown, bottom, 10 Dec 1881; **306** bottom, 30 Jan 1897; **307** top, middle left, middle right, 19 Apr 1890, bottom, 19 Nov 1887; **311** bottom, 13 Jun 1863; **312**, 23 Dec 1882; **317** top, 24 Aug 1872; **318** bottom, 2 Dec 1882; **320** top, 17 Apr 1875, bottom, 24 May, 1884; **321** top right, middle right, 17 Apr 1875; **324**, 18 Sep 1880; **325** bottom, 3 Jan 1885; **330** top, 10 Sep 1881, bottom, 14 Aug 1897; **331** bottom, 5 Aug 1893; **333**, 27 May 1893; **346**, 5 Aug 1882; **353** bottom, 18 Dec 1869; **358** top left, top right, 31 Jan 1891; **359**, 10 Jul 1886.

Spies, William B. *The Pennsylvania Railroad*, Philadelphia, 1875. **29** top; **175** top left, top right; **177** top, upper middle, middle, lower middle, bottom.

The St. Louis Lumberman, St. Louis, Jan 1893, adv. **76** top.

The Street Railway Gazette, Chicago, Aug 1889, adv. **300** top left, top right, middle center, middle right.

Street Railway Journal, McGraw Publishing Co., New York. **300** middle left, 1891 adv., bottom, adv.; **322** top, middle, bottom, adv.

Stretton, Clement E. *The Locomotive Engine and Its Development*, Crosby Lockwood & Son, London, 1892. **47** middle right; **56** top right; **58** top center; middle left, middle center; **78** bottom.

Ueber Land und Meer: Allgemeine Illustrirte Zeitung, Stuttgart. **116**, 1883; **117**, 1 Dec 1889; **119**, 1893; **159** top, 1876; **195** top; **265**, 1872.

L'Univers Illustré, Michel Freres, Editeurs, Paris. **131** bottom, 3 Sep 1864; **236**; **275** bottom, 23 Feb 1867; **290**, 1889; **349** bottom.

Williams, F.S., *Our Iron Roads*, Bemrose & Sons, London, 1883. **3** top left; **4** top left; **218** bottom left; **229** middle center.

Zimmerman, Dr. B. F. A., *Elektrizität, Magnetismus, Galvanismus im Dienste des Menschen*, Ferd. Dümmlers Berlags Buchhandlung, Berlin, 1888. **350** botom.

Zur Guten Stunde, Deutsches Verlagshaus, Bong & Co., Berlin, 1893. **120**, 1893.

Source unknown. **1** lower middle left, middle left, bottom left; **2** middle left; **6** top right; **31**; **40**; **46** bottom; **50** top, bottom; **52** top, middle left; **53** bottom; **56** top left, middle right; **60** top; **64** all pictures; **68** top right; **69** top; **70** top left, top right; **80** top right; **84** middle left; middle right; **97** middle; **100** top (courtesy Mary Evans Picture Library, London); **104** middle (group); **113** top left; **118**; **121**; **129**; **141** bottom; **173** bottom; **174** bottom left; **178** upper middle left; **194** bottom; **203** top; **206** top right; **218** top center; bottom center; **233** top left, upper middle right, bottom left; **235** top, bottom; **244**; **249** bottom; **262** top; **277**; **285** top; **299** middle left; **304** bottom; **316** top; **318** top; **326** middle right; **335** middle; **355** upper middle left; **356** bottom.